Ubiquitin Proteasome System in CNS Neurodegenerative Diseases

Sarika Singh
Editor

Ubiquitin Proteasome System in CNS Neurodegenerative Diseases

Springer

Editor
Sarika Singh
Department of Toxicology and Experimental
Medicine and Professor of Biology (AcSIR)
CSIR-Central Drug Research Institute
Lucknow, Uttar Pradesh, India

ISBN 978-981-95-6098-1 ISBN 978-981-95-6099-8 (eBook)
https://doi.org/10.1007/978-981-95-6099-8

This Springer imprint is published by the registered company Springer Nature Singapore Pte Ltd.
The registered company address is: 152 Beach Road, #21-01/04 Gateway East, Singapore 189721, Singapore

Preface

Welcome to a captivating exploration of the central nervous system (CNS), the epicenter of our body's intricate operations. Think of the brain as the command center, orchestrating thoughts and movements, while the spinal cord acts as the vital highway, connecting the brain to the rest of our body. Together, they play a crucial role in processing information, making decisions, and managing both voluntary and involuntary actions. In this book, we dive into the fascinating world of CNS neurodegenerative diseases—conditions that unveil the complex and multifaceted mysteries behind their origins and the signaling pathways that underlie them. Despite the strides we've made in understanding these tragic disorders, we are still grappling with significant challenges in developing effective treatments that can slow or halt their progression. This reality highlights the urgent need for dedicated research and action in our field. One of the most exciting areas of investigation is the role of protein aggregation and degradation. These processes are tightly woven together in a web of intricate pathways, demanding our attention. A key focus is the ubiquitin-proteasome system (UPS), which offers a rich ground for discovery due to the diverse functions of various E3 ubiquitin ligases. While we have a basic framework for categorizing these ligases, many are still shrouded in mystery regarding their roles in health and disease, hinting at a wealth of potential breakthroughs yet to come. This meticulously curated book aims to engage contributors and share valuable insights on how the UPS intertwines with the pathogenesis of neurodegenerative diseases, while also exploring its promise in developing therapeutic strategies. Organized into four dynamic sections, we present: (1) an engaging introduction to CNS neurodegenerative diseases, (2) a deep dive into the intricate signaling events that fuel neurodegeneration, (3) a critical review of current therapeutic options and the gaps that exist, and (4) innovative proposals for experimental models that will pave the way for detailed mechanistic studies. Ultimately, this book aspires to be your go-to resource, offering a broad perspective on the mechanisms that drive CNS neurodegeneration. It seeks to illuminate the remarkable advancements achieved through recent research, point out the areas that still need exploration, and inspire innovative action that is not only encouraged but essential for propelling this crucial field forward. Join us on this academic journey to uncover and understand the complexities of the CNS neurodegenerative disease research and their future.

Lucknow, Uttar Pradesh, India

Sarika Singh

Contents

Part I

Introduction

Need for the Ubiquitin-Proteasome System in Cellular Physiology

1

S. Sreenivasa Reddy

1.1 Introduction

The ubiquitin-proteasome system (UPS) is involved in a wide array of cellular functions like maintenance of protein quality by degrading misfolded proteins in the cytoplasm and endoplasmic reticulum (ER), preventing the accumulation of toxic aggregates mostly linked to diseases like Alzheimer's and Parkinson's. It also plays a vital role in mitochondrial homeostasis, regulating mitochondrial dynamics and mitophagy. The UPS processes antigens for MHC presentation in the immune system and regulates key signaling pathways like nuclear factor kappa B (NF-κB). Furthermore, the UPS modulates cell death pathways by controlling the levels of pro- and anti-apoptotic proteins such as p53 and Bcl-2 family members. It also regulates DNA replication and cell cycle progression by degrading cyclins and cyclin-dependent kinase inhibitors (CKIs), ensuring the unidirectional nature of cell division. Finally, the UPS is critical for neuronal physiology, impacting synaptic plasticity by regulating key synaptic proteins and contributing to neurodegenerative diseases when dysfunctional. The UPS, therefore, represents a crucial cellular machinery with implications for numerous physiological and pathological processes.

Proteins are fundamental components of every cell in the body, serving several critical functions. They are involved in numerous biological processes. Proteins function as enzymes that catalyze a wide range of biochemical reactions. Hormones, predominantly composed of proteins such as insulin and growth hormone, regulate essential metabolic functions. Certain proteins, such as actin and fibronectin, provide structural support to cells. Moreover, proteins are essential for muscle contraction. Some proteins act as receptors for various hormones, while others, including immunoglobulins, play significant roles in the immune response. Proteins also facilitate the transport of essential substances, such as oxygen and sex hormones.

S. Sreenivasa Reddy (✉)
Biochemistry Division, ICMR-National Institute of Nutrition, Hyderabad, Telangana, India

S. Singh (ed.), *Ubiquitin Proteasome System in CNS Neurodegenerative Diseases*, https://doi.org/10.1007/978-981-95-6099-8_1

Additionally, proteins monitor extracellular and intracellular conditions, effectively relaying important information to other cellular components.

Proteins must achieve their native conformations to perform their biological functions effectively. Misfolding and aggregation of proteins can occur due to various stressors, emphasizing the importance of proteostasis for cellular health and survival. Proteostasis encompasses all stages in a protein's lifecycle, from its translation to its functional activity and, finally, to its degradation. The UPS and autophagy-lysosome system represent the primary cellular pathways for protein degradation. Dysfunction of these pathways can result in proteotoxicity, leading to various diseases.

The integrity of cellular processes is contingent upon the appropriate balance of various proteins. Among the regulatory mechanisms that govern the activities of these molecules, the modulation of protein half-life has emerged as a predominant strategy. The UPS serves as the principal pathway for protein degradation in eukaryotic cells. This process is characterized by its stringent regulation, specificity, and dependence on adenosine triphosphate (ATP). The UPS is integral to maintaining protein homeostasis, facilitating the rapid removal of unwanted or damaged proteins and the regulation of numerous fundamental cellular functions. These functions include cell division and differentiation, responses to environmental stress, immune responses, DNA repair, and apoptosis. The UPS plays a critical role in modulating cellular processes by regulating the half-life of proteins integral to these processes. The UPS targets a diverse range of proteins, including regulators of the cell cycle and growth, hormones along with their receptors, components of signaling pathways, enzymes involved in housekeeping and cell-specific metabolic pathways, and proteins that are mutated or damaged through post-translational modifications. It encompasses proteins localized in the cytosol, nucleus, and membrane and those compartmentalized within the secretory pathway. Given the wide variety of cellular targets, it is unsurprising that the UPS regulates numerous fundamental cellular activities. These activities include DNA repair mechanisms, cell cycle progression and division, differentiation, and development, responses to stress and extracellular signals, formation of neuronal networks, biogenesis of organelles, modulation of receptors and ion channels, the secretory pathway, regulation of immune and inflammatory responses, and the process of apoptosis [1–3].

1.2 Functions of UPS

1.2.1 UPS is a Cellular Protein Degradation Machinery

Intracellular and numerous extracellular proteins are constantly being recycled; they are broken down into amino acid components and replaced by freshly produced proteins. This process is essential for several homeostatic functions. Individual proteins located in the cytosol, cellular organelles, and membranes are degraded at varying rates, ranging from minutes to days or weeks to months. Cells utilize various proteolytic systems to facilitate this degradation process, supported by intricate regulatory mechanisms that guarantee selectivity and avoid excessive breakdown of cellular components. Thus, achieving a precise balance between protein synthesis and degradation is vital for maintaining proteostasis. UPS effectively targets most of the cellular proteins in all tissues for degradation and modulates their half-lives according to the cell's specific needs in varying circumstances [4].

The Nobel Prize in Chemistry for 2004 was awarded to Aaron Ciechanover, Avram Hershko, and Irwin Rose in recognition of their remarkable contributions to the field of ubiquitin-mediated protein degradation. The process of protein degradation via the UPS comprises two principal steps: ubiquitination of the target protein and proteasomal degradation of the ubiquitinated protein.

Ubiquitin is a small polypeptide of 76 amino acids. Goldstein and colleagues initially isolated it during the purification of polypeptide hormones from the thymus [5]. The amino acid sequence of ubiquitin exhibits remarkable evolutionary conservation, with the primary structure being nearly identical across a wide range of organisms. This conservation suggests that ubiquitin plays a fundamental role in essential cellular functions.

Ubiquitin is added to lysine residues of the target protein as a post-translational modification known as ubiquitination. Ubiquitination is facilitated by a sequential process involving three distinct enzymes: an E1 ubiquitin-activating enzyme, an E2 conjugating enzyme, and an E3 ligase (Fig. 1.1). The E1 enzyme is responsible for activating free ubiquitin, which subsequently interacts with the cysteine of the E1 active site to form a Ub-E1 thioester. After activation, the ubiquitin is transferred to the E2 enzyme, which, with assistance from the E3 ligase, transfers ubiquitin to the lysine residue of the target protein. This transfer typically involves the covalent linkage of the C-terminal glycine residue (Gly-76) of ubiquitin to the ε-amino group of lysine on the target protein, forming an isopeptide bond. This process represents the monoubiquitination of the substrate, which triggers various non-proteolytic events, including endocytosis, histone regulation, DNA repair, and nuclear export.

Once an initial ubiquitin molecule is conjugated to a target protein, the process may be reiterated to add additional ubiquitin molecules to the earlier ubiquitin, thereby generating a polyubiquitin chain [6]. Ubiquitin comprises seven lysine residues (K6, K11, K27, K29, K33, K48, and K63) and *N*-terminus methionine (M1), all of which can serve as sites for ubiquitin-chain extension. These variations yield different types of polyubiquitin chain linkages, which ultimately influence the fate of the substrates. Specifically, protein ubiquitination entails the formation of tetramers linked through K48 ubiquitin signals, which are designated for degradation by the proteasome [7, 8].

A single E1 enzyme, or in some cases two E1 enzymes, is responsible for activating ubiquitin, which is essential for all modifications. This enzyme facilitates the transfer of ubiquitin to various E2 enzymes, of which approximately 40 exist. Each E2 enzyme interacts with one or several E3 ligases. E3 ligases, numbering around 600 in humans, are pivotal in imparting high specificity to the UPS. They serve as the primary determinants of substrate specificity in the process of protein ubiquitination [1, 9].

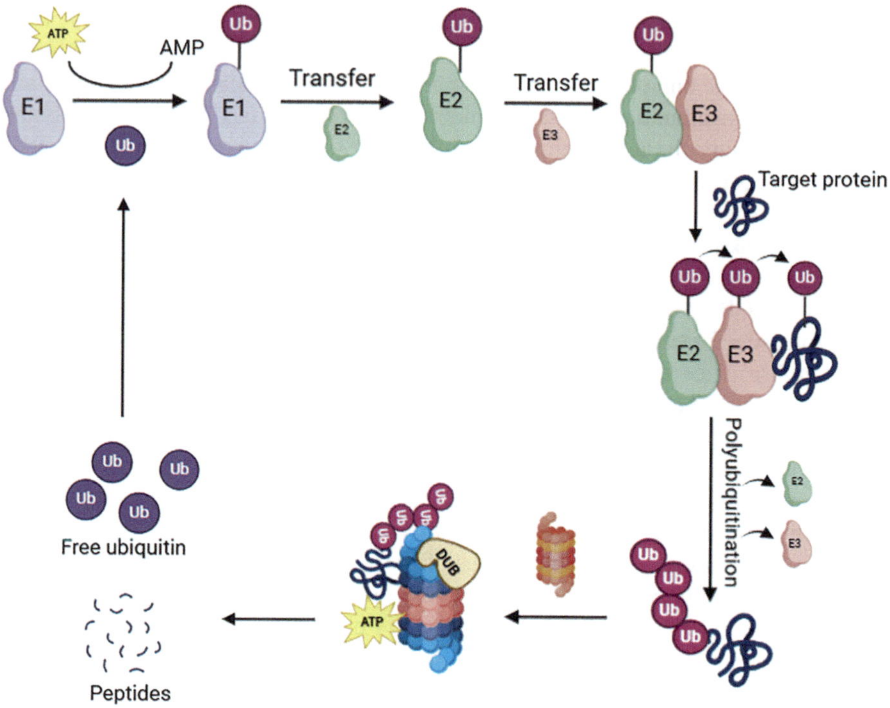

Fig. 1.1 The ubiquitin-proteasome pathway. In the first step, ubiquitin (Ub) is activated by E1, which transfers it to E2. Subsequently, Ub is transferred to the target protein with the help of E3 ligase. This process results in the monoubiquitination of the substrate, triggering various non-proteolytic events. The ubiquitination process can be repeated to add more ubiquitin molecules to the existing ubiquitin, forming a polyubiquitin chain that signals the protein for degradation by the proteasome. DUBs detach the polyubiquitin chain from the target protein, releasing free Ub for reuse. Finally, the target protein is cleaved into short peptides

E1s, E2s, and E3s are enzymes that help attach ubiquitin to target proteins, while deubiquitinases (DUBs) are proteolytic enzymes that specifically break the isopeptide bonds between two ubiquitin molecules or between ubiquitin and the target protein for proteasomal degradation. DUBs demonstrate specificity for unique types of ubiquitin linkages and play particular roles in ubiquitin cleavage, including the removal of entire polyubiquitin chains or the fragmentation of these chains into individual ubiquitin units. DUBs are integral to recycling ubiquitin, regulating ubiquitin chain lengths, and modulating the degradation rate of targets by the UPS. Both E3 ligases and DUBs play critical roles in catalyzing specific ubiquitination and deubiquitination events, which are essential for maintaining normal cellular function. Disruption of this balance has been linked to the progression of various diseases [10, 11].

The 26S proteasome is a molecular complex tasked with degrading ubiquitinated proteins. It comprises two principal components: the 20S core particle, which serves as the proteolytic core, and the 19S regulatory particle, commonly called the proteasome cap. The proteasome cap plays a crucial role in recognizing ubiquitinated substrates, removing ubiquitin, unfolding target proteins, and subsequently translocating the unfolded proteins into the proteolytic core for degradation. Within the regulatory subunit, DUBs facilitate the removal of ubiquitin from proteins, thereby enabling the recycling of ubiquitin for continued cellular utilization. The 20S subunit exhibits three different proteolytic activities: trypsin-like, chymotrypsin-like, and post-glutamyl peptide hydrolyzing activities, contributing to its functional diversity in protein turnover.

1.2.2 UPS Ensures Protein Quality Control

1.2.2.1 Cytoplasmic Protein Quality Control

The UPS is integral to protein quality control in conjunction with chaperones and the autophagic pathway. It accomplishes this by executing selective and timely degradation of misfolded or damaged proteins resulting from missense or nonsense mutations, biosynthetic errors, or oxidative damage. The accumulation of such defective proteins is linked to a variety of age-related diseases, including Parkinson's disease (PD), Huntington's disease (HD), Alzheimer's disease (AD), cataracts, and age-related macular degeneration.

Ribosomes are the primary sites for protein biosynthesis within cells. In the initial stage, nascent polypeptides undergo quality control checks to prevent the accumulation of faulty proteins. Several E3 ligases identify and ubiquitinate these defective nascent polypeptides for proteasomal degradation. Previous research shows that Not4p acts as an E3 ligase, mediating the ubiquitination of aberrant polypeptides on the ribosome and targeting them for degradation [12, 13].

The E3 ligase CHIP (C-terminus of Hsc70-interacting protein) plays a crucial role in removing cytosolic misfolded polypeptides [14]. CHIP directly interacts with both Hsp70 and Hsp90 chaperones, allowing for the ubiquitination of substrates bound to these chaperones, a process facilitated by the ubiquitin-conjugating enzyme UbcH5 [15]. Moreover, CHIP demonstrates chaperoning capabilities by binding directly to client proteins, which helps in preventing their aggregation. Ubiquitination is a crucial modification that ensures the solubility of substrates and guides them toward degradation. Another key proteolytic mechanism for targeting misfolded proteins, especially insoluble aggregates, is selective autophagy, which allows for their degradation. During this process, cytoplasmic components are enclosed by double-membrane vesicles known as autophagosomes, which later fuse with lysosomes (or vacuoles) for degradation. Autophagy assumes a critical role when the functionality of the UPS is compromised. Similar to the UPS, substrates that have been ubiquitinated (but with the K63-linked polyubiquitin chain) are recognized by two adaptor proteins, p62 and Nbr1. These adaptor proteins facilitate binding protein aggregates marked with K63 polyubiquitin chains, subsequently transporting the relevant substrates for lysosomal degradation [12]. The decision regarding whether a substrate is directed toward the proteasomal or autophagic pathway can be influenced by the specific E2 enzymes involved. Depending on the

E2 enzyme utilized, both E3 ligases CHIP and PARKIN can link K48 or K63 ubiquitin chains to their substrates. Proteins with K48 ubiquitin chains are marked for degradation by the proteasome, whereas those with K63 chains are targeted for the autophagic process [16–18].

1.2.2.2 Endoplasmic Reticular Protein Quality Control

The UPS also removes misfolded proteins accumulated in the ER, the primary site for synthesizing secretory and membrane proteins. Through the process known as ER-associated degradation (ERAD), numerous misfolded proteins within the ER are retro-translocated into the cytosol via a series of ubiquitin-conjugating enzymes associated with the ER membrane. Once in the cytosol, these proteins are subsequently targeted for degradation by cytosolic proteasomes.

The most extensively studied and well-understood aspect of UPS's protein quality control occurs in the ER. Approximately 30% of a cell's protein composition comprises proteins associated with the secretory pathway. These proteins are linked to the ER, the Golgi apparatus, lysosomes (or vacuoles), the plasma membrane, and the external surface of the cell. The predominant entry point for these proteins into the secretory pathway is the ER, where they are imported into an unfolded conformation. For proteins to acquire their functional conformation, they must properly fold within the ER before being exported to their respective sites of action. As a result, the ER functions as a highly concentrated folding factory for secretory proteins, incorporating a range of molecular chaperones, co-chaperones, oxidoreductases, glycan-modifying enzymes, and lectins. This complex repertoire of proteins facilitates the dynamic processes of folding and quality control, ensuring that proteins attain correct conformations before their subsequent trafficking.

Numerous modifications occur to nascent proteins during their entry into the ER and subsequent folding. These modifications include signal sequence cleavage, glycosylation, the formation of disulfide bonds, and lipid conjugation. Proteins in the ER and those associated with the secretory pathway are especially prone to errors arising from disulfide bond formation, glycosylation, and membrane insertion.

Mutations or errors in folding, which are significantly exacerbated by various cellular stresses, can lead to the accumulation of misfolded proteins within the ER, necessitating their clearance to avoid cytotoxic consequences. Misfolded proteins may form inactive aggregates that can precipitate, potentially binding to other proteins and disrupting their functions. These misfolded proteins are subsequently removed from the ER and degraded through the UPS via a mechanism known as ERAD. The UPS is primarily responsible for most degradation that occurs through ERAD, while autophagic processes utilizing ubiquitin systems also contribute to protein degradation in the ER.

GRP78 protects the newly synthesized proteins from aggregation and assists their folding into the native state. If the proper folding is not achieved, then it shuttles them for UPS-mediated degradation. The degradation process of these substrates takes place in the cytoplasm. All major E3 ligases involved in ERAD are located in the ER membrane, with their catalytic domains facing the cytoplasmic side. Therefore, ERAD substrates must be transported across the membrane to facilitate ubiquitination at the ER outer surface. These proteins undergo proteasomal degradation in the cytosol. HRD1 represents the most extensively researched E3 ubiquitin ligase implicated in ERAD. The insoluble protein aggregates undergo ubiquitination and are subsequently directed toward autophagic degradation [19, 20]. Impairment in the ERAD process can lead to the toxic accumulation of misfolded, unfolded, and aggregated proteins. This accumulation is linked to a variety of disorders, including cystic fibrosis, α1-antitrypsin insufficiency, diabetes, and several neurodegenerative disorders such as PD, AD, and HD, as well as viral infections and albinism. A dedicated chapter further examines the role of the UPS in maintaining proteostasis within the ER.

1.2.3 UPS Regulates Mitochondrial Homeostasis

Mitochondria are organelles characterized by a double membrane structure, and they play a vital role in various cellular processes, including oxidative phosphorylation, ATP synthesis, lipid biosynthesis, calcium signaling, and apoptosis. Within the mitochondrial matrix, a self-contained ATP-dependent system facilitates protein turnover. In contrast, UPS monitors the proteostasis at the outer mitochondrial membrane and recycles membrane-embedded and imported proteins. Mitochondria form an extensive tubular network, maintained through a delicate balance of fusion and fission events involving their outer and inner membranes. These dynamic processes are essential for cellular viability, as they enable the adaptation of mitochondrial morphology to the cell's energy requirements while preserving the redox potential, which is a primary function of mitochondria.

Mitochondrial ubiquitin ligase (MITOL) is an E3 ubiquitin ligase located in the mitochondrial outer membrane, facilitating the ubiquitination of various mitochondria-associated proteins. Two identified substrates, Fis1 and Drp1, are essential components of mitochondrial membrane fission, underscoring the roles of MITOL and the UPS in mitochondrial dynamics. Moreover, MITOL may play a broader role in maintaining quality control within the mitochondrial proteome by targeting mutated or damaged proteins for ubiquitination. Another E3 ubiquitin ligase, mitochondrial ubiquitin ligase activator of NF-kB (MULAN), is also positioned within the mitochondrial outer membrane and has been implicated in regulating mitochondrial dynamics. The silencing of MULAN using small interfering RNA has resulted in disruptions that led to the clustering of mitochondria.

In addition to targeting specific proteins for degradation by the proteasome, severe cellular damage can be addressed by removing mitochondrial fragments or entire mitochondria through the process known as mitophagy, thereby maintaining cellular homeostasis. A key early event and defining feature of mitophagy is the ubiquitination of proteins located on the outer mitochondrial membrane. This modification is thought to attract various autophagy receptors, effectively signaling which mitochondria must be removed. The crucial factors responsible for this process are the E3 ligase Parkin and the mitochondrial kinase Pink1 [21–23]. The subsequent chapter will further clarify the role of the UPS in maintaining mitochondrial stability.

1.2.4 UPS Involvement in Cell Death Pathways

The UPS serves a pivotal mechanistic function in regulating cell death by facilitating the degradation of key regulatory proteins that are integral to apoptosis. This encompasses aspects of cell cycle regulation and the DNA damage response, where stabilizing the p53 protein is particularly significant. The equilibrium between pro-apoptotic and anti-apoptotic proteins is critical, as it ultimately influences the survival of cells; therefore, meticulous regulation of these proteins is essential. The UPS plays a crucial role in preserving this finely balanced process by rigorously controlling the concentrations of both anti- and pro-apoptotic proteins. Through its modulation of apoptosis, the UPS may either trigger or inhibit cell death.

UPS is crucial in the inflammatory signaling of tumor necrosis factor (TNF) and the control of cell death. TNF is known to induce NF-κB and MAPK-mediated gene expression, as well as drive apoptotic and necroptotic cell death. Ubiquitination and deubiquitination processes of different proteins in TNF-signaling play a vital role in maintaining the stability of these complexes and directly affecting the outcomes. Cellular inhibitors of apoptosis 1 and 2 (c-IAP1/2) serve as E3 ligases that enable K11-, K48-, and K63-linked ubiquitination of various components, including receptor-interacting protein kinase 1 (RIP1). The ubiquitination of RIP1 is crucial for activating NF-κB. Besides stabilizing RIP1, c-IAP1 can also promote the degradation of RIP1 by attaching K48-linked ubiquitin chains, resulting in a reduced capacity for RIP1-mediated cell death [24].

p53 is a critical transcription factor that regulates senescence and apoptosis. It primarily promotes apoptosis by inducing pro-apoptotic protein transcription. The induction of these pro-apoptotic proteins enhances the outer mitochondrial membrane permeabilization, thereby amplifying the activation of caspases. The p53 protein remains at low levels in healthy cells and has a short half-life. However, in response to various stress signals, the half-life of p53 can significantly increase due to the inhibition of its E3 ligases, resulting in an accumulation of p53 protein. The main E3 ligase responsible for the ubiquitylation of p53 is MDM2. The N-terminal region of MDM2 hinders the transcriptional activity of p53. In contrast, its C-terminal region aids in the recruitment of the E2 ubiquitin-conjugating enzyme, which is crucial for the ubiquitylation and subsequent degradation of p53. MDM2 knockout animals show embryonic lethality due to the excessive buildup of p53. Furthermore, many tumors exhibit high levels of MDM2 alongside reduced levels of p53, thereby facilitating the evasion of apoptosis.

Bcl-2 protein encompasses pro-apoptotic and anti-apoptotic proteins that primarily regulate the apoptotic pathway. The balance between these proteins is crucial in determining cellular susceptibility to apoptosis. The Bcl-2 family is categorized into three distinct groups: anti-apoptotic (Bcl-2), pro-apoptotic (Bax), and pro-apoptotic BH3-only proteins. Importantly, Bcl-2 is a key regulator of cell death as it inhibits apoptosis. In cells that are undergoing apoptosis, the levels of Bcl-2 decrease due to its degradation through the UPS. The X-linked inhibitor of apoptosis protein (XIAP) serves as the E3 ligase responsible for Bcl-2 degradation. After apoptotic stimuli, Bcl-2 is positioned close to XIAP through the scaffolding function of

ARTS. The ternary complex formed by Bcl-2, ARTS, and XIAP facilitates the ubiquitylation of Bcl-2 by XIAP, leading to its degradation by the proteasome.

In normal conditions, Bax is primarily localized in the cytosol. However, Bax becomes activated in response to apoptotic signals through a conformational change that facilitates its translocation to the outer mitochondrial membrane. At this site, Bax interacts with BAK to initiate the permeabilization of the outer mitochondrial membrane, which permits the release of pro-apoptotic proteins. These proteins include Smac/Diablo and Cytochrome c, typically within the inner mitochondrial membrane space. Bax protein levels are regulated through UPS. Specifically, the IBRDC2 E3 ligase mediates the degradation of Bax in response to apoptosis facilitated by p53.

Neurodegenerative diseases ultimately result from the death of specific neuronal populations. However, the exact causes behind this neuronal death remain unclear, as several factors contribute, including protein aggregates, mitochondrial dysfunction, neuroinflammatory mediators, and oxidative stress. Apoptosis is widely regarded as the main pathway for neuronal death in several neurodegenerative diseases, including AD, PD, and frontotemporal dementia. XIAP ubiquitinates apoptotic caspases such as caspase-3, -7, and -9 to facilitate their degradation. Elevated nitric oxide (NO), known to inhibit XIAP's ligase activity, is found in neurons experiencing inflammation, and increased S-nitrosylated XIAP has been identified in the brain tissue of AD, HD, and PD patients, hinting at a possible role for XIAP in neurodegeneration. Other E3 ubiquitin ligases, like Parkin, are also S-nitrosylated in the brains of PD patients. Research involving animals suggests that pro-survival BCL-2 proteins, particularly MCL1, are vital for maintaining dopaminergic neurons under mitochondrial defects. Several E3 ubiquitin ligases, such as HUWE1 and Parkin, have been reported to promote MCL1's proteasomal degradation. Furthermore, two deubiquitinating enzymes, USP9X and USP13, serve to deubiquitinate and stabilize MCL1, with mutations in both linked to neurodegenerative conditions [25].

1.2.5 UPS Modulates Immune Response

A prime illustration of the role of the UPS in the immune response is the processing of MHC class I antigens by antigen-presenting cells. It has been well-documented that the UPS is responsible for producing the majority of peptide antigens presented on MHC I molecules. The 11S multimeric complex known as PA28 is critical in this process within the UPS. Research has shown that PA28 interacts with the ends of the 20S proteasome, significantly enhancing its capacity to hydrolyze oligopeptides. This interaction results in the generation of small oligopeptides that are suitable for presentation by MHC I. Specifically, the proteasome is essential for producing the precise C termini of many peptides that are presented by MHC molecules.

UPS plays a critical role in regulating T cell receptors and costimulatory CD28 signaling through the action of ubiquitin ligases belonging to the Cbl family.

Costimulation via CD28 induces the ubiquitination and subsequent degradation of Cbl-β, thereby removing negative regulatory influences and facilitating the expression of proinflammatory cytokines and their respective receptors. A particularly vital connection between the UPS and inflammatory responses pertains to the NF-κB pathway. NF-κB is a principal regulator for numerous inflammatory cytokine genes, and the UPS mediates its activation. Under resting conditions, NF-κB is inhibited while bound to its inhibitor IκB. The activation of NF-κB transpires following the degradation of IκB, a process contingent upon the ubiquitination of IκB followed by proteasomal degradation. Furthermore, it has been documented that the DUB, Cylindromatosis Tumor Suppressor Protein (CYLD) enhances proximal T cell receptor signaling in thymocytes by specifically binding to and deubiquitinating the active form of the kinase Lck. Consequently, any alterations in the UPS may significantly affect immune responses, particularly in regulating a diverse array of inflammatory cytokines. Additionally, proteasomes have been demonstrated to modulate inflammatory pathways by influencing macrophage functionality [26].

1.2.6 UPS Regulates DNA Replication and Cell Cycle Progression

The cell cycle is a tightly regulated process driven by the fluctuating activities of Cyclin-Dependent Kinase proteins (CDKs), which are activated by cyclins and inhibited by Cyclin-Dependent Kinase Inhibitors (CKIs). Various cyclins and CDKs have been discovered, with each cyclin partnering with one or more CDKs, while most CDKs engage with multiple cyclins. The key cyclins that propel cell cycle progress are cyclins A, B, and E. These cyclins are expressed according to the cell cycle phase and interact with Cdk1 or Cdk2 to facilitate subsequent processes, such as DNA replication and mitosis. Specifically, cyclin E-Cdk2 is activated during the late G1 phase to encourage the transition into the S-phase and initiate DNA replication. Cyclin A-Cdk2 and cyclin A-Cdk1 are involved in S-phase progression and the shift from G2 to M phase, respectively, whereas cyclin B-Cdk1 aids in the initiation of mitosis [27].

Multiple mechanisms, including the degradation of cyclins and CKIs through the UPS, control the activity of CDKs [28, 29]. Unlike reversible modifications such as phosphorylation, ubiquitin-mediated proteasomal degradation represents an irreversible mechanism that guarantees the strict unidirectionality of the cell cycle. This process is crucial for regulating the cell cycle by mediating the precise spatial and temporal proteolysis of the key components associated with the cell cycle machinery.

The E3 ligases most notably involved in regulating the cell cycle are the cullin RING ligase 1 (CRL1 or SCF) and the anaphase-promoting complex/cyclosome (APC/C). The APC/C acts as a key E3 ubiquitin ligase in cell cycle regulation and works with two co-activators, Cdc20 and Cdh1, defining substrate specificity. While SCF ligases are active throughout the cell cycle, the function of APC/C is limited to the period between metaphase and the end of the following G1 phase. During this time, APC/CCdc20 facilitates the start of anaphase and the exit from mitosis, whereas APC/CCdh1 is essential for exiting mitosis and establishing a stable G1 state [27, 30, 31].

1.3 Essential Role of UPS in Neuronal Physiology

The UPS is essential for the optimal functioning of the nervous system and is instrumental in regulating various facets of neuronal development, including neurogenesis, differentiation, and synapse formation. Research indicates that proteasome inhibition can negatively affect the central and peripheral nervous systems.

The UPS plays a role in synaptic plasticity, which encompasses strengthening and weakening synaptic connections between neurons. This mechanism is widely regarded as fundamental to learning and memory. Protein turnover is essential for the remodeling and signaling of postsynaptic structures. The UPS regulates critical synaptic proteins, including postsynaptic density protein 95 (PSD-95), Shank, guanylate kinase-associated protein, A-kinase anchoring proteins, spine-associated Rap guanosine triphosphatase-activating protein, Rab3-interacting molecule 1 (RIM1), and glutamate receptor-interacting protein 1. Increased synaptic activity leads to elevated ubiquitin conjugation of PSD proteins in hippocampal cultures. Upon the activation of *N*-methyl-D-aspartate (NMDA) receptors, PSD-95, which serves as a critical anchoring factor for AMPA receptors (AMPARs) at the postsynaptic density, is ubiquitylated by the E3 ligase Mdm2 and subsequently degraded by the proteasome. This NMDA receptor-induced reduction in PSD-95 levels results in the endocytosis of AMPARs. Furthermore, the inhibition of proteasome activity obstructs the recycling of AMPARs by stabilizing PSD-95, thereby attenuating long-term depression. At presynaptic sites, the E3 ligase SCRAPPER ubiquitylates RIM1, a key regulator of the vesicle priming process, to modulate the release of synaptic vesicles [32–35].

Numerous neurodegenerative disorders, including AD, PD, ALS, and HD, are characterized by neuronal damage potentially attributed to toxic, abnormal, and aggregation-prone proteins. It has become evident that dysfunction of the UPS contributes to a variety of neurodegenerative diseases. The transcription factor nuclear factor erythroid 2-related factor 2 (Nrf2) is a primary regulator of the antioxidant response. It facilitates the synthesis of lysosomal and anti-inflammatory proteins, thereby safeguarding neurons from oxidative stress. The translocation of Nrf2 to the nucleus is primarily stimulated by the activity of the UPS [36].

1.4 Conclusion

The UPS is an essential cellular mechanism for protein quality control that ensures the proper functioning of proteins, which is critical for maintaining proteostasis and overall cellular health. By facilitating the selective degradation of misfolded, damaged, or unnecessary proteins, the UPS plays a pivotal role in various biological processes, including cell division, DNA repair, stress responses, immune function, mitochondrial homeostasis, and neuronal integrity. The UPS is indispensable for the optimal functioning of the nervous system and is crucial in regulating numerous aspects of neuronal development and function. Its dynamic regulation is vital for preventing the accumulation of toxic protein aggregates and for adapting to changing cellular environments. Numerous neurodegenerative disorders, such as AD, PD, amyotrophic lateral sclerosis, and HD, are characterized by neuronal damage that may be associated with toxic, abnormal, and aggregation-prone proteins. It is evident that UPS dysfunction contributes to various neurodegenerative diseases. Given its implications in numerous physiological functions and its involvement in disease processes when dysregulated, the UPS represents a significant focus for therapeutic strategies aimed at enhancing neuronal health and addressing degenerative conditions.

References

1. Ciechanover A. The ubiquitin-proteasome pathway: on protein death and cell life. EMBO J. 1998;17(24):7151–60.
2. Rock KL, Gramm C, Rothstein L, Clark K, Stein R, Dick L, et al. Inhibitors of the proteasome block the degradation of most cell proteins and the generation of peptides presented on MHC class I molecules. Cell. 1994;78(5):761–71.
3. Myung J, Kim KB, Crews CM. The ubiquitin-proteasome pathway and proteasome inhibitors. Med Res Rev. 2001;21(4):245–73.
4. Lecker SH, Goldberg AL, Mitch WE. Protein degradation by the ubiquitin-proteasome pathway in normal and disease states. J Am Soc Nephrol. 2006;17(7):1807–19.
5. Hershko A. Ubiquitin: roles in protein modification and breakdown. Cell. 1983;34(1):11–2.
6. Guo HJ, Rahimi N, Tadi P. Biochemistry, ubiquitination. Treasure Island ineligible companies. Disclosure: Nader Rahimi declares no relevant financial relationships with ineligible companies. Disclosure: Prasanna Tadi declares no relevant financial relationships with ineligible companies: StatPearls; 2024.
7. Burslem GM. The chemical biology of ubiquitin. Biochim Biophys Acta Gen Subj. 2022;1866(3):130079.
8. Yau R, Rape M. The increasing complexity of the ubiquitin code. Nat Cell Biol. 2016;18(6):579–86.
9. Zheng N, Shabek N. Ubiquitin ligases: structure, function, and regulation. Annu Rev Biochem. 2017;86:129–57.
10. Mevissen TET, Komander D. Mechanisms of deubiquitinase specificity and regulation. Annu Rev Biochem. 2017;86:159–92.
11. Clague MJ, Urbe S, Komander D. Breaking the chains: deubiquitylating enzyme specificity begets function. Nat Rev Mol Cell Biol. 2019;20(6):338–52.
12. Amm I, Sommer T, Wolf DH. Protein quality control and elimination of protein waste: the role of the ubiquitin-proteasome system. Biochim Biophys Acta. 2014;1843(1):182–96.
13. Dimitrova LN, Kuroha K, Tatematsu T, Inada T. Nascent peptide-dependent translation arrest leads to Not4p-mediated protein degradation by the proteasome. J Biol Chem. 2009;284(16):10343–52.
14. Rosser MF, Washburn E, Muchowski PJ, Patterson C, Cyr DM. Chaperone functions of the E3 ubiquitin ligase CHIP. J Biol Chem. 2007;282(31):22267–77.
15. Jiang J, Ballinger CA, Wu Y, Dai Q, Cyr DM, Hohfeld J, et al. CHIP is a U-box-dependent E3 ubiquitin ligase: identification of Hsc70 as a target for ubiquitylation. J Biol Chem. 2001;276(46):42938–44.
16. Olzmann JA, Li L, Chudaev MV, Chen J, Perez FA, Palmiter RD, et al. Parkin-mediated K63-linked polyubiquitination targets misfolded DJ-1 to aggresomes via binding to HDAC6. J Cell Biol. 2007;178(6):1025–38.

17. Zhang M, Windheim M, Roe SM, Peggie M, Cohen P, Prodromou C, et al. Chaperoned ubiquitylation—crystal structures of the CHIP U box E3 ubiquitin ligase and a CHIP-Ubc13-Uev1a complex. Mol Cell. 2005;20(4):525–38.
18. Zhang Y, Gao J, Chung KK, Huang H, Dawson VL, Dawson TM. Parkin functions as an E2-dependent ubiquitin- protein ligase and promotes the degradation of the synaptic vesicle-associated protein, CDCrel-1. Proc Natl Acad Sci USA. 2000;97(24):13354–9.
19. Buchberger A, Bukau B, Sommer T. Protein quality control in the cytosol and the endoplasmic reticulum: brothers in arms. Mol Cell. 2010;40(2):238–52.
20. Mehrtash AB, Hochstrasser M. Ubiquitin-dependent protein degradation at the endoplasmic reticulum and nuclear envelope. Semin Cell Dev Biol. 2019;93:111–24.
21. Livnat-Levanon N, Glickman MH. Ubiquitin-proteasome system and mitochondria - reciprocity. Biochim Biophys Acta. 2011;1809(2):80–7.
22. Alsayyah C, Ozturk O, Cavellini L, Belgareh-Touze N, Cohen MM. The regulation of mitochondrial homeostasis by the ubiquitin proteasome system. Biochim Biophys Acta Bioenerg. 2020;1861(12):148302.
23. Escobar-Henriques M, Altin S, Brave FD. Interplay between the ubiquitin proteasome system and mitochondria for protein homeostasis. Curr Issues Mol Biol. 2020;35:35–58.
24. Cockram PE, Kist M, Prakash S, Chen SH, Wertz IE, Vucic D. Ubiquitination in the regulation of inflammatory cell death and cancer. Cell Death Differ. 2021;28(2):591–605.
25. Schmidt MF, Gan ZY, Komander D, Dewson G. Ubiquitin signalling in neurodegeneration: mechanisms and therapeutic opportunities. Cell Death Differ. 2021;28(2):570–90.
26. Wang J, Maldonado MA. The ubiquitin-proteasome system and its role in inflammatory and autoimmune diseases. Cell Mol Immunol. 2006;3(4):255–61.
27. Bassermann F, Eichner R, Pagano M. The ubiquitin proteasome system - implications for cell cycle control and the targeted treatment of cancer. Biochim Biophys Acta. 2014;1843(1):150–62.
28. Glotzer M, Murray AW, Kirschner MW. Cyclin is degraded by the ubiquitin pathway. Nature. 1991;349(6305):132–8.
29. Skaar JR, Pagano M. Control of cell growth by the SCF and APC/C ubiquitin ligases. Curr Opin Cell Biol. 2009;21(6):816–24.
30. Acquaviva C, Pines J. The anaphase-promoting complex/cyclosome: APC/C. J Cell Sci. 2006;119(Pt 12):2401–4.
31. Li M, Zhang P. The function of APC/CCdh1 in cell cycle and beyond. Cell Div. 2009;4:2.
32. Colledge M, Snyder EM, Crozier RA, Soderling JA, Jin Y, Langeberg LK, et al. Ubiquitination regulates PSD-95 degradation and AMPA receptor surface expression. Neuron. 2003;40(3):595–607.
33. Yao I, Takagi H, Ageta H, Kahyo T, Sato S, Hatanaka K, et al. SCRAPPER-dependent ubiquitination of active zone protein RIM1 regulates synaptic vesicle release. Cell. 2007;130(5):943–57.
34. Upadhyay A, Joshi V, Amanullah A, Mishra R, Arora N, Prasad A, et al. E3 ubiquitin ligases neurobiological mechanisms: development to degeneration. Front Mol Neurosci. 2017;10:151.
35. Turker F, Cook EK, Margolis SS. The proteasome and its role in the nervous system. Cell Chem Biol. 2021;28(7):903–17.
36. Petrucelli L, Dawson TM. Mechanism of neurodegenerative disease: role of the ubiquitin proteasome system. Ann Med. 2004;36(4):315–20.

Dysfunctional Ubiquitin-Proteasome System and Consequences: Disease Mechanism

2

Mohini Milmile, Shweta Singh, Ajay Pandey, Nilesh Ahire, and Ruchi Chawla

2.1 Introduction

2.1.1 Overview of the Ubiquitin-Proteasome System

The *ubiquitin-proteasome system (UPS)* is a critical intracellular mechanism that maintains protein homeostasis (proteostasis) and regulates numerous cell processes, such as cell cycle progression, signal transduction, and stress responses. It serves as the primary pathway for targeted protein degradation in eukaryotic cells and comprises two key components: the ubiquitination machinery and the 26S proteasome.

Ubiquitination is a biochemical process where ubiquitin, a small polypeptide composed of 76 amino acids, is covalently linked to target proteins through a series of enzymatic steps. This pathway involves three key enzyme types: E1 ubiquitin-activating enzymes, E2 ubiquitin-conjugating enzymes, and E3 ubiquitin ligases. The process begins with the ATP-dependent activation of ubiquitin by E1 enzymes, followed by its transfer to E2 enzymes. Finally, E3 ligases facilitate the attachment of ubiquitin to the target protein, ensuring substrate specificity [1–3].

Polyubiquitination refers to the process of attaching multiple ubiquitin molecules to the lysine residues of an initial ubiquitin, forming a chain. This modification plays a pivotal role in dictating the fate of proteins. For instance, K48-linked polyubiquitin chains target proteins for degradation via the proteasome, whereas K63-linked chains participate in non-degradative cellular processes such as DNA repair and signal transduction [4]. The proteasome is a large, ATP-dependent enzyme complex that facilitates the breakdown of polyubiquitinated proteins. The 26S proteasome is composed of two main components: the 20S core particle (CP),

M. Milmile · S. Singh · A. Pandey · N. Ahire · R. Chawla (✉)
Department of Pharmaceutical Engineering and Technology, Indian Institute of Technology, Banaras Hindu University, Varanasi, Uttar Pradesh, India
e-mail: rchawla.phe@iitbhu.ac.in

S. Singh (ed.), *Ubiquitin Proteasome System in CNS Neurodegenerative Diseases*, https://doi.org/10.1007/978-981-95-6099-8_2

which performs proteolytic activities, and the 19S regulatory particle (RP), which identifies polyubiquitinated proteins, unfolds them, and directs them into the 20S core for degradation [5].

The UPS plays an essential role in maintaining protein quality control by degrading misfolded or damaged proteins, thereby preventing toxic protein aggregation [6]. It also regulates various cellular pathways by controlling the turnover of regulatory proteins such as cyclins, transcription factors, and tumor suppressors [7]. Additionally, the UPS contributes to immune responses by processing antigenic peptides for presentation on MHC class I molecules [8]. It also fine-tunes cellular signaling cascades by removing active components, ensuring precise control of cellular processes [9].

Dysfunction in the UPS is implicated in various pathological conditions, including neurodegenerative disorders such as Alzheimer's disease (AD) and Parkinson's disease (PD), where the failure to clear aggregated proteins contributes to cellular toxicity [10]. In cancer, dysregulation of the UPS affects the balance between oncogene activation and tumor suppressor degradation, promoting tumor progression [11]. UPS dysfunction is also linked to cardiac diseases, with alterations contributing to conditions such as cardiomyopathy [12].

The UPS is a promising target for therapeutic intervention. Proteasome inhibitors, such as bortezomib, have shown efficacy in treating multiple myeloma by disrupting protein turnover in cancer cells [13]. Advances in understanding the UPS and its components are paving the way for novel drug development, offering hope for effective treatments for a wide range of UPS-associated diseases.

This chapter explores the UPS and its crucial role in maintaining cellular health. It explains how the UPS works, including the processes of tagging unwanted proteins with ubiquitin, recognizing these tagged proteins, and breaking them down using the proteasome. The focus is on how this system ensures proper protein turnover and prevents the accumulation of damaged or misfolded proteins, which can be harmful to cells.

This chapter highlights the importance of the UPS in key cellular activities such as regulating the cell cycle, managing stress responses, controlling immune functions, and maintaining protein quality. It discusses what happens when the UPS does not function properly, leading to various diseases, for example, AD and PD, cancers, and heart diseases. By understanding these disease mechanisms, the reader can appreciate the role of UPS dysfunction in human health.

2.1.2 Significance of UPS in Cellular Function

UPS is indispensable for cellular homeostasis and functionality. It serves as the primary pathway for targeted intracellular protein degradation, ensuring the regulated turnover of proteins and maintaining a dynamic proteome. The UPS maintains cellular health by specifically targeting and degrading short-lived, misfolded, or damaged proteins, thereby preventing the accumulation of toxic proteins. This functionality is critical for diverse cellular activities, including cell cycle control, gene transcription, DNA repair, and signal transduction [6]. The UPS regulates the levels of essential regulatory proteins, such as cyclins, transcription factors, and tumor suppressors, allowing cells to respond effectively to physiological and environmental changes [7].

One of the hallmark roles of the UPS is in the cell cycle. Regulatory proteins, such as cyclins and cyclin-dependent kinase inhibitors, undergo timed degradation through ubiquitination, ensuring orderly progression through cell cycle phases [14]. This precisely controlled proteolysis also governs processes such as mitosis and apoptosis, making the UPS essential for cellular proliferation and survival. For example, during DNA damage, the UPS regulates p53 and other repair proteins, orchestrating cell cycle arrest and DNA repair or inducing apoptosis when damage is irreparable [15].

Beyond proteostasis, the UPS contributes significantly to cellular signaling pathways by modulating protein activity and receptor turnover. Ubiquitination regulates protein localization, interaction, and activity, thereby influencing critical processes such as inflammatory responses, immune signaling, and hormonal regulation. For instance, in immune cells, the UPS processes antigenic peptides for presentation via the major histocompatibility complex (MHC) class I molecules, a function integral to adaptive immunity [8]. In inflammatory signaling, UPS-mediated degradation of inhibitors, such as IκBα (inhibitors of kappa Bα), is necessary for the activation of transcription factors like NF-κB (nuclear factor kappa-light-chain-enhancer of activated B cells), which control inflammatory and immune responses [16].

The UPS also interacts with other degradation pathways, such as autophagy, forming a coordinated network to maintain protein and organelle quality control. Dysregulation of this system can have profound consequences, contributing to diseases ranging from neurodegeneration to cancer and cardiovascular disorders [17]. Importantly, these interconnected roles of the UPS highlight its centrality to cellular function and survival.

2.2 Structure and Function of UPS

2.2.1 Ubiquitination: Key Molecular Players and Pathway

Ubiquitination is a highly regulated post-translational modification (PTM) process that tags proteins for degradation, signaling, or localization. This process is central to maintaining proteostasis and involves the covalent attachment of ubiquitin, a 76-amino-acid polypeptide, to substrate proteins via an enzymatic cascade. The ubiquitination pathway consists of three main enzymatic steps: ubiquitin activation by E1 enzymes, ubiquitin conjugation by E2 enzymes, and substrate ligation by E3 ligases [18].

Ubiquitin itself plays a critical role as the functional unit of this pathway. Structurally, it contains seven lysine residues (K6, K11, K27, K29, K33, K48, and K63) and an N-terminal methionine, which form the basis for its versatile chain architecture. The specific linkage types dictate the biological outcomes of ubiquitination, with K48-linked chains typically signaling proteasomal degradation and K63-linked chains involved in non-proteolytic functions such as DNA repair and signal transduction [19].

The ubiquitination process begins with E1 ubiquitin-activating enzymes, which utilize ATP to activate ubiquitin by creating a high-energy thioester bond between the C-terminal glycine of ubiquitin and the active-site cysteine of the E1 enzyme. This activation is essential for preparing ubiquitin for subsequent transfer. The activated ubiquitin is then handed off to E2 ubiquitin-conjugating enzymes, which act as carriers in the pathway. Finally, E3 ubiquitin ligases facilitate the attachment of ubiquitin from E2 enzymes to lysine residues on substrate proteins. These E3 ligases ensure specificity by identifying the substrate proteins and catalyzing the formation of the isopeptide bond [2].

E3 ligases can be categorized into two primary families: HECT (Homologous to the E6AP Carboxyl Terminus) and RING (Really Interesting New Gene) ligases. HECT ligases first form an intermediate thioester bond with ubiquitin before transferring it to the target protein. In contrast, RING ligases enable the direct transfer of ubiquitin from the E2 enzyme to the substrate. This step determines the substrate specificity and the type of ubiquitin linkage formed, which, in turn, dictates the cellular fate of the tagged protein [20, 21].

Understanding the roles of E1, E2, and E3 enzymes is critical as they are implicated in various cellular processes and diseases. Aberrations in ubiquitination contribute to conditions such as cancer, neurodegeneration, and immune disorders, making the pathway an important therapeutic target.

2.3 Structure and Function of the Proteasome

The proteasome is a highly conserved and essential multicatalytic protein complex responsible for degrading ubiquitin-tagged proteins within eukaryotic cells. It is central to the maintenance of cellular proteostasis and plays a pivotal role in the regulated turnover of misfolded, damaged, or short-lived regulatory proteins. The primary functional form is the 26S proteasome, composed of a 20S catalytic CP and one or two 19S RPs, which assemble into a dynamic and intricate molecular machinery [22].

2.3.1 20S CP

The 20S CP is responsible for the proteolytic function of the proteasome. It has a cylindrical architecture composed of four stacked heptameric rings. The outer rings, formed by α-subunits, act as gatekeepers that regulate substrate entry. The inner rings consist of β-subunits, which contain the proteolytic active sites. These active sites perform three distinct proteolytic functions: chymotrypsin-like, trypsin-like, and caspase-like activities, allowing for the degradation of substrate proteins into oligopeptides [23]. Substrate access to the catalytic chamber is regulated by the α-subunits, which form a gate that is opened upon interaction with the 19S RP.

2.3.2 19S RP

The 19S RP is crucial for substrate recognition, unfolding, deubiquitination, and translocation into the 20S CP. It consists of two major subcomplexes: the base and the lid. The base includes ATPase subunits that provide the energy required for substrate unfolding and channel opening in the CP, while the lid contains non-ATPase subunits responsible for binding polyubiquitin chains and processing them via deubiquitinating enzymes (DUBs) [24]. This ensures that only properly tagged substrates are directed into the proteasome.

2.4 Proteasome Function

The primary function of the proteasome is protein degradation, tightly regulated by ubiquitination to ensure the specific targeting of unwanted or regulatory proteins. By degrading cyclins, p53, and transcription factors, the proteasome regulates critical processes such as cell cycle progression, apoptosis, and gene expression [25]. It also prevents toxic protein aggregation by removing misfolded and oxidatively damaged proteins, thus maintaining cellular health under stress conditions [6].

In addition to proteostasis, proteasomes play specialized roles in immune responses, such as processing antigenic peptides presented on MHC class I molecules. Immunoproteasomes, variants of the standard 20S CP with specific subunit compositions, enhance the generation of peptides optimized for MHC class I binding [25].

Dysfunction or dysregulation of proteasome activity is implicated in various pathological states, including neurodegenerative diseases such as AD, cancers, and cardiovascular disorders. Consequently, targeting the proteasome, as with the proteasome inhibitor bortezomib in cancer therapy, represents a therapeutic strategy [26].

2.4.1 Role of UPS in Protein Homeostasis

The UPS is a critical cellular mechanism that ensures protein homeostasis (proteostasis), which refers to the balance of protein synthesis, folding, modification, and degradation within the cell. By maintaining proteostasis, the UPS helps prevent the accumulation of damaged, misfolded, or surplus proteins, which, if unregulated, could disrupt cellular function and contribute to disease development [27].

The UPS functions through two main processes: ubiquitination and proteasomal degradation. Ubiquitination involves marking damaged or surplus proteins with ubiquitin, a small regulatory protein. This tagging is achieved through the orchestrated actions of three enzyme types: E1 (activating enzyme), E2 (conjugating enzyme), and E3 (ligase). This mechanism ensures precise targeting of specific proteins for destruction. Following ubiquitination, the tagged proteins are transported to the 26S proteasome, a large proteolytic complex. Inside the proteasome, the proteins are unfolded, transported into a catalytic core, and degraded into short peptides. These peptides are subsequently broken down into amino acids, which are then recycled for cellular processes [28].

The UPS is particularly vital during stress conditions, such as oxidative stress or heat shock, when cells face an increased burden of misfolded and damaged proteins. Under these circumstances, the UPS, in conjunction with molecular chaperones, prevents the accumulation of toxic protein aggregates, protecting the cell from damage [29]. However, when UPS function is impaired, protein aggregates accumulate, a characteristic of neurodegenerative disorders such as Alzheimer's and Parkinson's [30].

In addition to its role in quality control, the UPS is crucial for regulating various cellular functions. It manages the degradation of regulatory proteins, such as cyclins, transcription factors, and tumor suppressors, thereby influencing processes such as cell cycle progression, DNA repair, and apoptosis.

Dysregulation of the UPS can result in cancer, immune disorders, and other diseases [31].

The UPS also interacts with autophagy, another protein degradation pathway, to ensure a coordinated response to proteotoxic stress. This interplay ensures efficient clearance of larger protein aggregates and damaged organelles, underscoring the system's critical role in sustaining cellular health [32].

2.5 Mechanisms Leading to Dysfunction in the UPS

The UPS, a key regulator of protein degradation and homeostasis, can malfunction through diverse mechanisms, profoundly affecting cellular function. Dysfunction in the UPS is associated with various pathological conditions, including neurodegenerative diseases, cancer, and immune disorders. In the following, we detail how genetic mutations, PTMs, and external factors disrupt UPS function (Fig. 2.1).

2.5.1 Genetic Mutations in UPS Components

Genetic mutations in UPS components compromise the system's functionality by altering protein stability, substrate recognition, or enzymatic activity. Mutations in the FBXO7 gene, encoding an F-box protein, disrupt SCF (Skp1-Cullin-F-box complex)-type E3 ubiquitin ligase activity, contributing to early-onset Parkinsonism. The loss of proper substrate specificity results in protein aggregation and cellular toxicity [33]. Similarly, mutations in CUL4B, a core component of cullin-RING E3 ligases, impair ubiquitin-mediated protein degradation, leading to developmental syndromes and impaired DNA damage repair [34].

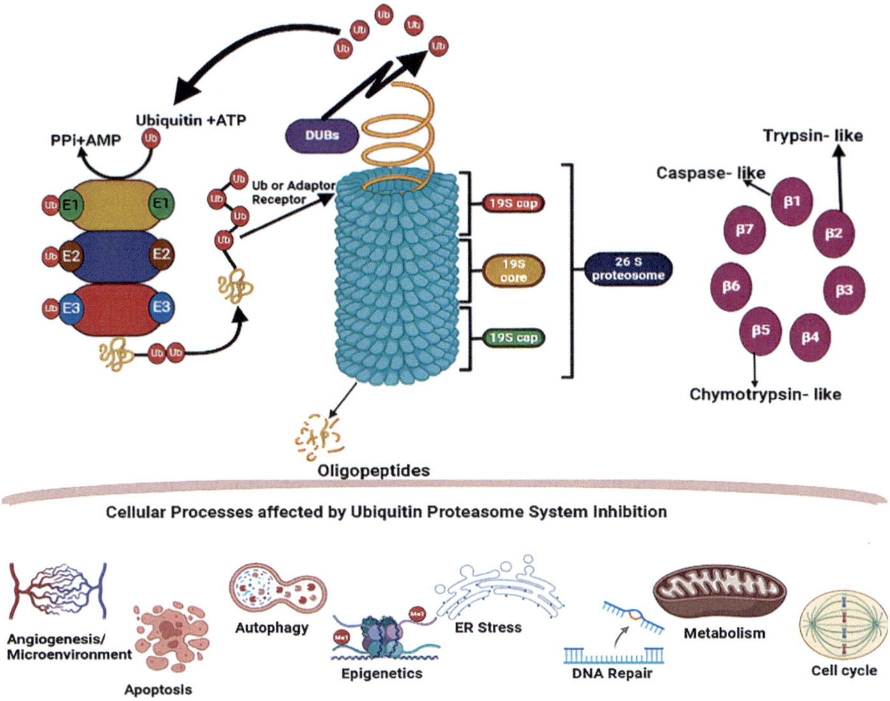

Fig. 2.1 UPS mechanism and its impact on cellular process

2.5.2 PTMs Disrupting UPS Function

PTMs of UPS components, phosphorylation, acetylation, and oxidation, can disrupt their activity, leading to proteostasis imbalance. Phosphorylation of E3 ligases like TRIM32 can alter substrate binding and ubiquitination efficiency, contributing to diseases like limb-girdle muscular dystrophy [35]. Moreover, oxidative stress modifies proteasomal subunits through carbonylation or nitration, reducing proteasomal activity and leading to the accumulation of ubiquitinated proteins, as observed in neurodegenerative conditions such as AD [36]. Acetylation of the 20S proteasome, which normally enhances degradation of oxidized proteins, may also result in an imbalance in protein turnover when dysregulated [37].

2.5.3 External Factors: Environmental Toxins, Stress, and Infections

External stressors, including environmental toxins, cellular stress, and infections, are potent disruptors of UPS function. Cadmium and arsenic inhibit proteasome activity by binding to its active sites, resulting in protein aggregation and oxidative stress. This has been linked to the pathogenesis of diseases such as cancer and cardiovascular disorders [38]. Infectious agents like viruses can hijack the UPS for their replication or suppress its function to evade immune responses. For example, human papillomavirus encodes E6 protein, which recruits host E3 ligases like E6AP to degrade tumor suppressors such as p53, fostering oncogenesis [39]. Additionally, chronic oxidative and metabolic stress induced by diabetes or aging impairs proteasome activity, enhancing vulnerability to disease [40].

These mechanisms underline the complexity of UPS dysregulation and its broad implications for health and disease. A deeper understanding of these processes could inform strategies to target specific disruptions therapeutically, providing new avenues for intervention.

2.6 Pathophysiological Consequences of UPS Dysfunction

The UPS is integral to maintaining proteostasis and regulating crucial cellular processes. Dysfunction in the UPS has profound pathophysiological consequences, including protein misfolding and accumulation, altered protein turnover, and the activation of compensatory pathways. These disruptions are associated with several diseases, including neurodegeneration, cancer, and immune disorders.

2.6.1 Protein Misfolding and Accumulation

One of the primary consequences of UPS dysfunction is the accumulation of misfolded or damaged proteins within the cell. In healthy conditions, the UPS selectively degrades misfolded proteins, preventing their aggregation and the subsequent generation of proteotoxic stress (Fig. 2.2) [41]. When the UPS is impaired, these proteins accumulate and form insoluble aggregates, which can disrupt cellular architecture and function. This phenomenon is particularly evident in neurodegenerative disorders such as AD, PD, and Huntington's disease (HD), where aggregated proteins such as amyloid-beta (Aβ), alpha-synuclein, and huntingtin (HTT) create cellular inclusions that impair neuronal function and survival [42].

2.6.2 Altered Protein Turnover

The UPS also plays a crucial role in regulating the turnover of proteins, including key signaling molecules and cell-cycle regulators. Dysfunction in this system results in the dysregulation of cell signaling pathways. For instance, impaired degradation of proteins such as beta-catenin and NF-κB can lead to enhanced proliferation and inflammation, contributing to oncogenesis and autoimmune diseases [43]. Additionally, improper clearance of cyclins or p53 results in disrupted cell-cycle regulation and apoptosis, further exacerbating pathological conditions such as cancer and tissue degeneration [44].

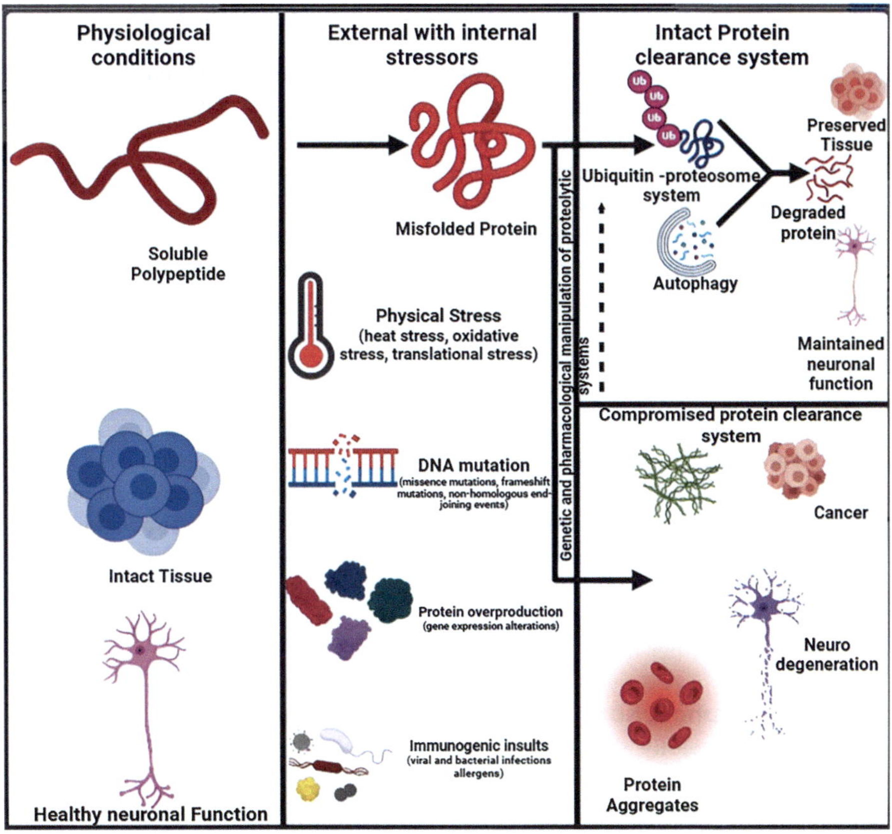

Fig. 2.2 Role of UPS in protein quality control and cellular outcomes

2.6.3 Activation of Compensatory Pathways

When UPS activity is compromised, cells activate compensatory pathways to manage protein homeostasis, most notably autophagy and stress response mechanisms. Autophagy, a lysosome-dependent degradation system, can partially compensate by removing protein aggregates and damaged organelles. For example, increased autophagic flux has been observed in response to UPS inhibition in experimental models of neurodegenerative diseases [45]. Furthermore, stress response pathways, such as the unfolded protein response and heat shock response, are upregulated to promote the expression of molecular chaperones that facilitate protein refolding or degradation [45]. However, these compensatory mechanisms are often insufficient or maladaptive in chronic conditions, contributing to cellular stress and disease progression.

In summary, UPS dysfunction leads to proteostasis failure, disrupting cellular function and promoting disease. Understanding these mechanisms highlights the critical role of UPS in cellular health and its potential as a therapeutic target.

2.7 Diseases Mechanisms Linked to UPS Dysfunction

2.7.1 Ubiquitination: Key Molecular Players and Pathway of Alzheimer's Disease

The UPS is crucial for maintaining protein homeostasis (proteostasis) by regulating the degradation of misfolded, damaged, or short-lived proteins. This tightly regulated system is particularly critical in neurons, which are highly susceptible to protein aggregation due to their post-mitotic nature. In AD, the UPS is often compromised, leading to the accumulation of toxic protein aggregates, including Aβ and hyperphosphorylated tau, two hallmark features of the disease pathology.

2.7.1.1 Mechanisms of UPS Dysfunction in AD

Proteasome Inhibition by Aβ

Aβ peptides, especially their oligomeric forms, directly inhibit proteasomal activity by binding to the catalytic subunit of the 26S proteasome, blocking substrate entry. This interaction reduces the clearance of ubiquitinated substrates, contributing to toxic protein buildup [46]. Intracellular Aβ accumulation disrupts the activity of DUBs like ubiquitin carboxyl-terminal hydrolase L1 (UCHL1), critical for recycling ubiquitin, further impairing the UPS [47].

2.7.1.2 Hyperphosphorylated Tau Protein Aggregation

Tau proteins, primarily stabilizers of microtubules, become hyperphosphorylated and aggregate into neurofibrillary tangles. Normally degraded by the UPS, these tau aggregates overwhelm the system. Dysregulation of specific E3 ligases such as CHIP (carboxy terminus of Hsc70-interacting protein) prevents effective ubiquitination of abnormal tau, contributing to its accumulation [48]. Tau tangles also indirectly impair proteasome function by physically disrupting the proteasomal machinery and inducing oxidative stress [49].

Effects of UBB+1

The mutant UBB+1, which arises due to mRNA frameshift mutations, is resistant to proteasomal degradation and acts as a dominant-negative inhibitor. It accumulates in AD brain regions, such as the hippocampus, and inhibits proteasome function, creating a cycle of protein aggregation and neuronal toxicity [50].

2.7.1.3 Altered APP and ERAD Dysfunction

Amyloid precursor protein (APP) is processed through endoplasmic reticulum (ER)-associated degradation (ERAD), which is closely linked to the UPS. Downregulation of HRD1, a key E3 ligase in ERAD, reduces APP clearance, increasing its cleavage by beta- and gamma-secretases, leading to enhanced production of Aβ peptides [51]. ER stress, common in AD, also disrupts ERAD activity, adding to the proteostatic imbalance.

2.7.1.4 Genetic Contributions

Mutations and altered expression of genes involved in the UPS exacerbate dysfunction in AD:

UCHL1 Reduced levels of this DUB lead to impaired ubiquitin recycling, accumulation of protein aggregates, and synaptic dysfunction [52].

Parkin (PARK2) Loss or downregulation of this E3 ligase limits the removal of damaged mitochondrial proteins, contributing to oxidative stress and Aβ accumulation [53].

FBXO2 and FBXO45 Components of the E3 ubiquitin ligase complexes are downregulated in AD, contributing to neuronal apoptosis and impaired synaptic signaling [54].

Oxidative Stress and Inflammation Oxidative damage alters UPS components, such as the 26S proteasome and E2/E3 enzymes, further reducing their activity. This feeds a vicious cycle where impaired protein degradation exacerbates oxidative damage [55]. Inflammation triggered by glial cells in AD promotes the release of cytokines, such as tumor necrosis factor-alpha (TNF-α), which impair proteasome function through PTMs like nitrosylation [56].

Autophagy

The UPS interacts with autophagy, another degradation pathway. When autophagy is impaired in AD, the proteasome is overwhelmed with substrates, amplifying protein aggregation and cellular stress. Defects in both systems synergistically worsen neurodegeneration [57].

2.7.2 Parkinson's Disease

PD is a chronic neurodegenerative condition marked by the degeneration of dopaminergic neurons in the substantia nigra. This neuronal loss results in motor symptoms, including slowed movement (bradykinesia), muscle rigidity, and tremors. One of the key molecular mechanisms implicated in Parkinson's pathology is dysfunction of the UPS. The UPS, responsible for degrading damaged, misfolded, or unnecessary proteins, fails in PD, resulting in protein aggregation, mitochondrial dysfunction, and neuronal death.

2.7.2.1 Mechanisms of UPS Dysfunction in PD

In PD, abnormalities in various components of the UPS result in defective protein degradation, leading to intracellular accumulation of toxic proteins like α-synuclein.

2.7.2.2 α-Synuclein Aggregation

α-Synuclein is the main component of Lewy bodies, a pathological hallmark of PD. Normally degraded by the UPS, misfolded or overexpressed α-synuclein overwhelms the UPS, reducing its ability to handle other substrates [58].

2.7.2.3 Feedback Inhibition

Aggregated α-synuclein directly binds to the proteasome, inhibiting its activity. It also impairs the E1–E2–E3 cascade by sequestering ubiquitin and UPS enzymes [59].

2.7.2.4 Mutations in UPS-Related Genes

Parkin (PARK2) Parkin is an E3 ubiquitin ligase involved in targeting damaged mitochondria for degradation via mitophagy. Mutations in the parkin gene reduce its ability to ubiquitinate mitochondrial substrates such as VDAC1 and Mfn2, leading to mitochondrial dysfunction and neurotoxicity [60, 61].

UCHL1 (Ubiquitin C-Terminal Hydrolase-L1) UCHL1 recycles ubiquitin by cleaving it from polyubiquitin chains. Mutations in UCHL1 reduce available free ubiquitin, impairing protein turnover and contributing to the accumulation of toxic proteins [62].

PINK1 (PTEN-Induced Kinase 1) PINK1 collaborates with Parkin to regulate mitophagy by tagging damaged mitochondria for degradation. Mutations disrupt this pathway, leading to mitochondrial accumulation and oxidative stress [63].

2.7.2.5 Proteasomal Inhibition by Oxidative Stress

Dopaminergic neurons in the substantia nigra are highly susceptible to oxidative damage, largely due to the metabolic processes involved in dopamine breakdown. Reactive oxygen species oxidize proteasomal subunits, reducing their activity and preventing efficient protein degradation. Oxidized α-synuclein and lipids further impair the proteasome, creating a cycle of proteostatic stress [64].

2.7.2.6 Dysregulation of Ubiquitin Levels

Role of Mono- and Polyubiquitination Disruption in ubiquitin supply due to mutations in UCHL1 or E1/E2 enzymes reduces the tagging efficiency of the UPS. Inadequate ubiquitination leads to persistence of misfolded proteins, which form aggregates that are toxic to neurons [65].

2.7.2.7 UPS-Mediated NeuroInflammation

Microglia activation in PD releases pro-inflammatory cytokines like TNF-α, which can modify UPS components through nitration or oxidation, impairing their function. Chronic neuroinflammation exacerbates proteasomal dysfunction [66].

2.7.2.8 Interaction with Other Pathways

Mitophagy and Autophagy While the UPS primarily degrades soluble proteins, insoluble protein aggregates and damaged organelles are degraded by autophagy.

Impairments in autophagy, often seen alongside UPS dysfunction in PD, intensify proteostatic stress, contributing to neurodegeneration [67].

ERAD The UPS and ERAD work together to clear misfolded proteins from the ER. In PD, impaired ERAD exacerbates proteostasis collapse, increasing cellular stress [68].

2.7.3 Huntington's Disease

HD is a hereditary neurodegenerative disorder with an autosomal dominant inheritance pattern. It is caused by an expanded CAG trinucleotide repeat in the HTT gene, which results in the production of an abnormal polyglutamine (polyQ) tract within the HTT protein. This abnormal HTT protein forms toxic aggregates that contribute to neuronal dysfunction and death, particularly in the striatum and cortex. A critical mechanism implicated in HD pathology is the dysfunction of the *UPS*, which normally degrades misfolded and aggregated proteins.

2.7.3.1 Inefficient Degradation of Mutant Huntingtin

The expanded polyQ tract in mutant huntingtin (mHTT) impairs its proper folding, leading to misfolded protein species that aggregate and resist proteasomal degradation [69]. mHTT fragments enter the proteasome and block its activity, impairing degradation of other cellular substrates and causing widespread proteostasis disruption [70]. These aggregates can inhibit the 19S regulatory subunit of the proteasome, preventing substrate recognition and processing [71].

2.7.4 Cancer

The UPS is a critical regulator of many cellular processes, including the degradation of damaged or misfolded proteins, cell cycle control, signaling pathways, and apoptosis. The proper functioning of UPS is vital for cellular homeostasis, and its dysregulation is a key contributor to cancer progression. UPS dysregulation in cancer often involves the overexpression of specific components such as E3 ubiquitin ligases, proteasomal subunits, and DUBs, which enhance tumor growth, survival, and resistance to therapies.

A key player in the regulation of the UPS in cancer is the tumor suppressor protein p53 (encoded by the TP53 gene). p53 governs critical cellular processes such as cell cycle arrest, DNA repair, and apoptosis. Under normal conditions, its levels are kept low by the E3 ubiquitin ligase MDM2 (Mouse Double Minute 2), which facilitates its degradation via the proteasome. In response to DNA damage, p53 becomes stabilized, triggering cell cycle arrest and programmed cell death. However, in many cancers, MDM2 is overexpressed, leading to the excessive degradation of p53 and promoting unregulated cell growth. This disruption of p53 regulation is implicated in cancer formation and progression. Targeting the MDM2/p53 interaction has become a therapeutic strategy, with several MDM2 inhibitors under investigation to restore p53 function in tumors that rely on its inhibition [72, 73].

Another key regulator of cancer development through UPS dysfunction is cyclin D1 (CCND1). Cyclins are regulatory proteins that control the progression of cells through the cell cycle. CCND1 is involved in the transition from the G1 to the S phase, and its stability is tightly regulated by UPS. Normally, CCND1 is ubiquitinated and degraded by the SCF complex (an E3 ubiquitin ligase complex), but in many cancers, the regulation of CCND1 is disrupted, leading to its accumulation and sustained cell cycle progression. CCND1 overexpression promotes cancer cell proliferation and contributes to tumorigenesis in cancers such as breast, liver, and esophageal carcinoma [74].

The UPS also plays a crucial role in regulating NF-κB, a transcription factor essential for controlling immune responses, inflammation, and cell survival. Under normal conditions, the IκB proteins prevent the nuclear translocation of NF-κB dimers. IκB degradation, mediated by E3 ligases, allows the release and nuclear translocation of NF-κB to initiate transcription of pro-inflammatory and survival genes. In cancer cells, aberrant activation of NF-κB is common and is often triggered by mutations in E3 ligases or DUBs, resulting in enhanced immune evasion, inflammation, and resistance to cell death. This dysregulated NF-κB signaling is particularly implicated in the progression of lymphoma, breast cancer, and prostate cancer [75, 76].

The proteasome itself, responsible for the degradation of polyubiquitinated substrates, also plays a significant role in cancer. In response to the increased burden of misfolded and mutated proteins in cancer cells, there is often upregulation of proteasomal activity. Elevated levels of proteasome subunits such as proteasomal subunit alpha and proteasomal subunit beta facilitate this process, helping the cancer cell cope with oxidative stress, altered protein expression, and apoptotic signals. Additionally, DUBs, such as ubiquitin-specific protease 7, further stabilize key oncogenic proteins by reversing the ubiquitination process, allowing them to accumulate. Elevated levels of DUBs in cancer can promote tumor growth, survival, and resistance to therapies [77, 78].

UPS-related mutations can activate oncogenes or impair tumor suppressors. For example, mutations in the E3 ligases responsible for regulating proteins such as p53 and CCND1 lead to improper degradation and stabilization of oncoproteins, contributing to unchecked cell division and tumor development. The inhibition of these mutations and the restoration of normal proteolysis mechanisms are currently major areas of cancer research. Pharmacological interventions targeting the proteasome and E3 ligases hold promise as a therapeutic avenue to correct UPS dysfunction in cancer [79].

2.7.5 Cardiovascular Disease

In heart failure, the UPS regulates the degradation of misfolded proteins, damaged proteins, and regulatory proteins involved in hypertrophic signaling. One of the key genes implicated in this process is *MuRF1* (muscle-specific ring finger protein 1), an E3 ubiquitin ligase. *MuRF1* plays a crucial role in the ubiquitination and degradation of contractile proteins in cardiac muscle cells, and its overexpression has been observed in pathological cardiac hypertrophy. The accumulation of abnormal proteins, due to impaired UPS function, leads to cardiac dysfunction, with a reduction in contractility and progression toward heart failure [80, 81].

The UPS also plays a role in regulating ion channels, which are vital for maintaining cellular excitability in the heart. For instance, the E3 ligase NEDD4 (neural precursor cell-expressed developmentally downregulated 4) has been found to ubiquitinate ion channels, including Na+/K+ ATPase and potassium (K+) channels. These channels regulate the electrochemical gradients essential for normal myocardial contraction and relaxation. Disruption of *NEDD4* activity and improper regulation of ion channels have been linked to arrhythmias and increased risk of cardiovascular events. Furthermore, other components such as *Parkin* and *PINK1*, which are involved in mitophagy, also play roles in maintaining mitochondrial integrity in cardiac cells, with mutations leading to a buildup of damaged mitochondria, ultimately contributing to cardiovascular dysfunction [82, 83].

In atherosclerosis, the UPS plays a critical role in controlling inflammation, which is one of the hallmarks of this disease. Proteins involved in immune response and cell adhesion, such as tumor necrosis factor receptor-associated factor 6 (TRAF6), are regulated by the UPS. *TRAF6* ubiquitinates several key signaling molecules, including *NF-κB*, a master regulator of inflammatory responses. Dysregulation of this pathway due to aberrant *E3* ligase activity or impaired proteasomal degradation of NF-κB inhibitors can lead to chronic inflammation, endothelial cell activation, and the formation of atherosclerotic plaques [84].

Another aspect of UPS function that affects cardiovascular diseases is its role in regulating smooth muscle cell differentiation and vascular remodeling. Homologous to E6AP carboxyl terminus 10, another E3 ligase, has been shown to play a crucial role in vascular smooth muscle cells (VSMCs) by regulating the levels of focal adhesion kinase, which is essential for VSMC contraction. The dysregulation of such regulatory proteins contributes to vascular stiffness and remodeling, phenomena that occur in various cardiovascular conditions, including aortic aneurysms and hypertension [85].

The UPS also regulates the degradation of apoptotic regulators, including *cIAP1* (cellular inhibitor of apoptosis protein 1) and *cFLIP* (FLICE-like inhibitory protein). These proteins, through their proteasomal degradation, prevent excessive apoptosis in endothelial cells. However, when the UPS becomes impaired, an accumulation of anti-apoptotic proteins can prevent the normal removal of damaged or senescent cells in the vascular system, contributing to vascular remodeling, fibrosis, and maladaptive cardiac responses. This aberrant regulation may be an underlying factor in the progression of heart failure and atherosclerosis [86].

2.7.6 Role of the UPS in Immune Dysregulation and Inflammatory Diseases

The UPS is essential for maintaining the balance of immune cell activation and function, thus regulating inflammatory responses. Dysregulation of the UPS has been linked to various immune dysregulation conditions, including autoimmune diseases and chronic inflammation. By controlling the degradation of key signaling molecules, transcription factors, and receptors, UPS plays a fundamental role in orchestrating both innate and adaptive immune responses. This regulation impacts a variety of immune processes such as cytokine production, immune cell differentiation, and antigen presentation.

The UPS plays a significant role in autoimmunity through its regulation of the NF-κB pathway. NF-κB is a key transcription factor involved in the immune response, controlling the production of pro-inflammatory cytokines such as TNF-α and interleukin-1 beta. In conditions of immune dysfunction, this pathway is often hyperactive. The UPS regulates NF-κB activity by degrading IκB proteins, which serve as inhibitors that prevent the nuclear translocation of NF-κB dimers. Alterations in E3 ligases such as *TRAF6* and Casitas B-lineage lymphoma can lead to an uncontrolled activation of NF-κB, resulting in chronic inflammation associated with diseases such as rheumatoid arthritis and systemic lupus erythematosus [87, 88].

The UPS regulates the function of antigen-presenting cells, including dendritic cells and macrophages, by controlling the degradation of MHC class II molecules and related signaling proteins. Proteins such as CIAP1 and E2 transcription factor 1, along with other ubiquitin-conjugating enzymes, play a crucial role in modulating antigen presentation to T-cells. Overactive UPS-mediated degradation of these components can impair antigen presentation, impairing the proper activation of T cells, leading to immune tolerance failure and autoimmunity [89, 90].

In autoimmune diseases like RA, the UPS plays a crucial role in regulating the degradation of the pro-inflammatory transcription factor *JUN* (AP-1) and cytokines, which contribute to synovial inflammation and cartilage destruction. Elevated levels of *MuRF1*, an E3 ligase, have been implicated in the degradation of myofilament proteins and other muscle regulatory components in RA, indicating the role of UPS in inflammation-induced tissue remodeling [91]. Furthermore, excessive degradation of apoptotic inhibitors like *cFLIP* in macrophages is related to increased cell survival and the persistence of inflammation in chronic conditions such as inflammatory bowel disease [92].

Chronic inflammation associated with UPS dysregulation can also contribute to the development of cancer, as the immune system's inability to regulate inflammatory signals effectively facilitates tumor promotion. E3 ligases such as *TRAF6* and *NEDD4* mediate the degradation of key modulators of immune cell function and inflammation. This alteration disrupts the normal balance of immune cell function and inflammatory mediators, creating an environment conducive to sustained inflammatory signaling that can exacerbate autoimmune and chronic diseases, including inflammatory skin conditions such as psoriasis [93].

Moreover, the UPS controls the degradation of forkhead box O transcription factors that regulate T-cell homeostasis and apoptosis. Altered UPS function can lead to abnormal immune responses by preventing the proper degradation of apoptosis-related proteins, which promotes the survival of autoreactive immune cells and contributes to inflammatory diseases. Dysregulation of the UPS has also been linked to malfunction in regulatory T-cells (*Tregs*), whose failure contributes to the breakdown of self-tolerance seen in autoimmune conditions [94].

2.8 Conclusion

The dysfunction of the UPS has emerged as a pivotal factor in the pathogenesis of numerous diseases, ranging from neurodegenerative disorders to cancers and immune dysregulation. The intricate balance maintained by the UPS in protein homeostasis underscores its critical role in cellular health. Advances in understanding the molecular underpinnings of UPS dysfunction have revealed novel therapeutic targets, such as proteasome inhibitors and E3 ligase modulators, which are being actively explored in clinical trials. Future research holds immense potential to further elucidate the complex interplay between UPS dysfunction and disease mechanisms, with the integration of systems biology, artificial intelligence, and high-throughput screening approaches promising to accelerate the discovery of precise biomarkers and tailored therapeutic strategies. Moreover, the development of innovative techniques such as targeted protein degradation and ubiquitin code editing could transform our ability to manipulate the UPS in a disease-specific manner. Such advances not only deepen our understanding of UPS-mediated pathologies but also pave the way for groundbreaking interventions in medicine.

References

1. Hershko A, Heller H, Elias S, Ciechanover A. Components of ubiquitin-protein ligase system. Resolution, affinity purification. J Biol Chem. 1983;258(13):8206–14. PMID: 6305978.
2. Zheng N, Shabek N. Ubiquitin ligases: structure, function, and regulation. Annu Rev Biochem. 2017;86:129–57. https://doi.org/10.1146/annurev-biochem-060815-014922.
3. Wenzel DM, Klevit RE. Following Ariadne's thread: a new perspective on RBR ubiquitin ligases. BMC Biol. 2012;10:24. https://doi.org/10.1186/1741-7007-10-24.
4. Komander D, Rape M. The ubiquitin code. Annu Rev Biochem. 2012;81:203–29. https://doi.org/10.1146/annurev-biochem-060310-170328.
5. Bard JAM, Goodall EA, Greene ER, Jonsson E, Dong KC, Martin A. Structure and function of the 26S proteasome. Annu Rev Biochem. 2018;87:697–724. https://doi.org/10.1146/annurev-biochem-062917-011931.
6. Goldberg AL. Protein degradation and protection against misfolded or damaged proteins. Nature. 2003;426:895–9. https://doi.org/10.1038/nature02263.
7. Pickart CM, Cohen RE. Proteasomes and their kin: proteases in the machine age. Nat Rev Mol Cell Biol. 2004;5:177–87. https://doi.org/10.1038/nrm1336.
8. Rock KL, Gramm C, Rothstein L, Clark K, Stein R, Dick L, Hwang D, Goldberg AL. Inhibitors of the proteasome block the degradation of most cell proteins and the generation of peptides presented on MHC class I molecules. Cell. 1994;78:761–71. https://doi.org/10.1016/S0092-8674(94)90462-6.
9. Scott DC, Monda JK, Grace CRR, Duda DM, Kriwacki RW, Kurz T, Schulman BA. A dual E3 mechanism for Rub1 ligation to Cdc53. Mol Cell. 2010;39:784–96. https://doi.org/10.1016/j.molcel.2010.08.030.
10. Ciechanover A, Kwon YT. Degradation of misfolded proteins in neurodegenerative diseases: therapeutic targets and strategies. Exp Mol Med. 2015;47:e147. https://doi.org/10.1038/emm.2014.117.
11. Bedford L, Lowe J, Dick LR, Mayer RJ, Brownell JE. Ubiquitin-like protein conjugation and the ubiquitin–proteasome system as drug targets. Nat Rev Drug Discov. 2011;10:29–46. https://doi.org/10.1038/nrd3321.

12. Willis MS, Patterson C. Proteotoxicity and cardiac dysfunction — Alzheimer's disease of the heart? N Engl J Med. 2013;368:455–64. https://doi.org/10.1056/NEJMra1106180.
13. Chauhan D, Bianchi G, Anderson KC. Targeting the UPS as therapy in multiple myeloma. BMC Biochem. 2008;9:S1. https://doi.org/10.1186/1471-2091-9-S1-S1.
14. King RW, Deshaies RJ, Peters J-M, Kirschner MW. How proteolysis drives the cell cycle. Science. 1996;274:1652–9. https://doi.org/10.1126/science.274.5293.1652.
15. Murray AW. Recycling the cell cycle. Cell. 2004;116:221–34. https://doi.org/10.1016/S0092-8674(03)01080-8.
16. Karin M, Ben-Neriah Y. Phosphorylation meets ubiquitination: the control of NF-κB activity. Annu Rev Immunol. 2000;18:621–63. https://doi.org/10.1146/annurev.immunol.18.1.621.
17. Ciechanover A, Brundin P. The ubiquitin proteasome system in neurodegenerative diseases. Neuron. 2003;40:427–46. https://doi.org/10.1016/S0896-6273(03)00606-8.
18. Guo HJ, Rahimi N, Tadi P. Biochemistry, ubiquitination. 2023. https://pubmed.ncbi.nlm.nih.gov/32310512/.
19. Dittmar G, Winklhofer KF. Linear ubiquitin chains: cellular functions and strategies for detection and quantification. Front Chem. 2020;7:915. https://doi.org/10.3389/fchem.2019.00915.
20. Lorenz S. Structural mechanisms of HECT-type ubiquitin ligases. Biol Chem. 2018;399:127–45. https://doi.org/10.1515/hsz-2017-0184.
21. Metzger MB, Pruneda JN, Klevit RE, Weissman AM. RING-type E3 ligases: master manipulators of E2 ubiquitin-conjugating enzymes and ubiquitination. Biochim Biophys Acta, Mol Cell Res. 2014;1843:47–60. https://doi.org/10.1016/j.bbamcr.2013.05.026.
22. Bard JAM, Bashore C, Dong KC, Martin A. The 26S proteasome utilizes a kinetic gateway to prioritize substrate degradation. Cell. 2019;177:286–298.e15. https://doi.org/10.1016/j.cell.2019.02.031.
23. Groll M, Ditzel L, Löwe J, Stock D, Bochtler M, Bartunik HD, Huber R. Structure of 20S proteasome from yeast at 2.4Å resolution. Nature. 1997;386:463–71. https://doi.org/10.1038/386463a0.
24. Glickman MH, Rubin DM, Coux O, Wefes I, Pfeifer G, Cjeka Z, Baumeister W, Fried VA, Finley D. A subcomplex of the proteasome regulatory particle required for ubiquitin-conjugate degradation and related to the COP9-signalosome and eIF3. Cell. 1998;94:615–23. https://doi.org/10.1016/S0092-8674(00)81603-7.
25. Collins GA, Goldberg AL. The logic of the 26S proteasome. Cell. 2017;169:792–806. https://doi.org/10.1016/j.cell.2017.04.023.
26. Kisselev AF, van der Linden WA, Overkleeft HS. Proteasome inhibitors: an expanding army attacking a unique target. Chem Biol. 2012;19:99–115. https://doi.org/10.1016/j.chembiol.2012.01.003.
27. Kim HT, Collins GA, Goldberg AL. Measurement of the multiple activities of 26S proteasomes. In: The ubiquitin proteasome system. New York: Humana Press; 2018. p. 289–308. https://doi.org/10.1007/978-1-4939-8706-1_19.
28. Loftus RM, Finlay DK. Immunometabolism: cellular metabolism turns immune regulator. J Biol Chem. 2016;291:1–10. https://doi.org/10.1074/jbc.R115.693903.
29. Benaroudj N, Zwickl P, Seemüller E, Baumeister W, Goldberg AL. ATP hydrolysis by the proteasome regulatory complex PAN serves multiple functions in protein degradation. Mol Cell. 2003;11:69–78. https://doi.org/10.1016/S1097-2765(02)00775-X.
30. Korhonen L, Lindholm D. The ubiquitin proteasome system in synaptic and axonal degeneration. J Cell Biol. 2004;165:27–30. https://doi.org/10.1083/jcb.200311091.
31. Hershko A, Ciechanover A. The ubiquitin system. Annu Rev Biochem. 1998;67:425–79. https://doi.org/10.1146/annurev.biochem.67.1.425.
32. Komatsu M, Waguri S, Koike M, Sou Y, Ueno T, Hara T, Mizushima N, Iwata J, Ezaki J, Murata S, Hamazaki J, Nishito Y, Iemura S, Natsume T, Yanagawa T, Uwayama J, Warabi E, Yoshida H, Ishii T, Kobayashi A, Yamamoto M, Yue Z, Uchiyama Y, Kominami E, Tanaka K. Homeostatic levels of p62 control cytoplasmic inclusion body formation in autophagy-deficient mice. Cell. 2007;131:1149–63. https://doi.org/10.1016/j.cell.2007.10.035.

33. Di Fonzo A, Dekker MCJ, Montagna P, Baruzzi A, Yonova EH, Guedes LC, Szczerbinska A, Zhao T, Dubbel-Hulsman LOM, Wouters CH, de Graaff E, Oyen WJG, Simons EJ, Breedveld GJ, Oostra BA, Horstink MW, Bonifati V. FBXO7 mutations cause autosomal recessive, early-onset parkinsonian-pyramidal syndrome. Neurology. 2009;72:240–5. https://doi.org/10.1212/01.wnl.0000338144.10967.2b.
34. Yin Y, Zhang L, Zeng Y, Chen D, Guan H, Ran G, Du K. Histone ubiquitination-related gene CUL4B promotes lung adenocarcinoma progression and cisplatin resistance. Front Genet. 2023;14:1242137. https://doi.org/10.3389/fgene.2023.1242137.
35. Kudryashova E, Kudryashov D, Kramerova I, Spencer MJ. Trim32 is a ubiquitin ligase mutated in limb girdle muscular dystrophy type 2H that binds to skeletal muscle myosin and ubiquitinates actin. J Mol Biol. 2005;354:413–24. https://doi.org/10.1016/j.jmb.2005.09.068.
36. Höhn A, Tramutola A, Cascella R. Proteostasis failure in neurodegenerative diseases: focus on oxidative stress. Oxidative Med Cell Longev. 2020;2020:1–21. https://doi.org/10.1155/2020/5497046.
37. Grune T, Catalgol B, Licht A, Ermak G, Pickering AM, Ngo JK, Davies KJA. HSP70 mediates dissociation and reassociation of the 26S proteasome during adaptation to oxidative stress. Free Radic Biol Med. 2011;51:1355–64. https://doi.org/10.1016/j.freeradbiomed.2011.06.015.
38. Yu X, Ponce RA, Faustman EM. Metals induced disruption of ubiquitin proteasome system, activation of stress signaling and apoptosis. In: Cellular effects of heavy metals. Dordrecht: Springer Netherlands; 2011. p. 291–311. https://doi.org/10.1007/978-94-007-0428-2_14.
39. Scheffner M, Huibregtse JM, Vierstra RD, Howley PM. The HPV-16 E6 and E6-AP complex functions as a ubiquitin-protein ligase in the ubiquitination of p53. Cell. 1993;75:495–505. https://doi.org/10.1016/0092-8674(93)90384-3.
40. Vilchez D, Saez I, Dillin A. The role of protein clearance mechanisms in organismal ageing and age-related diseases. Nat Commun. 2014;5:5659. https://doi.org/10.1038/ncomms6659.
41. Vabulas RM, Raychaudhuri S, Hayer-Hartl M, Hartl FU. Protein folding in the cytoplasm and the heat shock response. Cold Spring Harb Perspect Biol. 2010;2:a004390. https://doi.org/10.1101/cshperspect.a004390.
42. Dantuma NP, Bott LC. The ubiquitin-proteasome system in neurodegenerative diseases: precipitating factor, yet part of the solution. Front Mol Neurosci. 2014;7:70. https://doi.org/10.3389/fnmol.2014.00070.
43. Zhang L, Li Z, Skrzypczynska KM, Fang Q, Zhang W, O'Brien SA, He Y, Wang L, Zhang Q, Kim A, Gao R, Orf J, Wang T, Sawant D, Kang J, Bhatt D, Lu D, Li C-M, Rapaport AS, Perez K, Ye Y, Wang S, Hu X, Ren X, Ouyang W, Shen Z, Egen JG, Zhang Z, Yu X. Single-cell analyses inform mechanisms of myeloid-targeted therapies in colon cancer. Cell. 2020;181:442–459.e29. https://doi.org/10.1016/j.cell.2020.03.048.
44. Yuan L, Li P, Zheng Q, Wang H, Xiao H. The ubiquitin-proteasome system in apoptosis and apoptotic cell clearance. Front Cell Dev Biol. 2022;10:914288. https://doi.org/10.3389/fcell.2022.914288.
45. Menzies FM, Fleming A, Rubinsztein DC. Compromised autophagy and neurodegenerative diseases. Nat Rev Neurosci. 2015;16:345–57. https://doi.org/10.1038/nrn3961.
46. Myeku N, Duff KE. Targeting the 26S proteasome to protect against proteotoxic diseases. Trends Mol Med. 2018;24:18–29. https://doi.org/10.1016/j.molmed.2017.11.006.
47. Tramutola A, Di Domenico F, Barone E, Perluigi M, Butterfield DA. It is all about (U)biquitin: role of altered ubiquitin-proteasome system and UCHL1 in Alzheimer disease. Oxidative Med Cell Longev. 2016;2016:2756068. https://doi.org/10.1155/2016/2756068.
48. Petrucelli L. CHIP and Hsp70 regulate tau ubiquitination, degradation and aggregation. Hum Mol Genet. 2004;13:703–14. https://doi.org/10.1093/hmg/ddh083.
49. Klaips CL, Jayaraj GG, Hartl FU. Pathways of cellular proteostasis in aging and disease. J Cell Biol. 2018;217:51–63. https://doi.org/10.1083/jcb.201709072.
50. van Leeuwen FW, de Kleijn DPV, van den Hurk HH, Neubauer A, Sonnemans MAF, Sluijs JA, Köycü S, Ramdjielal RDJ, Salehi A, Martens GJM, Grosveld FG, Burbach JPH, Hol EM. Frameshift mutants of β amyloid precursor protein and ubiquitin-B in Alzheimer's and down patients. Science. 1998;279:242–7. https://doi.org/10.1126/science.279.5348.242.

51. Kaneko N, Marín O, Koike M, Hirota Y, Uchiyama Y, Wu JY, Lu Q, Tessier-Lavigne M, Alvarez-Buylla A, Okano H, Rubenstein JLR, Sawamoto K. New neurons clear the path of astrocytic processes for their rapid migration in the adult brain. Neuron. 2010;67:213–23. https://doi.org/10.1016/j.neuron.2010.06.018.
52. Mi Z, Graham SH. Role of UCHL1 in the pathogenesis of neurodegenerative diseases and brain injury. Ageing Res Rev. 2023;86:101856. https://doi.org/10.1016/j.arr.2023.101856.
53. Palacino JJ, Sagi D, Goldberg MS, Krauss S, Motz C, Wacker M, Klose J, Shen J. Mitochondrial dysfunction and oxidative damage in parkin-deficient mice. J Biol Chem. 2004;279:18614–22. https://doi.org/10.1074/jbc.M401135200.
54. Saiga T, Fukuda T, Matsumoto M, Tada H, Okano HJ, Okano H, Nakayama KI. Fbxo45 forms a novel ubiquitin ligase complex and is required for neuronal development. Mol Cell Biol. 2009;29:3529–43. https://doi.org/10.1128/MCB.00364-09.
55. Grimm S, Höhn A, Grune T. Oxidative protein damage and the proteasome. Amino Acids. 2012;42:23–38. https://doi.org/10.1007/s00726-010-0646-8.
56. Al-Ghraiybah NF, Wang J, Alkhalifa AE, Roberts AB, Raj R, Yang E, Kaddoumi A. Glial cell-mediated neuroinflammation in Alzheimer's disease. Int J Mol Sci. 2022;23:10572. https://doi.org/10.3390/ijms231810572.
57. Nixon RA, Yang D-S. Autophagy failure in Alzheimer's disease—locating the primary defect. Neurobiol Dis. 2011;43:38–45. https://doi.org/10.1016/j.nbd.2011.01.021.
58. Lee JE, Park B, Song SK, Sohn YH, Park H, Lee PH. A comparison of gray and white matter density in patients with Parkinson's disease dementia and dementia with Lewy bodies using voxel-based morphometry. Mov Disord. 2010;25:28–34. https://doi.org/10.1002/mds.22858.
59. Kim S, Kwon M, Hwang Y, Yoon J, Park S, Kang HC. Stress-induced NEDDylation promotes cytosolic protein aggregation through HDAC6 in a p62-dependent manner. iScience. 2021;24:102146. https://doi.org/10.1016/j.isci.2021.102146.
60. Shimura H, Hattori N, Kubo S, Mizuno Y, Asakawa S, Minoshima S, Shimizu N, Iwai K, Chiba T, Tanaka K, Suzuki T. Familial Parkinson disease gene product, parkin, is a ubiquitin-protein ligase. Nat Genet. 2000;25:302–5. https://doi.org/10.1038/77060.
61. Geisler S, Holmström KM, Skujat D, Fiesel FC, Rothfuss OC, Kahle PJ, Springer W. PINK1/Parkin-mediated mitophagy is dependent on VDAC1 and p62/SQSTM1. Nat Cell Biol. 2010;12:119–31. https://doi.org/10.1038/ncb2012.
62. Bishop P, Rocca D, Henley JM. Ubiquitin C-terminal hydrolase L1 (UCH-L1): structure, distribution and roles in brain function and dysfunction. Biochem J. 2016;473:2453–62. https://doi.org/10.1042/BCJ20160082.
63. Pickrell AM, Youle RJ. The roles of PINK1, parkin, and mitochondrial fidelity in Parkinson's disease. Neuron. 2015;85:257–73. https://doi.org/10.1016/j.neuron.2014.12.007.
64. Dias V, Junn E, Mouradian MM. The role of oxidative stress in Parkinson's disease. J Parkinsons Dis. 2013;3:461–91. https://doi.org/10.3233/JPD-130230.
65. Ebrahimi-Fakhari D, Cantuti-Castelvetri I, Fan Z, Rockenstein E, Masliah E, Hyman BT, McLean PJ, Unni VK. Distinct roles in vivo for the ubiquitin–proteasome system and the autophagy–lysosomal pathway in the degradation of α-synuclein. J Neurosci. 2011;31:14508–20. https://doi.org/10.1523/JNEUROSCI.1560-11.2011.
66. Reynolds AD, Kadiu I, Garg SK, Glanzer JG, Nordgren T, Ciborowski P, Banerjee R, Gendelman HE. Nitrated alpha-synuclein and microglial neuroregulatory activities. J Neuroimmune Pharmacol. 2008;3:59–74. https://doi.org/10.1007/s11481-008-9100-z.
67. Kocaturk NM, Gozuacik D. Crosstalk between mammalian autophagy and the ubiquitin-proteasome system. Front Cell Dev Biol. 2018;6:128. https://doi.org/10.3389/fcell.2018.00128.
68. Yoo YS, Han HG, Jeon YJ. Unfolded protein response of the endoplasmic reticulum in tumor progression and immunogenicity. Oxidative Med Cell Longev. 2017;2017:2969271. https://doi.org/10.1155/2017/2969271.
69. Jarosińska OD, Rüdiger SGD. Molecular strategies to target protein aggregation in Huntington's disease. Front Mol Biosci. 2021;8:769184. https://doi.org/10.3389/fmolb.2021.769184.
70. Harding RJ, Tong Y. Proteostasis in Huntington's disease: disease mechanisms and therapeutic opportunities. Acta Pharmacol Sin. 2018;39:754–69. https://doi.org/10.1038/aps.2018.11.

71. Ortega Z, Lucas JJ. Ubiquitin-proteasome system involvement in Huntington's disease. Front Mol Neurosci. 2014;7:77. https://doi.org/10.3389/fnmol.2014.00077.
72. Marine J-C. MDM2 and MDMX in cancer and development. Curr Top Dev Biol. 2011;94:45–75. https://doi.org/10.1016/B978-0-12-380916-2.00003-6.
73. Karni-Schmidt O, Lokshin M, Prives C. The roles of MDM2 and MDMX in cancer. Annu Rev Pathol. 2016;11:617–44. https://doi.org/10.1146/annurev-pathol-012414-040349.
74. Diehl JA. Cycling to cancer with cyclin D1. Cancer Biol Ther. 2002;1:226–31. https://doi.org/10.4161/cbt.72.
75. Ghosh S, Karin M. Missing pieces in the NF-κB puzzle. Cell. 2002;109:S81–96. https://doi.org/10.1016/S0092-8674(02)00703-1.
76. Hoesel B, Schmid JA. The complexity of NF-κB signaling in inflammation and cancer. Mol Cancer. 2013;12:86. https://doi.org/10.1186/1476-4598-12-86.
77. Colland F. The therapeutic potential of deubiquitinating enzyme inhibitors. Biochem Soc Trans. 2010;38:137–43. https://doi.org/10.1042/BST0380137.
78. Kouroukis CT, Baldassarre FG, Haynes AE, Imrie K, Reece DE, Cheung MC. Bortezomib in multiple myeloma: a practice guideline. Clin Oncol. 2014;26:110–9. https://doi.org/10.1016/j.clon.2013.11.022.
79. Glickman MH, Ciechanover A. The ubiquitin-proteasome proteolytic pathway: destruction for the sake of construction. Physiol Rev. 2002;82:373–428. https://doi.org/10.1152/physrev.00027.2001.
80. Li N, Wang H-X, Han Q-Y, Li W-J, Zhang Y-L, Du J, Xia Y-L, Li H-H. Activation of the cardiac proteasome promotes angiotension II-induced hypertrophy by down-regulation of ATRAP. J Mol Cell Cardiol. 2015;79:303–14. https://doi.org/10.1016/j.yjmcc.2014.12.007.
81. Maejima Y, Usui S, Zhai P, Takamura M, Kaneko S, Zablocki D, Yokota M, Isobe M, Sadoshima J. Muscle-specific RING finger 1 negatively regulates pathological cardiac hypertrophy through downregulation of calcineurin A. Circ Heart Fail. 2014;7:479–90. https://doi.org/10.1161/CIRCHEARTFAILURE.113.000713.
82. Tait SWG, Green DR. Mitochondrial regulation of cell death. Cold Spring Harb Perspect Biol. 2013;5:a008706. https://doi.org/10.1101/cshperspect.a008706.
83. Zhu J, Heidersbach A, Kathiriya IS, Garay BI, Ivey KN, Srivastava D, Han Z, King IN. The E3 ubiquitin ligase Nedd4/Nedd4L is directly regulated by microRNA 1. Development. 2017;144:866–75. https://doi.org/10.1242/dev.140368.
84. Rios FJ, de Ciuceis C, Georgiopoulos G, Lazaridis A, Nosalski R, Pavlidis G, Tual-Chalot S, Agabiti-Rosei C, Camargo LL, Dąbrowska E, Quarti-Trevano F, Hellmann M, Masi S, Lopreiato M, Mavraganis G, Mengozzi A, Montezano AC, Stavropoulos K, Winklewski PJ, Wolf J, Costantino S, Doumas M, Gkaliagkousi E, Grassi G, Guzik TJ, Ikonomidis I, Narkiewicz K, Paneni F, Rizzoni D, Stamatelopoulos K, Stellos K, Taddei S, Touyz RM, Virdis A. Mechanisms of vascular inflammation and potential therapeutic targets: a position paper from the ESH working group on small arteries. Hypertension. 2024;81:1218–32. https://doi.org/10.1161/HYPERTENSIONAHA.123.22483.
85. Shi N, Mei X, Chen S-Y. Smooth muscle cells in vascular Remodeling. Arterioscler Thromb Vasc Biol. 2019;39:e247–52. https://doi.org/10.1161/ATVBAHA.119.312581.
86. Sun H-J, Wu Z-Y, Nie X-W, Bian J-S. Role of endothelial dysfunction in cardiovascular diseases: the link between inflammation and hydrogen sulfide. Front Pharmacol. 2020;10:1568. https://doi.org/10.3389/fphar.2019.01568.
87. Barnabei L, Laplantine E, Mbongo W, Rieux-Laucat F, Weil R. NF-κB: at the borders of autoimmunity and inflammation. Front Immunol. 2021;12:716469. https://doi.org/10.3389/fimmu.2021.716469.
88. Miraghazadeh B, Cook MC. Nuclear factor-kappaB in autoimmunity: man and mouse. Front Immunol. 2018;9:613. https://doi.org/10.3389/fimmu.2018.00613.
89. Moffat JM, Mintern JD, Villadangos JA. Control of MHC II antigen presentation by ubiquitination. Curr Opin Immunol. 2013;25:109–14. https://doi.org/10.1016/j.coi.2012.10.008.

90. Zheng Q, Huang T, Zhang L, Zhou Y, Luo H, Xu H, Wang X. Dysregulation of ubiquitin-proteasome system in neurodegenerative diseases. Front Aging Neurosci. 2016;8:303. https://doi.org/10.3389/fnagi.2016.00303.
91. Balendran T, Lim K, Hamilton JA, Achuthan AA. Targeting transcription factors for therapeutic benefit in rheumatoid arthritis. Front Immunol. 2023;14:1196931. https://doi.org/10.3389/fimmu.2023.1196931.
92. Tsuchiya Y, Nakabayashi O, Nakano H. FLIP the switch: regulation of apoptosis and necroptosis by cFLIP. Int J Mol Sci. 2015;16:30321–41. https://doi.org/10.3390/ijms161226232.
93. Chen Y, Dai R, Cheng M, Wang W, Liu C, Cao Z, Ge Y, Wang Y, Zhang L. Status and role of the ubiquitin–proteasome system in renal fibrosis. Biomed Pharmacother. 2024;178:117210. https://doi.org/10.1016/j.biopha.2024.117210.
94. Wang Y, Zhou Y, Graves DT. FOXO transcription factors: their clinical significance and regulation. Biomed Res Int. 2014;2014:1–13. https://doi.org/10.1155/2014/925350.

Ubiquitin-Proteasome System: Collective Operation of Molecular Chaperones

3

Sunny Kumar and Mrinal K. Ghosh

3.1 Introduction

Ubiquitin-proteasome system (UPS) is a highly coordinated cellular machinery responsible for protein quality control, regulating degradation, and maintaining protein homeostasis. The UPS operates alongside molecular chaperones—specialized proteins that assist in protein folding, assembly, and disassembly processes. Together, these systems play a critical role in cellular health by preventing the accumulation of misfolded or damaged proteins, which is essential for preventing cellular stress, apoptosis, and diseases associated with protein aggregation, such as Huntington's disease (HD), Parkinson's disease (PD), and Alzheimer's disease (AD) [1].

The UPS itself is a multi-step pathway involving protein tagging by ubiquitin molecules, recognition by proteasomes, and subsequent proteolytic degradation. Proteins are tagged through a process involving ubiquitin-activating (E1), ubiquitin-conjugating (E2), and ubiquitin-ligase (E3) enzymes, which sequentially attach ubiquitin molecules to a substrate (Fig. 3.1). Once a protein is poly-ubiquitinated, it is recognized by the 26S proteasome, an Adenosine tri-phosphate (ATP)-dependent protease complex that unfolds and degrades proteins into short peptides. This system is tightly regulated, ensuring specific and timely degradation of misfolded, damaged, or excess proteins. This selective process not only eliminates defective proteins but also controls the levels of various regulatory proteins involved in cell

S. Kumar
Cancer Biology and Inflammatory Disorder Division, Council of Scientific and Industrial Research-Indian Institute of Chemical Biology (CSIR-IICB), Kolkata, West Bengal, India

Academy of Scientific and Innovative Research, Ghaziabad, Uttar Pradesh, India

M. K. Ghosh (✉)
Cancer Biology and Inflammatory Disorder Division, Council of Scientific and Industrial Research-Indian Institute of Chemical Biology (CSIR-IICB), Kolkata, West Bengal, India
e-mail: mrinalghosh@iicb.res.in

S. Singh (ed.), *Ubiquitin Proteasome System in CNS Neurodegenerative Diseases*, https://doi.org/10.1007/978-981-95-6099-8_3

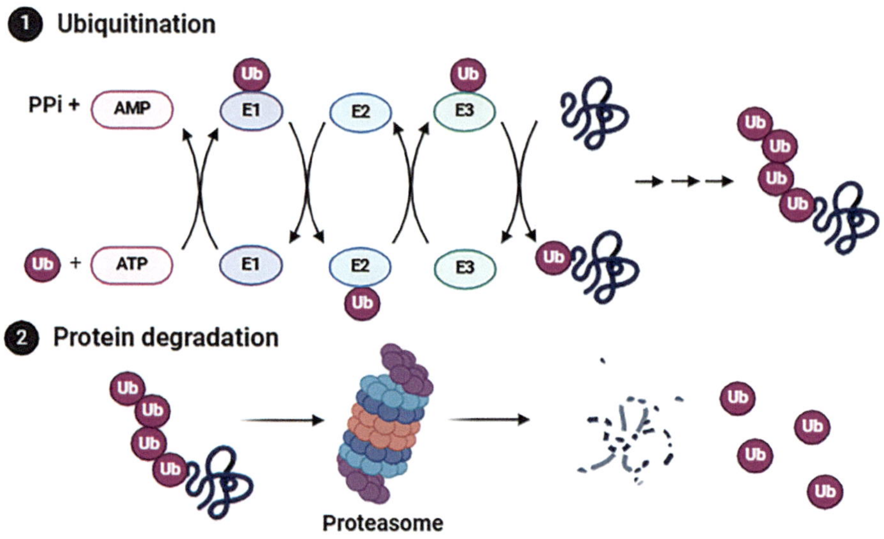

Fig. 3.1 Ubiquitin proteasome system. The figure illustrates the pathway through which proteins get ubiquitinated with the involvement of a series of enzymatic (E1, E2, and E3) steps and finally leads to 26S proteasome-mediated degradation. (Created with BioRender.com)

signaling, growth, and apoptosis, highlighting the UPS's broader roles in cellular homeostasis [1].

Molecular chaperones, including families such as HSP70, HSP90, and HSP40, play distinct but complementary roles in the UPS by recognizing misfolded proteins, preventing aggregation, and delivering these substrates to the proteasome for degradation. In cases where aggregated proteins accumulate, as in neurodegenerative diseases, additional chaperones such as small heat shock proteins (sHSPs) and chaperonins can act to either refold or sequester these toxic species, thereby reducing cellular stress. Additionally, the coordinated function of the UPS and chaperones is modulated by co-chaperones and other regulatory proteins that direct substrate specificity and determine the cellular response to misfolded proteins [2].

The UPS and molecular chaperones collaborate to manage cellular proteins throughout their life cycle, ensuring that only properly folded and functional proteins persist while damaged or unnecessary ones are degraded. This process begins with molecular chaperones, which recognize and stabilize newly synthesized or partially folded proteins. By preventing aggregation and facilitating correct folding, chaperones enhance the stability and solubility of these proteins. In cases where proteins cannot be correctly folded or are damaged beyond repair, chaperones mark them for degradation by the UPS, signaling that they are no longer viable for cellular functions [3]. Beyond simple degradation, the UPS and chaperones are essential in mediating stress responses, signaling pathways, and cellular adaptation to environmental changes. During cellular stress, such as heat shock or oxidative stress, chaperones are upregulated to enhance protein stabilization and prevent aggregation, while the UPS may prioritize the degradation of damaged proteins. This collaboration between the UPS and chaperones demonstrates a highly responsive and adaptive system designed to protect cells from proteotoxic stress and to maintain the integrity of the proteome under a range of physiological conditions [4].

In summary, the collective operation of the UPS and molecular chaperones forms a dynamic network crucial for proteostasis. By identifying, refolding, or degrading damaged proteins, this system helps maintain cellular health and prevents the onset of proteotoxic diseases. This chapter explores the mechanisms, key players, and regulatory pathways involved in the UPS and chaperone network, discussing their roles in cellular homeostasis, disease prevention, and potential therapeutic strategies targeting this system in various proteinopathies. Understanding this collaborative process provides insight into a fundamental aspect of cellular biology and opens avenues for interventions in diseases where proteostasis is disrupted [2].

3.2 Protein Quality Control System

UPS is a cornerstone of cellular quality control, overseeing the regulated degradation of proteins to maintain protein homeostasis, or proteostasis. Through selective targeting and elimination of misfolded, damaged, or unnecessary proteins, the UPS plays a pivotal role in cellular health. It's especially crucial in high-demand cells like neurons, where the buildup of damaged proteins can lead to cellular toxicity. Dysfunction within this system is implicated in a range of neurodegenerative diseases, where protein aggregation is a hallmark of disease progression. The UPS relies on two main components: ubiquitin, a small, conserved regulatory protein that labels proteins for degradation, and the proteasome, a large proteolytic complex that breaks down tagged proteins [5].

The process of targeting a protein for degradation begins with the addition of ubiquitin through a series of steps mediated by three enzymes. First, E1, the ubiquitin-activating enzyme, activates ubiquitin in an ATP-dependent reaction. This is followed by E2, the ubiquitin-conjugating enzyme, which transfers the activated ubiquitin to the substrate protein. Finally, E3 ubiquitin ligase confers specificity by recognizing the protein substrate and facilitating the transfer of ubiquitin from E2 to the substrate. This cascade of enzyme actions results in polyubiquitination, where a chain of ubiquitin molecules is attached to a target protein, marking it for destruction by the proteasome. The proteasome is composed of a 20S core particle, which performs proteolytic cleavage, and two 19S regulatory particles, which recognize ubiquitinated proteins, remove their ubiquitin tags, and unfold them for translocation into the core for degradation. Once a protein is marked for degradation, it binds to the 19S regulatory particle of the proteasome, where ubiquitin chains are removed and recycled. The 19S particle then unfolds the protein and translocates it into the 20S core particle, where proteolytic enzymes cleave the protein into smaller peptides. These peptides are subsequently degraded or recycled, allowing the amino acids to be reused by the cell. This tightly regulated process is energy-intensive, requiring ATP for ubiquitin activation, protein unfolding, and translocation steps.

The UPS does not just remove damaged proteins; it also regulates the levels of functional proteins involved in essential cellular processes such as cell cycle progression, transcription, and signal transduction, enabling the cell to rapidly respond to changing conditions [2].

As a quality control mechanism, the UPS prevents the accumulation of potentially toxic, misfolded proteins by recognizing abnormal conformations and tagging them for degradation. Molecular chaperones assist in refolding misfolded proteins, reducing the load on the degradation machinery. However, when protein misfolding is extensive, as in neurodegenerative diseases, the UPS can become overwhelmed. In HD, for example, the mutant huntingtin protein (mHtt) misfolds and aggregates, and though the UPS attempts to degrade it, the high levels of misfolding strain the system's capacity. When overwhelmed, mHtt aggregates accumulate, leading to cellular dysfunction and neuronal death. Similar processes are observed in AD and PD, where the proteins beta-amyloid and alpha-synuclein, respectively, aggregate and overwhelm the UPS, contributing to disease pathology. Beyond managing protein quality under stress, the UPS also responds to cellular damage. In times of oxidative stress or environmental insult, cells upregulate UPS components to handle the increased load of damaged proteins. This adaptation is vital for neurons, where long-term survival depends on efficient protein turnover. Yet, even under optimal conditions, the UPS declines with age, making cells more susceptible to accumulating damage and, thus, more prone to neurodegenerative conditions. Disruptions in UPS function can be due to mutations in E3 ligases or proteasome subunits, oxidative damage, and even the natural decline of proteasome activity with aging, which diminishes the cell's ability to manage damaged proteins effectively [6, 7].

Therapeutic strategies aimed at enhancing UPS activity hold promise for treating neurodegenerative diseases. By increasing proteasome activity or facilitating the clearance of aggregated proteins, scientists hope to alleviate some of the cellular toxicity that results from protein misfolding. In neurodegenerative diseases, particularly HD, the UPS is crucial as it attempts to control the accumulation of toxic protein aggregates, yet its limited capacity highlights the need for interventions that can bolster its activity. The UPS thus remains central to cellular health, providing insights into both the mechanisms of disease and potential avenues for treatment [8].

3.3 Molecular Chaperones in Protein Refolding

Different molecular chaperones are expressed by neurons, which work together to build a complex network of PQC to stop aggregation. The main functions of these chaperones are to help fold new polypeptides and refold damaged or misfolded proteins. They fall into groups such as Hsp70, Hsp90, Hsp60, Hsp40 (DnaJ), and tiny HSPs and are categorized according to size and function. Although many chaperones have similar processes, including ATP-driven substrate binding and substrate recognition, their specific mechanisms, location, and substrate specificity vary [2, 9] (Fig. 3.2).

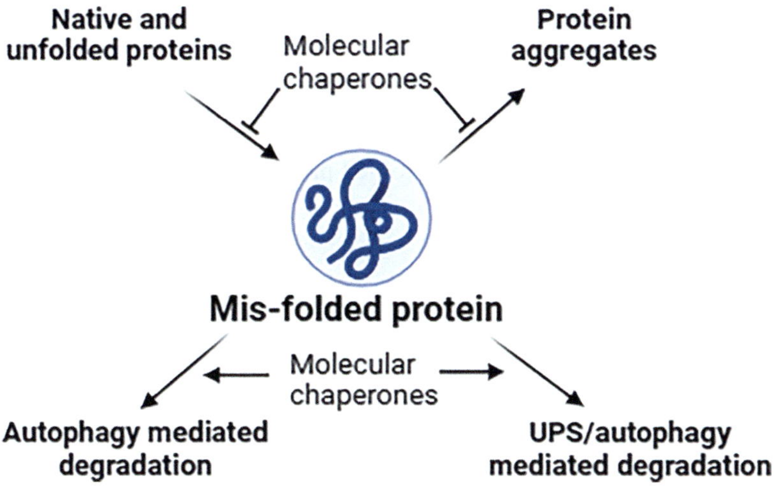

Fig. 3.2 Role of molecular chaperones. The figure illustrates the role of molecular chaperones in proteolysis (UPS or autophagy mediated) of misfolded proteins and their role in the inhibition of aggregate formation, as well as in protein misfolding of native or unfolded proteins

3.3.1 HSP70 Family

Hsp70, a highly conserved and abundant cytosolic chaperone, plays a critical role across cellular compartments, including the cytosol (as Hsc70) and the ER (as BiP/GRP78). Hsp70 assists in folding new polypeptides, refolding misfolded proteins, degrading irreversibly misfolded proteins, and resolving protein aggregates. In order to identify misfolded proteins, Hsp70 binds to brief hydrophobic amino acid sequences that are visible on their surface and usually occur every 30–40 residues. The capacity of Hsp70 to alternate between open and closed conformations in the substrate-binding domain is essential to its activity. Hsp70 has a low substrate affinity when it is ATP-bound in its open form. However, it changes to an ADP-bound state with a high substrate affinity when ATP hydrolysis occurs, which is frequently brought on by co-chaperones. As a result, unfolded proteins can be bound and stabilized by Hsp70 until they can refold on their own. After folding, proteins are freed from Hsp70 and lose their exposed hydrophobic areas. According to research, Hsp70 facilitates the refolding and possible reduction of aggregation of a number of misfolded proteins linked to neurodegenerative diseases, including mHtt in HD, α-synuclein in PD, amyloid-β and hyperphosphorylated tau in AD, and mutant SOD1 in ALS [8, 10].

3.3.2 HSP40 Family

Hsp40 proteins, or J-proteins, are a large family of 49 cochaperones, including neuron-expressed members DnaJB6 and DnaJB8, which help prevent polyglutamine aggregation and toxicity in neurodegeneration. These cochaperones assist Hsp70 by binding misfolded proteins and triggering Hsp70's ATP hydrolysis through their J domain. This process increases Hsp70's affinity for the misfolded substrate, facilitating its transfer from Hsp40 to Hsp70 for refolding. Beyond the J domain, Hsp40 proteins contain specialized domains for specific functions, like targeting proteins for degradation, playing a vital role in stabilizing and refolding misfolded proteins in neurodegenerative diseases [11, 12].

3.3.3 HSP90 Family

Hsp90, an ATP-dependent chaperone dimer found in the cytosol, nucleus, endoplasmic reticulum (ER), and mitochondria, plays a key role in protein stability and function. Constituting 1–2% of cellular proteins (rising to 4–6% under stress), Hsp90 is regulated by HSF1. Neurons express both stress-inducible Hsp90α and constitutive Hsp90β, which together interact with roughly 2000 client proteins, including kinases, receptors, and transcription factors. Its structure includes an N-terminal ATP-binding domain, a middle substrate-binding domain, and a C-terminal dimerization domain. During its ATP-driven cycle, Hsp90 shifts between open and closed conformations, aided by cochaperones such as Hop, p23, and Cdc37, to bind and release substrates. While crucial in stabilizing essential proteins, Hsp90's activity can inadvertently promote toxic aggregate formation in neurodegenerative diseases [11, 13–15].

3.3.4 HSP60 Family

Hsp60, a mitochondrial chaperone, forms a double-ring complex that sequesters hydrophobic residues of misfolded proteins during refolding. It works with cochaperones like Hsp10 and releases the client after ATP hydrolysis. Hsp60 also collaborates with Hsp70 in protein folding. In neurons, the TCP-1 Ring Complex, similar to Hsp60, functions without cochaperones. Though Hsp60 interacts with mutant α-synuclein in PD, its role in neurodegenerative diseases remains poorly understood [16–20].

3.3.5 Small HSP Family

sHSPs are ATP-independent chaperones, with 10 known types in humans, ranging from 12 to 42 kDa. In the mouse brain, five are prominently expressed, including stress-inducible Hsp27 and αB crystallin in neurons. These proteins contain a 100-residue α-crystallin domain and form oligomers that bind misfolded proteins, preventing aggregation. They act as holding factors until proteins are passed to other chaperones such as Hsp70 and Hsp40. Hsp27 is the most abundant and well-studied, induced by stress that disrupts proteostasis [3, 21–23].

3.4 Molecular Chaperones in Protein Degradation

Molecular chaperones assist in refolding proteins and degrade irreversibly misfolded ones via the UPS or autophagy. Misfolded proteins tagged with ubiquitin go to the proteasome, while aggregation-prone proteins are directed to autophagy. Chaperones play a key role in recognizing and delivering these proteins for clearance [24].

3.4.1 Ub-Dependent Proteostasis

The UPS manages selective degradation of short-lived proteins, responsible for over 80% of intracellular proteolysis. The process begins with ATP-dependent activation of Ub by the E1 enzyme, which then transfers Ub to E2. E3 ligases attach Ub to substrate lysines, creating an isopeptide bond. With over 500 types, E3 ligases are categorized into RING finger, HECT, and U-box classes. E4 enzymes can further extend Ub chains, with K48-linked chains marking proteins for degradation and K63-linked chains participating in non-degradative signaling. K11 and K27 linkages assist in cell cycle and ER-associated processes. Polyubiquitinated proteins are degraded by the 26S proteasome, while DUB enzymes can reverse or modify ubiquitination. Protein breakdown is managed by the 2.5-MDa proteasome complex, which is made up of two 19S regulatory particles and a 20S core (700 kDa). RPN10 and RPN13 identify Ub-tagged substrates in the 19S unit, where Ub chains are eliminated by deubiquitinating enzymes such as RPN11, USP14, and UCHL5. The substrate is unfolded by ATP hydrolysis in the 19S particle and then fed into the 20S core to be broken down into peptides of 8–12 amino acids. Amino peptidases further degrade these peptides for immune presentation or recycling in protein synthesis [25–27].

3.4.2 Collective Operation of Molecular Chaperones and Ub E3 Ligases

To break down misfolded proteins, a number of cytosolic and nuclear E3 ligases work in tandem with molecular chaperones, including UBR1, UBR2, San1, Hul5, E6-AP, CHIP, and Parkin. The adaptable 35-kDa protein CHIP functions as an E3 ligase and a cochaperone for Hsp70 and Hsp90. It interacts with the E2 enzyme UBCH5 to tag misfolded proteins with Ub via its U-box domain. Carboxy terminus of Hsp70-Interacting Protein (CHIP) recognizes misfolded clients and ubiquitinates them for proteasomal destruction when it binds to the Hsp70- or Hsp90-client complex. By attaching itself to Rpn10 on the 19S proteasome, CHIP further aids in this process. Through its tetratricopeptide repeat domain, which interacts with Hsp70 or Hsp90, CHIP detects misfolded proteins indirectly. Mutant SOD1 and hyperphosphorylated tau are important CHIP substrates. By serving as a nucleotide exchange factor for Hsp70, the co-chaperone BAG-1 facilitates this process by releasing misfolded substrates and guiding them to the proteasome via its Ub-like (UBL) domain. Further encouraging the breakdown of ubiquitinated proteins is the ability of Hsp27, a member of the small HSP family, to bind Ub chains on clients [13–15, 28, 29].

The UPS may designate these dissociated clients for destruction when Hsp90 is unable to stabilize its client proteins any longer, frequently as a result of client misfolding. These misfolded proteins can be ubiquitinated by E3 ligases like CHIP, which are usually connected to Hsp70, in order to be cleared by proteasomes. For targeted degradation, Hsp90 clients most likely depend on E3 ligases other than CHIP. Triad3A, an E3 ligase that combines with receptor-interacting protein 1 (RIP-1) and Hsp90, is one such possibility. When geldanamycin, for example, inhibits Hsp90 activity, Triad3A ubiquitinates RIP-1, designating it for proteasomal destruction. This demonstrates how Triad3A may be used to care for clients who are dependent on Hsp90 in situations where Hsp90 is unable to provide for them [30–33].

Based on the exposed N-terminal residues of proteins, which serve as N-degrons—signals for recognition by certain E3 ligases, such as those in the N-recognin family—the N-end rule route guides the degradation of proteins. This family includes Ub ligases such as p62, UBR1, UBR2, UBR4, and UBR5 in mammals. Using chaperones such as Hsp70 and Hsp110 to identify misfolded proteins, UBR1 and UBR2 in particular are essential for tagging misfolded cytosolic proteins for proteasomal destruction. N-recognins, such as UBR1 and UBR2, bind to hydrophobic residues and, in some situations, to N-terminal arginine residues to particularly target misfolded proteins such as TDP43, tau, and amyloid-β. These proteins become N-recognin substrates when these residues are added post-translationally by conjugation by Arg-tRNA, which is facilitated by the enzyme ATE1. Mammalian cells, such as yeast, may use a cooperative strategy when misfolded proteins overwhelm the cytosolic quality control system. The nuclear E3 ligase San1 is used by yeast cells. Its flexible, disordered regions immediately bind misfolded proteins and carry them to the nucleus for destruction. It is hypothesized that UHRF2 in mammals serves a similar purpose, acting as a nuclear backup mechanism to help handle misfolded proteins when cytosolic routes are exhausted [34–36].

Different protein degradation mechanisms, each tailored to deal with certain kinds of misfolded proteins, are used by eukaryotic cells. The HECT-domain E3 ligase Hul5 in yeast marks heat-shock-denatured proteins for destruction. Together with the BAG6 chaperone complex, the RING finger E3 ligase RNF126 in mammals marks mislocalized membrane proteins. The ribosome-associated E3 ligase Listerin (Ltn1) ubiquitinates unfinished polypeptides that are translated from defective mRNAs that do not include stop codons. Depending on where they are located in the ER, distinct E3 ligases are responsible for managing misfolded proteins. DOA10 ubiquitinates those in the ER membrane, whereas Hrd1 and Gp78 induce ER-associated degradation for misfolded proteins that reside in the ER lumen. Most E3 ligases, in contrast to Hul5 and San1, rely on chaperones to recognize and deliver misfolded protein substrates for ubiquitination rather than directly identifying misfolded proteins. Although there are many different degradation processes, little is known about their precise functions in neurodegenerative illnesses. This information gap might be crucial for the development of new treatments [37–41].

3.5 Conclusion and Future Perspectives

UPS plays a crucial role in cellular homeostasis with the involvement of molecular chaperones, which assist in protein folding, prevent aggregation, and guide misfolded proteins towards the UPS for degradation or repair. The collaboration between chaperones and the UPS is vital in preventing the accumulation of toxic aggregates in neurodegenerative diseases. Chaperones not only aid in protein folding but also modulate UPS activity, enhancing substrate recognition and proteasomal degradation. However, this interaction is complex and regulated by various cellular signals. A deeper understanding of the chaperone–UPS interplay, including the role of post-translational modifications, could lead to novel therapies for diseases related to protein misfolding, such as cancer and neurodegeneration. Research into dysregulated chaperones or UPS components may identify biomarkers for disease detection and prognosis. Enhancing chaperone or UPS function could offer therapeutic strategies, while proteasome inhibitors could be useful in diseases like cancer. In summary, the chaperone–UPS partnership is essential for proteostasis and offers potential for therapeutic interventions in protein-misfolding diseases. Further interdisciplinary research is needed to unlock its full therapeutic potential.

Authors' Contributions SK and MKG conceived the idea, reviewed the structure, and drafted the paper. The authors have agreed to the final draft of the manuscript.

Funding Statement This work is jointly supported by the Department of Science and Technology (NanoMission: DST/NM/NT/2018/105(G); SERB: EMR/2017/000992) and Focused Basic Research (FBR), HCT, CSIR, Govt. of India.

Declarations

Conflicts of Interest There is no competing interest to disclose.

References

1. Soto C, Pritzkow S. Protein misfolding, aggregation, and conformational strains in neurodegenerative diseases. Nat Neurosci. 2018;21(10):1332–40.
2. Batko J, Antosz K, Miśków W, Pszczołowska M, Walczak K, Leszek J. Chaperones—a new class of potential therapeutic targets in Alzheimer's disease. Int J Mol Sci. 2024;25(6):3401.
3. Kumar S, Basu M, Ghosh MK. Chaperone-assisted E3 ligase CHIP: a double agent in cancer. Genes Dis. 2021;9(6):1521–55. https://doi.org/10.1016/j.gendis.2021.08.003.
4. Seyoum Tola F. The role of ubiquitin-proteasome system in the pathogenesis of severe acute respiratory syndrome coronavirus-2 disease. Int J Inflamm. 2023;2023:6698069. https://www.hindawi.com/journals/iji/2023/6698069/. Accessed 10 Mar 2024.
5. Koszła O, Sołek P. Misfolding and aggregation in neurodegenerative diseases: protein quality control machinery as potential therapeutic clearance pathways. Cell Commun Signal. 2024;22(1):421. https://doi.org/10.1186/s12964-024-01791-8.
6. Giusti V, Kaur G, Giusto E, Civiero L. Brain clearance of protein aggregates: a close-up on astrocytes. Mol Neurodegener. 2024;19(1):5. https://doi.org/10.1186/s13024-024-00703-1.

7. Limanaqi F, Biagioni F, Mastroiacovo F, Polzella M, Lazzeri G, Fornai F. Merging the multi-target effects of phytochemicals in neurodegeneration: from oxidative stress to protein aggregation and inflammation. Antioxidants. 2020;9(10):1022.
8. Upaganlawar AB, Wankhede NL, Kale MB, et al. Interweaving epilepsy and neurodegeneration: vitamin E as a treatment approach. Biomed Pharmacother. 2021;143:112146.
9. Pillai RN, Fennell DA, Kovcin V, et al. Randomized phase III study of ganetespib, a heat shock protein 90 inhibitor, with docetaxel versus docetaxel in advanced non–small-cell lung cancer (GALAXY-2). J Clin Oncol. 2020;38(6):613–22.
10. Rai P. Conformational dynamics of Hsp90 and Hsp70 chaperones in treating neurodegenerative diseases: insights from the drosophila model. Proc Indian Natl Sci Acad. 2024;90(3):628–37. https://doi.org/10.1007/s43538-024-00325-7.
11. Acosta-Alvear D, Harnoss JM, Walter P, Ashkenazi A. Homeostasis control in health and disease by the unfolded protein response. Nat Rev Mol Cell Biol. 2025;26(3):193–212.
12. Kim Chiaw P, Hantouche C, Wong MJH, et al. Hsp70 and DNAJA2 limit CFTR levels through degradation. Brodsky JL, ed. PLoS One. 2019;14(8):e0220984. https://doi.org/10.1371/journal.pone.0220984.
13. Gavilán E, Medina-Guzman R, Bahatyrevich-Kharitonik B, Ruano D. Protein quality control systems and ER stress as key players in SARS-CoV-2-induced neurodegeneration. Cells. 2024;13(2):123.
14. Limcharoensuk T, Chusuth P, Utaisincharoen P, Auesukaree C. Protein quality control systems in the endoplasmic reticulum and the cytosol coordinately prevent alachlor-induced proteotoxic stress in Saccharomyces cerevisiae. J Hazard Mater. 2024;471:134270.
15. Wickramaratne AC, Wickner S, Kravats AN. Hsp90, a team player in protein quality control and the stress response in bacteria. Detweiler CS, ed. Microbiol Mol Biol Rev. 2024;88(2):e00176–22. https://doi.org/10.1128/mmbr.00176-22.
16. Esser C, Alberti S, Höhfeld J. Cooperation of molecular chaperones with the ubiquitin/proteasome system. Biochim Biophys Acta. 2004;1695(1–3):171–88.
17. Kumar S, Chatterjee M, Ghosh P, Ganguly KK, Basu M, Ghosh MK. Targeting PD-1/PD-L1 in cancer immunotherapy: an effective strategy for treatment of triple-negative breast cancer (TNBC) patients. Genes Dis. 2023;10(4):1318–50.
18. Ghosh MK, Kumar S, Ganguly KK, et al. COVID-19 and cancer: insights into their association and influence on genetic and epigenetic landscape. Epigenomics. 2023;15(4):227–48. https://doi.org/10.2217/epi-2023-0052.
19. Kumar S, Basu M, Ghosh MK. E3 ubiquitin ligases and deubiquitinases in colorectal cancer: emerging molecular insights and therapeutic opportunities. Biochim Biophys Acta Mol Cell Res. 2024;1871(8):119827.
20. Ghosh MK, Kumar S, Begam S, Ghosh S, Basu M. GBM immunotherapy: exploring molecular and clinical frontiers. Life Sci. 2024;356:123018. https://doi.org/10.1016/j.lfs.2024.123018.
21. Kurop MK, Huyen CM, Kelly JH, Blagg BS. The heat shock response and small molecule regulators. Eur J Med Chem. 2021;226:113846.
22. Kumar S, Basu M, Ghosh P, Pal U, Ghosh MK. COVID-19 therapeutics: clinical application of repurposed drugs and futuristic strategies for target-based drug discovery. Genes Dis. 2023;10(4):1402–28. https://doi.org/10.1016/j.gendis.2022.12.019.
23. Kumar S, Basu M, Ghosh P, Ansari A, Ghosh MK. COVID-19: clinical status of vaccine development to date. Br J Clin Pharmacol. 2023;89(1):114–49. https://doi.org/10.1111/bcp.15552.
24. Howell RA, Wang S, Khambete M, McDonald DM, Spiegel DA. Bifunctional molecules that induce both targeted degradation and transcytosis of extracellular proteins in brain cells. J Am Chem Soc. 2024;146(24):16404–11. https://doi.org/10.1021/jacs.3c13320.
25. Kandel R, Jung J, Neal S. Proteotoxic stress and the ubiquitin proteasome system. In: Seminars in cell & developmental biology, vol. 156. Elsevier; 2024. p. 107–20. https://www.sciencedirect.com/science/article/pii/S1084952123001611. Accessed 10 Nov 2024.
26. Ahammed MS, Wang X. Promoting proteostasis by cAMP/PKA and cGMP/PKG. Trends Mol Med. 2025;31:224–39. https://www.cell.com/trends/molecular-medicine/abstract/S1471-4914(24)00273-9. Accessed 8 Nov 2024.

27. Xu P, Yang JC, Ning S, et al. Allosteric inhibition of HSP70 in collaboration with STUB1 augments enzalutamide efficacy in antiandrogen resistant prostate tumor and patient-derived models. Pharmacol Res. 2023;189:106692.
28. Scheidt T, Lemke EA. Chaperone quality control in liquid-phase separated organelles. bioRxiv. Published online 2024:2024-01.
29. Jirout M. Mechanisms of membrane protein quality control in the endoplasmic reticulum. Published online 2024. https://dspace.cuni.cz/handle/20.500.11956/190254. Accessed 10 Nov 2024.
30. Sharma A, Shah OP, Sharma L, et al. Molecular chaperones as therapeutic target: Hallmark of neurodegenerative disorders. Mol Neurobiol. 2024;61(7):4750–67. https://doi.org/10.1007/s12035-023-03846-2.
31. Baker HA, Bernardini JP, Csizmok V, et al. The co-chaperone DNAJA2 buffers proteasomal degradation of cytosolic proteins with missense mutations. bioRxiv. Published online 2024:2024-02.
32. Perez RF, Mochi G, Khan A, Woodford M. Mitochondrial chaperone code: just warming up. Cell Stress Chaperones. 2024;29(3):483–96.
33. Que Y, Qiu Y, Ding Z, et al. The role of molecular chaperone CCT/TRiC in translation elongation: a literature review. Heliyon. 2024;10(7):e29029. https://www.cell.com/heliyon/fulltext/S2405-8440(24)05060-6. Accessed 10 Nov 2024.
34. Varshavsky A. N-degron pathways. Proc Natl Acad Sci. 2024;121(39):e2408697121. https://doi.org/10.1073/pnas.2408697121.
35. Wu Z, Huang Y, Liu K, Min J. N/C-degron pathways and inhibitor development for PROTAC applications. Biochim Biophys Acta Gene Regul Mech. 2024;1867(1):194952.
36. Kim SH, Park JS, Lee MH, et al. The N-degron pathway governs autophagy to promote thermotolerance in Arabidopsis. bioRxiv. Published online 2024:2024-07.
37. Abaeva IS, Bulakhov AG, Hellen CU, Pestova TV. The ribosome-associated quality control factor TCF25 imposes K48 specificity on Listerin-mediated ubiquitination of nascent chains by binding and specifically orienting the acceptor ubiquitin. bioRxiv. Published online 2024:2024-10.
38. Qin F, Cai B, Wang P, et al. LTN1 promotes RLR degradation to inhibit immune response to RNA virus through the ESCRT pathway. Autophagy. 2024;20(6):1270–85. https://doi.org/10.1080/15548627.2023.2291939.
39. Dos Santos OAL, Carneiro RL, Requião RD, Ribeiro-Alves M, Domitrovic T, Palhano FL. Transcriptional profile of ribosome-associated quality control components and their associated phenotypes in mammalian cells. Sci Rep. 2024;14(1):1439.
40. Svetlov MS, Dunand CF, Nakamoto JA, et al. Peptidyl-tRNA hydrolase is the nascent chain release factor in bacterial ribosome-associated quality control. Mol Cell. 2024;84(4):715–26.
41. Aslani S, Jacob J. Utilisation of deep learning for COVID-19 diagnosis. Clin Radiol. 2023;78(2):150–7.

Cellular Bioenergetics in Ubiquitin-Proteasome System Functionality

4

Akhil Sharma, Ashi Mannan, and Thakur Gurjeet Singh

4.1 Introduction

The ubiquitin-proteasome system (UPS) is the primary proteolytic system that manages the degradation of proteins, which plays an important function in cells by modulating various functions such as deoxyribonucleic acid (DNA) repair, response to stress, and cell division [1]. The main function of ubiquitin was thought to be related to protein degradation; it is now recognized as a key modulator of protein functioning, influencing several cellular processes. These include signal transduction pathways, gene expression, and apoptosis [2]. As a result, during UPS malfunctions, it may contribute to the development of various human diseases [3]. Research using proteasome inhibitors has shown that 26S and 20S proteasomes play a role in breaking down proteins for cellular functions, including proteins that are associated with various processes such as transcription, apoptosis, and cell cycle control. The proteasomes degrade short-term as well as long-term proteins, by using a pathway that may or may not be ubiquitin-dependent [4].

Cellular bioenergetics is a fundamental aspect of cell biology that encompasses the processes by which cells obtain, convert, and use energy. Abnormal cellular

A. Sharma · A. Mannan
Chitkara College of Pharmacy, Chitkara University, Rajpura, Punjab, India
e-mail: ashi.mannan@chitkara.edu.in

T. G. Singh (✉)
Chitkara College of Pharmacy, Chitkara University, Rajpura, Punjab, India

School of Public Health, Faculty of Health, University of Technology Sydney, Broadway, NSW, Australia
e-mail: gurjeet.singh@chitkara.edu.in

S. Singh (ed.), *Ubiquitin Proteasome System in CNS Neurodegenerative Diseases*, https://doi.org/10.1007/978-981-95-6099-8_4

bioenergetics plays an essential function in various health diseases, including obesity, diabetes, tumors, neurodegenerative diseases, and cardiomyopathy [5].

Mitochondria are widely recognized as cytoplasmic organelles that serve a key function in sustaining the energy balance of cells via generating adenosine 5′-triphosphate (ATP) [6]. In healthy replicative eukaryotic cells, mitochondria control key cellular functions including growth, programmed cell death, metabolism adjustments, and calcium (Ca^{2+}) balance [7]. Mitochondria serve an important function in producing energy for the cell via the process of oxidative phosphorylation (OXPHOS) while also coordinating various synthetic processes and intracellular communication pathways. As a consequence, mitochondrial dysfunction can lead to an oversupply of harmful reactive oxygen species (ROS) that triggers the disruption of cell balance, triggering programmed cell death, and consequently is at the root cause of various human diseases [8]. Mitochondrial dysfunction is also linked with many health issues such as metabolic problems, neurological diseases, tumors, cardiovascular diseases, infectious diseases, and inflammatory diseases [9]. The primary mechanism by which eukaryotic cells eliminate damaged or unnecessary mitochondria is through selective autophagy, known as mitophagy. Increasing evidence suggests that specific mitochondrial proteins undergo degradation by the UPS rapidly before remaining mitochondrial components are cleared through mitophagy [10].

4.2 Mitochondrial Structure, Function, and Bioenergetics

Mitochondria are widespread intracellular organelles present in nearly the entire eukaryotes and play an important role in a variety of cellular activities, primarily in the generation of energy [11]. The significance of mitochondria has become more evident, as researchers have uncovered their association with numerous diseases, including aging, neurodegenerative diseases, and cancer [12].

4.2.1 Structure and Function of Mitochondria

Mitochondria consist of the mitochondrial lumen, referred to as "matrix," which hosts key processes such as the Krebs cycle, mitochondrial DNA (mtDNA) replication, and protein synthesis. This matrix is enclosed by two distinct membranes: an external membrane is the outer mitochondrial membrane (OMM), and an internal membrane is the inner mitochondrial membrane (IMM) [13]. OMM serves as a transmission wall while, moreover, facilitating the transmission of signals from the mitochondria. IMM is divided into two key sub-compartments: inner boundary membrane (IBM) and mitochondrial cristae [14]. IBM is equivalent to OMM and consists of various channel transporters that facilitate the movement of small metabolites, ADP, ATP, and ions between the cytoplasm and mitochondrial matrix. Mitochondrial cristae are folds of the membrane that act as key locations for OXPHOS, accommodation of respiratory chain complexes, and ATP synthase. They are also involved in the iron–sulphur cluster synthesis and the maintenance of mtDNA [15]. Cristae are linked with IBM through cristae junctions, which are fine gaps to enclose the cristae and prevent their contents from leaking into the intermembrane space (IMS) [16]. The formation of cristae in the IMM relies on various membrane-shaping proteins. One key player is the mitochondrial contact site and cristae organizing system, which is situated at the cristae junctions. This system associates IMM with the OMM via interacting with protein complexes in the outer membrane. Additionally, the dynamin-related protein optic atrophy protein 1 (OPA1) makes oligomerization on cristae junctions and causes interruption of these oligomers, which are linked to the cavity of cristae. ATP synthase molecules also aggregate and then form extended rows that run along the edges of the cristae [17]. The OMM forms a web of connections with different organelles. These include contact points with the plasma membrane, lysosomes, endoplasmic reticulum (ER), endosomes, peroxisomes, melanosomes, and lipid droplets [18]. Contact sites of these membranes play a role in regulating mitochondrial functions. Additionally, mitochondria also produce mitochondrial-derived compartments (MDCs) and mitochondrial-derived vesicles (MDVs), which contain mitochondrial components [19]. MDVs facilitate communication with other organelles, such as peroxisomes and lysosomes, although MDCs serve a crucial role in maintaining the quality of mitochondria via removing misfolded proteins from mitochondrial membranes. The shape and structure of mitochondria are essential for metabolic activity, such as the effectiveness of ATP generation is enhanced, and the exchange of matrix content is favored in fused organelles [20]. In contrast, fragmented organelles generate higher levels of ROS and are effectively removed through mitophagy. Nevertheless, mitochondrial fragmentation is necessary to ensure the equal division of mitochondria into daughter cells during cell division [21].

4.2.2 Role of Mitochondria as Bioenergetics

Mitochondria play a key function in cellular bioenergetics, helping as the major site of energy generation in cells [22]. In most tissues, mitochondria supply the energy for cell development and various functions. It is estimated that around 90% of oxygen expenditure in mammals occurs in mitochondria, mainly for ATP synthesis, though the amount varies depending on the tissue and the organism's level of activity [23]. Mitochondria play a function in the final stage of cellular catabolism, following intermediate metabolic reactions that break down proteins, carbohydrates, and fats into smaller molecules such as amino acids, pyruvate, and fatty acids, respectively. Mitochondria also convert these energy-rich molecules into Flavin adenine dinucleotide (FADH2) and nicotinamide adenine dinucleotide hydrogen (NADH), via the Krebs cycle and ß-oxidation [24]. They also regulate calcium homeostasis, are a source of ROS, and are involved in apoptosis. The ubiquitin-dependent regulation of the mitochondrial proteome is a vital quality control (QC) system that maintains mitochondrial health [25].

4.2.3 Mitochondria and ATP Production Through OXPHOS

ATP production through OXPHOS is a multi-step membrane-based method to generate the majority of cellular energy in eukaryotic cells and in many prokaryotic cells [26]. There are two key aspects involved: Oxidative reactions catalyzed by respiratory enzymes embedded in the membrane must create an electrochemical gradient of protons (H^+) across the membrane. ATP is then generated from ADP and inorganic phosphate (Pi) by the FIFO-type proton-ATPase, also known as ATP synthase, which is embedded in the membrane. This enzyme, which can work in both directions, harnesses the energy stored in the proton gradient to power the energy-consuming process of ATP synthesis. These types of respiratory enzymes found in mitochondria are mainly consistent across different species, where prokaryotes exhibit significant diversity [27].

OXPHOS occurs in the IMM wherever four large multi-protein complexes—Complex I, II, III, and IV are organized. These complexes contain heme proteins, iron-sulfur clusters, flavins, and copper structures, which work together in the ETC to produce water and a H^+ gradient in the occurrence of oxygen [28]. Energy stored as an electrochemical H^+ gradient is utilized by F0–F1 ATP synthase (also known as Complex V)-generate ATP from ADP and phosphate [29]. Complex I (NADH–CoQ reductase) is the biggest complex in the respiratory chain and transfers electrons from NADH to the CoQ (ubiquinone), thereby regenerating NAD^+ levels by oxidizing the NADH, which is produced during the tricarboxylic acid (TCA) cycle and the β-oxidation of fatty acids [30]. Complex II (succinate dehydrogenase), a membrane-bound enzyme of the Krebs cycle, facilitates the oxidation of succinate to fumarate while transferring two protons to CoQ, via the FAD/FADH2 coenzymes, thereby linking the ETC and TCA cycle [31]. Complex III (coenzyme Q: cytochrome c-oxidoreductase) facilitates the reduction of cytochrome c, supplying two protons to the CoQH2, which is derived from complex I and II, to regenerate the CoQ. Additionally, it pumps $4H^+$ from the mitochondrial matrix into the IMS [32]. The resulting *Q cycle* enables the development of an H^+ gradient across the membrane: the oxidation of CoQ pumps four H^+ into the IMS (positive side), while two H^+ are absorbed from the matrix (negative side). Additionally, two electrons are transferred from the ubiquinol to ubiquinone, through two cytochrome c intermediates, completing the reduction of cytochrome c [33]. Complex IV (cytochrome c oxidase) transfers electrons from cytochrome c to oxygen, resulting in the generation of water and a H^+ gradient [34]. Complex V is composed of two primary subunits: F0, which functions as a transmembrane unit resembling a driven turbine; F1 is the catalytic subunit situated upon the side of the mitochondrial matrix [35].

4.2.4 Role of Mitochondrial Membrane Potential

The mitochondrial membrane potential ($\Delta\Psi m$) is the chief bioenergetic factor that regulates the respiratory rate, generation of ATP, and production of ROS, and is itself regulated by electron transport and proton leaks [36]. The $\Delta\Psi m$ drives three primary functions: ATP production, Ca^{2+} uptake and storage, and the production and detoxification of ROS. Additionally, numerous apoptotic pathways trigger apoptosis by disturpting mitochondrial function adequately to cause cytochrome c release [37]. $\Delta\Psi m$ arises from redox reactions linked to the Krebs cycle activity and acts as a transitional energy reservoir that is utilized by ATP synthase to produce ATP. These transformations produce not only an electrical potential due to charge separation but also a proton gradient, which mutually constitute the transmembrane potential of hydrogen ions [38]. $\Delta\Psi m$ is the most consistent indicator of mitochondrial function. The $\Delta\Psi m$ sort lies between −136 and −140 mV and is regarded as ideal for optimal ATP generation across living organisms [39]. Even slight deviations from this range lead to a significant reduction in ATP generation and a substantial rise in ROS generation. This bioenergetic imbalance is measured as the cause of many human diseases [40]. Numerous experimental models have demonstrated a strong affirmative relationship between $\Delta\Psi m$ and ROS generation [41].

4.3 UPS Functionality

4.3.1 Overview of UPS

UPS is the main signaling pathway responsible for the degradation of proteins in the cell. UPS removes misfolded, aggregated, or excess cellular proteins. Autophagy, on the other hand, handles lysosome-mediated degradation of protein aggregates, long-lived proteins, cellular organelles, and intracellular pathogens. Both degradation systems, UPS and autophagy, play a crucial role in stabilizing cellular homeostasis. The proteasome-mediated degradation of intracellular proteins is crucial for cell growth as well as the functioning of the cell [42].

4.3.2 Ubiquitination Function as a Protein Modification

Ubiquitin, a protein made up of 76 amino acids, is highly conserved across species. It links itself to other proteins by forming a chemical bond between its own end (C terminus) and specific sites (lysine residues) on the target proteins. This attachment can happen to proteins that already have ubiquitin attached, creating chains of ubiquitin molecules of varying sizes [43]. This reaction is facilitated via a series of enzymes: Ubiquitin-activating enzyme (E1, Uba1 in yeast) creates a thioester with ubiquitin and then transports it to the target protein through one of numerous cellular ubiquitin-conjugating enzymes (E2s) [44]. Ubiquitin ligases (E3s), also known as E3s, play a crucial role in determining which proteins get tagged with ubiquitin. They do this by often binding to both E2 enzymes and the target proteins simultaneously, acting as a bridge that ensures that ubiquitin is attached to the specific substrate. Cells possess a wide variety of E3s for different groups of substrates. Deubiquitylating enzymes inhibit ubiquitination enzymes, modify the interactions in the ubiquitin chain, and eliminate ubiquitin from proteins to protect them from degradation [45]. In a corresponding process, ubiquitination also contributes to the selection of damaged mitochondria for degradation lysozome/vacuole through autophagy [46].

4.4 Role of UPS in Mitochondrial Functionality

The UPS is an individual key method through which cells control the proteome. Most of mitochondrial proteins are expressed by nuclear DNA and, after being translated into the cytosol, near the OMM, are imported into the organelle. Various proteomic analyses of mitochondria across various organisms, such as yeast, mouse, and plants, have identified ubiquitinated proteins [25]. Intriguingly, recent research has discovered ubiquitin-mediated protein degradation signaling that regulates various processes of mitochondrial functionality. Dysfunction in mitochondrial surveillance, supported by the UPS and mitophagy, is generally associated with various diseases [47]. Significantly, further age-related diseases, such as diabetes mellitus and cancer, are also associated with a decrease in mitochondrial metabolism function.

4.4.1 Ubiquitin-Dependent Observation of the Mitochondrial Proteome

The ubiquitin-dependent regulation of the mitochondrial proteome is a vital QC system that maintains mitochondrial health. This process involves ubiquitin, a small protein tag, working with E1, E2, and E3 enzymes to mark specific mitochondrial proteins for various fates [48]. Breakdown of mitochondrial proteins is controlled by two mechanisms that are the same as the ubiquitin-dependent pathways. These processes are called ER-associated degradation and mitochondria-associated degradation [49]. Like the cytoplasm, mitochondrial "ubiquitinome" comprises proteins from different biological mechanisms such as the TCA cycle, OXPHOS, and mitochondrial dynamics [50]. The UPS may target mitochondrial proteins for ubiquitination when they are improperly folded. This process could act as a QC mechanism, allowing the cell to identify and eliminate abnormal proteins intended for mitochondria [51]. The discovery of these proteins with multiple ubiquitinated in the cytoplasm may replicate the function of (cytosolic) UPS in regulating the levels, which plays a crucial function in regulating both the amount and quality of proteins designed for mitochondria [48]. The presence of multiple ubiquitination sites on proteins in the cytoplasm reflects the complex regulatory role of the cytosolic UPS, which controls both quality and quantity of proteins through various mechanisms, including selective protein degradation through 26s proteasome, modification of protein activity or localization, regulation of protein–protein interactions, and control of protein trafficking to and from mitochondria [52].

4.4.2 UPS-Mediated Control of Mitochondrial Fission and Fusion

Accumulating research indicates that the UPS is crucial for modifying mitochondrial dynamics. Mitofusion1 (Mfn1)/2, dynamin-related protein 1(Drp1), and mitochondria fission protein (Fis1) are key players for modulating the mitochondrial fission and fusion and are degraded by the proteasome [53].

4.4.2.1 Mechanism of Drp1-Dependent Mitochondrial Fission and Fusion

The Drp1-dependent mitochondrial fission process has four steps: first, Drp1 moves to OMM, second, it forms a complex structure, third, guanosine triphosphate (GTP) hydrolysis, and finally, it disassembles. Drp1 binds to a specific receptor site of OMM and forms a functional unit, and a bigger oligomer, then subsequently composed and transported from the cytosol to the locations where mitochondrial division will occur [54]. Additionally, ER plays a supporting function in transferring the Ca^{2+} within the mitochondria, resulting in recruitment of Drp1 proteins to the outer surface of mitochondria [55]. Nevertheless, ROS also triggers mitochondrial fission, in which Drp1 may be stimulated by various signaling pathways such as calcineurin, Cdk1, and PKCδ [56]. Activation of Drp1, accompanied by narrowing of the mitochondrial diameter, results in the mitochondrion undergoing an asymmetric division and two daughter mitochondria with unequal membrane potential [57].

Typically, the process begins when the ER starts folding around the mitochondria. This triggers an immediate division of mitochondria, and mitochondrial fission factor (Mff)-dependent Drp1 assembly is then triggered [58]. While multiple isoforms of Drp1 are produced through careful fusion of pre-mRNA transcripts, Mff distinctively as well as efficiently controls Drp1 activity via a supportive GTPase-dependent mechanism [59]. Mff knockdown has been demonstrated to cause mitochondrial elongation in mammalian cells, while the overexpression of Mff promotes the requirement of Drp1 for mitochondrial fragmentation [60]. Premature self-assembly of Drp1 disrupts its connections with Mff, which leads to the formation of Drp1–Mff heterodimers, which enhance the elimination of the insert B domain of Drp1. Mainly, two key receptors are responsible for Drp1-dependent mitochondrial fission: mitochondrial elongation factor 1 (MiD51), which attaches with ADP cofactor, and mitochondrial elongation factor 2 (MiD49), which engages with Drp1 to the mitochondrial surface [61]. MiD51 has already been shown to facilitate the enrollment of Drp1 by lacking regulatory elements such as mitochondrial fission 1 (Fis1), mitofusion 2 (Mfn2), and Mff [62]. This is still controversial regarding the role of human Fis1, which consistently localizes across OMM and enhances Drp1 action by competitively binding to MiD51 [63]. Research demonstrated that Fis1 knockdown enhances the association of mitochondrial tubules, while its over-activation causes an enhancement in the process of mitochondrial fragmentation [64].

Beyond the normal process of mitochondrial fission, there are several corresponding pathways and regulatory elements that control Drp1-dependent fission. GTP hydrolysis and Drp1 activity are inhibited during mitochondrial division once

Drp1 identifies the phosphatidic acid head group and two saturated acyl chains of another phospholipid present in OMM [65]. Actin filaments can bind with Drp1 and work together with Mff, enhancing GTPase activity, while blocking actin polymerization considerably decreases the Drp1-dependent mitochondrial fission [66]. Inhibiting the receptor-interacting protein kinase 1 reduces the activation of c-Jun NH_2-terminal kinase and efficiently prevents the translocation of Drp1 [67].

For a better understanding of the regulatory mechanisms of mitochondrial fission, we need to gain a basic understanding of the processes opposing Drp1-dependent mitochondrial fission. In mitochondrial fusion, the membranes of two neighboring mitochondria merge by forming hetero- and homo-oligomeric complexes of protein [68]. While the OMM is merged independently of the IMM, both control the mitochondrial redistribution process. Mitochondrial fusion proteins such as Mfn1 and Mfn2 are situated on OMM, while OPA1 is found on IMM. Together, these three proteins considerably manage mitochondrial morphology and function [69]. Although overactivation of Mfn1 and Mfn2 does not repair mitochondrial fragmentation and improve cellular viability, ubiquitylation of Mfn1 is important for increasing cell life in stress conditions [70].

4.4.3 Degradation of Mitochondrial Proteins by the UPS

UPS manages the selective degeneration of intracellular proteins in eukaryotes. Elimination of dysfunctional proteins is essential for preserving the tissue integrity and overall health of an organism because dysfunctional proteins have the ability to accumulate. Mitochondrial proteins are removed through various degradation systems, such as mitophagy and proteolytic enzymes [71]. Primary components that ensure the specificity of abnormal protein degradation within the QC system are E3 ubiquitin ligases of UPS. The QC system is also important for refolding damaged stress proteins and preventing the accumulation of proteotoxic misfolded species that expose hydrophobic surfaces [72]. Ubiquitin serves as a marker for protein degradation when it is covalently linked to a substrate protein. E3 ubiquitin ligases are essential modulators of UPS, directly targeting the substrate proteins for degradation in both regulatory and protein QC degradation (PQCD) pathways. UPS-dependent PQCD signaling specifically targets the dysfunctional proteins. Notably, while every protein has the ability to be damaged, just a limited number of E3 ubiquitin ligases have been recognized as participated in PQCD [73]. Therefore, degradation signals of dysfunctional proteins, known as PQC degrons, are recognized by a small group of specific E3 ubiquitin ligases, indicating a specialization in identifying a broad and potentially various array of substrates [74]. We highlight that numerous regulatory degradation pathways that focus on the N-terminal and C-terminal degrons also encourage PQC functions through destroying the protein fragments that are produced by protease cleavage during metabolic pathways [73]. These fragments primarily participate in detrimental protein interactions, disrupting cellular processes, and are thus degraded. Likewise, proteins that are local but do not play a role in physiological function, such as those affected by mislocalization, stoichiometric ratios of their protein complex partners, and are degraded [75]. Impairment of the UPS is associated with various disorders such as chronic inflammatory conditions, neurodegenerative disorders, and cancer. For instance, mutations in the UPS machinery can result in the buildup of dysfunctional proteins, which contribute to the progression of various diseases [76].

4.5 Mitochondrial Dysfunction and UPS Impairment

Impairment in mitochondrial dysfunction and UPS are two key characteristics linked to numerous ageing-related diseases such as neurodegenerative diseases and certain cancers [77]. Even though mitochondria are involved in numerous processes for free radical neutralization, mitochondria remain susceptible to oxidative damage and depend on UPS, among other QC processes, to eliminate dysfunctional mitochondrial proteins. Therefore, the functioning of UPS is essential for preserving mitochondrial health, and vice versa; healthy mitochondria are necessary to support an effective UPS [78]. If mitochondrial dysfunction occurs, a vicious cycle may begin, resulting in the progressive failure of both systems. Research has shown that mitochondrial dysfunction can influence proteasomal activity in various systems, including mammalian cells, yeast, and *C. elegans* [79].

4.5.1 Oxidative Stress and Protein Damage

Oxidative stress arises from an imbalance between the production and accumulation of ROS, which are unavoidable by-products of metabolism. In biological systems, it acts as a double-edged sword: under carefully controlled conditions, it also plays a crucial role in signaling pathways, but can also damage cells when produced in extreme amounts since it can harm cells by oxidizing key biomolecules, such as lipids, nucleic acids (DNA and RNA), and proteins [80]. Oxidative damage can severely affect the structure and function of proteins and may even result in cell death. As the mitochondria are a primary source of ROS, mitochondrial proteins are especially exposed to oxidative modification, and removal of oxidized proteins is essential for maintaining the integrity of the organelle [81]. Therefore, enzymatic repair of protein oxidation and degradation of damaged protein is crucial for maintaining homeostasis, while a failure in protein maintenance has been associated with the age-related accumulation of oxidized proteins [82].

4.5.2 MDVs and UPS

Both MDVs and UPS are interconnected in preserving mitochondrial function, controlling protein degradation, and responding to cellular stress. In 2008, Neuspiel and colleagues were the first to discover MDVs budding from the mitochondria, which included single-membrane vesicles originating from OMM and double-membrane vesicles formed from the OMM, inner membrane, and matrix [83]. MDVs are characterized based on three divergent criteria: configuration in the absence of DRP1, length ranging between 70 and 150 nm, and cargo selectivity [84]. However, MDVs may serve a major function as a conserved mechanism, adapted for various functions in eukaryotes [19]. Unlike mitophagy, mitochondrial-derived vesicles (MDVs) form rapidly during mitochondrial stress and selectively deliver damaged or oxidized cargo to lysosomes or extracellular vesicles, enabling faster regulation of mitochondrial quality and mass. Additionally, a large amount of research has suggested that MDVs play diverse functions, such as resistance to oxidative stress, peroxisome biogenesis, mitochondrial fission, anti-infection, and innate immunity [85]. Nevertheless, dysfunction of MDVs is linked with various diseases, such as ischemic-hypoxic myocardial damage and neurodegenerative diseases. Furthermore, MDVs are also linked with infections, cancers, aging, and autoimmune diseases [86]. Proper functioning of MDVs plays a positive role in the treatment of numerous diseases.

UPS provides both a specific and a common mechanism for the cell to eliminate the unusual proteins such as misfolded, terminally modified, mutant, excessively accumulated, and damaged proteins [87]. The UPS has capacity for rapid, substrate-specific proteolysis. Also, it acts as a molecular signaling in various mechanistic pathways [88]. UPS plays an important function in sustaining protein degradation and mitochondrial function in addition to regulating dynamics in mitochondrial vesicles [89]. Both the MDVs and UPS work together to maintain mitochondrial homeostasis by facilitating the elimination of damaged mitochondrial elements, regulating protein QC, and enhancing mitochondrial function, dynamics, and biogenesis. This ultimately leads to maintaining the cellular energy metabolism and preventing mitochondrial-related diseases.

4.5.3 Mitophagy and Its Relation to UPS

Mitophagy is a specific type of macroautophagy that particularly targets the mitochondria to the lysosome for degradation [90]. Mitophagy plays a key function in maintaining the mitochondrial fission, fusion, homeostasis, and biogenesis for MQC within cells [91]. Latest investigations show that mitophagy can take place under normal conditions, which is referred to as basal mitophagy [92]. Removal of dysfunctional mitochondria is a crucial procedure for maintaining cellular homeostasis and is exactingly managed by the UPS and lysosomal functions [93]. Ubiquitin is a tiny regulatory protein made up of 76 amino acids that can be linked to the target substrates. This protein is extremely conserved across eukaryotes, and the method of attaching ubiquitin to substrates is called ubiquitination [94]. In the context of mitophagy, there are two distinct subtypes: receptor-mediated mitophagy and ubiquitin-mediated mitophagy [95]. The receptor-mediated mitophagy signaling serves critical and often very specific functions, such as the role of NIX in the maturation of reticulocytes or the function of BNIP3 in hypoxic conditions (Fig. 4.1).

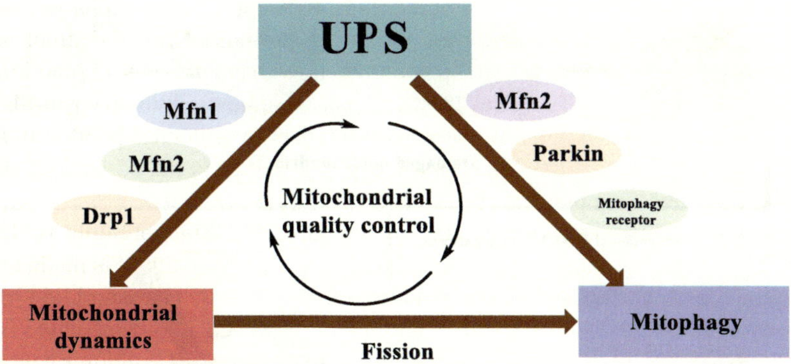

Fig. 4.1 Complex interactions between the UPS, mitophagy, and mitochondrial dynamics. Mfn1 Mitofusion1, Drp1 dynamin-related protein1

4.5.4 UPS-Mediated PINK1/Parkin Pathway in Regulation of Mitophagy

The PINK1/Parkin pathway is a UPS-mediated pathway that regulates mitophagy and helps maintain MQC. In the process of preserving mitochondrial health and function, the PINK1/Parkin-signaling mechanism is regarded as a key player for eliminating dysfunctional mitochondria (Fig. 4.2) [96]. PINK1-Parkin-mediated mitophagy signaling mechanism involves a series of distinct stages 1–4. These stages can be disconnected from each other due to mutations linked to PD or through chemical/genetic disruption within several critical components [97]. When cells are in a stable state, two proteins work together to maintain the shape of mitochondria: Ser/Thr kinase PINK1 and cytosolic E3 ubiquitin ligase Parkin control the morphology of mitochondria by engaging with fission/fusion machineries. In this context, PINK1 and Parkin function within a similar genetic pathway, with PINK1 acting as a regulator of Parkin [98]. In normal physiology, the functioning of mitochondrial PINK1 protein levels is typically very low. This is likely due to possible mechanisms: continuous breakdown via unidentified voltage-sensitive protease and through non-canonical import signaling [99]. On the contrary, the Parkin protein is comparatively steady as well as uniformly dispersed through the cytoplasm. (1) Oxidative stress causes mitochondrial membrane depolarization. (2) Chemical uncoupling, mtDNA mutations, and elevated ROS levels lead to a loss of $\Delta\Psi m$, which consequently stabilizes PINK1 on the OMM of damaged mitochondria [100]. Accumulating PINK1 facilitates PINK1 autophosphorylation and activation. After that, Parkin and ubiquitin are recruited and phosphorylated by PINK1. (3)

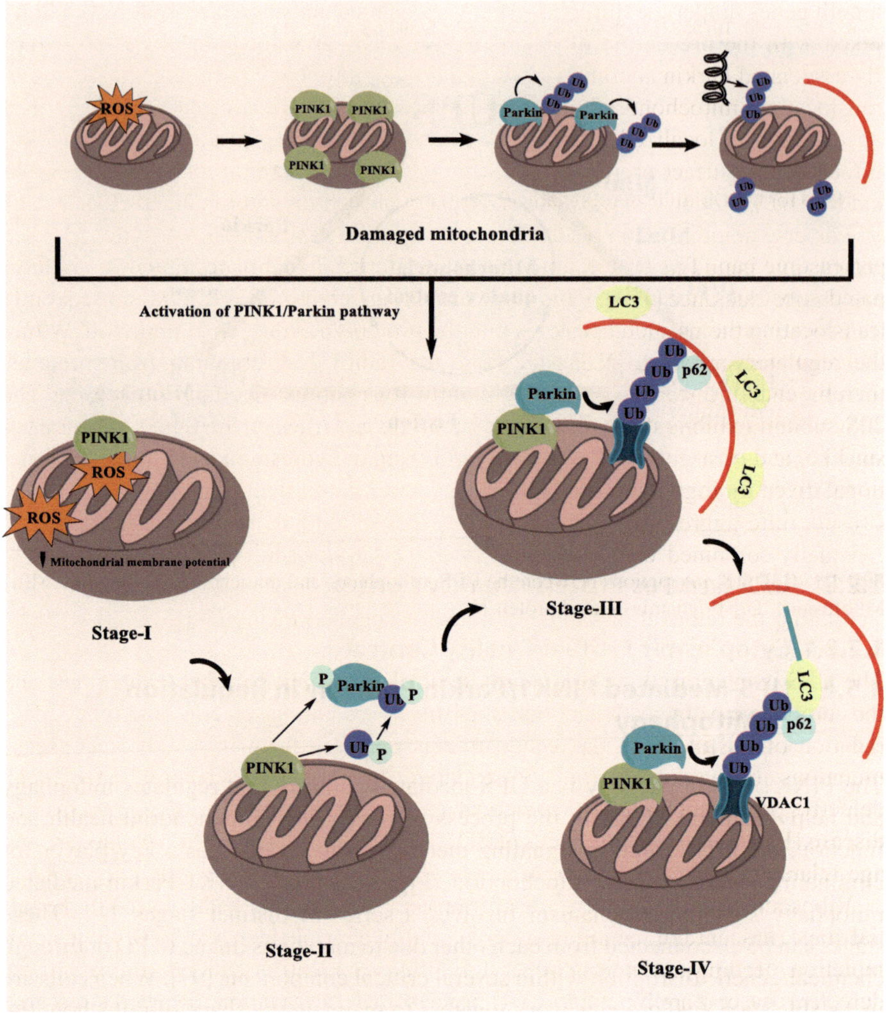

Fig. 4.2 PINK1/Parkin and UPS-mediated pathway that regulates mitophagy. (A) Oxidative stress causes mitochondrial membrane depolarization, (B) Elevated ROS levels lead to a loss of ΔΨm, which consequently stabilizes PINK1 on OMM of damaged mitochondria. (C) Parkin localization leads to the malfunction of mitochondria, as it ubiquitinates specific mitochondrial surface proteins such as VDAC1 and Mfn1/2. (D) LC3 and autophagosomal membranes are co-recruited through the transport and accumulation of damaged mitochondria. ROS reactive oxygen species, OMM outer mitochondrial membrane, VDACI voltage-dependent anion channel-1, Mnf1/2 mitofusion1/2, LC3 light chain, Ub ubiquitin

Subsequently, PINK1 on the mitochondria triggers the movement of cytosolic E3 ubiquitin ligase Parkin to dysfunctional mitochondria. Sub-cellular rearrangement of Parkin necessitates mitochondrial localization and kinase activity of PINK1, although the specific phosphorylation targets are still unidentified [101]. Alterations in both genes hinder the mitophagy via various mechanisms, like PINK1 mutations linked with the prevention of PD, the preliminary translocation of Parkin, while PD-associated Parkin alterations also disturb the mitophagy during the processes of translocation, mitochondrial clustering, and autophagic clearance of mitochondria. Once Parkin is localized to dysfunctional mitochondria, it ubiquitinates specific mitochondrial surface proteins such as voltage-dependent anion channel-1 (VDAC1) and Mfn1/2 [102]. In this way, Parkin, along with various sets of cofactors, can catalyze diverse ubiquitination reactions that are possibly essential for specific stages of mitophagy, including the hindrance of refusion, mitochondrial clustering, and autophagic clearance [103]. Parkin-dependent mitochondrial ubiquitination requires the adapter proteins such as p62/SQSTM1 and HDAC6, which connect with ubiquitinated substrates to cellular transport and autophagic signaling [104]. Mitochondrial clustering may help to check the cellular division of dysfunctional mitochondria and the stress they cause, while also preparing them for well-organized autophagic clearance. (4) Light chain 3 (LC3) and autophagosomal membranes are co-recruited through the transport and accumulation of damaged mitochondria. As a result, mito-aggresomes are engulfed by the developing autophagosomes, which eventually combined with lysosomes in a HDAC6-cortactin-actin-dependent manner. UPS-mediated PINK1/Parkin signaling is a critical regulator of mitophagy, facilitating the removal of damaged mitochondria and maintaining cellular energy metabolism.

4.6 Cellular Bioenergetics, UPS, and Disease

The UPS and mitochondria systems are closely interconnected. Mitochondria are semi-independent organelles in human cells that play a key function in the generation of energy and as well as metabolism, cell signaling, and apoptosis [105]. UPS is also accepted for the generation of RNS and ROS free radicals during ATP production [106]. Excessive amounts of free radicals can harm the mtDNA and the proteome, leading to the buildup of misfolded proteins that can negatively impact their function and are linked with the development of numerous diseases [107].

4.6.1 Neurodegenerative Disorders

The brain has disproportionately high energy demand compared to other organs; it comprises only 2% of body weight while consuming 20% of the total body's energy [108]. These require large amounts of energy to perform their unique tasks, of neurons including the synthesis of neurotransmitters, neurotransmission, proper ionic gradients, and membrane integrity [109]. Neurons are mainly dependent on glucose to generate ATP through various processes: glycolysis and the citric acid cycle. Impaired glucose transport functions, insulin signaling, reduced glycolysis, defective mitochondrial function, including diminished Krebs cycle, and elevated oxidative stress are critical factors that disrupt neuronal bioenergetics, ultimately leading to neuronal dysfunction and cell death [110]. Possible involvement of inadequate energy production in neurons is a contributing factor for AD, PD, and HD. In AD, beta-amyloid peptides, along with other risk factors, lead to energy failure by various mechanisms. In PD, ecological and genetic factors can interact with α-synuclein to result in energy failure [111]. Similarly, in HD, a genetic disorder, the mutant protein mHtt can also cause mitochondrial dysfunction and energy deficits [112]. Failure to identify and remove misfolded proteins is a significant factor in the development of the pathogenesis of neurodegeneration [113]. Studies showed that the association between the two genes Parkin and ubiquitin-C-terminal hydrolase-L1 within the UPS in heritable PD suggests that the UPS is involved in the development of PD and has high significance of the UPS in the neurological disorders, including HD, AD, and ALS. The UPS, a multi-component pathway, primarily degrades abnormal and misfolded proteins, eliminating unwanted proteins [114].

4.6.2 Cancer and Metabolic Reprogramming

Metabolic reprogramming typically originates from the tumor microenvironment. Based on the philosophy of the "Pasteur Effect," Warburg recommended that cancer cells and embryonic tissues exhibit heightened aerobic glycolysis due to compromised mitochondrial bioenergetic function [115]. Nevertheless, the variation of the bioenergetic function of mitochondria in carcinomas remains a topic of discussion, even though there is evidence connecting the downward regulation of mitochondrial metabolism and bioenergetics with cancer progression [116]. According to an analysis involving 859 patients across 20 different cancer types, 66% of cancers contained at least one somatic point mutation in mtDNA, indicating that the somatic point mutation in mtDNA is a general incident in human cancer progression. However, the specific role of these mtDNA alterations in tumorigenesis and cancer progression is still unclear [117]. Furthermore, reduced mtDNA copy number in cancer may be linked to impairment in mitochondrial biogenesis or other mitochondria-localized proteins such as p53 or SIRT3 [118]. The proteasome is a multiprotein cellular compound that regulates/degrades most of the cellular proteins, playing a role in nearly all cellular functions. Angiogenesis, cell proliferation, and motility processes are of particular significance for carcinogenesis, which are coordinated through the UPS [119]. Dysfunction of UPS can facilitate cancer development and progression by interfering with the normal cellular function, resulting in uncontrolled cell growth, genomic instability, and inflammation.

4.6.3 Cardiovascular Diseases

Dysfunction of UPS can also contribute to the progression of cardiovascular diseases by generating reactive species that harm the nDNA and mtDNA, while simultaneously regulating energy availability [120]. Proper mitochondrial function is essential for cells, as well as organs with high energy requirements, like the heart [121]. The uncoupling of the ETC in malfunctioning mitochondria leads to an enhanced production of ROS, a reduction in the cellular ATP pool, significant cellular damage, and apoptosis of cardiomyocytes [122]. Dysregulation of mitophagy significantly contributes to the development of cardiovascular disorders by promoting cell survival through the removal of long-lived organelles and proteins [123]. The first report that demonstrated the proteasome plays an important function in degrading certain stress-sensitive transcriptional regulators, including hypoxia-inducible factor-1 alpha (HIF-1α), nuclear factor kappa-light-chain-enhancer of activated b cells (NF-κB), and inducible cyclic adenosine monophosphate (cAMP) early repressor in the heart [124]. In contrast, the study outlines the various mechanisms that contribute to the dysregulation of UPS in the context of cardiac ischemia [125]. Dysfunction of UPS is pivotal in the onset development of cardiovascular diseases such as heart failure, hypertension, and atherosclerosis by promoting protein aggregation, inflammation, and oxidative stress, which ultimately result in cardiac dysfunction and increase mortality.

4.7 Therapeutic Approaches Targeting Bioenergetics and UPS

Latest advances in understanding the intricate relationship between bioenergetics and UPS have unveiled novel therapeutic strategies for combating complex diseases.

4.7.1 Mitochondria-Targeted Therapies in Various Diseases

4.7.1.1 Mitochondria-Targeted Drug Therapies in Cancer (Table 4.1)

Table 4.1 Pre-clinical and clinical mitochondria-targeted drug therapies in cancer

Complications	Sr. no.	Mitochondria-targeted	Pre-clinical/clinical studies	Methods of action	References
Cancer	1.	SkQ1	HT1080 cells	Inhibit cell multiplication as well as the cell division process Increased the activity of Aurora family kinases	[126]
	2.		Benzopyrene-induced carcinogenesis in SHR mice	By inhibiting the tumor growth	[127]
	3.		N-Nitrosodiethylamine-induced hepatocarcinogenesis in BALB/c mice	Reduced tumor occurrence multiplicity Restored gap junctions and intercellular communication of tumor cells	[128]
CVDs	4.	Mito-TEMPO	THP-1 cell model induced by ox-LDL; HFD rats	Reduced foam cell formation by enhancing the autophagic flux Enhanced cholesterol efflux through the upregulation of ABCA1 and ABCG1 in an autophagy-dependent manner Reduced the buildup of TC and LDL-c	[129]
	5.	MitoQ	Pressure overload-induced heart failure in rats	Decreased ventricular hypertrophy and pulmonary congestion Recovered membrane potential in IFM Enhanced retention capability of mitochondrial calcium in the SSM and IFM	[130]
	6.	SS-31/elamipretide/MTP-131	Clinical trials on heart failure patients	Enhanced left ventricular capacity Enhanced oxygen flow linked with super-complexes and enhanced the activity of Complex-I	[131], NCT02388464

Neurodegenerative diseases	7.	MitoQ	Cellular MPP^+ model Mouse MPTP model	Blocked the reduction in dopamine levels through MPP^+ Sheltered the nigrostriatal axis against MPTP toxicity Block the mitochondrial aconitase inactivation	[132, 133]
	8.		Clinical trial	Slowed the development of PD as measured through the UPDRS	NCT00329056
	9.	Mito-TEMPO	Isolated MCAs from rats with traumatic injury	Reduced myogenic constriction Scavenged H_2O_2 (PEG-catalase) by inhibiting both BKCa channels and TRPV4 channels	[134]
	10.	SkQR1	Rat model by brain surgery	Reduced the volume of the lesion in the cortex By decreasing the mitochondrial ROS and GSK-3β activity	[135]

4.7.2 Proteasome Inhibitors and Activators Targeted Therapies in Various Diseases

4.7.2.1 Proteasome Inhibitors Targeted Therapies in Various Diseases

Sr. no.	Name	Clinical trials (Phase)	Patient population	Adverse effects	References
1.	Bortezomib	Phase-II	Patients with newly diagnosed light-chain amyloidosis	Peripheral neuropathy, nausea, vomiting, diarrhoea, constipation, cytopenias infection	https://clinicaltrials.gov/
2.	Lxazomib	III	Patients of relapsed and refractory AL amyloidosis	Skin rash, peripheral edema, peripheral neuropathy, fatigue, cytopenia infection	NCT01850524, NCT01659658
3.	Carfilzomib	II	Patients with relapsed Waldenström's macroglobulinaemia	Skin rash, peripheral edema, peripheral neuropathy, fatigue, cytopenia infection	NCT01850524, NCT01813227

4.7.2.2 Proteasome Activator Targeted Therapies in Neurodegenerative Diseases

Sr. no.	Name of drug	Mechanism of action	Action of drug	Experimental model	References
1.	Betulinic acid	Activation of chymotrypsin proteasome activity	Increased the clearance of tau and amyloid β	MT4 human T cells	[136]
2.	Oleuropein	Activation of trypsin, chymotrypsin, and caspase activity of proteasome	Inhibition of tau aggregation	In in-vitro aggregation assay with P301L tau protein	[137]

4.8 Conclusion

The complex relationship among cellular bioenergetics, UPS, and mitochondrial function, including membrane potential, dynamics, and mitophagy, plays a key role in cellular homeostasis and disease pathogenesis. This complex interplay is implicated in various diseases such as neurodegenerative diseases, cardiovascular diseases, metabolic reprogramming, and cancer. Understanding these interconnected processes provides valuable insights into potential therapeutic approaches. Such strategies may include enhancing mitophagy, modulating UPS functioning, and regulating mitochondrial fission and fusion to maintain cellular energy production and protein QC. By addressing these fundamental cellular processes, we may uncover novel treatment modalities for a diverse range of pathologies. The challenge lies in fine-tuning these interventions to achieve therapeutic benefits while minimizing adverse effects, ultimately leading to more effective and personalized treatment strategies for complex diseases.

Acknowledgments The authors are grateful to the Chitkara College of Pharmacy, Chitkara University, Rajpura, Patiala, Punjab, India, for providing the necessary facilities to carry out the research work.

Credit Author Statement *Conceptualization*: Thakur Gurjeet Singh. *Analyzed the data*: Ashi Mannan, Thakur Gurjeet Singh; *Wrote the manuscript*: Akhil Sharma, Ashi Mannan. *Visualization*: Ashi Mannan. *Editing of the Manuscript*: Akhil Sharma, Ashi Mannan, and Thakur Gurjeet Singh. *Critically reviewed the article*: Thakur Gurjeet Singh. *Supervision*: Thakur Gurjeet Singh.

All authors read and approved the final manuscript.

Funding Statement Nil.

Availability of Data and Materials Not applicable.

Declarations

Competing Interests There are no conflicts of interest.

Ethics Approval Not applicable.

Compliance with Ethical Standards Not applicable.

Consent to Participate Not applicable.

Consent to Publish The paper is submitted with the consent of all listed authors.

References

1. Mata-Cantero L, Lobato-Gil S, Aillet F, Lang V, Rodriguez MS. The ubiquitin-proteasome system (UPS) as a cancer drug target: emerging mechanisms and therapeutics. In: Wondrak GT, editor. Stress response pathways in cancer. Dordrecht: Springer Netherlands; 2015. p. 225–64. Available from: http://link.springer.com/10.1007/978-94-017-9421-3_11.
2. Abbas R, Larisch S. Killing by degradation: regulation of apoptosis by the ubiquitin-proteasome-system. Cells. 2021;10(12):3465.
3. Höhn A, Tramutola A, Cascella R. Proteostasis failure in neurodegenerative diseases: focus on oxidative stress. Oxidative Med Cell Longev. 2020;2020:1–21.

4. Račková L, Csekes E. Proteasome biology: chemistry and bioengineering insights. Polymers. 2020;12(12):2909.
5. Ferrick DA, Neilson A, Beeson C. Advances in measuring cellular bioenergetics using extracellular flux. Drug Discov Today. 2008;13(5–6):268–74.
6. Harrington JS, Ryter SW, Plataki M, Price DR, Choi AMK. Mitochondria in health, disease, and aging. Physiol Rev. 2023;103(4):2349–422.
7. Monzel AS, Enríquez JA, Picard M. Multifaceted mitochondria: moving mitochondrial science beyond function and dysfunction. Nat Metab. 2023;5(4):546–62.
8. Casanova A, Wevers A, Navarro-Ledesma S, Pruimboom L. Mitochondria: it is all about energy. Front Physiol. 2023;14:1114231.
9. Sulkshane P, Ram J, Thakur A, Reis N, Kleifeld O, Glickman MH. Ubiquitination and receptor-mediated mitophagy converge to eliminate oxidation-damaged mitochondria during hypoxia. Redox Biol. 2021;45:102047.
10. Wei Y, Chiang WC, Sumpter R, Mishra P, Levine B. Prohibitin 2 is an inner mitochondrial membrane mitophagy receptor. Cell. 2017;168(1–2):224–238.e10.
11. Protasoni M, Zeviani M. Mitochondrial structure and bioenergetics in normal and disease conditions. Int J Mol Sci. 2021;22(2):586.
12. Reddy PH, Reddy TP. Mitochondria as a therapeutic target for aging and neurodegenerative diseases. Curr Alzheimer Res. 2011;8(4):393–409.
13. Giacomeli R, de Gomes MG, Reolon JB, Haas SE, Colomé LM, Jesse CR. Chrysin loaded lipid-core nanocapsules ameliorates neurobehavioral alterations induced by β-amyloid1-42 in aged female mice. Behav Brain Res. 2020;390:112696.
14. Busch KB. Inner mitochondrial membrane compartmentalization: dynamics across scales. Int J Biochem Cell Biol. 2020;120:105694.
15. Ježek P, Jabůrek M, Holendová B, Engstová H, Dlasková A. Mitochondrial cristae morphology reflecting metabolism, superoxide formation, redox homeostasis, and pathology. Antioxid Redox Signal. 2023;39(10–12):635–83.
16. Mohanraj K, Nowicka U, Chacinska A. Mitochondrial control of cellular protein homeostasis. Biochem J. 2020;477(16):3033–54.
17. Caron C, Bertolin G. Cristae shaping and dynamics in mitochondrial function. J Cell Sci. 2024;137(1):jcs260986.
18. Peruzzo R, Costa R, Bachmann M, Leanza L, Szabò I. Mitochondrial metabolism, contact sites and cellular calcium signaling: implications for tumorigenesis. Cancers (Basel). 2020;12(9):2574.
19. König T, Nolte H, Aaltonen MJ, Tatsuta T, Krols M, Stroh T, Langer T, McBride HM. MIROs and DRP1 drive mitochondrial-derived vesicle biogenesis and promote quality control. Nat Cell Biol. 2021;23(12):1271–86.
20. Bennett CF, Latorre-Muro P, Puigserver P. Mechanisms of mitochondrial respiratory adaptation. Nat Rev Mol Cell Biol. 2022;23(12):817–35.
21. Madan S, Uttekar B, Chowdhary S, Rikhy R. Mitochondria lead the way: mitochondrial dynamics and function in cellular movements in development and disease. Front Cell Dev Biol. 2021;9:781933.
22. Chandel NS. Mitochondria. Cold Spring Harb Perspect Biol. 2021;13(3):a040543.
23. Arunachalam E, Ireland W, Yang X, Needleman D. Dissecting flux balances to measure energetic costs in cell biology: techniques and challenges. Annu Rev Condens Matter Phys. 2023;14(1):211–35.
24. Schirrmacher V. Mitochondria at work: new insights into regulation and dysregulation of cellular energy supply and metabolism. Biomedicine. 2020;8(11):526.
25. Ravanelli S, den Brave F, Hoppe T. Mitochondrial quality control governed by ubiquitin. Front Cell Dev Biol. 2020;8:270.
26. Morelli AM, Ravera S, Panfoli I. The aerobic mitochondrial ATP synthesis from a comprehensive point of view. Open Biol. 2020;10(10):200224.
27. Senior AE. ATP synthesis by oxidative phosphorylation. Physiol Rev. 1988;68(1):177–231.

28. Coetzer J. A metabolomics and biochemical investigation of selected brain regions from Ndufs4 knockout mice [Internet]. PhD thesis, North-West University, South Africa. 2020 [cited 2024 Oct 2]. Available from: https://repository.nwu.ac.za/handle/10394/34742.
29. Barbato A, Scandura G, Puglisi F, Cambria D, La Spina E, Palumbo GA, Lazzarino G, Tibullo D, Di Raimondo F, Giallongo C, Romano A. Mitochondrial bioenergetics at the onset of drug resistance in Hematological malignancies: an overview. Front Oncol. 2020;10:604143.
30. Rai NK, Venugopal H, Rajesh R, Ancha P, Venkatesh S. Mitochondrial complex-1 as a therapeutic target for cardiac diseases. Mol Cell Biochem. 2024;480(2):869–90.
31. Markevich NI, Markevich LN. Computational modeling analysis of kinetics of fumarate reductase activity and ROS production during reverse electron transfer in mitochondrial respiratory complex II. Int J Mol Sci. 2023;24(9):8291.
32. González García P. Pathological and therapeutic mechanisms in CoQ deficiency: the role of the proteins involved in the Q-Junction. 2023 [cited 2024]. Available from: https://digibug.ugr.es/handle/10481/81257.
33. Scialo F, Sanz A. Coenzyme Q redox signalling and longevity. Free Radic Biol Med. 2021;164:187–205.
34. Vercellino I, Sazanov LA. The assembly, regulation and function of the mitochondrial respiratory chain. Nat Rev Mol Cell Biol. 2022;23(2):141–61.
35. Kühlbrandt W. Structure and mechanisms of F-type ATP synthases. Annu Rev Biochem. 2019;88:515–49.
36. Zorov DB, Andrianova NV, Babenko VA, Pevzner IB, Popkov VA, Zorov SD, Zorova LD, Plotnikov EY, Sukhikh GT, Silachev DN. Neuroprotective potential of mild uncoupling in mitochondria. Pros and Cons. Brain Sci. 2021;11(8):1050.
37. Nicholls DG. Mitochondrial membrane potential and aging. Aging Cell. 2004;3(1):35–40.
38. Zorova LD, Popkov VA, Plotnikov EY, Silachev DN, Pevzner IB, Jankauskas SS, Babenko VA, Zorov SD, Balakireva AV, Juhaszova M, Sollott SJ, Zorov DB. Mitochondrial membrane potential. Anal Biochem. 2018;552:50–9.
39. Bagkos G, Koufopoulos K, Piperi C. A new model for mitochondrial membrane potential production and storage. Med Hypotheses. 2014;83(2):175–81.
40. Bhatt S, Puli L, Patil CR. Role of reactive oxygen species in the progression of Alzheimer's disease. Drug Discov Today. 2021;26(3):794–803.
41. Suski JM, Lebiedzinska M, Bonora M, Pinton P, Duszynski J, Wieckowski MR. Relation between mitochondrial membrane potential and ROS formation. Methods Mol Biol. 2012;810:183–205.
42. Kodroń A, Mussulini BH, Pilecka I, Chacińska A. The ubiquitin-proteasome system and its crosstalk with mitochondria as therapeutic targets in medicine. Pharmacol Res. 2021;163:105248.
43. Eletr ZM, Wilkinson KD. Regulation of proteolysis by human deubiquitinating enzymes. Biochim Biophys Acta. 2014;1843(1):114–28.
44. Lacoursiere RE, Shaw GS. Acetylated ubiquitin modulates the catalytic activity of the E1 enzyme Uba1. Biochemistry. 2021;60(16):1276–85.
45. Snyder NA, Silva GM. Deubiquitinating enzymes (DUBs): regulation, homeostasis, and oxidative stress response. J Biol Chem. 2021;297(3):101077.
46. Livnat-Levanon N, Glickman MH. Ubiquitin-proteasome system and mitochondria - reciprocity. Biochim Biophys Acta. 2011;1809(2):80–7.
47. Franz A, Kevei É, Hoppe T. Double-edged alliance: mitochondrial surveillance by the UPS and autophagy. Curr Opin Cell Biol. 2015;37:18–27.
48. Song J, Herrmann JM, Becker T. Quality control of the mitochondrial proteome. Nat Rev Mol Cell Biol. 2021;22(1):54–70.
49. Escobar-Henriques M, Anton V. Mitochondrial surveillance by Cdc48/p97: MAD vs. membrane fusion. Int J Mol Sci. 2020;21(18):6841.
50. Maitland MER, Kuljanin M, Wang X, Lajoie GA, Schild-Poulter C. Proteomic analysis of ubiquitination substrates reveals a CTLH E3 ligase complex-dependent regulation of glycolysis. FASEB J. 2021;35(9):e21825.

51. Hegde RS, Zavodszky E. Recognition and degradation of mislocalized proteins in health and disease. Cold Spring Harb Perspect Biol. 2019;11(11):a033902.
52. Zavileyskiy L, Bunik V. Regulation of p53 function by formation of non-nuclear heterologous protein complexes. Biomolecules. 2022;12(2):327.
53. Heo JM, Rutter J. Ubiquitin-dependent mitochondrial protein degradation. Int J Biochem Cell Biol. 2011;43(10):1422–6.
54. Ren L, Chen X, Chen X, Li J, Cheng B, Xia J. Mitochondrial dynamics: fission and fusion in fate determination of mesenchymal stem cells. Front Cell Dev Biol. 2020;8:580070.
55. Hu C, Huang Y, Li L. Drp1-dependent mitochondrial fission plays critical roles in physiological and pathological progresses in mammals. Int J Mol Sci. 2017;18(1):144.
56. Li RL, Wang LY, Duan HX, Zhang Q, Guo X, Wu C, Peng W. Regulation of mitochondrial dysfunction induced cell apoptosis is a potential therapeutic strategy for herbal medicine to treat neurodegenerative diseases. Front Pharmacol. 2022;13:937289.
57. Shi W, Tan C, Liu C, Chen D. Mitochondrial fission mediated by Drp1-Fis1 pathway and neurodegenerative diseases. Rev Neurosci. 2023;34(3):275–94.
58. Adachi Y, Kato T, Yamada T, Murata D, Arai K, Stahelin RV, Chan DC, Iijima M, Sesaki H. Drp1 tubulates the ER in a GTPase-independent manner. Mol Cell. 2020;80(4):621–632.e6.
59. Chowdhury SR. The role of MicroRNAs in chlamydia infection [Internet]. Bayerische Julius-Maximilians-Universitaet Wuerzburg, Germany. 2017 [cited 2024]. Available from: https://search.proquest.com/openview/bf923fd3152218c44938e2075498703e/1?pq-origsite=gscholar&cbl=2026366.
60. Yu R, Liu T, Jin SB, Ning C, Lendahl U, Nistér M, Zhao J. MIEF1/2 function as adaptors to recruit Drp1 to mitochondria and regulate the association of Drp1 with Mff. Sci Rep. 2017;7(1):880.
61. Richter V, Singh AP, Kvansakul M, Ryan MT, Osellame LD. Splitting up the powerhouse: structural insights into the mechanism of mitochondrial fission. Cell Mol Life Sci. 2015;72(19):3695–707.
62. Samangouei P, Crespo-Avilan GE, Cabrera-Fuentes H, Hernández-Reséndiz S, Ismail NI, Katwadi KB, Boisvert WA, Hausenloy DJ. MiD49 and MiD51: new mediators of mitochondrial fission and novel targets for cardioprotection. Cond Med. 2018;1(5):239–46.
63. Zhu T, Hu Q, Yuan Y, Yao H, Zhang J, Qi J. Mitochondrial dynamics in vascular remodeling and target-organ damage. Front Cardiovasc Med. 2023;10:1067732. Available from: https://www.frontiersin.org/journals/cardiovascular-medicine/articles/10.3389/fcvm.2023.1067732/full.
64. Mai S, Klinkenberg M, Auburger G, Bereiter-Hahn J, Jendrach M. Decreased expression of Drp1 and Fis1 mediates mitochondrial elongation in senescent cells and enhances resistance to oxidative stress through PINK1. J Cell Sci. 2010;123(Pt 6):917–26.
65. Stepanyants N, Macdonald PJ, Francy CA, Mears JA, Qi X, Ramachandran R. Cardiolipin's propensity for phase transition and its reorganization by dynamin-related protein 1 form a basis for mitochondrial membrane fission. Mol Biol Cell. 2015;26(17):3104–16.
66. Liu A, Kage F, Higgs HN. Mff oligomerization is required for Drp1 activation and synergy with actin filaments during mitochondrial division. Mol Biol Cell. 2021;32(20):ar5.
67. Chen D, Ni HM, Wang L, Ma X, Yu J, Ding WX, Zhang L. p53 up-regulated modulator of apoptosis induction mediates acetaminophen-induced necrosis and liver injury in mice. Hepatology. 2019;69(5):2164–79.
68. Westermann B. Mitochondrial fusion and fission in cell life and death. Nat Rev Mol Cell Biol. 2010;11(12):872–84.
69. Di Nottia M, Verrigni D, Torraco A, Rizza T, Bertini E, Carrozzo R. Mitochondrial dynamics: molecular mechanisms, related primary mitochondrial disorders and therapeutic approaches. Genes (Basel). 2021;12(2):247.
70. Dong L, Li P, Yang K, Liu L, Gao H, Zhou G, Zhao Q, Xia T, Wang A, Zhang S. Promotion of mitochondrial fusion protects against developmental PBDE-47 neurotoxicity by restoring mitochondrial homeostasis and suppressing excessive apoptosis. Theranostics. 2020;10(3):1245–61.

71. Lavie J, De Belvalet H, Sonon S, Ion AM, Dumon E, Melser S, Lacombe D, Dupuy JW, Lalou C, Bénard G. Ubiquitin-dependent degradation of mitochondrial proteins regulates energy metabolism. Cell Rep. 2018;23(10):2852–63.
72. Papsdorf K, Richter K. Protein folding, misfolding and quality control: the role of molecular chaperones. Essays Biochem. 2014;56:53–68.
73. Müller L, Hoppe T. UPS-dependent strategies of protein quality control degradation. Trends Biochem Sci. 2024;49:859–74.
74. Hu C, Tian Y, Xu H, Pan B, Terpstra EM, Wu P, Wang H, Li F, Liu J, Wang X. Inadequate ubiquitination-proteasome coupling contributes to myocardial ischemia-reperfusion injury. J Clin Invest. 2018;128(12):5294–306.
75. Swamy KBS, Schuyler SC, Leu JY. Protein complexes form a basis for complex hybrid incompatibility. Front Genet. 2021;12:609766.
76. Forloni G, Terreni L, Bertani I, Fogliarino S, Invernizzi R, Assini A, Ribizzi G, Negro A, Calabrese E, Volonté MA, Mariani C, Franceschi M, Tabaton M, Bertoli A. Protein misfolding in Alzheimer's and Parkinson's disease: genetics and molecular mechanisms. Neurobiol Aging. 2002;23(5):957–76.
77. Ross JM, Olson L, Coppotelli G. Mitochondrial and ubiquitin proteasome system dysfunction in ageing and disease: two sides of the same coin? Int J Mol Sci. 2015;16(8):19458–76.
78. Sedlackova L, Korolchuk VI. Mitochondrial quality control as a key determinant of cell survival. Biochim Biophys Acta, Mol Cell Res. 2019;1866(4):575–87.
79. Livnat-Levanon N, Kevei É, Kleifeld O, Krutauz D, Segref A, Rinaldi T, Erpapazoglou Z, Cohen M, Reis N, Hoppe T, Glickman MH. Reversible 26S proteasome disassembly upon mitochondrial stress. Cell Rep. 2014;7(5):1371–80.
80. Misrani A, Tabassum S, Yang L. Mitochondrial dysfunction and oxidative stress in Alzheimer's disease. Front Aging Neurosci. 2021;13:617588.
81. Ugarte N, Petropoulos I, Friguet B. Oxidized mitochondrial protein degradation and repair in aging and oxidative stress. Antioxid Redox Signal. 2010;13(4):539–49.
82. Reeg S, Grune T. Protein oxidation in aging: does it play a role in aging progression? Antioxid Redox Signal. 2015;23(3):239–55.
83. Neuspiel M, Schauss AC, Braschi E, Zunino R, Rippstein P, Rachubinski RA, Andrade-Navarro MA, McBride HM. Cargo-selected transport from the mitochondria to peroxisomes is mediated by vesicular carriers. Curr Biol. 2008;18(2):102–8.
84. Picca A, Guerra F, Calvani R, Coelho-Júnior HJ, Landi F, Bucci C, Marzetti E. Mitochondrial-derived vesicles: the good, the bad, and the ugly. Int J Mol Sci. 2023;24(18):13835.
85. Peng T, Xie Y, Sheng H, Wang C, Lian Y, Xie N. Mitochondrial-derived vesicles: gatekeepers of mitochondrial response to oxidative stress. Free Radic Biol Med. 2022;188:185–93.
86. Picca A, Guerra F, Calvani R, Coelho-Junior HJ, Bossola M, Landi F, Bernabei R, Bucci C, Marzetti E. Generation and release of mitochondrial-derived vesicles in health, aging and disease. J Clin Med. 2020;9(5):1440.
87. Lehman NL. The ubiquitin proteasome system in neuropathology. Acta Neuropathol. 2009;118(3):329–47.
88. Zhang X, Linder S, Bazzaro M. Drug development targeting the ubiquitin-proteasome system (UPS) for the treatment of human cancers. Cancers (Basel). 2020;12(4):902.
89. Picca A, Guerra F, Calvani R, Bucci C, Lo Monaco MR, Bentivoglio AR, Coelho-Júnior HJ, Landi F, Bernabei R, Marzetti E. Mitochondrial dysfunction and aging: insights from the analysis of extracellular vesicles. Int J Mol Sci. 2019;20(4):805.
90. Song Y, Xu Y, Liu Y, Gao J, Feng L, Zhang Y, Shi L, Zhang M, Guo D, Qi B, Zhang M. Mitochondrial quality control in the maintenance of cardiovascular homeostasis: the roles and interregulation of UPS, mitochondrial dynamics and mitophagy. Oxidative Med Cell Longev. 2021;2021:3960773.
91. De R, Mazumder S, Bandyopadhyay U. Mediators of mitophagy that regulate mitochondrial quality control play crucial role in diverse pathophysiology. Cell Biol Toxicol. 2021;37(3):333–66.

92. Doblado L, Lueck C, Rey C, Samhan-Arias AK, Prieto I, Stacchiotti A, Monsalve M. Mitophagy in human diseases. Int J Mol Sci. 2021;22(8):3903.
93. Quarato G, Mari L, Barrows NJ, Yang M, Ruehl S, Chen MJ, Guy CS, Low J, Chen T, Green DR. Mitophagy restricts BAX/BAK-independent, Parkin-mediated apoptosis. Sci Adv. 2023;9(21):eadg8156.
94. Park GH, Park JH, Chung KC. Precise control of mitophagy through ubiquitin proteasome system and deubiquitin proteases and their dysfunction in Parkinson's disease. BMB Rep. 2021;54(12):592–600.
95. Czegle I, Gray AL, Wang M, Liu Y, Wang J, Wappler-Guzzetta EA. Mitochondria and their relationship with common genetic abnormalities in hematologic malignancies. Life (Basel). 2021;11(12):1351.
96. Vizziello M, Borellini L, Franco G, Ardolino G. Disruption of mitochondrial homeostasis: the role of PINK1 in Parkinson's disease. Cells. 2021;10(11):3022.
97. Springer W, Kahle PJ. Regulation of PINK1-Parkin-mediated mitophagy. Autophagy. 2011;7(3):266–78.
98. Yu R, Lendahl U, Nistér M, Zhao J. Regulation of mammalian mitochondrial dynamics: opportunities and challenges. Front Endocrinol (Lausanne). 2020;11:374.
99. Needs HI, Protasoni M, Henley JM, Prudent J, Collinson I, Pereira GC. Interplay between mitochondrial protein import and respiratory complexes assembly in neuronal health and degeneration. Life (Basel). 2021;11(5):432.
100. Ammal Kaidery N, Thomas B. Current perspective of mitochondrial biology in Parkinson's disease. Neurochem Int. 2018;117:91–113.
101. Barazzuol L, Giamogante F, Brini M, Calì T. PINK1/parkin mediated mitophagy, Ca2+ signalling, and ER-mitochondria contacts in Parkinson's disease. Int J Mol Sci. 2020;21(5):1772.
102. Li J, Yang D, Li Z, Zhao M, Wang D, Sun Z, Wen P, Dai Y, Gou F, Ji Y, Zhao D, Yang L. PINK1/Parkin-mediated mitophagy in neurodegenerative diseases. Ageing Res Rev. 2023;84:101817.
103. Bayne AN, Trempe JF. Mechanisms of PINK1, ubiquitin and Parkin interactions in mitochondrial quality control and beyond. Cell Mol Life Sci. 2019;76(23):4589–611.
104. Jeong SJ, Zhang X, Rodriguez-Velez A, Evans TD, Razani B. p62/SQSTM1 and selective autophagy in cardiometabolic diseases. Antioxid Redox Signal. 2019;31(6):458–71.
105. De I, Dogra N, Singh S. The mitochondrial unfolded protein response: role in cellular homeostasis and disease. Curr Mol Med. 2017;17(9):587–97.
106. Ifeanyi OE. A review on free radicals and antioxidants. Int J Curr Res Med Sci. 2018;4(2):123–33.
107. Bhatia V, Sharma S. Role of mitochondrial dysfunction, oxidative stress and autophagy in progression of Alzheimer's disease. J Neurol Sci. 2021;421:117253.
108. Bosy-Westphal A, Hägele FA, Müller MJ. What is the impact of energy expenditure on energy intake? Nutrients. 2021;13(10):3508.
109. Pereira CD, Martins F, Marques F, Sousa JC, Rebelo S. Beyond brain signaling. In: Silva JV, Freitas MJ, Fardilha M, editors. Tissue-specific cell signaling. Cham: Springer International Publishing; 2020. p. 1–32. Available from: https://link.springer.com/10.1007/978-3-030-44436-5_1.
110. Jové M, Mota-Martorell N, Torres P, Ayala V, Portero-Otin M, Ferrer I, Pamplona R. The causal role of lipoxidative damage in mitochondrial bioenergetic dysfunction linked to Alzheimer's disease pathology. Life (Basel). 2021;11(5):388.
111. Saxena U. Bioenergetics failure in neurodegenerative diseases: back to the future. Expert Opin Ther Targets. 2012;16(4):351–4.
112. Kabir MT, Uddin MS, Abdeen A, Ashraf GM, Perveen A, Hafeez A, Bin-Jumah MN, Abdel-Daim MM. Evidence linking protein misfolding to quality control in progressive neurodegenerative diseases. Curr Top Med Chem. 2020;20(23):2025–43.
113. Kumar D, Ambasta RK, Kumar P. Ubiquitin biology in neurodegenerative disorders: from impairment to therapeutic strategies. Ageing Res Rev. 2020;61:101078.

114. Petrucelli L, Dawson TM. Mechanism of neurodegenerative disease: role of the ubiquitin proteasome system. Ann Med. 2004;36(4):315–20.
115. Warburg O. On respiratory impairment in cancer cells. Science. 1956;124(3215):269–70.
116. Torresano L, Nuevo-Tapioles C, Santacatterina F, Cuezva JM. Metabolic reprogramming and disease progression in cancer patients. Biochim Biophys Acta Mol Basis Dis. 2020;1866(5):165721.
117. Lee HC, Huang KH, Yeh TS, Chi CW. Somatic alterations in mitochondrial DNA and mitochondrial dysfunction in gastric cancer progression. World J Gastroenterol. 2014;20(14):3950–9.
118. Li H, Cai Z. SIRT3 regulates mitochondrial biogenesis in aging-related diseases. J Biomed Res. 2022;37(2):77–88.
119. Voutsadakis IA. The ubiquitin-proteasome system in colorectal cancer. Biochim Biophys Acta. 2008;1782(12):800–8.
120. Mercer JR. Mitochondrial bioenergetics and therapeutic intervention in cardiovascular disease. Pharmacol Ther. 2014;141(1):13–20.
121. Boengler K, Kosiol M, Mayr M, Schulz R, Rohrbach S. Mitochondria and ageing: role in heart, skeletal muscle and adipose tissue. J Cachexia Sarcopenia Muscle. 2017;8(3):349–69.
122. Chistiakov DA, Shkurat TP, Melnichenko AA, Grechko AV, Orekhov AN. The role of mitochondrial dysfunction in cardiovascular disease: a brief review. Ann Med. 2018;50(2):121–7.
123. Ajoolabady A, Aslkhodapasandhokmabad H, Aghanejad A, Zhang Y, Ren J. Mitophagy receptors and mediators: therapeutic targets in the management of cardiovascular ageing. Ageing Res Rev. 2020;62:101129.
124. de Moissac D, Mustapha S, Greenberg AH, Kirshenbaum LA. Bcl-2 activates the transcription factor NFkappaB through the degradation of the cytoplasmic inhibitor IkappaBalpha. J Biol Chem. 1998;273(37):23946–51.
125. Powell SR, Divald A. The ubiquitin-proteasome system in myocardial ischaemia and preconditioning. Cardiovasc Res. 2010;85(2):303–11.
126. Dhanasekaran A, Kotamraju S, Kalivendi SV, Matsunaga T, Shang T, Keszler A, Joseph J, Kalyanaraman B. Supplementation of endothelial cells with mitochondria-targeted antioxidants inhibit peroxide-induced mitochondrial iron uptake, oxidative damage, and apoptosis. J Biol Chem. 2004;279(36):37575–87.
127. Yang Y, Karakhanova S, Hartwig W, D'Haese JG, Philippov PP, Werner J, Bazhin AV. Mitochondria and mitochondrial ROS in cancer: novel targets for anticancer therapy. J Cell Physiol. 2016;231(12):2570–81.
128. Morgan MJ, Liu Z, gang. Crosstalk of reactive oxygen species and NF-κB signaling. Cell Res. 2011;21(1):103–15.
129. Ma Y, Huang Z, Zhou Z, He X, Wang Y, Meng C, Huang G, Fang N. A novel antioxidant Mito-Tempol inhibits ox-LDL-induced foam cell formation through restoration of autophagy flux. Free Radic Biol Med. 2018;129:463–72.
130. Ribeiro Junior RF, Dabkowski ER, Shekar KC, Connell KAO, Hecker PA, Murphy MP. MitoQ improves mitochondrial dysfunction in heart failure induced by pressure overload. Free Radic Biol Med. 2018;117:18–29.
131. Chatfield KC, Sparagna GC, Chau S, Phillips EK, Ambardekar AV, Aftab M, Mitchell MB, Sucharov CC, Miyamoto SD, Stauffer BL. Elamipretide improves mitochondrial function in the failing human heart. JACC Basic Transl Sci. 2019;4(2):147–57.
132. Ghosh A, Chandran K, Kalivendi SV, Joseph J, Antholine WE, Hillard CJ, Kanthasamy A, Kanthasamy A, Kalyanaraman B. Neuroprotection by a mitochondria-targeted drug in a Parkinson's disease model. Free Radic Biol Med. 2010;49(11):1674–84.
133. Jin H, Kanthasamy A, Ghosh A, Anantharam V, Kalyanaraman B, Kanthasamy AG. Mitochondria-targeted antioxidants for treatment of Parkinson's disease: preclinical and clinical outcomes. Biochim Biophys Acta. 2014;1842(8):1282–94.
134. Szarka N, Pabbidi MR, Amrein K, Czeiter E, Berta G, Pohoczky K, Helyes Z, Ungvari Z, Koller A, Buki A, Toth P. Traumatic brain injury impairs myogenic constriction of cerebral arteries: role of mitochondria-derived H2O2 and TRPV4-dependent activation of BKca channels. J Neurotrauma. 2018;35(7):930–9.

135. Isaev NK, Novikova SV, Stelmashook EV, Barskov IV, Silachev DN, Khaspekov LG, Skulachev VP, Zorov DB. Mitochondria-targeted plastoquinone antioxidant SkQR1 decreases trauma-induced neurological deficit in rat. Biochemistry (Mosc). 2012;77(9):996–9.
136. Friedman LG, Price K, Lane RF, Carman AJ, Dacks PA, Shineman DW, Fillit HM. Meeting report on the Alzheimer's drug discovery foundation 14th international conference on Alzheimer's drug discovery. Alzheimers Res Ther. 2014;6(2):22.
137. Daccache A, Lion C, Sibille N, Gerard M, Slomianny C, Lippens G, Cotelle P. Oleuropein and derivatives from olives as Tau aggregation inhibitors. Neurochem Int. 2011;58(6):700–7.

Part II

Signaling Mechanisms

Triggers of UPS: Disease Aspects/ Initiator and Signaling Factors for Ubiquitin Proteasome System: Disease Aspect

5

Angel P. Godad, Tanmay L. Vasant, Shubham S. Kurmi, and Maheshkumar R. Borkar

5.1 Introduction

The ubiquitin-proteasome system (UPS) plays a critical role in maintaining cellular homeostasis by degrading misfolded, damaged, or excess proteins, thus preventing the accumulation of potentially toxic protein aggregates. This system is essential for the regulation of protein quality control (PQC) through a process known as PQC degradation, where E3 ubiquitin ligases selectively target damaged proteins for degradation based on specific degradation signals, or degrons, which are typically exposed hydrophobic sites in misfolded proteins [1]. As organisms age, the efficiency of these PQC mechanisms declines, leading to the accumulation of aggregation-prone proteins, which are implicated in age-related diseases such as Alzheimer's disease (AD), Parkinson's disease (PD), Huntington's disease (HD), amyotrophic lateral sclerosis (ALS), and multiple sclerosis (MS) [2–5]. UPS is the primary machinery for protein turnover in eukaryotic cells, mediating the ubiquitylation and subsequent degradation of unwanted proteins, which is crucial for various cellular processes, including cell cycle control, DNA damage response, and apoptosis [6, 7]. It operates in conjunction with autophagy, another degradation pathway, to ensure efficient protein and organelle turnover, thereby maintaining cellular integrity and nutrient management [8]. Genetic variations in UPS components can lead to substrate-specific effects on protein degradation, highlighting the system's complexity and its role in post-translational regulation [9]. Central to this system is the 26S

A. P. Godad · T. L. Vasant · S. S. Kurmi
Department of Pharmacology, SVKM's Dr. Bhanuben Nanavati College of Pharmacy, Mumbai, India

M. R. Borkar (✉)
Department of Pharmaceutical Chemistry, SVKM's Dr. Bhanuben Nanavati College of Pharmacy, Mumbai, India
e-mail: maheshkumar.borkar@bncp.ac.in

S. Singh (ed.), *Ubiquitin Proteasome System in CNS Neurodegenerative Diseases*, https://doi.org/10.1007/978-981-95-6099-8_5

proteasome, a complex composed of the 20S core particle and the 19S regulatory particle. The 20S core is a cylindrical structure responsible for the proteolytic activity, where proteins are degraded into peptides. It is composed of four stacked rings, each containing seven subunits, forming a barrel-like structure that encloses the proteolytic sites, thus preventing uncontrolled protein degradation [10]. The 19S regulatory particle, attached to one or both ends of the 20S core, is responsible for recognizing ubiquitin-tagged proteins, unfolding them, and translocating them into the 20S core for degradation. This process is ATP-dependent, highlighting the energy requirement for substrate processing and unfolding [11, 12]. The UPS is not only crucial for degrading misfolded or damaged proteins but also regulates various cellular processes, including cell cycle progression, signal transduction, and immune responses [13, 14]. Ubiquitination is a critical post-translational modification process that involves the tagging of proteins for degradation, primarily through the UPS. This process is orchestrated by a cascade of enzymatic reactions involving three main types of enzymes: ubiquitin-activating enzymes (E1), ubiquitin-conjugating enzymes (E2), and ubiquitin ligases (E3) [15, 16]. The E1 enzyme initiates the process by activating ubiquitin, which is then transferred to the E2 enzyme. The E2 enzyme, in conjunction with an E3 ligase, facilitates the transfer of ubiquitin to the target protein, forming a covalent bond with a lysine residue on the substrate [15, 17]. The E3 ligase is crucial for substrate recognition and specificity, as it determines which proteins are tagged for degradation [16]. There are over 600 E3 ligases in humans, each responsible for ubiquitinating specific target proteins in a spatio-temporal manner, thereby regulating various cellular processes such as DNA repair, cell division, and signaling pathways [18]. The ubiquitination process can result in different types of modifications, including mono-ubiquitination, multi-mono-ubiquitination, and polyubiquitination, which can signal for diverse cellular outcomes, including protein degradation, cellular localization, and activity modulation [19]. DUBs are a diverse group of proteases, classified into seven subfamilies, that reverse the ubiquitination process, which is essential for protein degradation, DNA repair, apoptosis, and immune responses [20]. By reversing the ubiquitination of proteins, DUBs play a critical role in the degradation and stability in neurodegenerative diseases characterized by protein aggregation, such as PD and AD [21, 22]. For instance, the DUB YOD1 has been shown to interact with α-synuclein, a protein implicated in PD, by deubiquitinating specific polyubiquitin chains, thus destabilizing α-synuclein and promoting its degradation through the E3 ligase NEDD4 pathway [23]. The regulation of DUBs is complex and involves various mechanisms, including protein–protein interactions, post-translational modifications, and subcellular localization, which ensure precise substrate targeting and prevent aberrant deubiquitination [24, 25].

5.2 Molecular Mechanisms of the UPS in Neurodegenerative Diseases

5.2.1 Ubiquitination Pathways

The ubiquitination pathway is a highly regulated and sequential process involving three main types of enzymes: E1 (ubiquitin-activating enzyme), E2 (ubiquitin-conjugating enzyme), and E3 (ubiquitin ligases) enzymes, which work together to tag proteins for degradation [26]. The process begins with the E1 enzyme, which activates ubiquitin in an ATP-dependent manner, forming a high-energy thiol-ester bond with ubiquitin. This activated ubiquitin is then transferred to the E2 conjugating enzyme, which acts as a carrier in the ubiquitin transfer process. The final step involves the E3 ligases, which facilitate the transfer of ubiquitin from the E2 enzyme to the substrate protein, often forming a polyubiquitin chain that signals the protein for degradation by the 26S proteasome [27–29]. The specificity of the ubiquitination process is largely determined by the E3 ligases, which are responsible for recognizing and binding to specific substrate proteins. E3 ligases are diverse and can be classified into different families, such as HECT and RING finger families, each with distinct mechanisms of action. For instance, the RING-type E3 ligases act as scaffolds that bring the E2–ubiquitin complex and the substrate into proximity, facilitating the direct transfer of ubiquitin from E2 to the substrate [30, 31]. The specificity of E3 ligases is crucial for the regulation of numerous cellular processes, including cell cycle progression, signal transduction, and apoptosis. For example, the degradation of cyclin E is a key regulator of the G1/S transition in the cell cycle, is mediated by specific E3 ligases that recognize and ubiquitinate cyclin E, thereby controlling its levels and ensuring proper cell cycle progression [29]. Additionally, the ubiquitin system is involved in the regulation of inflammatory responses and programmed cell death, with E3 ligases playing a pivotal role in modulating these processes by targeting key signaling proteins for degradation [32]. The complexity and specificity of the ubiquitin system are further enhanced by the presence of DUBs, which can remove ubiquitin from substrate, thus providing an additional layer of regulation [33]. The ubiquitination pathway, through the coordinated action of E1, E2, and E3 enzymes, and the specificity imparted by the E3 ligases, plays a fundamental role in maintaining cellular homeostasis and regulating diverse biological processes [27, 28, 34].

5.2.2 Types of Ubiquitination

Ubiquitination is a versatile post-translational modification that involves the attachment of ubiquitin, a small regulatory protein to substrate proteins, influencing their fate and function. This process can occur as mono-ubiquitination, where a single ubiquitin molecule is attached to a substrate, or poly-ubiquitination, where a chain of ubiquitin molecules is formed. Mono-ubiquitination regulates processes such as endocytosis, DNA repair, and histone modifications, acting as a signal for non-degradative cellular processes [35]. Poly-ubiquitination typically targets proteins for degradation via the 26S proteasome, a process crucial for maintaining protein homeostasis and regulating various cellular pathways [36, 37]. The type of ubiquitin chain linkage plays a significant role in determining the functional outcome of ubiquitination. K48-linked polyubiquitin chains are the most well-known and are primarily associated with proteasomal degradation, marking proteins for destruction and thus regulating protein levels within the cell [38]. K63-linked polyubiquitin chains are involved in non-degradative signaling pathways such as DNA repair, signal transduction, and receptor endocytosis, highlighting their role in cellular signaling rather than protein degradation [39, 40]. The diversity of ubiquitin chain linkages, including less common ones such as K6, K11, K27, K29, and K33, further expands the functional repertoire of ubiquitination, influencing processes such as cell cycle regulation, DNA damage response, and immune signaling [38]. The specificity of ubiquitination is largely determined by the E3 ubiquitin ligases, which confer substrate specificity and dictate the type of ubiquitin modifications, whether mono- or poly-ubiquitination, and the specific lysine residues involved. This specificity is crucial for the precise regulation of cellular processes, as different ubiquitin linkages can lead to vastly different outcomes for the substrate proteins [37, 41].

5.2.3 Proteasomal Degradation

Proteasomal degradation is a critical ATP-dependent process in neurons, involving the recognition, unfolding, and translocation of proteins into the proteasome for degradation. The 26S proteasome, a complex multi-subunit protease, plays a pivotal role in maintaining cellular homeostasis by degrading misfolded, damaged, or unnecessary proteins, which is essential for neuronal function and survival [42]. The process begins with the recognition of substrates tagged with ubiquitin chains, which serve as signals for degradation. The ubiquitin chains are recognized by ubiquitin receptors on the proteasome, triggering conformational changes that facilitate substrate engagement and processing [43, 44]. The ATPase motor of the proteasome then unfolds the substrate and translocates it into the proteolytic chamber for degradation [42, 45]. In neurons, the proteasome is involved in various processes, including synaptic plasticity, memory function, and the regulation of circadian rhythms. Dysregulation of proteasomal degradation in neurons can lead to neurodegenerative diseases such as AD, PD, and HD, where impaired proteasome function contributes to the accumulation of toxic protein aggregates. Age-related decline in proteasome activity is also observed, often characterized by a shift from 26S to the 20S proteasome, which lacks the regulatory particle necessary for ubiquitin-dependent degradation [46, 47]. This decline can exacerbate neurodegenerative conditions by impairing the clearance of damaged proteins. Regulation of proteasome activity is complex and involves various factors, including the TORC1, which modulates proteasome assembly and activity in response to cellular needs [47].

5.2.4 Deubiquitination and Regulation

DUBs play a crucial role in the regulation of ubiquitin chains, serving as editors, proofreaders, and recyclers of ubiquitin, which is essential for maintaining cellular homeostasis. DUBs are specialized proteases that remove ubiquitin from substrates or cleave within ubiquitin chains, thereby regulating the ubiquitination process and influencing various cellular pathways [48, 49]. The diversity of DUBs, with over 90 identified enzymes, underscores their significance in the ubiquitin system, where they modulate protein degradation, trafficking, and signal transduction [48, 50]. DUB-1 is a member of the UBP superfamily and has been shown to regulate cellular growth by modulating ubiquitin-dependent proteolysis, highlighting the role of DUBs in cell cycle control [51]. DUBs are integral to apoptosis, where they balance ubiquitination and deubiquitination to ensure proper cell death and survival responses, thus maintaining cellular health [52]. Ubiquitination and deubiquitination are critical, as dysregulation can lead to neurodegenerative disorders (NDDs) [25]. DUBs also influence protein trafficking by determining the fate of ubiquitinated substrates, which is vital for cellular localization and function [50]. The regulation of DUB activity is complex, involving protein–protein interaction, posttranslational modifications, and responses to oxidative stress, which can directly affect their enzymatic activity [25]. The deubiquitination complex platform is proposed as a mechanism for DUB substrate recognition, further illustrating the complexity of DUB regulation [53]. They exhibit specificity for different ubiquitin linkage types, which is critical for their function in various cellular processes, including DNA repair, PQC, and immune signaling [54, 55]. For instance, Ubp2 and Ubp3 are DUBs that facilitate ribosome-associated quality control (RQC) by removing mixed K48/K63 ubiquitin chains from collided ribosomes, thus promoting the recruitment of the RQC–trigger complex [56]. Similarly, Ubp14 in *Saccharomyces cerevisiae* is primarily involved in recycling free ubiquitin chains, which is vital for maintaining ubiquitin homeostasis and regulating biological processes such as amino acid metabolism and stress responses [57]. Disruptions in the balance between ubiquitination and deubiquitination due to mutations in components of the ubiquitin system, such as E3 ligases and DUBs, are linked to various neurodegenerative diseases, including PD, AD, and ALS [58, 59]. In PD, for instance, DUBs regulate mitophagy by deubiquitinating mitochondrial substrates, thus influencing the degradation of dysfunctional mitochondria and the aggregation of α-synuclein, a hallmark of the disease [60]. Similarly, in ALS and frontotemporal dementia (FTD), the ubiquitin-positive aggregation of TDP-43 is modulated by the interplay between ubiquitinating and deubiquitinating enzymes (DUBs), affecting protein stability and inclusion formation [61]. Moreover, the aging process exacerbates these issues by reducing the efficiency of proteasomal degradation and autophagy, thereby increasing the risk of neurodegenerative diseases through the accumulation of misfolded proteins [62].

5.3 Signaling Pathways Involved in UPS Regulation and Dysfunction

5.3.1 NF-κB Pathway

The NF-κB pathway plays a crucial role in inflammation and stress responses, acting as a central mediator in various biological processes, including immune responses, cell growth, and apoptosis. It is activated by a range of stimuli such as proinflammatory cytokines, oxidative stress, and mechanical stress, which trigger the transcription of genes involved in inflammation and immune regulation [63–65]. NF-κB signaling is implicated in the inhibition of neurogenesis in the adult hippocampus, contributing to depressive behaviors by decreasing the proliferation of neural stem-like cells [66]. Dysregulation of the NF-κB pathway has significant implications for the UPS function, particularly in neurodegenerative diseases. NF-κB is a transcription factor involved in regulating immune responses, inflammation, and cell survival, which is activated by various stressors, including oxidative and inflammatory stresses, which are prevalent in neurodegenerative conditions [63, 64]. NF-κB activation is linked to the expression of proinflammatory cytokines and stress response genes, which can further impair UPS function by promoting an inflammatory environment that overwhelms the proteasome's capacity to degrade proteins [66, 67].

5.3.2 PI3K/Akt/mTOR Pathway

The PI3K/Akt/mTOR pathway is critical for cellular processes such as protein synthesis, degradation, and cell survival, playing a pivotal role in both normal cellular function and disease states. This pathway is activated by various stimuli, including growth factors and stress signals, leading to the activation of Akt, which, in turn, phosphorylates and regulates multiple downstream targets involved in cell survival and growth. Akt activation can inhibit apoptosis in neuronal cells, as demonstrated in studies where activated forms of Akt enhanced survival in differentiating neuronal cells, independent of Bcl-2 induction [68]. Dysregulation of the PI3K/Akt/mTOR pathway is implicated in various neurodegenerative diseases (NDDs). In AD and PD, the PI3K/Akt pathway influences neuronal survival and neurotoxicity through substrates such as GSK-3β and caspase-9, with natural products showing potential in modulating this pathway for therapeutic benefits [69]. Altered PI3K/Akt signaling contributes to dysfunction in the UPS, a critical pathway for protein degradation, which is often observed in neurodegenerative diseases. The PI3K/Akt//mTOR pathway, when dysregulated, can affect the UPS by altering the expression and activity of components involved in protein degradation [70].

5.3.3 p53 Pathway

The p53 protein is a pivotal regulator of cellular stress responses, particularly in the context of DNA damage and cellular stress, where it orchestrates a variety of cellular outcomes such as DNA repair, cell cycle arrest, apoptosis, and senescence [71, 72]. The regulation of the UPS by p53 is crucial in maintaining protein homeostasis, especially under stress conditions. p53 modulates the UPS by transcribing its negative regulator, Mdm2, an E3 ligase that ubiquitinates p53, targeting it for degradation via the proteasome pathway [72, 73]. This regulation is essential for controlling p53 levels and activity, ensuring that p53 can effectively respond to DNA damage by either promoting repair to initiating apoptosis if the damage is irreparable. The degradation of p53 is not limited to the cytoplasm; it also occurs in the nucleus, which adds a layer of control in down-regulating the p53 response once DNA damage is repaired [74]. Mutations can lead to the accumulation of misfolded proteins and proteotoxic stress, which are hallmarks of neurodegenerative disease [75].

5.3.4 Wnt/β-Catenin Signaling

Wnt/β-catenin signaling plays a crucial role in neuronal development and synaptic function in neurodegenerative diseases such as PD and AD. The signaling pathway is essential for the development and maintenance of midbrain dopaminergic (DA) neurons, which are critical for motor functions and are notably affected in PD [76, 77]. The pathway's involvement in neurogenesis is highlighted by its regulation of progenitor cell proliferation and differentiation in the ventral midbrain, where it interacts with Sonic Hedgehog signaling to influence dopamine neuron development [78]. Dysregulation of the Wnt/β-catenin signaling pathway can lead to synaptic degeneration in the striatum, leading to motor coordination deficits [79]. In AD, the pathway's dysregulation is linked to synaptotoxicity, where a decrease in β-catenin levels is associated with amyloid-beta (Aβ) induced neurotoxicity [80]. The dysregulation of Wnt/β-catenin signaling also affects the UPS, which is crucial for protein degradation and cellular homeostasis. The UPS is responsible for the non-lysosomal degradation of proteins crucial for regulating protein levels and ensuring that cellular functions such as growth, division, and stress response are maintained [81]. The Wnt signaling pathway is tightly regulated by ubiquitination and deubiquitination process, where E3 ligases and DUBs plays pivotal roles in modulating the stability of key signaling factors, including β-catenin, Axin, GSK3, and Dishevelled (Dvl) [82]. This regulation involves both proteolytic and non-proteolytic functions of ubiquitin, which control protein activity and localization, thereby influencing the pathway's activation and cellular outcomes [83].

5.3.5 c-Jun Terminal Kinase Pathway

The c-Jun terminal kinase (JNK) pathway plays a crucial role in cellular stress responses and apoptotic signaling, acting as a mediator of various cellular processes in response to both internal and external stress signals. JNK, a member of the mitogen-activated protein kinase (MAPK) family, is activated by a range of stressors, including UV radiation, growth factors, and viral infections, and is involved in regulating cell differentiation, proliferation, survival, and apoptosis [84, 85]. The pathway's complexity is underscored by its interaction with numerous protein partners, which allow it to mediate diverse cellular responses. In stress responses, JNK activation leads to the phosphorylation of substrates such as c-Jun, which modulates gene expression and influences cell fate decisions [86, 87]. JNK's role in apoptosis is particularly significant, as it can either promote or inhibit cell death depending on the context and duration of activation [88]. For instance, JNK-mediated phosphorylation of nSMase1 leads to ceramide generation, a critical step in stress-induced apoptosis [89]. Additionally, JNK interacts with mitochondrial proteins like SAB, which sustains its activity and promotes ROS production, further influencing apoptotic pathways [90]. In the CNS, JNK is highly active and implicated in pathogenesis such as neurodegeneration and glioblastoma, where it regulates processes such as cell death and regeneration [85]. Furthermore, JNK activation is associated with dysfunction and neuronal apoptosis, contributing to memory deficits and neurodegeneration, as seen in AD and other neurodegenerative conditions. The JNK pathway, particularly the JNK3 isoform, is involved in the formation of neurofibrillary tangles and senile plaques through the phosphorylation of tau and amyloid precursor protein (APP) [91, 92].

5.4 Signaling Cascades Contributing to Neurodegeneration

5.4.1 Role of Key Signaling Pathway in AD

5.4.1.1 NF-κB Pathway

The NF-κB pathway plays a significant role in regulating the UPS in AD by influencing various molecular mechanisms that contribute to the disease's pathogenesis. NF-κB is a transcription factor involved in inflammation, oxidative stress, and apoptosis, all of which are critical in AD progression [93]. The UPS is crucial for maintaining proteostasis by degrading misfolded proteins, and its impairment is linked to the accumulation of amyloid plaques and neurofibrillary tangles, hallmark features of AD [94]. NF-κB signaling is intricately connected to the UPS, as it regulates the degradation of its inhibitors through ubiquitination, thereby modulating its own activity [95]. This regulation is essential because dysregulated NF-κB activity can exacerbate neuroinflammation and Aβ production, further contributing to AD pathology. Moreover, NF-κB activation can upregulate BACE1 expression, a key enzyme in Aβ production, linking NF-κB directly to amyloidogenic processes [96]. The ubiquitin system also influences NF-κB signaling through linear ubiquitination, which is mediated by the linear ubiquitin chain assembly complex, affecting immune and inflammatory responses relevant to AD [97]. Additionally, NF-κB's role in neuroprotection is highlighted by its ability to upregulate neuroglobin, which can protect against Aβ-induced neurotoxicity, suggesting a dual role in both promoting and mitigating AD pathology [98]. Therefore, the NF-κB pathway's regulation of the UPS is multifaceted, involving both the promotion of pathological processes and potential protective mechanisms, making it a complex but promising target for therapeutic intervention in AD [99].

5.4.1.2 PI3K/Akt/mTOR Pathway

The PI3K/Akt/mTOR pathway plays a multifaceted role in regulating the UPS in AD, primarily through its influence on autophagy and protein homeostasis. This pathway is crucial in maintaining cellular homeostasis by modulating autophagy, a process essential for the degradation and recycling of damaged proteins and organelles, which is often impaired in AD. Dysregulation of the PI3K/Akt/mTOR pathway has been linked to the accumulation of Aβ and tau proteins, key pathological features of AD. For instance, hyperactivation of mTOR signaling, as observed in NS-Pten knockout mice, correlates with increased levels of Aβ and tau, suggesting that mTOR dysregulation exacerbates AD pathology by impairing protein degradation pathways [100]. Conversely, inhibition of mTOR, such as through rapamycin treatment, has been shown to enhance autophagy, thereby reducing Aβ and tau accumulation and improving cognitive deficits in AD models [101]. This is further supported by studies demonstrating that rapamycin-induced autophagy activation can reverse oxidative stress and synaptic dysfunction in Aβ-insulted neurons, highlighting the protective role of autophagy in AD [102]. Additionally, the PI3K/Akt/mTOR pathway is involved in Aβ-induced autophagy, where its inhibition can mitigate Aβ-induced neuronal cell cycle events and synaptic damage, further underscoring its regulatory role in UPS and neuroprotection [103]. Moreover, the pathway's modulation affects tau phosphorylation, a process linked to mitochondrial dysfunction and cognitive impairment, which can be alleviated by mTOR inhibitors [104].

5.4.1.3 p53 Pathway

The p53 pathway plays a multifaceted role in regulating the UPS in the context of AD, influencing both protein degradation and transcriptional regulation. The p53 protein, a key regulator of apoptosis and DNA damage response, is stabilized and activated in AD, often in response to Aβ peptides, which are central to AD pathology. This stabilization is partly due to the phosphorylation by Cdk5, leading to neuronal apoptosis [105]. Furthermore, p53 aggregation and its interaction with tau proteins in AD brains disrupt its normal function, contributing to neurodegeneration [106]. The UPS is involved in regulating p53-mediated transcription, as evidenced by the requirement of ubiquitylation and proteasomal functions for efficient transcription at the p21(waf1) promoter, a process that is disrupted in AD [107]. In AD models, proteasome inhibition has been shown to restore synaptic plasticity, suggesting that modulating UPS activity can ameliorate some AD-related deficits [108]. Additionally, mutations in the Trp53 gene, which encodes p53, are associated with altered p53 function and increased Aβ levels, further linking p53 dysregulation to AD pathology [109].

5.4.1.4 Wnt/β-Catenin Signaling Pathway

The Wnt/β-catenin signaling pathway plays a crucial role in regulating the UPS in AD by influencing several key processes associated with the disease's pathogenesis. This signaling pathway is essential for neuronal survival, synaptic plasticity, and neurogenesis, and its dysfunction is linked to the progression of AD [110]. Specifically, Wnt/β-catenin signaling is involved in the processing of APP and the modulation of Aβ peptide production, which are central to AD pathology [111]. Activation of this pathway has been shown to repress the transcription of the β-amyloid precursor protein cleaving enzyme (BACE1) through the binding of T-cell factor-4 to the BACE1 promoter, thereby reducing Aβ levels [112]. Conversely, inhibition of Wnt signaling accelerates amyloidogenic processing of APP, leading to increased production and aggregation of Aβ peptides, which are toxic to neurons [113]. Furthermore, the suppression of Wnt/β-catenin signaling is associated with increased tau phosphorylation, another hallmark of AD [114]. The pathway's dysfunction is exacerbated by factors such as the overexpression of CXXC5, a negative regulator of Wnt/β-catenin signaling, which is observed in AD patients and models, leading to cognitive deficits and increased Aβ plaque formation. Therapeutic strategies that aim to restore Wnt/β-catenin signaling, such as the use of lithium or small molecules like KY19334, have shown promise in reducing AD-related neuropathological markers and improving cognitive functions in animal models [115].

5.4.1.5 JNK Pathway

The JNK pathway plays a multifaceted role in regulating the UPS in AD, primarily through its involvement in synaptic dysfunction, Aβ production, and tau pathology. JNK activation is a critical factor in the early synaptic dysfunction observed in AD, as it is activated at synapses before cognitive impairment becomes apparent, and its inhibition can prevent synaptic degeneration and loss of postsynaptic proteins in AD models [116]. JNK, particularly JNK3, is implicated in the amyloidogenic processing of APP, where it phosphorylates APP, facilitating its cleavage by γ-secretase, thus increasing Aβ production and plaque formation [117]. This pathway is also linked to tau pathology, as JNK activation leads to tau phosphorylation and aggregation, processes that are exacerbated by additional phosphorylation by GSK-3β, contributing to neurofibrillary tangle formation [118]. Furthermore, JNK activation is associated with oxidative stress and endoplasmic reticulum stress, which are prevalent in AD, and its inhibition can mitigate these stress responses, thereby promoting neuronal survival [119]. The therapeutic potential of targeting JNK is underscored by studies showing that JNK inhibition can reduce amyloid plaque burden, tau hyperphosphorylation, and synaptic loss, while improving cognitive deficits in AD models [120].

5.4.2 Role of Signaling Pathways in PD

5.4.2.1 NF-κB Pathway

The NF-κB pathway plays a significant role in regulating the UPS in PD by influencing both the transcriptional regulation of DUBs and the modulation of inflammatory responses. NF-κB is a transcription factor that, when activated, translocates to the nucleus and influences the expression of various genes, including those involved in inflammation and apoptosis, which are critical in PD pathogenesis [121]. Specifically, NF-κB regulates the transcription of the USP24 gene, a DUB implicated in PD, suggesting that NF-κB may modulate UPS activity by controlling the expression of enzymes that remove ubiquitin from proteins, thereby affecting protein degradation and cellular homeostasis. Additionally, NF-κB activation is linked to the inflammatory processes in PD, where it mediates the expression of pro-inflammatory cytokines, exacerbating neurodegeneration [122]. The ubiquitin system is integral to NF-κB signaling, as ubiquitination and deubiquitination processes regulate the activation and inhibition of NF-κB, respectively [123]. Parkin, an E3 ubiquitin ligase associated with PD, interacts with NF-κB signaling by mediating the ubiquitination of RIPK1, which enhances NF-κB activation, further linking UPS dysfunction to PD pathology [124]. Moreover, NF-κB's role in PD is underscored by its increased nuclear translocation in DA neurons of PD patients, suggesting its involvement in oxidative stress-induced apoptosis [125].

5.4.2.2 PI3K/Akt/mTOR Pathway

The PI3K/Akt/mTOR pathway plays a significant role in regulating the UPS in the context of PD, primarily through its influence on cellular processes such as autophagy and proteostasis. The mTOR signaling pathway, a central component of this axis, is crucial for modulating autophagy, which is a key process in the degradation and recycling of cellular components, including dysfunctional proteins and organelles like mitochondria, which are implicated in PD pathogenesis [126]. The PI3K/Akt/mTOR pathway is known to be dysregulated in PD, affecting cell survival and contributing to neurodegeneration by impairing the balance between protein synthesis and degradation [127]. Specifically, mTORC1, a complex within the mTOR pathway, has dual roles in muscle proteostasis by promoting protein breakdown through the UPS, highlighting its paradoxical role in both muscle atrophy and maintenance of proteome integrity [128]. In PD, the UPS is activated by Parkin, an E3 ubiquitin ligase, which facilitates the degradation of dysfunctional mitochondria, a process critical for maintaining neuronal health. Furthermore, the PI3K/Akt pathway is involved in cell survival and metabolism, and its impairment in PD, often due to genetic factors like PINK1 mutations, leads to reduced Akt activation, contributing to neurodegeneration [129]. The inhibition of the PI3K/Akt/mTOR pathway, as demonstrated by the neuroprotective effects of sinomenine in PD models, enhances autophagy and supports DA neuron survival, suggesting therapeutic potential in modulating this pathway to regulate UPS activity and mitigate PD progression [126].

5.4.2.3 p53 Pathway

The p53 pathway plays a multifaceted role in regulating the UPS in the context of PD, primarily through its interactions with parkin and its influence on cellular stress responses. Parkin, an E3 ubiquitin ligase, is known to repress p53 transcriptionally, creating a feedback loop that maintains cellular homeostasis. This balance is disrupted in PD, where mutations in parkin lead to elevated p53 levels, contributing to increased neuronal cell death [130]. The UPS is crucial for p53-mediated transcriptional regulation, as evidenced by the requirement of proteasomal functions for efficient p53 transcriptional activity at the p21 (waf1) promoter [107]. In PD, proteasomal dysfunction is a hallmark, with decreased 26S proteasomal activity observed in the substantia nigra, which may exacerbate p53-related apoptotic pathways [131]. Additionally, p53 activation is linked to autophagy induction in response to proteasomal inhibition, suggesting a compensatory mechanism to mitigate protein aggregation and neurotoxicity [132]. The interplay between p53 and parkin is further complicated by oxidative stress, where S-nitrosylation of parkin impairs its ability to repress p53, leading to increased p53-mediated neuronal apoptosis [133]. Moreover, p53 activation following oxidative stress, such as exposure to 6-hydroxydopamine, initiates apoptotic pathways in DA neurons, highlighting its role in neurodegeneration in PD [134]. The deletion of p53 specifically in DA neurons has been shown to be neuroprotective, reducing neuronal loss and motor deficits in PD models, further underscoring the detrimental role of p53 activation in PD pathology [135].

5.4.2.4 Wnt/β-Catenin Signaling Pathway

The Wnt/β-catenin signaling pathway plays a crucial role in regulating the UPS in the context of PD, primarily through its influence on DA neuron survival, neuroprotection, and neurorepair. This pathway is essential for the development and maintenance of midbrain DA neurons, which are significantly affected in PD [136]. In PD models, such as the MPTP mouse model, Wnt/β-catenin signaling has been shown to promote the survival and regeneration of DA neurons by engaging in neurorestorative programs [76]. The pathway's activation leads to the accumulation of β-catenin in the cytosol, which then translocates to the nucleus to activate transcription factors like TCF/LEF, promoting the expression of genes crucial for neurogenesis and neuronal survival [137]. Furthermore, Wnt/β-catenin signaling is involved in the dynamic interplay between DA neurons and astrocytes, where astrocyte-derived factors modulate DAergic neurodegeneration and repair [138]. This interaction is vital for maintaining the neurogenic potential of neural stem cells (NSCs) in the subventricular zone and mesencephalic aqueduct-periventricular region, which are critical for DA neuron regeneration [137]. Aging and oxidative stress, which are exacerbated by environmental toxins like paraquat, can downregulate Wnt/β-catenin signaling, leading to increased oxidative damage and DA neuron loss [136].

5.4.2.5 JNK Pathway

The JNK pathway plays a significant role in regulating the UPS in PD through several molecular mechanisms. JNK is activated in response to cellular stress and is involved in the phosphorylation of c-Jun, a component of the transcription factor AP-1, which is crucial for neuronal apoptosis in PD models [139]. The phosphorylation of c-Jun by JNK not only activates transcriptional programs leading to cell death but also influences its ubiquitination. Specifically, JNK can enhance c-Jun ubiquitination, marking it for degradation via the proteasome pathway, although phosphorylation at certain sites can protect c-Jun from ubiquitination, thus prolonging its half-life [140]. This dual role of JNK in modulating c-Jun stability highlights its regulatory capacity over the UPS. Furthermore, JNK activation is implicated in proteasome inhibition-induced DA neuron degeneration, as seen with the proteasome inhibitor lactacystin, which induces JNK phosphorylation and subsequent apoptotic pathways involving caspase activation [141]. The inhibition of JNK has been shown to protect against neuron loss, suggesting that JNK-mediated pathways are critical in UPS-related neuronal degeneration in PD [142]. Additionally, JNK's interaction with mitochondrial proteins like SH3BP5/SAB suggests a role in sustaining JNK activity through mitochondrial signaling, which may further influence UPS function by modulating reactive oxygen species (ROS) production and mitochondrial health [90].

5.4.3 Role of Signaling Pathways in HD

5.4.3.1 NF-κB Pathway

The NF-κB pathway plays a multifaceted role in regulating the UPS in HD, primarily through its involvement in inflammatory responses and protein degradation processes. In HD, the mutant huntingtin (mHTT) protein interacts with the NF-κB pathway, leading to dysregulation that contributes to disease pathogenesis. Specifically, mHTT affects the NF-κB pathway by interacting with The full name is IkappaB kinase gamma (IKKγ), resulting in increased degradation of IκB and subsequent nuclear translocation of RelA, which enhances inflammatory cytokine production in myeloid cells [143]. This dysregulation is further evidenced in astrocytes, where enhanced NF-κB-p65 activation leads to prolonged inflammatory responses, exacerbating neuronal damage and accelerating disease progression [144]. The NF-κB pathway's role extends to the UPS, as the presence of ubiquitin in neuronal inclusions suggests a link between NF-κB activity and UPS impairment, although the exact mechanistic implications remain unresolved [145]. Moreover, the NF-κB pathway's involvement in oxidative stress and cell death is highlighted by the down-regulation of NF-κB-p65 in the presence of mHTT, which is associated with decreased antioxidant levels and increased oxidative stress [146]. Additionally, the NF-κB pathway is implicated in the transport of NF-κB from synapses to the nucleus, a process impaired by the polyglutamine expansion in mHTT, potentially affecting neuronal transcription and contributing to HD pathology [147]. The interplay between NF-κB and UPS is further complicated by the role of ubiquilin-2 (UBQLN2), which enhances NF-κB activation and is involved in protein aggregation, suggesting a convergence of pathways that may exacerbate neurodegeneration [148].

5.4.3.2 PI3K/Akt/mTOR Pathway

The PI3K/Akt/mTOR pathway plays a significant role in regulating the UPS in HD by influencing protein degradation and cellular survival mechanisms. The PI3K/Akt pathway is known for its pro-survival effects, and in the context of HD, it is dysregulated, contributing to the disease's pathophysiology. Specifically, mHTT downregulates PHLPP1, a phosphatase that dephosphorylates Akt, leading to increased Akt activation in the striatum, which may delay cell death and support neuronal viability [149]. The Akt pathway also interacts with the UPS by modulating the activity of PTEN, a negative regulator of Akt. In HD, USP11, a DUB, is upregulated, and its knockdown activates the Akt pathway by promoting PTEN ubiquitination, thereby alleviating mitochondrial dysfunction and neuronal damage. Furthermore, the mTOR component of the pathway is sequestered in polyglutamine aggregates, impairing its kinase activity and inducing autophagy, which is a compensatory mechanism to clear mHTT fragments and reduce toxicity [150]. This sequestration suggests a dual role where mTOR inhibition can be beneficial by promoting autophagy, yet its dysregulation can impair cellular clearance mechanisms. Additionally, the PI3K/Akt/mTOR pathway's regulation of TFEB, a master regulator of lysosomal pathways, further underscores its role in cellular clearance processes. Akt inhibition can activate TFEB independently of mTORC1, enhancing the clearance of proteolipid aggregates, which is crucial in neurodegenerative diseases like HD [151].

5.4.3.3 p53 Pathway

The p53 pathway plays a multifaceted role in regulating the UPS in HD, primarily through its interactions with mHTT and its influence on protein degradation and transcriptional regulation. In HD, the stabilization of p53 is facilitated by the disruption of its interaction with the E3 ligase MDM2, leading to increased transcription of protein kinase CK2 alpha prime and E3 ligase FBXW7, which are responsible for the degradation of heat shock factor 1 [152]. This stabilization of p53 contributes to the degradation of proteins via the UPS, as p53-mediated transcription is positively regulated by the UPS, which is essential for the transcriptional activation of p53 target genes like p21 [107]. Furthermore, p53 is upregulated in HD, enhancing the expression and activation of caspase-6, which is implicated in both neuronal and peripheral tissue degeneration, such as muscle atrophy [153]. The interaction between mHtt and p53 also leads to transcriptional repression of p53-regulated promoters, such as p21, which may contribute to neuronal dysfunction and cell death in HD [154]. The UPS's role in HD is further complicated by its involvement in synaptic function, where mHtt decreases synaptic UPS activity, potentially contributing to selective neuronal dysfunction [155].

5.4.3.4 Wnt/β-Catenin Signaling Pathway

The Wnt/β-catenin signaling pathway plays a significant role in regulating UPS in the context of HD, an NDD characterized by the accumulation of mHTT protein aggregates. The Wnt/β-catenin pathway is crucial for maintaining cellular homeostasis and is implicated in neurogenesis and cell proliferation, processes that are disrupted in HD [156]. In HD, mHTT interferes with the degradation of β-catenin, leading to its toxic stabilization, which exacerbates neurotoxicity. This interference occurs because mHTT associates with the β-catenin destruction complex, altering its normal function and preventing the proper degradation of β-catenin [157]. The UPS, which is responsible for the degradation of misfolded proteins, is also affected in HD. Although the UPS is not entirely inhibited, there are alterations in its activity, such as increased chymotrypsin-like and PGPH-like activities and decreased trypsin-like activity, which may be due to changes in proteasome regulation and maturation [158]. The presence of mHTT aggregates, which are marked for degradation by the UPS, suggests a dysfunction in the recognition or degradation of ubiquitylated substrates [159]. Furthermore, the role of ubiquitin-specific proteases, such as USP34, in regulating components of the Wnt pathway highlights the complex interplay between Wnt signaling and the UPS. USP34, for instance, regulates the stability of axin, a component of the β-catenin destruction complex, thereby influencing β-catenin-mediated transcription [160]. This intricate relationship suggests that targeting the Wnt/β-catenin pathway could offer therapeutic potential in restoring UPS function and mitigating neurodegeneration in HD [161].

5.4.3.5 JNK Pathway

The JNK pathway plays a significant role in regulating the UPS in the context of HD, primarily through its involvement in neuronal cell death and dysfunction. In HD, the pathogenic huntingtin protein (polyQ-Htt) activates JNK3, a neuron-specific isoform of JNK, which in turn phosphorylates kinesin-1, a motor protein essential for fast axonal transport (FAT). This phosphorylation reduces kinesin-1's binding to microtubules, thereby inhibiting FAT, a process critical for neuronal function and survival, and contributing to the selective vulnerability of neurons in HD [162]. The JNK pathway's involvement in UPS regulation is further highlighted by its dual role in ubiquitination processes. JNK can target proteins like c-Jun for ubiquitination, marking them for degradation, while also protecting them from ubiquitination through phosphorylation, thus modulating protein stability and turnover [140]. In HD, the UPS is disrupted, as evidenced by decreased recognition or degradation of ubiquitylated substrates, despite normal ubiquitylation processes. This dysfunction is linked to the presence of mHTT aggregates, which may result from or cause UPS impairment [159]. The sustained activation of JNK, particularly through its interaction with mitochondrial proteins like SH3BP5/SAB, exacerbates oxidative stress and cell death pathways, further impairing neuronal function [90].

5.4.4 Roles of Key Signaling Pathways in ALS

5.4.4.1 NF-κB Pathway

The NF-κB pathway plays a significant role in the pathogenesis of ALS and is intricately linked with the UPS. In ALS, NF-κB is activated in response to various stressors, including oxidative stress and protein aggregation, which are common in neurodegenerative diseases. This activation is observed in different cell types within the central nervous system, such as microglia and astrocytes, contributing to neuroinflammation and potentially exacerbating motor neuron degeneration [163, 164]. The NF-κB pathway is also implicated in the regulation of PQC mechanisms, such as autophagy, which are crucial for clearing protein aggregates that accumulate due to UPS dysfunction [165]. Mutations in genes such as optineurin (OPTN) and UBQLN2, which are associated with ALS, further illustrate the connection between NF-κB and UPS. OPTN mutations lead to a loss of NF-κB suppression, resulting in increased neuronal cell death, while UBQLN2 mutations enhance NF-κB activation and promote TDP-43 aggregation, a hallmark of ALS pathology [166]. Additionally, TDP-43 itself can act as a co-activator of NF-κB, increasing the production of pro-inflammatory cytokines and neurotoxic mediators, thereby contributing to the disease's progression [167].

5.4.4.2 PI3K/Akt/mTOR Pathway

The PI3K/Akt/mTOR pathway plays a significant role in the pathophysiology of ALS and its interaction with the UPS. In ALS, the PI3K/Akt/mTOR pathway is crucial for cell survival and is activated in response to growth factors, yet its dysregulation can contribute to neurodegeneration. Specifically, increased PI3K activity has been observed in the spinal cord of ALS patients, suggesting its involvement in the disease, although the downstream signaling through Akt and mTOR may be impaired, as indicated by unchanged activities of PKB (Akt) and S6K in ALS tissues [168]. The neuroprotective potential of the PI3K/Akt pathway is further supported by studies showing that activation of Akt3 in motor neurons can prevent neuronal loss in ALS models, highlighting its therapeutic potential [169]. Additionally, the UPS, responsible for protein degradation, is impaired in ALS, leading to the accumulation of ubiquitinated proteins and contributing to motor neuron death. This impairment is evident in ALS models where proteasome subunit expression decreases, and immunoproteasome expression increases, correlating with glial activation and inflammation [170]. The p53 pathway plays a significant role in the pathophysiology of both the UPS and ALS, with distinct mechanisms involved in each context. In the UPS, p53 is intricately linked to protein degradation processes. Proteasomal inhibition leads to increased p53 levels and autophagy activity, suggesting a compensatory mechanism where p53-mediated autophagy is induced to manage protein aggregation when the UPS is compromised [132]. In ALS, p53 is implicated in neurodegeneration through several pathways. The C9orf72 poly-PR dipeptide repeats, associated with ALS, stabilize p53 by impairing its degradation via the UPS, leading to its accumulation and activation of downstream apoptotic pathways [171]. The p53 pathway is also activated in response to inflammatory stress, such as prolonged IFNγ exposure, which leads to TDP-43 dysfunction and subsequent motor neuron degeneration in ALS [172]. Moreover, the long non-coding RNA NR3C2-8:1 promotes p53-mediated apoptosis in ALS by modulating the miR-129-5p/USP10 axis, highlighting the role of non-coding RNAs in p53 activation and neuronal cell death [173].

5.4.4.3 Wnt/β-Catenin Signaling Pathway

The Wnt/β-catenin signaling pathway plays a significant role in the pathogenesis of ALS and is considered a potential therapeutic target for this neurodegenerative disease. In ALS, the canonical Wnt/β-catenin pathway is involved in various cellular processes, including neurogenesis, cell proliferation, and differentiation, which are crucial for maintaining neuronal health and function [161]. Abnormal activation or dysregulation of this pathway has been linked to neuronal degeneration and glial cell proliferation, contributing to the progression of ALS [174]. In particular, the downregulation of miR-139 in ALS leads to the activation of Wnt/β-catenin signaling, which is a major driver of motor neuron degeneration [175]. Additionally, alterations in Wnt signaling components, such as increased expression of Wnt3, Wnt4, and Fz2, have been observed in the spinal cords of ALS patients, suggesting their involvement in the disease pathology [176]. The pathway's role in energy metabolism is also noteworthy, as its activation can induce aerobic glycolysis, potentially affecting neuronal viability and contributing to ALS progression [177]. Furthermore, in ALS models, the Wnt/β-catenin pathway's transcriptional activity is decreased, possibly due to cytosolic aggregation of β-catenin, which impairs motor neuron differentiation and survival [178]. The pathway's involvement in maintaining the blood–brain barrier (BBB) integrity is also critical, as its downregulation in ALS patient-derived brain microvascular endothelial cells leads to increased permeability and immune cell adhesion, highlighting its role in BBB dysfunction in ALS [179].

5.4.4.4 JNK Pathway

The JNK pathway plays a multifaceted role in the pathogenesis of ALS and related neurodegenerative conditions, such as FTD. In ALS, particularly in cases associated with C9orf72 mutations, JNK is activated by toxic dipeptide repeats via the ER-stress response protein IRE1, leading to stress granule assembly and neurodegeneration. This process involves JNK-induced histone modifications that increase the expression of stress granule proteins like G3BP1, thereby exacerbating neurotoxicity [180]. JNK's involvement in ALS is further highlighted by its role in oxidative stress responses, where it is overexpressed in astrocytes but not in motor neurons, suggesting a protective mechanism in glial cells that is deficient in neurons [164]. Additionally, JNK signaling is implicated in synaptic dysfunction and neuronal apoptosis, contributing to the neurodegenerative processes observed in ALS [181]. The pathway's activation is linked to the phosphorylation of proteins such as tau, which, when combined with other kinases like GSK-3β, leads to neurofibrillary tangle formation, a hallmark of several neurodegenerative diseases [118]. Moreover, JNK's presynaptic localization and its interaction with SNARE proteins suggest a role in synaptic plasticity and vesicle release, which may be disrupted in ALS [182]. The dual role of JNK in both promoting cell death and potentially offering neuroprotection through its interaction with mitochondrial proteins like SAB further complicates its involvement in ALS pathogenesis [90].

5.4.5 The Role of Signaling Pathway in MS

5.4.5.1 NF-κB Pathway

The NF-κB pathway plays a crucial role in the pathogenesis of MS and is intricately linked with the UPS. In MS, NF-κB signaling is dysregulated, leading to increased activation and translocation of NF-κB to the nucleus, which is partly due to reduced levels of the inhibitor IκB-α in peripheral blood mononuclear cells of patients [183]. This dysregulation is associated with chronic inflammation, a hallmark of MS, and contributes to the disease's progression by affecting immune responses and promoting neurodegeneration [184]. The NF-κB pathway consists of canonical and non-canonical pathways, both of which are tightly regulated by ubiquitination processes. The canonical pathway is rapidly activated and involves the degradation of IκB proteins via the UPS, which releases NF-κB dimers to translocate to the nucleus and initiate transcription of pro-inflammatory genes [123]. Ubiquitination, particularly through K63- and K48-linked polyubiquitin chains, plays a positive role in NF-κB activation, while deubiquitination acts as a negative feedback mechanism to balance the pathway's activity [185]. In MS, aberrant NF-κB activation is linked to the immune system's dysregulation, as evidenced by the identification of NF-κB p65 target genes involved in immune functions and MS pathogenesis [186]. Furthermore, therapeutic strategies targeting NF-κB pathways, such as fingolimod and dimethyl fumarate, have shown promise in modulating immune responses in MS patients by interfering with NF-κB signaling [187].

5.4.5.2 PI3K/Akt/mTOR Pathway

The PI3K/Akt/mTOR pathway is a critical signaling cascade involved in various cellular processes, including cell growth, proliferation, metabolism, and survival, and plays a significant role in both MS. In the context of MS, the pathway is implicated in the regulation of immune responses and neuroinflammation, which are central to the disease's pathogenesis. The mTOR component of the pathway is particularly important in modulating T-cell activation and glial responses, which are crucial for the progression of MS. Inhibitors of mTOR, such as rapamycin, have shown promise in experimental models of MS by reducing T-cell activation and enhancing T-regulatory cell function, although their clinical application is complicated by side effects and the need for cell-selective targeting to improve efficacy [188]. Additionally, the PI3K/Akt/mTOR pathway is involved in the regulation of oligodendrocyte development and myelination, suggesting potential therapeutic avenues for promoting remyelination in MS [189]. Dysregulation of the PI3K/Akt/mTOR pathway is linked to various disorders, including neurodegenerative diseases, where it plays a role in apoptosis, inflammation, and oxidative stress, highlighting its potential as a therapeutic target [190].

5.4.5.3 p53 Pathway

The p53 pathway is intricately involved in both the UPS and the unfolded protein response (UPR), playing crucial roles in cellular stress responses and tumor suppression. The UPS regulates p53 stability and activity through ubiquitination, primarily mediated by the E3 ligase Mdm2, which tags p53 for proteasomal degradation. This process is dynamic, with Mdm2 capable of both mono- and polyubiquitination, and is counteracted by deubiquitinases like HAUSP, which stabilize p53 by removing ubiquitin tags, thus enhancing its tumor suppressor functions [191]. The UPS also influences p53-mediated transcriptional activation, as evidenced by the recruitment of proteasomal components like Sug1 to the p21(waf1) promoter, which is crucial for p53's role in cell cycle regulation [107]. In the context of the UPR, p53 is activated in response to nuclear protein misfolding, engaging a transcriptional response that is distinct from the cytosolic UPR, highlighting the unique PQC mechanisms in the nucleus [192]. This nuclear UPR involves p53-driven transcriptional responses and transient cell cycle delays, underscoring p53's role in maintaining genomic stability under stress conditions [193]. Additionally, p53's activity is modulated by post-translational modifications such as S-glutathionylation, which can inhibit its DNA-binding ability during oxidative stress, further illustrating the complexity of its regulation under different cellular conditions [194].

5.4.5.4 Wnt/β-Catenin Signaling Pathway

The Wnt/β-catenin signaling pathway plays a crucial role in both the UPS and MS, influencing various cellular processes and disease mechanisms. In the context of MS, Wnt/β-catenin signaling is implicated in neuroinflammation and demyelination, with its upregulation associated with impaired remyelination and oligodendrocyte differentiation failure. However, downstream targets such as PI3K/Akt and TCF7L2 can promote remyelination, suggesting a complex role in MS pathology [195]. The pathway's involvement in MS is further highlighted by its contribution to chronic pain, as seen in experimental autoimmune encephalomyelitis models, where upregulation of Wnt signaling proteins in the spinal cord dorsal horn correlates with pain development [196]. In terms of cellular aging, Wnt/β-catenin signaling induces mesenchymal stem cell aging through ROS production and the DNA damage response, involving the p53/p21 pathway [197]. This signaling pathway is also crucial for maintaining NSC homeostasis, influencing proliferation and differentiation, which are vital for brain development and function [198]. In the UPS, Wnt/β-catenin signaling regulates β-catenin stability through a destruction complex involving proteins such as GSK3β and CK1, which phosphorylate β-catenin for degradation in the absence of Wnt signals [199]. The pathway's regulation of cellular processes such as cell growth, differentiation, and migration underscores its importance in both normal physiology and disease states, including neurodegenerative diseases and MS [161].

5.4.5.5 JNK Pathway

The JNK pathway plays a multifaceted role in both the UPS and MS, primarily through its involvement in stress signaling, synaptic plasticity, and neurodegeneration. JNK, a member of the MAPK family, is activated by various stress stimuli, including inflammatory cytokines and environmental stressors, and is crucial in regulating cellular processes such as proliferation, differentiation, and apoptosis [200]. In the context of the UPS, JNK is implicated in the regulation of protein degradation, as it interacts with non-canonical Wnt components and ubiquitin ligases like Fbw7, which are essential for maintaining cellular homeostasis and preventing NDDs [201]. JNK's role in synaptic plasticity is highlighted by its presynaptic localization, where it modulates vesicle release through phosphorylation of T-SNARE proteins, thereby influencing synaptic transmission and plasticity [182]. In MS, JNK contributes to neuroinflammation and neuronal death, processes that are central to the disease's pathogenesis. The pathway's involvement in inflammatory responses and its regulation of pro-inflammatory cytokines make it a potential therapeutic target for MS and other autoimmune diseases [202].

5.5 Conclusion

This chapter emphasizes the critical role of the UPS in maintaining cellular homeostasis and its implications in various diseases, particularly neurodegenerative disease. The UPS is essential for degrading misfolded or damaged proteins, thereby regulating numerous cellular processes such as cell cycle progression, signal transduction, and immune responses. Dysregulation of the UPS is linked to several neurodegenerative diseases, including AD, PD, HD, ALS, and MS, due to the accumulation of neurotoxic proteins. The chapter also highlights the complexity of the ubiquitination process involves a cascade of enzymatic reactions facilitated by E1, E2, and E3 enzymes. E3 ligases in particular are crucial for substrate specificity, determining which proteins are tagged for degradation. DUBs add another layer of regulation by reversing ubiquitination, playing a significant role in protein degradation and stability, especially in neurodegenerative diseases. The chapter also advocates the interplay between UPS and other cellular pathways, such as autophagy, which together ensure efficient protein and organelle turnover. Dysregulation of signaling factors associated with UPS, such as NF-kB and PI3K/Akt/mTOR, can lead to illness, highlighting the system's complexity and its role in post-translational regulation. The chapter underscores the importance of understanding the triggers and dysregulation of the UPS to develop novel therapeutic approaches for treating neurodegenerative diseases. Addressing UPS-related issues could lead to better management of these conditions, potentially improving outcomes for affected individuals.

References

1. Müller L, Hoppe T. UPS-dependent strategies of protein quality control degradation. Trends Biochem Sci [Internet]. 2024;49:859–74. Available from: https://linkinghub.elsevier.com/retrieve/pii/S096800042400149X
2. Pras A, Nollen EAA. Regulation of age-related protein toxicity. Front Cell Dev Biol [Internet]. 2021:9. Available from: https://www.frontiersin.org/articles/10.3389/fcell.2021.637084/full
3. Ciechanover A, Kwon YT. Protein quality control by molecular chaperones in neurodegeneration. Front Neurosci [Internet]. 2017:11. Available from: http://journal.frontiersin.org/article/10.3389/fnins.2017.00185/full

4. Kim J, Mook-Jung I. Special issue on neurodegenerative diseases and their therapeutic approaches. Exp Mol Med [Internet]. 2015;47:e146. Available from: https://www.nature.com/articles/emm201513
5. Khanam H, Ali A, Asif M, Shamsuzzaman. Neurodegenerative diseases linked to misfolded proteins and their therapeutic approaches: a review. Eur J Med Chem [Internet]. 2016;124:1121–41. Available from: https://linkinghub.elsevier.com/retrieve/pii/S0223523416306493
6. Chen Y, Xue H, Jin J. Applications of protein ubiquitylation and deubiquitylation in drug discovery. J Biol Chem [Internet]. 2024;300:107264. Available from: https://linkinghub.elsevier.com/retrieve/pii/S0021925824017654
7. Roy PK, Biswas A, Deepak K, Mandal M. An insight into the ubiquitin-proteasomal axis and related therapeutic approaches towards central nervous system malignancies. Biochim Biophys Acta – Rev Cancer [Internet]. 2022;1877:188734. Available from: https://linkinghub.elsevier.com/retrieve/pii/S0304419X22000592
8. Yu P, Hua Z. To kill or to be killed: how does the Battle between the UPS and autophagy maintain the intracellular homeostasis in eukaryotes? Int J Mol Sci [Internet]. 2023;24:2221. Available from: https://www.mdpi.com/1422-0067/24/3/2221
9. Collins MA, Mekonnen G, Albert FW. Variation in ubiquitin system genes creates substrate-specific effects on proteasomal protein degradation. Elife [Internet]. 2022:11. Available from: https://elifesciences.org/articles/79570
10. Zhang X, Linder S, Bazzaro M. Drug development targeting the Ubiquitin–Proteasome System (UPS) for the treatment of human cancers. Cancers (Basel) [Internet]. 2020;12:902. Available from: https://www.mdpi.com/2072-6694/12/4/902
11. Braganca CE, Kraut DA. Mode of protein targeting to the proteasome determines substrate fate. FASEB J [Internet]. 2020;34:1–1. Available from: https://faseb.onlinelibrary.wiley.com/doi/10.1096/fasebj.2020.34.s1.04547
12. Huang L. Structural analysis of the 26S proteasome complex to understand its function and regulation. FASEB J [Internet]. 2017:31. Available from: https://faseb.onlinelibrary.wiley.com/doi/10.1096/fasebj.31.1_supplement.124.1
13. Kudriaeva AA, Belogurov AA. Proteasome: a nanomachinery of creative destruction. Biochem Int. 2019;84:159–92. Available from: http://link.springer.com/10.1134/S0006297919140104
14. Meyer-Schwesinger C. The ubiquitin–proteasome system in kidney physiology and disease. Nat Rev Nephrol [Internet]. 2019;15:393–411. Available from: https://www.nature.com/articles/s41581-019-0148-1
15. Johnson J-AK, Sumner I. On the possibility that bond strain is the mechanism of RING E3 activation in the E2-catalyzed ubiquitination reaction. J Chem Inf Model [Internet]. 2022;62:6475–81. Available from: https://pubs.acs.org/doi/10.1021/acs.jcim.2c00423
16. dos Passos PMS, Spagnol V, de Correia CRST, Teixeira FR. Evaluation of substrate Ubiquitylation by E3 ubiquitin-ligase in mammalian cell lysates. J Vis Exp [Internet]. 2022; Available from: https://app.jove.com/v/63561
17. Middleton AJ, Day CL. From seeds to trees: how E2 enzymes grow ubiquitin chains. Biochem Soc Trans [Internet]. 2023;51:353–62. Available from: https://portlandpress.com/biochemsoctrans/article/51/1/353/232445/From-seeds-to-trees-how-E2-enzymes-grow-ubiquitin
18. Mukhopadyay U, Levantovsky S, Gharbi S, Stein F, Behrends C, Bhogaraju S. A ubiquitin-specific, proximity-based labeling approach for the identification of ubiquitin ligase substrates [Internet]. 2023. Available from: http://biorxiv.org/lookup/doi/10.1101/2023.09.04.556194
19. Liu CH, Rao Y, Gao D, Wan L, Zhang L. Editorial: ubiquitin code: from cell biology to translational medicine. Front Cell Dev Biol [Internet]. 2021:9. Available from: https://www.frontiersin.org/articles/10.3389/fcell.2021.791967/full
20. Ren J, Yu P, Liu S, Li R, Niu X, Chen Y, et al. Deubiquitylating enzymes in cancer and immunity. Adv Sci [Internet]. 2023;10(36):e2303807. Available from: https://onlinelibrary.wiley.com/doi/10.1002/advs.202303807

21. Bello AI, Goswami R, Brown SL, Costanzo K, Shores T, Allan S, et al. Deubiquitinases in neurodegeneration. Cells [Internet]. 2022;11:556. Available from: https://www.mdpi.com/2073-4409/11/3/556
22. Liu B, Ruan J, Chen M, Li Z, Manjengwa G, Schlüter D, et al. Deubiquitinating enzymes (DUBs): decipher underlying basis of neurodegenerative diseases. Mol Psychiatry [Internet]. 2022;27:259–68. Available from: https://www.nature.com/articles/s41380-021-01233-8
23. Park S-S, Do H-A, Park H-B, Choi H-S, Baek K-H. Deubiquitinating enzyme YOD1 deubiquitinates and destabilizes α-synuclein. Biochem Biophys Res Commun [Internet]. 2023;645:124–31. Available from: https://linkinghub.elsevier.com/retrieve/pii/S0006291X23000566
24. Estavoyer B, Messmer C, Echbicheb M, Rudd CE, Milot E, Affar EB. Mechanisms orchestrating the enzymatic activity and cellular functions of deubiquitinases. J Biol Chem [Internet]. 2022;298:102198. Available from: https://linkinghub.elsevier.com/retrieve/pii/S0021925822006408
25. Snyder NA, Silva GM. Deubiquitinating enzymes (DUBs): regulation, homeostasis, and oxidative stress response. J Biol Chem [Internet]. 2021;297:101077. Available from: https://linkinghub.elsevier.com/retrieve/pii/S0021925821008802
26. Faccenda E, Layfield R. E3 ubiquitin ligase components in GtoPdb v.2023.2. IUPHAR/BPS Guid to Pharmacol CITE [Internet]. 2023;2023. Available from: http://journals.ed.ac.uk/gtopdb-cite/article/view/9042
27. Hershko A, Ciechanover A. The ubiquitin system. Annu Rev Biochem [Internet]. 1998;67:425–79. Available from: https://www.annualreviews.org/doi/10.1146/annurev.biochem.67.1.425
28. Ciechanover A. The ubiquitin-proteasome pathway: on protein death and cell life. EMBO J [Internet]. 1998;17:7151–60. Available from: http://emboj.embopress.org/cgi/doi/10.1093/emboj/17.24.7151
29. Winston JT, Chu C, Harper JW. Culprits in the degradation of cyclin E apprehended. Genes Dev [Internet]. 1999;13:2751–7. Available from: http://www.genesdev.org/cgi/doi/10.1101/gad.13.21.2751
30. Li W, Tu D, Brunger AT, Ye Y. A ubiquitin ligase transfers preformed polyubiquitin chains from a conjugating enzyme to a substrate. Nature [Internet]. 2007;446:333–7. Available from: https://www.nature.com/articles/nature05542
31. Tyers M, Willems AR. One ring to rule a superfamily of E3 ubiquitin ligases. Science Internet. 1999;284:601–4. Available from: https://www.science.org/doi/10.1126/science.284.5414.601
32. Cockram PE, Kist M, Prakash S, Chen S-H, Wertz IE, Vucic D. Ubiquitination in the regulation of inflammatory cell death and cancer. Cell Death Differ [Internet]. 2021;28:591–605. Available from: https://www.nature.com/articles/s41418-020-00708-5
33. Love KR, Pandya RK, Spooner E, Ploegh HL. Ubiquitin C-terminal electrophiles are activity-based probes for identification and mechanistic study of ubiquitin conjugating machinery. ACS Chem Biol [Internet]. 2009;4:275–87. Available from: https://pubs.acs.org/doi/10.1021/cb9000348
34. Rolfe M, Chiu MI, Pagano M. The ubiquitin-mediated proteolytic pathway as a therapeutic area. J Mol Med [Internet]. 1997;75:5–17. Available from: http://link.springer.com/10.1007/s001090050081
35. Sadowski M, Suryadinata R, Tan AR, Roesley SNA, Sarcevic B. Protein monoubiquitination and polyubiquitination generate structural diversity to control distinct biological processes. IUBMB Life [Internet]. 2012;64:136–42. Available from: https://iubmb.onlinelibrary.wiley.com/doi/10.1002/iub.589
36. Robinson PA, Ardley HC. Ubiquitin-protein ligases. J Cell Sci [Internet]. 2004;117:5191–4. Available from: https://journals.biologists.com/jcs/article/117/22/5191/27982/Ubiquitin-protein-ligases

37. Joazeiro CAP, Hunter T. Ubiquitination–more than two to Tango. Science Internet. 2000;289:2061–2. Available from: https://www.science.org/doi/10.1126/science.289.5487.2061
38. Tracz M, Bialek W. Beyond K48 and K63: non-canonical protein ubiquitination. Cell Mol Biol Lett [Internet]. 2021;26:1. Available from: https://cmbl.biomedcentral.com/articles/10.1186/s11658-020-00245-6
39. Wang H, Matsuzawa A, Brown SA, Zhou J, Guy CS, Tseng P-H, et al. Analysis of non-degradative protein ubiquitylation with a monoclonal antibody specific for lysine-63-linked polyubiquitin. Proc Natl Acad Sci [Internet]. 2008;105:20197–202. Available from: https://pnas.org/doi/full/10.1073/pnas.0810461105
40. Seo J, Kim MW, Bae K-H, Lee SC, Song J, Lee E-W. The roles of ubiquitination in extrinsic cell death pathways and its implications for therapeutics. Biochem Pharmacol [Internet]. 2019;162:21–40. Available from: https://linkinghub.elsevier.com/retrieve/pii/S0006295218304817
41. Marfany G, Denuc A. To ubiquitinate or to deubiquitinate: it all depends on the partners. Biochem Soc Trans [Internet]. 2008;36:833–8. Available from: https://portlandpress.com/biochemsoctrans/article/36/5/833/65682/To-ubiquitinate-or-to-deubiquitinate-it-all
42. Jonsson E, Htet ZM, Bard JAM, Dong KC, Martin A. Ubiquitin modulates 26S proteasome conformational dynamics and promotes substrate degradation [Internet]. 2021. Available from: http://biorxiv.org/lookup/doi/10.1101/2021.08.18.456915
43. Sahu I, Glickman MH. Proteasome in action: substrate degradation by the 26S proteasome. Biochem Soc Trans [Internet]. 2021;49:629–44. Available from: https://portlandpress.com/biochemsoctrans/article/49/2/629/228091/Proteasome-in-action-substrate-degradation-by-the
44. Saeki Y. Ubiquitin recognition by the proteasome. J Biochem [Internet]. 2017:mvw091. Available from: https://academic.oup.com/jb/article-lookup/doi/10.1093/jb/mvw091
45. Majumder P, Baumeister W. Proteasomes: unfoldase-assisted protein degradation machines. Biol Chem [Internet]. 2019;401:183–99. Available from: https://www.degruyter.com/document/doi/10.1515/hsz-2019-0344/html
46. Davidson K, Pickering AM. The proteasome: a key modulator of nervous system function, brain aging, and neurodegenerative disease. Front Cell Dev Biol [Internet]. 2023:11. Available from: https://www.frontiersin.org/articles/10.3389/fcell.2023.1124907/full
47. Rousseau A, Bertolotti A. Regulation of proteasome assembly and activity in health and disease. Nat Rev Mol Cell Biol [Internet]. 2018;19:697–712. Available from: https://www.nature.com/articles/s41580-018-0040-z
48. Chung CH, Baek SH. Deubiquitinating enzymes: their diversity and emerging roles. Biochem Biophys Res Commun [Internet]. 1999;266:633–40. Available from: https://linkinghub.elsevier.com/retrieve/pii/S0006291X99918806
49. Lange SM, Armstrong LA, Kulathu Y. Deubiquitinases: from mechanisms to their inhibition by small molecules. Mol Cell [Internet]. 2022;82:15–29. Available from: https://linkinghub.elsevier.com/retrieve/pii/S1097276521009473
50. Millard SM, Wood SA. Riding the DUBway: regulation of protein trafficking by deubiquitylating enzymes. J Cell Biol [Internet]. 2006;173:463–8. Available from: https://rupress.org/jcb/article/173/4/463/53652/Riding-the-DUBway-regulation-of-protein
51. Zhu Y, Carroll M, Papa FR, Hochstrasser M, D'Andrea AD. DUB-1, a deubiquitinating enzyme with growth-suppressing activity. Proc Natl Acad Sci [Internet]. 1996;93:3275–9. Available from: https://pnas.org/doi/full/10.1073/pnas.93.8.3275
52. Ramakrishna S, Suresh B, Baek K-H. The role of deubiquitinating enzymes in apoptosis. Cell Mol Life Sci [Internet]. 2011;68:15–26. Available from: http://link.springer.com/10.1007/s00018-010-0504-6
53. Fang Y-Z, Jiang L, He Q, Cao J, Yang B. Deubiquitination complex platform: a plausible mechanism for regulating the substrate specificity of deubiquitinating enzymes. Acta Pharm Sin B [Internet]. 2023;13:2955–62. Available from: https://linkinghub.elsevier.com/retrieve/pii/S2211383523000667

54. Clague MJ, Urbé S, Komander D. Breaking the chains: deubiquitylating enzyme specificity begets function. Nat Rev Mol Cell Biol [Internet]. 2019;20:338–52. Available from: https://www.nature.com/articles/s41580-019-0099-1
55. Suresh HG, Pascoe N, Andrews B. The structure and function of deubiquitinases: lessons from budding yeast. Open Biol [Internet]. 2020;10. Available from: https://royalsocietypublishing.org/doi/10.1098/rsob.200279
56. Tomomatsu S, Matsuo Y, Ohtake F, Saeki Y, Inada T. Proofreading of the unbranched mixed-linkage ubiquitin chain on the collided ribosome promotes RQC [Internet]. 2023. Available from: https://www.ssrn.com/abstract=4530811
57. Li Z, Lan Q, Li Y, Xu C, Chang L, Xu P, et al. Quantitative proteomics reveal the potential biological functions of the deubiquitinating enzyme Ubp14 in Saccharomyces cerevisiae. Sheng Wu Gong Cheng Xue Bao [Internet]. 2022;38:3901–13. Available from: http://www.ncbi.nlm.nih.gov/pubmed/36305417
58. Zenge C, Ordureau A. Ubiquitin system mutations in neurological diseases. Trends Biochem Sci [Internet]. 2024;49:875–87. Available from: https://linkinghub.elsevier.com/retrieve/pii/S0968000424001543
59. Atkin G, Paulson H. Ubiquitin pathways in neurodegenerative disease. Front Mol Neurosci [Internet]. 2014:7. Available from: http://journal.frontiersin.org/article/10.3389/fnmol.2014.00063/abstract
60. Nielsen PYØ, Okarmus J, Meyer M. Role of Deubiquitinases in Parkinson's disease—therapeutic perspectives. Cells [Internet]. 2023;12:651. Available from: https://www.mdpi.com/2073-4409/12/4/651
61. Tran N-N, Lee B-H. Functional implication of ubiquitinating and deubiquitinating mechanisms in TDP-43 proteinopathies. Front Cell Dev Biol [Internet]. 2022;10:931968. Available from: https://www.frontiersin.org/articles/10.3389/fcell.2022.931968/full
62. Schmidt MF, Gan ZY, Komander D, Dewson G. Ubiquitin signalling in neurodegeneration: mechanisms and therapeutic opportunities. Cell Death Differ [Internet]. 2021;28:570–90. Available from: https://www.nature.com/articles/s41418-020-00706-7
63. Mitchell S, Vargas J, Hoffmann A. Signaling via the NFκB system. WIREs Syst Biol Med [Internet]. 2016;8:227–41. Available from: https://wires.onlinelibrary.wiley.com/doi/10.1002/wsbm.1331
64. Tilstra JS, Clauson CL, Niedernhofer LJ, Robbins PD. NF-κB in aging and disease. Aging Dis [Internet]. 2011;2:449–65. Available from: http://www.ncbi.nlm.nih.gov/pubmed/22396894
65. Barnabei L, Laplantine E, Mbongo W, Rieux-Laucat F, Weil R. NF-κB: at the borders of autoimmunity and inflammation. Front Immunol [Internet]. 2021:12. Available from: https://www.frontiersin.org/articles/10.3389/fimmu.2021.716469/full
66. Koo JW, Russo SJ, Ferguson D, Nestler EJ, Duman RS. Nuclear factor-κB is a critical mediator of stress-impaired neurogenesis and depressive behavior. Proc Natl Acad Sci [Internet]. 2010;107:2669–74. Available from: https://pnas.org/doi/full/10.1073/pnas.0910658107
67. Lukiw WJ. NF-κB-regulated, proinflammatory miRNAs in Alzheimer's disease. Alzheimers Res Ther [Internet]. 2012;4:47. Available from: http://alzres.biomedcentral.com/articles/10.1186/alzrt150
68. Eves EM, Xiong W, Bellacosa A, Kennedy SG, Tsichlis PN, Rosner MR, et al. Akt, a target of phosphatidylinositol 3-kinase, inhibits apoptosis in a differentiating neuronal cell line. Mol Cell Biol [Internet]. 1998;18:2143–52. Available from: https://www.tandfonline.com/doi/full/10.1128/MCB.18.4.2143
69. Long H-Z, Cheng Y, Zhou Z-W, Luo H-Y, Wen D-D, Gao L-C. PI3K/AKT signal pathway: a target of natural products in the prevention and treatment of Alzheimer's disease and Parkinson's disease. Front Pharmacol [Internet]. 2021:12. Available from: https://www.frontiersin.org/articles/10.3389/fphar.2021.648636/full
70. Popova NV, Jücker M. The role of mTOR signaling as a therapeutic target in cancer. Int J Mol Sci [Internet]. 2021;22:1743. Available from: https://www.mdpi.com/1422-0067/22/4/1743

71. Liu Y. p53 protein at the hub of cellular DNA damage response pathways through sequence-specific and non-sequence-specific DNA binding. Carcinogenesis [Internet]. 2001;22:851–60. Available from: https://academic.oup.com/carcin/article-lookup/doi/10.1093/carcin/22.6.851
72. Rozan LM, El-Deiry WS. p53 downstream target genes and tumor suppression: a classical view in evolution. Cell Death Differ [Internet]. 2007;14:3–9. Available from: https://www.nature.com/articles/4402058
73. Joseph TW, Zaika A, Moll UM. Nuclear and cytoplasmic degradation of endogenous p53 and HDM2 occurs during down-regulation of the p53 response after multiple types of DNA damage. FASEB J [Internet]. 2003;17:1622–30. Available from: https://onlinelibrary.wiley.com/doi/10.1096/fj.02-0931com
74. Zavileyskiy L, Bunik V. Regulation of p53 function by formation of non-nuclear heterologous protein complexes. Biomolecules [Internet]. 2022;12:327. Available from: https://www.mdpi.com/2218-273X/12/2/327
75. Chatoo W, Abdouh M, Bernier G. p53 pro-oxidant activity in the central nervous system: implication in aging and neurodegenerative diseases. Antioxid Redox Signal [Internet]. 2011;15:1729–37. Available from: http://www.liebertpub.com/doi/10.1089/ars.2010.3610
76. L'Episcopo F, Tirolo C, Testa N, Caniglia S, Morale MC, Serapide MF, et al. Wnt/β-catenin signaling is required to rescue midbrain dopaminergic progenitors and promote Neurorepair in ageing mouse model of Parkinson's disease. Stem Cells [Internet]. 2014;32:2147–63. Available from: https://academic.oup.com/stmcls/article/32/8/2147-2163/6444718
77. Arenas E. Wnt signaling in midbrain dopaminergic neuron development and regenerative medicine for Parkinson's disease. J Mol Cell Biol [Internet]. 2014;6:42–53. Available from: https://academic.oup.com/jmcb/article-lookup/doi/10.1093/jmcb/mju001
78. Tang M, Villaescusa JC, Luo SX, Guitarte C, Lei S, Miyamoto Y, et al. Interactions of Wnt/-catenin signaling and sonic hedgehog regulate the neurogenesis of ventral midbrain dopamine neurons. J Neurosci [Internet]. 2010;30:9280–91. Available from: https://www.jneurosci.org/lookup/doi/10.1523/JNEUROSCI.0860-10.2010
79. Galli S, Lopes DM, Ammari R, Kopra J, Millar SE, Gibb A, et al. Deficient Wnt signalling triggers striatal synaptic degeneration and impaired motor behaviour in adult mice. Nat Commun [Internet]. 2014;5:4992. Available from: https://www.nature.com/articles/ncomms5992
80. Inestrosa NC, Varela-Nallar L, Grabowski CP, Colombres M. Synaptotoxicity in Alzheimer's disease: the Wnt signaling pathway as a molecular target. IUBMB Life [Internet]. 2007;59:316–21. Available from: https://iubmb.onlinelibrary.wiley.com/doi/10.1080/15216540701242490
81. Fhu CW, Ali A. Dysregulation of the ubiquitin proteasome system in human malignancies: a window for therapeutic intervention. Cancers (Basel) [Internet]. 2021;13:1513. Available from: https://www.mdpi.com/2072-6694/13/7/1513
82. Park H-B, Kim J-W, Baek K-H. Regulation of Wnt signaling through ubiquitination and Deubiquitination in cancers. Int J Mol Sci [Internet]. 2020;21:3904. Available from: https://www.mdpi.com/1422-0067/21/11/3904
83. Tauriello DVF, Maurice MM. The various roles of ubiquitin in Wnt pathway regulation. Cell Cycle [Internet]. 2010;9:3724–33. Available from: http://www.tandfonline.com/doi/abs/10.4161/cc.9.18.13204
84. Yan H, He L, Lv D, Yang J, Yuan Z. The role of the dysregulated JNK signaling pathway in the pathogenesis of human diseases and its potential therapeutic strategies: a comprehensive review. Biomolecules [Internet]. 2024;14:243. Available from: https://www.mdpi.com/2218-273X/14/2/243
85. de los Reyes Corrales T, Losada-Pérez M, Casas-Tintó S. JNK pathway in CNS pathologies. Int J Mol Sci [Internet]. 2021;22:3883. Available from: https://www.mdpi.com/1422-0067/22/8/3883
86. Jedariforoughi A. Complex interplay of MAPK signaling in cellular stress responses and apoptosis regulation. Int J Sci Res Arch [Internet]. 2024;11:1673–5. Available from: https://ijsra.net/content/complex-interplay-mapk-signaling-cellular-stress-responses-and-apoptosis-regulation

87. Zeke A, Misheva M, Reményi A, Bogoyevitch MA. JNK signaling: regulation and functions based on complex protein-protein partnerships. Microbiol Mol Biol Rev [Internet]. 2016;80:793–835. Available from: https://journals.asm.org/doi/10.1128/MMBR.00043-14
88. Dhanasekaran DN, Reddy EP. JNK-signaling: a multiplexing hub in programmed cell death. Genes Cancer [Internet]. 2017;8:682–94. Available from: https://www.genesandcancer.com/lookup/doi/10.18632/genesandcancer.155
89. Yabu T, Shiba H, Shibasaki Y, Nakanishi T, Imamura S, Touhata K, et al. Stress-induced ceramide generation and apoptosis via the phosphorylation and activation of nSMase1 by JNK signaling. Cell Death Differ [Internet]. 2015;22:258–73. Available from: https://www.nature.com/articles/cdd2014128
90. Win S, Than TA, Kaplowitz N. The regulation of JNK signaling pathways in cell death through the interplay with mitochondrial SAB and upstream post-translational effects. Int J Mol Sci [Internet]. 2018;19:3657. Available from: https://www.mdpi.com/1422-0067/19/11/3657
91. Zhao Y, Kuca K, Wu W, Wang X, Nepovimova E, Musilek K, et al. Hypothesis: JNK signaling is a therapeutic target of neurodegenerative diseases. Alzheimer's Dement [Internet]. 2022;18:152–8. Available from: https://alz-journals.onlinelibrary.wiley.com/doi/10.1002/alz.12370
92. Ardanaz CG, Bejarano A, Echarte B, Smerdou C, Martisova E, Martínez-Valbuena I, et al. JNK3 overexpression in the entorhinal cortex disseminates to the hippocampus and induces cognitive deficiencies and tau misfolding [Internet]. 2022. Available from: https://www.researchsquare.com/article/rs-1381483/v2
93. Sivamaruthi BS, Raghani N, Chorawala M, Bhattacharya S, Prajapati BG, Elossaily GM, et al. NF-κB pathway and its inhibitors: a promising frontier in the management of Alzheimer's disease. Biomedicines [Internet]. 2023;11:2587. Available from: https://www.mdpi.com/2227-9059/11/9/2587
94. Sulistio YA, Heese K. The ubiquitin-proteasome system and molecular chaperone deregulation in Alzheimer's disease. Mol Neurobiol [Internet]. 2016;53:905–31. Available from: http://link.springer.com/10.1007/s12035-014-9063-4
95. Xu H, You M, Shi H, Hou Y. Ubiquitin-mediated NFκB degradation pathway. Cell Mol Immunol [Internet]. 2015;12:653–5. Available from: https://www.nature.com/articles/cmi201499
96. Chen C-H, Zhou W, Liu S, Deng Y, Cai F, Tone M, et al. Increased NF-κB signalling upregulates BACE1 expression and its therapeutic potential in Alzheimer's disease. Int J Neuropsychopharmacol [Internet]. 2012;15:77–90. Available from: https://academic.oup.com/ijnp/article-lookup/doi/10.1017/S1461145711000149
97. Tokunaga F. Linear ubiquitination-mediated NF- B regulation and its related disorders. J Biochem [Internet]. 2013;154:313–23. Available from: https://academic.oup.com/jb/article-lookup/doi/10.1093/jb/mvt079
98. Liu N, Yu Z, Xun Y, Shu P, Yue Y, Yuan S, et al. Amyloid-β 25–35 upregulates endogenous Neuroprotectant Neuroglobin via NF κ B activation in vitro. J Alzheimer's Dis [Internet]. 2018;64:1163–74. Available from: https://journals.sagepub.com/doi/10.3233/JAD-180163
99. Marwarha G, Ghribi O. Nuclear factor kappa-light-chain-enhancer of activated B cells (NF-κB)- a friend, a foe, or a bystander – in the neurodegenerative Cascade and pathogenesis of Alzheimer's disease. CNS Neurol Disord – Drug Targets [Internet]. 2018;16:1050–65. Available from: http://www.eurekaselect.com/154444/article
100. Hodges SL, Reynolds CD, Smith GD, Jefferson TS, Nolan SO, Lugo JN. Molecular interplay between hyperactive mammalian target of rapamycin signaling and Alzheimer's disease neuropathology in the NS-Pten knockout mouse model. Neuroreport [Internet]. 2018;29:1109–13. Available from: https://journals.lww.com/00001756-201809010-00009
101. Tramutola A, Lanzillotta C, Barone E, Arena A, Zuliani I, Mosca L, et al. Intranasal rapamycin ameliorates Alzheimer-like cognitive decline in a mouse model of down syndrome. Transl Neurodegener [Internet]. 2018;7:28. Available from: https://translationalneurodegeneration.biomedcentral.com/articles/10.1186/s40035-018-0133-9

102. Singh AK, Kashyap MP, Tripathi VK, Singh S, Garg G, Rizvi SI. Neuroprotection through rapamycin-induced activation of autophagy and PI3K/Akt1/mTOR/CREB signaling against amyloid-β-induced oxidative stress, synaptic/neurotransmission dysfunction, and neurodegeneration in adult rats. Mol Neurobiol [Internet]. 2017;54:5815–28. Available from: http://link.springer.com/10.1007/s12035-016-0129-3
103. Fan S, Zhang B, Luan P, Gu B, Wan Q, Huang X, et al. PI3K/AKT/mTOR/p70S6K pathway is involved in a β 25-35-induced autophagy. Biomed Res Int [Internet]. 2015;2015:1–9. Available from: http://www.hindawi.com/journals/bmri/2015/161020/
104. Tang Z, Guo M, Ding Y, Wang L, Chen H, Li B, et al. mTOR inhibition alleviated tau phosphorylation-induced mitochondrial impairment, oxidative stress, and cognitive impairment [Internet]. 2024. Available from: http://biorxiv.org/lookup/doi/10.1101/2024.11.24.625068
105. Lapresa R, Agulla J, Sánchez-Morán I, Bolaños JP, Almeida A. Amyloid-ß promotes neurotoxicity by Cdk5-induced p53 stabilization [Internet]. 2018. Available from: http://biorxiv.org/lookup/doi/10.1101/365437
106. Farmer KM, Ghag G, Puangmalai N, Montalbano M, Bhatt N, Kayed R. P53 aggregation, interactions with tau, and impaired DNA damage response in Alzheimer's disease. Acta Neuropathol Commun [Internet]. 2020;8:132. Available from: https://actaneurocomms.biomedcentral.com/articles/10.1186/s40478-020-01012-6
107. Zhu Q, Wani G, Yao J, Patnaik S, Wang Q-E, El-Mahdy MA, et al. The ubiquitin–proteasome system regulates p53-mediated transcription at p21waf1 promoter. Oncogene [Internet]. 2007;26:4199–208. Available from: https://www.nature.com/articles/1210191
108. Krishna-K K, Behnisch T, Sajikumar S. Modulation of the ubiquitin-proteasome system restores plasticity in hippocampal pyramidal neurons of the APP/PS1 Alzheimer's disease-like mice. J Alzheimer's Dis [Internet]. 2022;86:1611–6. Available from: https://journals.sagepub.com/doi/full/10.3233/JAD-215718
109. Dorszewska J, Oczkowska A, Suwalska M, Rozycka A, Florczak-Wyspianska J, Dezor M, et al. Original article mutations in the exon 7 of Trp53 gene and the level of p53 protein in double transgenic mouse model of Alzheimer's disease. Folia Neuropathol [Internet]. 2014;1:30–40. Available from: http://www.termedia.pl/doi/10.5114/fn.2014.41742
110. Kostes WW, Brafman DA. The multifaceted role of WNT signaling in Alzheimer's disease onset and age-related progression. Cells [Internet]. 2023;12:1204. Available from: https://www.mdpi.com/2073-4409/12/8/1204
111. Inestrosa NC, Tapia-Rojas C, Cerpa W, Cisternas P, Zolezzi JM. WNT signaling is a key player in Alzheimer's disease. 2021. p. 357–382. Available from: https://link.springer.com/10.1007/164_2021_532
112. Parr C, Mirzaei N, Christian M, Sastre M. Activation of the Wnt/β-catenin pathway represses the transcription of the β-amyloid precursor protein cleaving enzyme (BACE1) via binding of T-cell factor-4 to BACE1 promoter. FASEB J [Internet]. 2015;29:623–35. Available from: https://onlinelibrary.wiley.com/doi/abs/10.1096/fj.14-253211
113. Tapia-Rojas C, Burgos PV, Inestrosa NC. Inhibition of Wnt signaling induces amyloidogenic processing of amyloid precursor protein and the production and aggregation of Amyloid-β (Aβ) 42 peptides. J Neurochem [Internet]. 2016;139:1175–91. Available from: https://onlinelibrary.wiley.com/doi/10.1111/jnc.13873
114. Tapia-Rojas C, Inestrosa NC. Wnt signaling loss accelerates the appearance of neuropathological hallmarks of Alzheimer's disease in J20- APP transgenic and wild-type mice. J Neurochem [Internet]. 2018;144:443–65. Available from: https://onlinelibrary.wiley.com/doi/10.1111/jnc.14278
115. Yoon M, Kim H, Shin H, Lee H, Kang M-J, Park S-H, et al. Inhibition of CXXC5 function rescues Alzheimer's disease phenotypes by restoring Wnt/β-catenin signaling pathway. Pharmacol Res [Internet]. 2023;194:106836. Available from: https://linkinghub.elsevier.com/retrieve/pii/S1043661823001925
116. Priori EC, Musi CA, Giani A, Colnaghi L, Milic I, Devitt A, et al. JNK activation correlates with cognitive impairment and alteration of the post-synaptic element in the 5xFAD

AD mouse model. Cells [Internet]. 2023;12:904. Available from: https://www.mdpi.com/2073-4409/12/6/904
117. Mazzitelli S, Xu P, Ferrer I, Davis RJ, Tournier C. The loss of c-Jun N-terminal protein kinase activity prevents the Amyloidogenic cleavage of amyloid precursor protein and the formation of amyloid plaques in vivo. J Neurosci [Internet]. 2011;31:16969–76. Available from: https://www.jneurosci.org/lookup/doi/10.1523/JNEUROSCI.4491-11.2011
118. Sahara N, Murayama M, Lee B, Park J, Lagalwar S, Binder LI, et al. Active c-Jun N-terminal kinase induces caspase cleavage of tau and additional phosphorylation by GSK-3β is required for tau aggregation. Eur J Neurosci [Internet]. 2008;27:2897–906. Available from: https://onlinelibrary.wiley.com/doi/10.1111/j.1460-9568.2008.06258.x
119. Yenki P, Khodagholi F, Shaerzadeh F. Inhibition of phosphorylation of JNK suppresses aβ-induced ER stress and upregulates Prosurvival mitochondrial proteins in rat hippocampus. J Mol Neurosci [Internet]. 2013;49:262–9. Available from: http://link.springer.com/10.1007/s12031-012-9837-y
120. Zhou Q, Wang M, Du Y, Zhang W, Bai M, Zhang Z, et al. Inhibition of c-Jun N-terminal kinase activation reverses Alzheimer disease phenotypes in APPswe/PS1dE9 mice. Ann Neurol [Internet]. 2015;77:637–54. Available from: https://onlinelibrary.wiley.com/doi/10.1002/ana.24361
121. Singh NK, Singh A, Mayank. Nuclear factor Kappa B: a nobel therapeutic target of flavonoids against parkinson's disease. Comb Chem High Throughput Screen [Internet]. 2024;27:2062–77. Available from: https://www.eurekaselect.com/225798/article
122. Ghosh A, Roy A, Liu X, Kordower JH, Mufson EJ, Hartley DM, et al. Selective inhibition of NF-κB activation prevents dopaminergic neuronal loss in a mouse model of Parkinson's disease. Proc Natl Acad Sci [Internet]. 2007;104:18754–9. Available from: https://pnas.org/doi/full/10.1073/pnas.0704908104
123. Düwel M, Hadian K, Krappmann D. Ubiquitin conjugation and deconjugation in NF-κB signaling. 2010. p. 88–99. Available from: http://link.springer.com/10.1007/978-1-4419-6676-6_7
124. Wang Y, Shan B, Liang Y, Wei H, Yuan J. Parkin regulates NF-κB by mediating site-specific ubiquitination of RIPK1. Cell Death Dis [Internet]. 2018;9:732. Available from: https://www.nature.com/articles/s41419-018-0770-z
125. Hunot S, Brugg B, Ricard D, Michel PP, Muriel M-P, Ruberg M, et al. Nuclear translocation of NF-κB is increased in dopaminergic neurons of patients with Parkinson disease. Proc Natl Acad Sci [Internet]. 1997;94:7531–6. Available from: https://pnas.org/doi/full/10.1073/pnas.94.14.7531
126. Bao X, He Y, Huang L, Li H, Li Q, Huang Y. Sinomenine exerts a neuroprotective effect on PD mouse model through inhibiting PI3K/AKT/mTOR pathway to enhance autophagy. Int J Neurosci [Internet]. 2024;134:301–9. Available from: https://www.tandfonline.com/doi/full/10.1080/00207454.2022.2100780
127. Khwanraj K, Madlah S, Grataitong K, Dharmasaroja P. Comparative mRNA expression of eEF1A isoforms and a PI3K/Akt/mTOR pathway in a cellular model of Parkinson's disease. Parkinsons Dis [Internet]. 2016;2016:1–11. Available from: http://www.hindawi.com/journals/pd/2016/8716016/
128. Kaiser MS, Milan G, Ham DJ, Lin S, Oliveri F, Chojnowska K, et al. Dual roles of mTORC1-dependent activation of the ubiquitin-proteasome system in muscle proteostasis. Commun Biol [Internet]. 2022;5:1141. Available from: https://www.nature.com/articles/s42003-022-04097-y
129. Furlong RM, Lindsay A, Anderson KE, Hawkins PT, Sullivan AM, O'Neill C. The Parkinson's disease gene PINK1 activates Akt via PINK1 kinase-dependent regulation of the phospholipid PI(3,4,5)P3. J Cell Sci [Internet]. 2019;132. Available from: https://journals.biologists.com/jcs/article/132/20/jcs233221/224770/The-Parkinson-s-disease-gene-PINK1-activates-Akt
130. Checler F, Alves da Costa C. Interplay between Parkin and p53 governs a physiological homeostasis that is disrupted in Parkinson's disease and cerebral cancer. Neurodegener Dis [Internet]. 2014;13:118–21. Available from: https://karger.com/NDD/article/doi/10.1159/000354075

131. Jenner P. Dysfunction of the ubiquitin–proteasome system in Parkinson's disease. J Neurochem [Internet]. 2003;85:1–1. Available from: https://onlinelibrary.wiley.com/doi/10.1046/j.1471-4159.85.s2.1_2.x
132. Du Y, Yang D, Li L, Luo G, Li T, Fan X, et al. An insight into the mechanistic role of p53-mediated autophagy induction in response to proteasomal inhibition-induced neurotoxicity. Autophagy [Internet]. 2009;5:663–75. Available from: http://www.tandfonline.com/doi/abs/10.4161/auto.5.5.8377
133. Sunico CR, Nakamura T, Rockenstein E, Mante M, Adame A, Chan SF, et al. S-Nitrosylation of parkin as a novel regulator of p53-mediated neuronal cell death in sporadic Parkinson's disease. Mol Neurodegener [Internet]. 2013;8:29. Available from: https://molecularneurodegeneration.biomedcentral.com/articles/10.1186/1750-1326-8-29
134. Nair VD. Activation of p53 signaling initiates apoptotic death in a cellular model of Parkinson's disease. Apoptosis [Internet]. 2006;11:955–66. Available from: http://link.springer.com/10.1007/s10495-006-6316-3
135. Qi X, Davis B, Chiang Y, Filichia E, Barnett A, Greig NH, et al. Dopaminergic neuron-specific deletion of p53 gene is neuroprotective in an experimental Parkinson's disease model. J Neurochem [Internet]. 2016;138:746–57. Available from: https://onlinelibrary.wiley.com/doi/10.1111/jnc.13706
136. Zhao F, Siedlak SL, Torres SL, Xu Q, Tang B, Zhu X. Conditional Haploinsufficiency of β-catenin aggravates neuronal damage in a Paraquat-based mouse model of Parkinson disease. Mol Neurobiol [Internet]. 2019;56:5157–66. Available from: http://link.springer.com/10.1007/s12035-018-1431-z
137. Marchetti B, Tirolo C, L'Episcopo F, Caniglia S, Testa N, Smith JA, et al. Parkinson's disease, aging and adult neurogenesis: Wnt/β-catenin signalling as the key to unlock the mystery of endogenous brain repair. Aging Cell [Internet]. 2020;19. Available from: https://onlinelibrary.wiley.com/doi/10.1111/acel.13101
138. Marchetti B, L'Episcopo F, Morale MC, Tirolo C, Testa N, Caniglia S, et al. Uncovering novel actors in astrocyte–neuron crosstalk in Parkinson's disease: the Wnt/β-catenin signaling cascade as the common final pathway for neuroprotection and self-repair. Eur J Neurosci [Internet]. 2013;37:1550–63. Available from: https://onlinelibrary.wiley.com/doi/10.1111/ejn.12166
139. Ries V, Silva RM, Oo TF, Cheng H, Rzhetskaya M, Kholodilov N, et al. JNK2 and JNK3 combined are essential for apoptosis in dopamine neurons of the substantia nigra, but are not required for axon degeneration. J Neurochem [Internet]. 2008;107:1578–88. Available from: https://onlinelibrary.wiley.com/doi/10.1111/j.1471-4159.2008.05713.x
140. Fuchs SY, Dolan L, Davis RJ, Ronai Z. Phosphorylation-dependent targeting of c-Jun ubiquitination by Jun N-kinase. Oncogene [Internet]. 1996;13:1531–5. Available from: http://www.ncbi.nlm.nih.gov/pubmed/8875991
141. Li X, Du Y, Fan X, Yang D, Luo G, Le W. c-Jun N-terminal kinase mediates Lactacystin-induced dopamine neuron degeneration. J Neuropathol Exp Neurol [Internet]. 2008;67:933–44. Available from: https://academic.oup.com/jnen/article-lookup/doi/10.1097/NEN.0b013e318186de64
142. Chambers JW, Pachori A, Howard S, Ganno M, Hansen D, Kamenecka T, et al. Small molecule c-Jun-N-terminal kinase inhibitors protect dopaminergic neurons in a model of Parkinson's disease. ACS Chem Neurosci [Internet]. 2011;2:198–206. Available from: https://pubs.acs.org/doi/10.1021/cn100109k
143. Träger U, Andre R, Lahiri N, Magnusson-Lind A, Weiss A, Grueninger S, et al. HTT-lowering reverses Huntington's disease immune dysfunction caused by NFκB pathway dysregulation. Brain [Internet]. 2014;137:819–33. Available from: https://academic.oup.com/brain/article-lookup/doi/10.1093/brain/awt355
144. Hsiao H-Y, Chen Y-C, Chen H-M, Tu P-H, Chern Y. A critical role of astrocyte-mediated nuclear factor-κB-dependent inflammation in Huntington's disease. Hum Mol Genet [Internet]. 2013;22:1826–42. Available from: https://academic.oup.com/hmg/article-lookup/doi/10.1093/hmg/ddt036

145. Ortega Z, Lucas JJ. Ubiquitin-proteasome system involvement in Huntington's disease. Front Mol Neurosci [Internet]. 2014:7. Available from: http://journal.frontiersin.org/article/10.3389/fnmol.2014.00077/abstract
146. Reijonen S, Kukkonen JP, Hyrskyluoto A, Kivinen J, Kairisalo M, Takei N, et al. Downregulation of NF-κB signaling by mutant huntingtin proteins induces oxidative stress and cell death. Cell Mol Life Sci [Internet]. 2010;67:1929–41. Available from: http://link.springer.com/10.1007/s00018-010-0305-y
147. Marcora E, Kennedy MB. The Huntington's disease mutation impairs huntingtin's role in the transport of NF-κB from the synapse to the nucleus. Hum Mol Genet [Internet]. 2010;19:4373–84. Available from: https://academic.oup.com/hmg/article-lookup/doi/10.1093/hmg/ddq358
148. Picher-Martel V, Dutta K, Phaneuf D, Sobue G, Julien J-P. Ubiquilin-2 drives NF-κB activity and cytosolic TDP-43 aggregation in neuronal cells. Mol Brain [Internet]. 2015;8:71. Available from: http://www.molecularbrain.com/content/8/1/71
149. Saavedra A, García-Martínez JM, Xifró X, Giralt A, Torres-Peraza JF, Canals JM, et al. PH domain leucine-rich repeat protein phosphatase 1 contributes to maintain the activation of the PI3K/Akt pro-survival pathway in Huntington's disease striatum. Cell Death Differ [Internet]. 2010;17:324–35. Available from: https://www.nature.com/articles/cdd2009127
150. Ravikumar B, Vacher C, Berger Z, Davies JE, Luo S, Oroz LG, et al. Inhibition of mTOR induces autophagy and reduces toxicity of polyglutamine expansions in fly and mouse models of Huntington disease. Nat Genet [Internet]. 2004;36:585–95. Available from: https://www.nature.com/articles/ng1362
151. Palmieri M, Pal R, Nelvagal HR, Lotfi P, Stinnett GR, Seymour ML, et al. mTORC1-independent TFEB activation via Akt inhibition promotes cellular clearance in neurodegenerative storage diseases. Nat Commun [Internet]. 2017;8:14338. Available from: https://www.nature.com/articles/ncomms14338
152. Mansky RH, Greguske EA, Yu D, Zarate N, Intihar TA, Tsai W, et al. Tumor suppressor p53 regulates heat shock factor 1 protein degradation in huntington's disease. SSRN Electron J [Internet]. 2022; Available from: https://www.ssrn.com/abstract=4119001
153. Ehrnhoefer DE, Skotte NH, Ladha S, Nguyen YTN, Qiu X, Deng Y, et al. p53 increases caspase-6 expression and activation in muscle tissue expressing mutant huntingtin. Hum Mol Genet [Internet]. 2014;23:717–29. Available from: https://academic.oup.com/hmg/article-lookup/doi/10.1093/hmg/ddt458
154. Steffan JS, Kazantsev A, Spasic-Boskovic O, Greenwald M, Zhu Y-Z, Gohler H, et al. The Huntington's disease protein interacts with p53 and CREB-binding protein and represses transcription. Proc Natl Acad Sci [Internet]. 2000;97:6763–8. Available from: https://pnas.org/doi/full/10.1073/pnas.100110097
155. Wang J, Wang C-E, Orr A, Tydlacka S, Li S-H, Li X-J. Impaired ubiquitin–proteasome system activity in the synapses of Huntington's disease mice. J Cell Biol [Internet]. 2008;180:1177–89. Available from: https://rupress.org/jcb/article/180/6/1177/45268/Impaired-ubiquitin-proteasome-system-activity-in
156. Anand AA, Khan M, Monica V, Kar D. The molecular basis of Wnt/β-Catenin signaling pathways in neurodegenerative diseases. Int J Cell Biol [Internet]. 2023;2023:1–13. Available from: https://www.hindawi.com/journals/ijcb/2023/9296092/
157. Godin JD, Poizat G, Hickey MA, Maschat F, Humbert S. Mutant huntingtin-impaired degradation of β-catenin causes neurotoxicity in Huntington's disease. EMBO J [Internet]. 2010;29:2433–45. Available from: http://emboj.embopress.org/cgi/doi/10.1038/emboj.2010.117
158. Schipper-Krom S, Juenemann K, Reits EAJ. The ubiquitin-proteasome system in Huntington's disease: are proteasomes impaired, initiators of disease, or coming to the rescue? Biochem Res Int [Internet]. 2012;2012:1–12. Available from: http://www.hindawi.com/journals/bri/2012/837015/

159. Hunter JM, Lesort M, Johnson GVW. Ubiquitin-proteasome system alterations in a striatal cell model of Huntington's disease. J Neurosci Res [Internet]. 2007;85:1774–88. Available from: https://onlinelibrary.wiley.com/doi/10.1002/jnr.21287
160. Lui TTH, Lacroix C, Ahmed SM, Goldenberg SJ, Leach CA, Daulat AM, et al. The ubiquitin-specific protease USP34 regulates Axin stability and Wnt/β-catenin signaling. Mol Cell Biol [Internet]. 2011;31:2053–65. Available from: https://www.tandfonline.com/doi/full/10.1128/MCB.01094-10
161. Ramakrishna K, Nalla LV, Naresh D, Venkateswarlu K, Viswanadh MK, Nalluri BN, et al. WNT-β catenin signaling as a potential therapeutic target for neurodegenerative diseases: current status and future perspective. Diseases [Internet]. 2023;11:89. Available from: https://www.mdpi.com/2079-9721/11/3/89
162. Morfini GA, You Y-M, Pollema SL, Kaminska A, Liu K, Yoshioka K, et al. Pathogenic huntingtin inhibits fast axonal transport by activating JNK3 and phosphorylating kinesin. Nat Neurosci [Internet]. 2009;12:864–71. Available from: https://www.nature.com/articles/nn.2346
163. Sako W, Ito H, Yoshida M, Koizumi H, Kamada M, Fujita K, et al. Nuclear factor ? B expression in patients with sporadic amyotrophic lateral sclerosis and hereditary amyotrophic lateral sclerosis with optineurin mutations. Clin Neuropathol [Internet]. 2012;31:418–23. Available from: http://www.dustri.com/article_response_page.html?artId=9849&doi=10.5414/NP300493&L=0
164. Migheli A, Piva R, Atzori C, Troost D, Schiffer D. c-Jun, JNK/SAPK kinase and transcription factor NF-κB are selectively activated in astrocytes, but not motor neurons, in amyotrophic lateral Sclerosis. J Neuropathol Exp Neurol [Internet]. 1997;56:1314–22. Available from: https://academic.oup.com/jnen/article-lookup/doi/10.1097/00005072-199712000-00006
165. Nivon M, Fort L, Muller P, Richet E, Simon S, Guey B, et al. NFκB is a central regulator of protein quality control in response to protein aggregation stresses via autophagy modulation. Mol Biol Cell [Internet]. 2016;27:1712–27. Available from: https://www.molbiolcell.org/doi/10.1091/mbc.e15-12-0835
166. Akizuki M, Yamashita H, Uemura K, Maruyama H, Kawakami H, Ito H, et al. Optineurin suppression causes neuronal cell death via NF-κB pathway. J Neurochem [Internet]. 2013;126:699–704. Available from: https://onlinelibrary.wiley.com/doi/10.1111/jnc.12326
167. Swarup V, Phaneuf D, Dupré N, Petri S, Strong M, Kriz J, et al. Deregulation of TDP-43 in amyotrophic lateral sclerosis triggers nuclear factor κB–mediated pathogenic pathways. J Exp Med [Internet]. 2011;208:2429–47. Available from: https://rupress.org/jem/article/208/12/2429/41090/Deregulation-of-TDP-43-in-amyotrophic-lateral
168. Wagey R, Pelech SL, Duronio V, Krieger C. Phosphatidylinositol 3-kinase: increased activity and protein level in amyotrophic lateral Sclerosis. J Neurochem [Internet]. 1998;71:716–22. Available from: https://onlinelibrary.wiley.com/doi/10.1046/j.1471-4159.1998.71020716.x
169. Peviani M, Tortarolo M, Battaglia E, Piva R, Bendotti C. Specific induction of Akt3 in spinal cord motor neurons is neuroprotective in a mouse model of familial amyotrophic lateral Sclerosis. Mol Neurobiol [Internet]. 2014;49:136–48. Available from: http://link.springer.com/10.1007/s12035-013-8507-6
170. Cheroni C, Marino M, Tortarolo M, Veglianese P, De Biasi S, Fontana E, et al. Functional alterations of the ubiquitin-proteasome system in motor neurons of a mouse model of familial amyotrophic lateral sclerosis†. Hum Mol Genet [Internet]. 2009;18:82–96. Available from: https://academic.oup.com/hmg/article-lookup/doi/10.1093/hmg/ddn319
171. Yang F, Mahaman YAR, Zhang B, Wang J, Liu R, Liu F, et al. C9orf72 poly-PR helps p53 escape from the ubiquitin-proteasome system and promotes its stability. J Neurochem [Internet]. 2023;166:389–402. Available from: https://onlinelibrary.wiley.com/doi/10.1111/jnc.15872
172. Chun C, Lee JH, Bothwell M, Nghiem P, Smith AST, Mack DL. Human motor neurons elicit pathological hallmarks of ALS and reveal potential biomarkers of the disease in response to prolonged IFNγ exposure. J Neurosci [Internet]. 2024:e1787232024. Available from: https://www.jneurosci.org/lookup/doi/10.1523/JNEUROSCI.1787-23.2024

173. Pang D, Yu Y, Zhao B, Huang J, Cui Y, Li T, et al. The Long non-coding RNA NR3C2-8:1 promotes p53-mediated apoptosis through the miR-129-5p/USP10 Axis in amyotrophic lateral Sclerosis. Mol Neurobiol [Internet]. 2024;61:7466–80. Available from: https://link.springer.com/10.1007/s12035-024-04059-x
174. Jiang X, Guan Y, Zhao Z, Meng F, Wang X, Gao X, et al. Potential roles of the WNT signaling pathway in amyotrophic lateral Sclerosis. Cells [Internet]. 2021;10:839. Available from: https://www.mdpi.com/2073-4409/10/4/839
175. Hawkins S, Namboori SC, Tariq A, Blaker C, Flaxman C, Dey NS, et al. Upregulation of β-catenin due to loss of miR-139 contributes to motor neuron death in amyotrophic lateral sclerosis. Stem Cell Reports [Internet]. 2022;17:1650–65. Available from: https://linkinghub.elsevier.com/retrieve/pii/S2213671122002727
176. González-Fernández C, Gonzalez P, Andres-Benito P, Ferrer I, Rodríguez FJ. Wnt signaling alterations in the human spinal cord of amyotrophic lateral Sclerosis cases: spotlight on Fz2 and Wnt5a. Mol Neurobiol [Internet]. 2019;56:6777–91. Available from: http://link.springer.com/10.1007/s12035-019-1547-9
177. Vallée A. Activation de la glycolyse aérobie par la voie canonique WNT/β-caténine. médecine/sciences [Internet]. 2018;34:326–30. Available from: https://www.medecinesciences.org/10.1051/medsci/20183404013
178. Pinto C, Cárdenas P, Osses N, Henríquez JP. Characterization of Wnt/β-catenin and BMP/Smad signaling pathways in an in vitro model of amyotrophic lateral sclerosis. Front Cell Neurosci [Internet]. 2013:7. Available from: http://journal.frontiersin.org/article/10.3389/fncel.2013.00239/abstract
179. Matsuo K, Nagamatsu J, Nagata K, Umeda R, Shiota T, Morimoto S, et al. Establishment of a novel amyotrophic lateral sclerosis patient (TARDBPN345K/+)-derived brain microvascular endothelial cell model reveals defective Wnt/β-catenin signaling: investigating diffusion barrier dysfunction and immune cell interaction. Front Cell Dev Biol [Internet]. 2024:12. Available from: https://www.frontiersin.org/articles/10.3389/fcell.2024.1357204/full
180. Sahana T, Chase KJ, Liu F, Lloyd TE, Rossoll W, Zhang K. c-Jun N-terminal kinase promotes stress granule assembly and neurodegeneration in C9orf72-mediated ALS and FTD [internet]. 2022. Available from: http://biorxiv.org/lookup/doi/10.1101/2022.04.28.489917
181. Yadav RK, Minz E, Mehan S. Understanding abnormal c-JNK/p38MAPK signaling in amyotrophic lateral Sclerosis: potential drug targets and influences on neurological disorders. CNS Neurol Disord - Drug Targets [Internet]. 2021;20:417–29. Available from: https://www.eurekaselect.com/190729/article
182. Biggi S, Buccarello L, Sclip A, Lippiello P, Tonna N, Rumio C, et al. Evidence of presynaptic localization and function of the c-Jun N-terminal kinase. Neural Plast [Internet]. 2017;2017:1–14. Available from: https://www.hindawi.com/journals/np/2017/6468356/
183. Yan J, McCombe PA, Pender MP, Greer JM. Reduced IκB-α protein levels in peripheral blood cells of patients with multiple sclerosis—a possible cause of constitutive NF-κB activation. J Clin Med [Internet]. 2020;9:2534. Available from: https://www.mdpi.com/2077-0383/9/8/2534
184. Srinivasan M, Lahiri DK. Significance of NF-κB as a pivotal therapeutic target in the neurodegenerative pathologies of Alzheimer's disease and multiple sclerosis. Expert Opin Ther Targets [Internet]. 2015;19:471–87. Available from: http://www.tandfonline.com/doi/full/10.1517/14728222.2014.989834
185. Tokunaga F. Ubiquitination-mediated NF-κB regulation in inflammatory response. Protein Modif Pathog dysregulation signal [internet]. Tokyo: Springer Japan; 2015. p. 177–96. Available from: https://link.springer.com/10.1007/978-4-431-55561-2_12
186. Satoh J. Molecular network of ChIP-Seq-based NF-κB p65 target genes involves diverse immune functions relevant to the immunopathogenesis of multiple sclerosis. Mult Scler Relat Disord [Internet]. 2014;3:94–106. Available from: https://linkinghub.elsevier.com/retrieve/pii/S2211034813000473

187. Leibowitz SM, Yan J. NF-κB pathways in the pathogenesis of multiple sclerosis and the therapeutic implications. Front Mol Neurosci [Internet]. 2016:9. Available from: http://journal.frontiersin.org/article/10.3389/fnmol.2016.00084/full
188. Mammana S, Bramanti P, Mazzon E, Cavalli E, Basile MS, Fagone P, et al. Preclinical evaluation of the PI3K/Akt/mTOR pathway in animal models of multiple sclerosis. Oncotarget [Internet]. 2018;9:8263–77. Available from: https://www.oncotarget.com/lookup/doi/10.18632/oncotarget.23862
189. Dello RC, Navarra P, Lisi L. mTOR in Multiple Sclerosis. Mol to Med with mTOR [Internet]. 2016:331–43. Available from: https://linkinghub.elsevier.com/retrieve/pii/B9780128027332000049
190. Evangelisti C, Martelli AM. The PI3K/Akt/mTOR pathway. Encycl Cell Biol [Internet] Elsevier. 2016:128–35. Available from: https://linkinghub.elsevier.com/retrieve/pii/B9780123944474300153
191. Brooks CL, Gu W. Dynamics in the p53-Mdm2 ubiquitination pathway. Cell Cycle [Internet]. 2004;3:893–7. Available from: http://www.tandfonline.com/doi/abs/10.4161/cc.3.7.997
192. Park JH, Wandless TJ. p53 engagement is a hallmark of an unfolded protein response in the nucleus of mammalian cells [Internet]. 2024. Available from: http://biorxiv.org/lookup/doi/10.1101/2024.11.08.622663
193. Park JH, Wandless TJ. p53 engagement is a hallmark of an unfolded protein response in the nucleus of mammalian cells [internet]. 2025. Available from: https://elifesciences.org/reviewed-preprints/105021v1
194. Velu CS, Niture SK, Doneanu CE, Pattabiraman N, Srivenugopal KS. Human p53 is inhibited by Glutathionylation of Cysteines present in the proximal DNA-binding domain during oxidative stress. Biochem Int. 2007;46:7765–80. Available from: https://pubs.acs.org/doi/10.1021/bi700425y
195. Vallée A, Vallée J-N, Guillevin R, Lecarpentier Y. Interactions between the canonical WNT/Beta-catenin pathway and PPAR gamma on neuroinflammation, demyelination, and remyelination in multiple sclerosis. Cell Mol Neurobiol [Internet]. 2018;38:783–95. Available from: http://link.springer.com/10.1007/s10571-017-0550-9
196. Yuan S, Shi Y, Tang S-J. Wnt signaling in the pathogenesis of multiple sclerosis-associated chronic pain. J Neuroimmune Pharmacol [Internet]. 2012;7:904–13. Available from: https://link.springer.com/10.1007/s11481-012-9370-3
197. Zhang D, Pan Y, Zhang C, Yan B, Yu S, Wu D, et al. Wnt/β-catenin signaling induces the aging of mesenchymal stem cells through promoting the ROS production. Mol Cell Biochem [Internet]. 2013;374:13–20. Available from: http://link.springer.com/10.1007/s11010-012-1498-1
198. Gao J, Liao Y, Qiu M, Shen W. Wnt/β-catenin signaling in neural stem cell homeostasis and neurological diseases. Neurosci [Internet]. 2021;27:58–72. Available from: https://journals.sagepub.com/doi/10.1177/1073858420914509
199. Groenendyk J, Michalak M. A genome-wide siRNA screen identifies novel Phosphoenzymes affecting Wnt/β-catenin signaling in mouse embryonic stem cells. Stem Cell Rev Reports [Internet]. 2011;7:910–26. Available from: http://link.springer.com/10.1007/s12015-011-9265-3
200. Chen YR, Tan TH. The c-Jun N-terminal kinase pathway and apoptotic signaling (review). Int J Oncol [Internet]. 2000; Available from: http://www.spandidos-publications.com/10.3892/ijo.16.4.651
201. Castro-Torres RD, Olloquequi J, Parcerisas A, Ureña J, Ettcheto M, Beas-Zarate C, et al. JNK signaling and its impact on neural cell maturation and differentiation. Life Sci [Internet]. 2024;350:122750. Available from: https://linkinghub.elsevier.com/retrieve/pii/S0024320524003400
202. Mehan S, Meena H, Sharma D, Sankhla R. JNK: a stress-activated protein kinase therapeutic strategies and involvement in Alzheimer's and various neurodegenerative abnormalities. J Mol Neurosci [Internet]. 2011;43:376–90. Available from: http://link.springer.com/10.1007/s12031-010-9454-6

Ubiquitin-Proteasome System in the Cellular Proteostasis

6

Sadhna Soni, Ashwini Nath Tiwari, and Arun Upadhyay

6.1 Introduction

Protein homeostasis (proteostasis) is a highly regulated process that maintains the healthy proteome in an organism. A "healthy proteome" refers to the correct concentration, conformation, and localization of all available cellular proteins. To regulate cellular proteostasis, a network of molecular players, collectively referred to as a proteostasis network (PN), ensures the tight regulation of protein synthesis, proper folding by molecular chaperones, and elimination of misfolded and old proteins [110, 113]. The PN oversees every aspect of the protein's lifecycle, which encompasses efficient protein synthesis, tightly controlled post-translational modifications (PTMs), proper folding, transport, interaction, localization, and degradation in a precisely regulated manner [13, 55, 88]. In fact, translation, folding, and degradation systems collectively orchestrate cellular proteostasis. In other words, PN is an intricate collection of components functioning consistently to acquire proteostasis, which is vital for cellular health. Key elements of the PN include protein synthesis machinery, chaperones/co-chaperones, ubiquitin-proteasome system (UPS), and autophagy [47, 75].

The translational machinery, including ribosomes, facilitates the synthesis of proteins from messenger RNA. Efficient and accurate translation is crucial for producing properly folded proteins [85]. Molecular chaperones play an essential role in assisting newly synthesized polypeptides in achieving their correct conformations. Additionally, chaperones bind to misfolded proteins, preventing aggregation, while additional proteins called co-chaperones regulate their activity, allowing tightly regulated protein folding under varying intracellular conditions [41, 109].

S. Soni · A. N. Tiwari · A. Upadhyay (✉)
Department of Bioscience and Biomedical Engineering, Indian Institute of Technology Bhilai, Durg, Chhattisgarh, India
e-mail: arun@iitbhilai.ac.in

S. Singh (ed.), *Ubiquitin Proteasome System in CNS Neurodegenerative Diseases*, https://doi.org/10.1007/978-981-95-6099-8_6

Degradation of proteins is performed by two major proteolytic systems: the proteasome for short-lived proteins and the lysosome for long-lived proteins [148]. While lysosomal (autophagy) pathways constitute bulk degradation processes of misfolded and aggregated proteins, the proteasome performs targeted proteolysis of proteins via the canonical ubiquitination pathway. The UPS can post-translationally modify damaged or superfluous proteins by attaching small (8 kDa) ubiquitin protein, marking them for proteasome-mediated degradation. This selective degradation process is indispensable for cellular protein quality control (PQC) and prevents the accumulation of unwanted proteins. If the proteins are degraded in such a timely and precise manner, they may lead to cellular distress [37, 58]. Additionally, the lysosomal machinery performs the degradation of cellular components and recycles macromolecules, including proteins, using a tightly regulated autophagy pathway. Autophagy-lysosomal degradation pathways protect cells under extreme proteotoxic stress and contribute to overall cellular viability [2, 57].

Beyond these core components, several stress response pathways modify the PN significantly. The heat shock response is triggered by stresses like elevated temperature, leading to the upregulation of heat shock proteins (HSPs) that assist in protein refolding and prevent aggregation [1, 50]. Similarly, the unfolded protein response (UPR) is triggered when proteins start accumulating in the endoplasmic reticulum (ER), activating a signaling cascade that enhances protein folding capacity and restores homeostasis within the ER [44, 45]. Moreover, various auxiliary factors are essential for maintaining proteostasis effectively. These include signaling pathways that regulate cellular responses to stress and metabolic changes, or metabolic factors that affect nutrient availability and impact protein synthesis and degradation [49, 139]. Chromatin remodelers that alter the expression of genes related to proteostasis, and regulators of transcription and PTMs, may fine-tune the activities of proteostasis-related proteins. Dysfunction in proteostasis can lead to several pathological conditions or diseases, collectively referred to as proteinopathies. Most common among these conditions are neurodegenerative disorders, for example, Alzheimer's disease (AD), Parkinson's disease (PD), Huntington's disease (HD), etc. [65, 74].

6.2 Ubiquitin-Proteasome System

The UPS is the primary cellular system for targeted protein degradation. It maintains proteostasis by eliminating unwanted, old, or misfolded proteins [18, 64]. The UPS removes non-functional proteins from cells to avoid their accumulation and prevents proteotoxicity. In the ubiquitination reaction, the ubiquitin (Ub) molecules are covalently attached to the target protein on the ε-amine group of a lysine (K) residue. The UPS consists of a series of enzymatic reactions that perform recognition and ubiquitination of target proteins. The ubiquitinated proteins are directed toward a multiprotease complex called the 26S proteasome. Before entering the proteasome, all proteins are deubiquitylated and unfolded in an adenosine triphosphate (ATP)-dependent manner [17, 24, 43]. Ubiquitin molecules are small proteins made of 76 amino-acid chain and are highly conserved, that is, its amino acid sequence and structure remain nearly unchanged throughout evolution. The process of ubiquitination involves multiple steps catalyzed by three successive enzymes: E1 ubiquitin-activating enzyme, E2 ubiquitin-conjugating enzyme, and E3 ubiquitin ligase [43, 99]. As shown in Fig. 6.1, ubiquitin is activated first in an ATP-dependent manner by the E1 enzyme. The E1 binds to ubiquitin and forms a thioester bond between its own cysteine residue and the carboxyl group of ubiquitin [32, 58]. In the second step, activated ubiquitin is transferred from the enzyme E1 to E2 cysteine, forming a new thioester bond [58]. The final step is catalyzed by the enzyme E3 ubiquitin ligase [22]. The E3s facilitate the transfer of ubiquitin from enzyme E2 to a lysine residue of the target protein and form an isopeptide bond between the ε-amine group of a lysine residue on the target protein and the α-carboxyl group of the C-terminus of the Ub molecule [15, 120].

In our proteome, multiple genes code for Ub: UBB, UBC, UBA52, and UBA80 (Rps27a). Expression of UBB and UBC genes is enhanced in stress conditions and encodes protein molecules for poly-ubiquitination [81, 131]. Gene UBA1 encodes for the E1 activation enzyme, while more than 40 E2 enzymes are known so far. In addition, approximately 600 E3 ubiquitin ligases are mapped in the human proteome that can ubiquitinate cellular proteins to modify their functions or target those for proteasomal degradation [96, 113]. E3 ubiquitin ligases are classified based on their structural motifs. The major classes are HECT domain-containing, RING finger type, U box-containing, and PHD finger type E3 ubiquitin ligase [23, 132]. In mammals, several E3s are present as multiprotein complexes, for example, SCF (SKP1-CUL1-F-Box-Protein) and APC/C (anaphase-promoting-complex/cyclosome) [99]. For poly-ubiquitination, the next Ub molecules are attached to one of the seven lysine (K6, K11, K27, K29, K33, K48, and K63) residues present on the previous Ub attached to the substrate [23, 131]. Ubiquitination on different lysine residues of target proteins may act as distinct signals to cells; for example, Lys-63 signals for the repair of protein and transcription activation factors, and

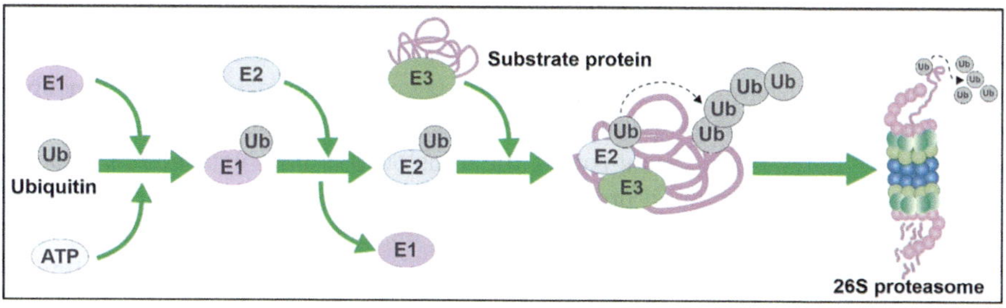

Fig. 6.1 Proteasomal degradation starts with the activation of the Ub molecule by the enzyme E1, which requires ATP. Activated Ub is transferred to the next enzyme E2 that interacts with another enzyme, E3 ligase. The E3s identify substrate proteins and facilitate the transfer of Ub from E2 to substrates. The proteasomal degradation often involves the subsequent addition of multiple ubiquitin molecules. The proteasome degrades polyubiquitinated proteins into smaller fragments after ubiquitin molecules are detached by DUBs

mono-ubiquitination for endocytosis, histone regulation, and so on. On the other hand, ubiquitination of Lys-48 and Lys-29 may lead to proteasomal degradation [58, 70, 81].

Proteasomes are large, cylindrical, multi-subunit protease complexes. In eukaryotes, proteasomes are found in both: cytosol and nucleus. Proteasomes are made up of two sub-complexes: one is the 20S catalytic core particle (CP), and the second is the 19S regulatory particle (RP). 20S CP may be present with a single RP and forms the 26S proteasome. Alternatively, two 19S RP could be present with one CP constituting the 30S proteasome. In eukaryotes, we have both 26S proteasome and 30S proteasome [58, 123]. The 20S catalytic CP has 28 subunits, stacked in barrel shaped structure, containing two outer rings of seven distinct alpha (α) subunits with no known proteolytic activity and two inner rings of seven distinct beta (β) subunits [7, 38]. Among the seven β subunits, three have proteolytic functions. 20S CP possesses trypsin-like (cleaves after basic residues), chymotrypsin-like (cuts after large hydrophobic residues), caspase, or peptidylglutamyl-peptide hydrolyzing-like (cuts after acidic residues) proteolytic activities in β2, β5, and β1 subunits of both the internal heptameric rings of CP [14, 87, 121].

The RP is responsible for receiving and preparing the target proteins for degradation. The most common form of 19S RP is PA700 in mammals. It is made up of 19 individual proteins, arranged in two distinct structures called base and lid [36, 118]. The base surrounds the 20S particle, and the lid is present above the base. The 19S RP functions to identify the substrate proteins, binding, de-ubiquitination, and translocation of proteins, while the 20S CP degrades the proteins [9, 31]. The lid contains nine non-ATPase (Rpn3, Rpn5, Rpn6, Rpn7, Rpn8, Rpn9, Rpn11, Rpn12, and Rpn15), while the base is composed of six AAA-ATPases (Rpt1–6), and four non-ATPase (Rpn1, Rpn2, Rpn10, and Rpn13) proteins [69, 78]. The ATPases form a hexameric ring that links RP and CP and performs ATP-dependent unfolding of target proteins. They are responsible for the gated translocation of substrate into the 20S particle after de-ubiquitination [7, 126]. Before entering the central proteolytic chamber of the 20S core proteasomal subcomplex, all proteins undergo a deubiquitylation reaction [84]. The ubiquitin molecules are taken off by one of the deubiquitinating (DUB) enzymes. Multiple proteins (Rpn11, Usp14, and Uch37) with DUB activity join the 19S regulatory proteasome subcomplex to remove ubiquitin chains from the substrate proteins [25, 81]. These enzymes belong to ubiquitin carboxy-terminal hydrolases and ubiquitin processing families [81]. Another important subunit, Rpn10, plays an important role in the translocation of target proteins into the proteasome, following which proteins are chewed up into small oligopeptides of 2–24 amino acids using multiple enzymatic digestions inside CP [114, 117]. Overall, the ubiquitination involves the identification, modification, and targeting of cellular proteins for proteasomal degradation.

6.3 Cellular Proteostasis: Multiple Players Orchestrating the Proteostasis Network

Proteostasis co-evolved with the proteome in all organisms to respond to changing environmental conditions and maintain cellular health. It primarily comprises chaperones and protein degradation machinery that collectively orchestrate the folding, maintenance, and degradation of polypeptides [47, 104, 107]. The process is maintained by PN in several stages of the life of a protein from synthesis to degradation (see Fig. 6.2). The PN is a multifaceted system comprising an array of components and cellular subsystems that together ensure proper maintenance, folding, and degradation of proteins [6, 55]. Understanding these components and their interactions is crucial for insights into cellular health and the implications of their dysfunction in various diseases [46]. To maintain a balanced proteome, regulated protein synthesis plays a crucial role, as it maintains the copy number and required quantities of all the cellular proteins. Soon after the synthesis, proteins go through PTM, as per the requirement of the cells [62, 85, 129]. Protein folding is an important aspect of

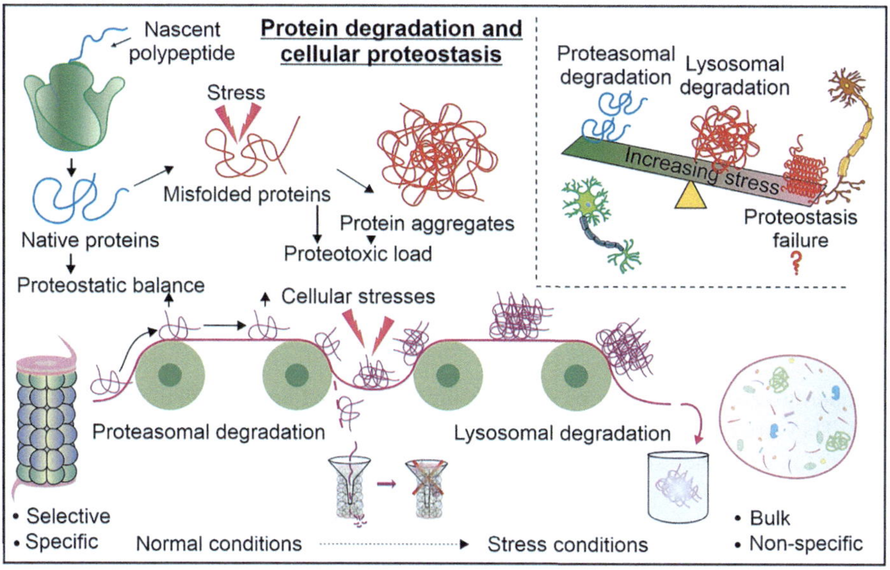

Fig. 6.2 Protein degradation is an essential component of cellular proteostasis. Under normal conditions, newly synthesized proteins fold into their functional shapes, while cellular stresses may lead to protein misfolding and the formation of aggregates. The protein aggregates generate proteotoxic stress, which can disrupt cellular proteostasis. To clear misfolded protein aggregates, mainly two pathways are responsible: the UPS and the lysosome. The proteasome acts as the first line of defense, selectively degrading misfolded proteins. However, when protein aggregates accumulate in large amounts, they can block the proteasome, making it less effective. The lysosome is capable of breaking down larger aggregates in a less specific manner, serving as a backup to prevent the buildup of toxic proteins. Together, these systems work to maintain cellular balance; however, their failure can lead to pathological conditions, such as neurodegeneration

maintaining a healthy proteome and is performed by molecular chaperones; they help proteins achieve their functional conformation [63]. Proteins fold into a unique three-dimensional structure by a not-so-well-understood mechanism. Attaining native conformation is a prerequisite for cellular proteins to perform functions.

The folding of proteins is guided by several chaperones and cofactors. Chaperones and many other stress-regulatory proteins also monitor the cellular *milieu* for any possible protein misfolding event and attempt to prevent further misfolding and aggregation by refolding any misfolded protein [109, 115, 116]. As explained in previous sections, the PN contains multiple proteins that contribute to the stability of the cellular proteome by maintaining their three-dimensional structures [63]. Notably, a well-regulated protein synthesis and degradation is required for maintaining the essential concentration and turnover of all cellular proteins [48]. As stated earlier, UPS consistently maintains cellular proteostasis by degrading the misfolded, old, unused proteins of a cell. However, many of the cellular proteins are degraded by the autosomal-lysosomal pathway (ALP) [76]. Most common substrates of lysosomal degradation are large-sized aggregated proteins that cannot enter into proteasome and hence start accumulating in or outside of the cells [148]. Such bulky protein aggregates, including amyloids, are targeted to ALP for degradation. Before being targeted for degradation, several unfolded/misfolded proteins are refolded back to their native three-dimensional structure. However, under overwhelming conditions, misfolded proteins may not enter into the refolding pathway; instead, they are recognized by chaperones and targeted for autophagic or proteasomal degradation [3, 63]. Such chaperone-mediated degradation pathways are termed chaperone-mediated autophagy (CMA) and chaperone-mediated proteasomal, respectively [59, 61]. Here we are briefly describing important components of PN, other than UPS, that contribute to the maintenance of the healthy proteome.

6.3.1 Molecular Chaperones

Chaperons are involved in protein folding events and are a crucial component of PN. They are classified into multiple protein families based on their size and functions, for example, small heat shock proteins (sHSP), HSP40, HSP60, HSP70, HSP90, and HSP110 [5]. Among these, Hsp70 and Hsp90 are the most well-studied classes playing a major part in protein folding and refolding events. Both Hsp70 and Hsp90 work with co-factors for protein recognition and proper folding of proteins in an ATP-dependent manner. Multiple small HSPs act as co-chaperones and assist the large HSPs in performing the folding process. For example, co-chaperones interrupt the HSP90 activation cycle at multiple points and control the activity/function of HSP90 [116]. These chaperones protect the cells from protein-conformational diseases and maintain a healthy cellular proteome inside the cell [106]. Overall, HSPs act as protein protectants, prevent their denaturation, and help the cell avoid protein misfolding and aggregation [127]. The smaller chaperone proteins of ~12–45 kDa molecular weight, called sHSPs, are ATP-independent and help prevent aggregation of the proteins [93].

Hsp90 plays a crucial role in protein folding, conformation, and stability. The conserved domains are responsible for protein binding and ATPase activity, and they also modulate various downstream processes [30]. Hsp90 ensures both intra- and extracellular proteome stability [51, 119]. The human HSP90 sequence is approximately 50% similar to bacterial high-temperature protein G, containing an evolutionary conserved structure [12, 97]. The highly conserved structure contains an N-terminal nucleotide-binding domain, the middle domain for client binding, and the C-terminal dimerization domain [140]. A highly charged linker region is also present in the conserved domain of eukaryotic Hsp90 and is crucial for its function and flexibility [128]. The Hsp70 family is the other chaperone family essential for folding, refolding, translocation, and degradation of proteins [50]. These chaperones are also conserved and comprise an N-terminal nucleotide-binding domain and a C-terminal substrate-binding domain, which work together to assist in the proper folding of nascent proteins, preventing aggregation of misfolded proteins. Hsp70 family members, such as Hsp72 and Hsc70, are expressed in multiple cellular compartments, performing diverse functions, including correct folding of nascent polypeptides and recovering misfolded proteins by refolding during stress responses. Hsp70 chaperones also facilitate protein transport across membranes and orchestrate targeted protein degradation via specialized pathways [11, 16, 33, 90].

6.3.2 Autosomal-Lysosomal Pathways

Lysosomal degradation is an important cellular process of removing cellular garbage, including old cell organelles, via a regulated mechanism called autophagy [83, 94]. The lysosomes contain several hydrolase enzymes responsible for the degradation of debris, macromolecules, aggregates, and proteins. These are targeted for degradation via multiple pathways. For example, macroautophagy involves lysosomal engulfment of organelles and molecules by membranous separation followed by fusion with the lysosome and digestion by hydrolases. In this most common form of autophagy, the target protein/substrates are first circumvented by an isolation membrane, followed by the formation of a membranous vacuole called autophagosomes [79]. The autophagosomes are labeled by Atg8 (a Ub-like small molecule). Later, autophagosomes combine with lysosomes to form autophagolysosomes [98]. Hydrolases present in the lumen of the lysosome lyse proteins into smaller polypeptides and free amino acids, which are further used for cellular metabolism [29]. Thus, ALP plays an important role in the recycling of cellular macromolecules. In addition, autophagy is crucial for the maintenance of cellular proteostasis, as proteasomal degradation may not be effective for the removal of large protein aggregates [95, 101].

Microautophagy involves direct engulfment of cytosolic material by membranous invagination, which could be selective or non-selective and does not involve autophagosome formation [80]. It occurs in several steps, starting with lysosomal membrane invagination, development of autophagic tubes, vesicle formation, their expansion, scission, and degradation, followed by recycling [73]. Another pathway that can target proteins for selective degradation via autophagy is called CMA. This pathway is responsible for the proteolysis of aberrant proteins containing a pentapeptide KFERQ-like motif [26, 35, 60]. This process occurs in multiple steps: recognition of the target protein by heat shock cognate 70 (Hsc70) protein, attachment of the protein to the outer lysosomal membrane, protein unfolding, and internalization into the lysosome, followed by degradation in the lysosomal lumen [20, 60]. In summary, autophagy is an important player in cellular proteostasis and is regulated by multiple proteins, for example, mammalian target of rapamycin (mTOR). As the name suggest, mTOR signaling pathway is inhibited by rapamycin, which is used as a potent inducer of autophagy [10, 147].

6.4 What If Proteostasis Machinery Fails

Human cells comprise approximately 10,000 unique proteins all the time [71]. To regulate their turnover, a well-coordinated network of proteins consistently works to ensure that a balance between synthesis and degradation is maintained. The PN comprises molecular chaperone machinery to maintain the correct shape of proteins and constantly checks and balances the proteome. Protein's life from synthesis to degradation continuously remains under surveillance by different players of PN. If found disturbed or out of order, the PN machinery attempts to repair proteins and fold them back to their native structure. Failing in doing so, a disturbed proteostasis condition may lead to multiple diseases collectively called proteinopathies, such as cystic fibrosis, type 2 diabetes, eye disorders, and neurological disorders [82, 100]. Aging is one of the driving factors that causes a decline in proteostasis functioning. While aging, a compromised proteostasis machinery may lead to significant negative consequences in cellular function and contribute to various diseases [86, 105]. When the mechanisms responsible for maintaining protein homeostasis fail, one most obvious changes in the cells is the accumulation of damaged or misfolded proteins, resulting in detrimental cellular outcomes [68, 124].

Perturbed proteostasis in aging cells leads to inefficient protein quality control. Therefore, during aging, misfolded proteins can aggregate, leading to cellular dysfunction and increased cytotoxic stress, which can further exacerbate the failure of the PN [46]. The failure of proteostasis is linked to the development of neurodegenerative diseases, such as AD, PD, amyotrophic lateral sclerosis (ALS), frontotemporal dementia (FTD), and cystic fibrosis [133]. The accumulation of misfolded proteins is a hallmark of neurodegeneration, contributing to neuronal dysfunction and death. For example, AD patient brains are characterized by the accumulation of two proteins: amyloid-beta and tau in plaques and tangles, respectively, both of which disrupt normal cellular functions and lead to cognitive decline [67]. Cells that suffer from proteostasis failure exhibit reduced viability due to the stress caused by accumulating toxic protein aggregates [136]. This stress can trigger a myriad of cellular dysfunctions, ultimately resulting in the initiation of cell death pathways and degeneration, particularly in tissues that are highly dependent on protein homeostasis, such as neurons [27]. Over time, the diminished ability to manage misfolded proteins contributes to the overall decline in cellular and tissue functions associated with aging.

Dysfunctional UPS and proteostasis are linked to various diseases, as failed proteostasis cannot remove unwanted protein aggregates, and these toxic protein species start accumulating in or outside the cell, which is a common cause of various neurodegenerative diseases [145]. When PN fails to remove the accumulated protein inclusions, it starts generating enormous proteotoxicity and deteriorating the cell's normal functions [13]. Here, in Table 6.1, we are enlisting the most common proteinopathies caused due to proteostasis failure and abnormal accumulation of proteins.

Many cancers have been found to be affected by the crosstalk of UPS and other PN components. Major changes occurring in tumor microenvironment and variety of cancers include, but are not limited to, activation of oncogenes, changes in expression of tumor suppressor genes, metabolic reprogramming, and hypoxia [21, 137]. Concentration and functioning of many of the genes responsible for causing these changes are regulated by targeted proteasomal degradation [42, 56, 122]. Activation of the UPS is one of the first-line responses inside a cell against the

Table 6.1 Most common pathological conditions caused due to proteostasis failure

Name of the disease	Genes/proteins	Primary target organs	References
AD	APP, MAPT, ApoE	Brain, nervous system	Cozachenco et al. [19]
Tauopathies	MAPT	Brain, nervous system	Papanikolopoulou and Skoulakis [103]
PD	SNCA, Parkin	Brain, nervous system	Kulkarni et al. [72]
Synucleinopathies	SNCA	Brain, nervous system	Santos and Outeiro [112]
Prion diseases	Prion protein	Brain, nervous system	Thellung et al. [125]
HD	Huntingtin	Brain, nervous system	Harding and Tong [39]
Spinocerebeller ataxias	Ataxin proteins	Brain, nervous system	Klockgether et al. [66]
FTD	C9orf72	Brain, nervous system	Mahali et al. [89]
ALS	SOD1, TDP43	Brain, nervous system	Lambert-Smith et al. [77]
Retinitis pigmentosa	Rhodopsin	Eyes	Saliba et al. [111]
Cystic fibrosis	CFTR	Lungs	McDonald et al. [91]
Diabetes mellitus	IAPP	Pancreas	Jaisson and Gillery [53]
Collagenopathies	COL1A1, COL1A2	Bones	Wong and Shoulders [142]
Cardiac diseases	Transthyretin, desmin	Heart	Wiersma et al. [141]

unwanted accumulation and generation of proteotoxicity [102]. Under extreme stress conditions, UPS and other PN components may not suffice the cellular requirements, leading to activation of the apoptosis machinery that helps avoid the spread of many detrimental changes [52, 108]. Among many, p53, Rap1GAP, p27, NF-κB or TGF-β receptor, Myc, and so on are the most common tumor-associated proteins regulated by targeted proteasomal degradation [34, 54, 135]. In addition to maintaining the cellular proteostasis, UPS involvement has been studied in many other physiological pathways, for example, DNA repair, cell cycle control, metabolism, signaling, cellular transport, and so on [92].

6.5 Therapeutic Benefits of UPS

Understanding the mechanisms behind proteostasis failure has led to the identification of potential therapeutic targets. Interventions aimed at enhancing the function of the PN or alleviating the burden of misfolded proteins could help in the prevention or treatment of age-associated diseases. Therapeutic modulation of proteasome-mediated degradation has gained significant attention in recent years, particularly due to its implications in various diseases, especially cancer, neurodegeneration, and aging [130]. Bortezomib and many other proteasome inhibitory drugs are already in clinics. Many others are under clinical trials. Not only the proteasome, but also other components of UPS machinery are used as promising therapeutic targets. For example, several inhibitors of E1 activating enzymes, like, Largazole, PYR-41, and TAK243, etc. have been identified and have shown potential benefits under experimental conditions [8, 143]. Similarly, multiple drugs, such as CC0651, NSC697923, Leucettamol A, IJ-5, and so on, have shown inhibitory potential against various E2 conjugating enzymes and have shown symptomatic benefits against inflammation, melanoma, and other cancers [28, 120].

More importantly, E3 ligases that provide high specificity to the system are also targeted for therapeutic benefits. Inhibitors of Mdm2 (nutlin-3a, APG-115, navtemadlin, Siremadlin, and alrizomadlin) are used against myelodysplastic syndrome and cancers, while inhibition of CRBN by thalidomide and Skp2 by lovastatin shows therapeutic benefits against the growth of multiple cancers [134, 138, 144, 146]. Many studies have shown that modulating DUB activity may further help in targeting several pathological conditions. Drugs such as ML323, 15-oxospiramilactone, pimozide, vialinin A, and mitoxantrone ameliorate symptoms in conditions such as inflammation, Tourette's syndrome, cancer, and neurodegeneration [4, 40]. Discoveries of all these small molecules have presented many clinically proven therapeutic benefits and have paved the way for multiple ongoing clinical trials for diseases and pathological conditions.

6.6 Conclusion

The proteasome plays a significant role in maintaining cellular proteostasis by degrading misfolded, damaged, or unneeded proteins. Research is ongoing to explore small molecules and other therapeutic strategies that may improve proteostasis and extend healthy lifespan by targeting the underlying causes of proteostasis decline. Overall, exploitation of UPS and its components has shown tremendous therapeutic benefits. Modulation of the proteasomal degradation pathway presents a promising avenue for treating a wide spectrum of diseases by leveraging its fundamental role in protein homeostasis and cellular regulation. Failure of proteostasis machinery is a critical factor in aging and the onset of various diseases. It leads to protein misfolding, the development of neurodegenerative conditions, and reduced cellular viability. Many FDA-approved drugs and multiple ongoing clinical trials have inspired therapeutic research aimed at enhancing protein homeostasis mechanisms. Understanding these processes is vital for developing approaches to mitigate the impacts of aging and related pathologies.

Acknowledgments S.S. thanks the Indian Institute of Technology Bhilai for providing a junior research fellowship. A.N.T. thanks the Department of Biotechnology, Government of India, for providing him junior research fellowship. A.U. thanks the Indian Institute of Technology Bhilai for providing the Research Initiation Grant (IITBH/R&D/2024-25/386).

Conflict of Interest The authors have no conflict of interest.

References

1. Alagar Boopathy LR, Jacob-Tomas S, Alecki C, Vera M. Mechanisms tailoring the expression of heat shock proteins to proteostasis challenges. J Biol Chem. 2022;298:101796.
2. Aman Y, Schmauck-Medina T, Hansen M, Morimoto RI, Simon AK, et al. Autophagy in healthy aging and disease. Nat Aging. 2021;1:634–50.
3. Amanullah A, Upadhyay A, Joshi V, Mishra R, Jana NR, Mishra A. Progressing neurobiological strategies against proteostasis failure: challenges in neurodegeneration. Prog Neurobiol. 2017;159:1–38.
4. Antao AM, Tyagi A, Kim KS, Ramakrishna S. Advances in deubiquitinating enzyme inhibition and applications in cancer therapeutics. Cancers (Basel). 2020;12:1579.
5. Bakthisaran R, Tangirala R, Rao CM. Small heat shock proteins: role in cellular functions and pathology. Biochim Biophys Acta. 2015;1854:291–319.
6. Balch WE, Morimoto RI, Dillin A, Kelly JW. Adapting proteostasis for disease intervention. Science. 2008;319:916–9.
7. Bard JAM, Goodall EA, Greene ER, Jonsson E, Dong KC, Martin A. Structure and function of the 26S proteasome. Annu Rev Biochem. 2018;87:697–724.
8. Barghout SH, Schimmer AD. E1 enzymes as therapeutic targets in cancer. Pharmacol Rev. 2021;73:1–58.
9. Bar-Nun S, Glickman MH. Proteasomal AAA-ATPases: structure and function. Biochim Biophys Acta. 2012;1823:67–82.
10. Benjamin D, Colombi M, Moroni C, Hall MN. Rapamycin passes the torch: a new generation of mTOR inhibitors. Nat Rev Drug Discov. 2011;10:868–80.

11. Bercovich B, Stancovski I, Mayer A, Blumenfeld N, Laszlo A, et al. Ubiquitin-dependent degradation of certain protein substrates in vitro requires the molecular chaperone Hsc70. J Biol Chem. 1997;272:9002–10.
12. Biebl MM, Buchner J. Structure, function, and regulation of the Hsp90 machinery. Cold Spring Harb Perspect Biol. 2019;11(9):a034017.
13. Brehme M, Sverchkova A, Voisine C. Proteostasis network deregulation signatures as biomarkers for pharmacological disease intervention. Curr Opin Syst Biol. 2019;15:74–81.
14. Britton M, Lucas MM, Downey SL, Screen M, Pletnev AA, et al. Selective inhibitor of proteasome's caspase-like sites sensitizes cells to specific inhibition of chymotrypsin-like sites. Chem Biol. 2009;16:1278–89.
15. Buetow L, Huang DT. Structural insights into the catalysis and regulation of E3 ubiquitin ligases. Nat Rev Mol Cell Biol. 2016;17:626–42.
16. Chiang H-L, Terlecky SR, Plant CP, Dice JF. A role for a 70-kilodalton heat shock protein in lysosomal degradation of intracellular proteins. Science. 1989;246:382–5.
17. Ciechanover A, Brundin P. The ubiquitin proteasome system in neurodegenerative diseases: sometimes the chicken, sometimes the egg. Neuron. 2003;40:427–46.
18. Ciechanover A, Schwartz AL. The ubiquitin-proteasome pathway: the complexity and myriad functions of proteins death. Proc Natl Acad Sci USA. 1998;95:2727–30.
19. Cozachenco D, Ribeiro FC, Ferreira ST. Defective proteostasis in Alzheimer's disease. Ageing Res Rev. 2023;85:101862.
20. Cuervo AM. Chaperone-mediated autophagy: selectivity pays off. Trends Endocrinol Metab. 2010;21:142–50.
21. Dakal TC, Dhabhai B, Pant A, Moar K, Chaudhary K, et al. Oncogenes and tumor suppressor genes: functions and roles in cancers. MedComm (2020). 2024;5:e582.
22. Damgaard RB. The ubiquitin system: from cell signalling to disease biology and new therapeutic opportunities. Cell Death Differ. 2021;28:423–6.
23. Dang F, Nie L, Wei W. Ubiquitin signaling in cell cycle control and tumorigenesis. Cell Death Differ. 2021;28:427–38.
24. Dantuma NP, Bott LC. The ubiquitin-proteasome system in neurodegenerative diseases: precipitating factor, yet part of the solution. Front Mol Neurosci. 2014;7:70.
25. Deol KK, Crowe SO, Du J, Bisbee HA, Guenette RG, Strieter ER. Proteasome-bound UCH37/UCHL5 debranches ubiquitin chains to promote degradation. Mol Cell. 2020;80:796–809.e9.
26. Dice JF. Peptide sequences that target cytosolic proteins for lysosomal proteolysis. Trends Biochem Sci. 1990;15:305–9.
27. Douglas PM, Dillin A. Protein homeostasis and aging in neurodegeneration. J Cell Biol. 2010;190:719–29.
28. Du X, Song H, Shen N, Hua R, Yang G. The molecular basis of ubiquitin-conjugating enzymes (E2s) as a potential target for cancer therapy. Int J Mol Sci. 2021;22(7):3440.
29. Duan X, Tong Q, Fu C, Chen L. Lysosome-targeted fluorescent probes: design mechanism and biological applications. Bioorg Chem. 2023;140:106832.
30. Eckl JM, Daake M, Schwartz S, Richter K. Nucleotide-free sB-Raf is preferentially bound by Hsp90 and Cdc37 in vitro. J Mol Biol. 2016;428:4185–96.
31. Ehlinger A, Walters KJ. Structural insights into proteasome activation by the 19S regulatory particle. Biochemistry. 2013;52:3618–28.
32. Escobar-Henriques M, Altin S, Brave FD. Interplay between the ubiquitin proteasome system and mitochondria for protein homeostasis. Curr Issues Mol Biol. 2020;35:35–58.
33. Evans CG, Chang L, Gestwicki JE. Heat shock protein 70 (hsp70) as an emerging drug target. J Med Chem. 2010;53:4585–602.
34. Farrell AS, Sears RC. MYC degradation. Cold Spring Harb Perspect Med. 2014;4(3):a014365.
35. Feder ME, Hofmann GE. Heat-shock proteins, molecular chaperones, and the stress response: evolutionary and ecological physiology. Annu Rev Physiol. 1999;61:243–82.

36. Gadhave K, Bolshette N, Ahire A, Pardeshi R, Thakur K, et al. The ubiquitin proteasomal system: a potential target for the management of Alzheimer's disease. J Cell Mol Med. 2016;20:1392–407.
37. Glickman MH, Ciechanover A. The ubiquitin-proteasome proteolytic pathway: destruction for the sake of construction. Physiol Rev. 2002;82:373–428.
38. Groll M, Ditzel L, Löwe J, Stock D, Bochtler M, et al. Structure of 20S proteasome from yeast at 2.4 A resolution. Nature. 1997;386:463–71.
39. Harding RJ, Tong Y-f. Proteostasis in Huntington's disease: disease mechanisms and therapeutic opportunities. Acta Pharmacol Sin. 2018;39:754–69.
40. Harrigan JA, Jacq X, Martin NM, Jackson SP. Deubiquitylating enzymes and drug discovery: emerging opportunities. Nat Rev Drug Discov. 2018;17:57–78.
41. Hartl FU, Bracher A, Hayer-Hartl M. Molecular chaperones in protein folding and proteostasis. Nature. 2011;475:324–32.
42. Haupt Y, Maya R, Kazaz A, Oren M. Mdm2 promotes the rapid degradation of p53. Nature. 1997;387:296–9.
43. Hershko A, Ciechanover A. The ubiquitin system. Annu Rev Biochem. 1998;67:425–79.
44. Hetz C, Papa FR. The unfolded protein response and cell fate control. Mol Cell. 2018;69:169–81.
45. Hetz C, Chevet E, Oakes SA. Proteostasis control by the unfolded protein response. Nat Cell Biol. 2015;17:829–38.
46. Hipp MS, Park S-H, Hartl FU. Proteostasis impairment in protein-misfolding and-aggregation diseases. Trends Cell Biol. 2014;24:506–14.
47. Hipp MS, Kasturi P, Hartl FU. The proteostasis network and its decline in ageing. Nat Rev Mol Cell Biol. 2019;20:421–35.
48. Hirano H, Kimura Y, Kimura A. Biological significance of co- and post-translational modifications of the yeast 26S proteasome. J Proteome. 2016;134:37–46.
49. Hotamisligil GS, Davis RJ. Cell signaling and stress responses. Cold Spring Harb Perspect Biol. 2016;8(10):a006072.
50. Hu C, Yang J, Qi Z, Wu H, Wang B, et al. Heat shock proteins: biological functions, pathological roles, and therapeutic opportunities. MedComm (2020). 2022;3:e161.
51. Hunter MC, O'Hagan KL, Kenyon A, Dhanani KCH, Prinsloo E, Edkins AL. Hsp90 binds directly to fibronectin (FN) and inhibition reduces the extracellular fibronectin matrix in breast cancer cells. PLoS One. 2014;9:e86842.
52. Iuliano L, Dalla E, Picco R, Mallavarapu S, Minisini M, et al. Proteotoxic stress-induced apoptosis in cancer cells: understanding the susceptibility and enhancing the potency. Cell Death Discov. 2022;8:407.
53. Jaisson S, Gillery P. Impaired proteostasis: role in the pathogenesis of diabetes mellitus. Diabetologia. 2014;57:1517–27.
54. Jang HH. Regulation of protein degradation by proteasomes in cancer. J Cancer Prev. 2018;23:153–61.
55. Jayaraj GG, Hipp MS, Hartl FU. Functional modules of the proteostasis network. Cold Spring Harb Perspect Biol. 2020;12(1):a033951.
56. Joshi V, Upadhyay A, Chhangani D, Amanullah A, Sharan RN, Mishra A. Gp78 involvement in cellular proliferation: can act as a promising modulator for cell cycle regulatory proteins? J Cell Physiol. 2018;233:6352–68.
57. Joshi V, Upadhyay A, Prajapati VK, Mishra A. How autophagy can restore proteostasis defects in multiple diseases? Med Res Rev. 2020;40:1385–439.
58. Kandel R, Jung J, Neal S. Proteotoxic stress and the ubiquitin proteasome system. Semin Cell Dev Biol. 2024;156:107–20.
59. Kaushik S, Cuervo AM. The coming of age of chaperone-mediated autophagy. Nat Rev Mol Cell Biol. 2018;19:365–81.
60. Kaushik S, Bandyopadhyay U, Sridhar S, Kiffin R, Martinez-Vicente M, et al. Chaperone-mediated autophagy at a glance. J Cell Sci. 2011;124:495–9.

61. Kettern N, Dreiseidler M, Tawo R, Höhfeld J. Chaperone-assisted degradation: multiple paths to destruction. Biol Chem. 2010;391:481–9.
62. Kikuchi J, Iwafune Y, Akiyama T, Okayama A, Nakamura H, et al. Co- and post-translational modifications of the 26S proteasome in yeast. Proteomics. 2010;10:2769–79.
63. Kim YE, Hipp MS, Bracher A, Hayer-Hartl M, Hartl FU. Molecular chaperone functions in protein folding and proteostasis. Annu Rev Biochem. 2013;82:323–55.
64. Kim Y, Kim EK, Chey Y, Song MJ, Jang HH. Targeted protein degradation: principles and applications of the proteasome. Cells. 2023;12:1846.
65. Klaips CL, Jayaraj GG, Hartl FU. Pathways of cellular proteostasis in aging and disease. J Cell Biol. 2018;217:51–63.
66. Klockgether T, Mariotti C, Paulson HL. Spinocerebellar ataxia. Nat Rev Dis Prim. 2019;5:24.
67. Knopman DS, Amieva H, Petersen RC, Chételat G, Holtzman DM, et al. Alzheimer disease. Nat Rev Dis Prim. 2021;7:33.
68. Koga H, Kaushik S, Cuervo AM. Protein homeostasis and aging: the importance of exquisite quality control. Ageing Res Rev. 2011;10:205–15.
69. Köhler A, Cascio P, Leggett DS, Woo KM, Goldberg AL, Finley D. The axial channel of the proteasome core particle is gated by the Rpt2 ATPase and controls both substrate entry and product release. Mol Cell. 2001;7:1143–52.
70. Komander D, Rape M. The ubiquitin code. Annu Rev Biochem. 2012;81:203–29.
71. Kulak NA, Geyer PE, Mann M. Loss-less nano-fractionator for high sensitivity, high coverage proteomics. Mol Cell Proteomics. 2017;16:694–705.
72. Kulkarni A, Preeti K, Tryphena KP, Srivastava S, Singh SB, Khatri DK. Proteostasis in Parkinson's disease: recent development and possible implication in diagnosis and therapeutics. Ageing Res Rev. 2023;84:101816.
73. Kunz JB, Schwarz H, Mayer A. Determination of four sequential stages during microautophagy in vitro. J Biol Chem. 2004;279:9987–96.
74. Kurtishi A, Rosen B, Patil KS, Alves GW, Møller SG. Cellular proteostasis in neurodegeneration. Mol Neurobiol. 2019;56(5):3676–89.
75. Labbadia J, Morimoto RI. The biology of proteostasis in aging and disease. Annu Rev Biochem. 2015;84:435–64.
76. Lamb CA, Yoshimori T, Tooze SA. The autophagosome: origins unknown, biogenesis complex. Nat Rev Mol Cell Biol. 2013;14:759–74.
77. Lambert-Smith IA, Saunders DN, Yerbury JJ. Proteostasis impairment and ALS. Prog Biophys Mol Biol. 2022;174:3–27.
78. Lander GC, Estrin E, Matyskiela ME, Bashore C, Nogales E, Martin A. Complete subunit architecture of the proteasome regulatory particle. Nature. 2012;482:186–91.
79. Levine B, Kroemer G. Autophagy in the pathogenesis of disease. Cell. 2008;132:27–42.
80. Li WW, Li J, Bao JK. Microautophagy: lesser-known self-eating. Cell Mol Life Sci. 2012;69:1125–36.
81. Li Y, Li S, Wu H. Ubiquitination-proteasome system (UPS) and autophagy two main protein degradation machineries in response to cell stress. Cells. 2022;11(5):851.
82. Lim SM, Nahm M, Kim SH. Proteostasis and ribostasis impairment as common cell death mechanisms in neurodegenerative diseases. J Clin Neurol. 2023;19:101.
83. Liu S, Yao S, Yang H, Liu S, Wang Y. Autophagy: regulator of cell death. Cell Death Dis. 2023;14:648.
84. Livnat-Levanon N, Kevei É, Kleifeld O, Krutauz D, Segref A, et al. Reversible 26S proteasome disassembly upon mitochondrial stress. Cell Rep. 2014;7:1371–80.
85. Llewellyn J, Hubbard SJ, Swift J. Translation is an emerging constraint on protein homeostasis in ageing. Trends Cell Biol. 2024;34:646–56.
86. López-Otín C, Blasco MA, Partridge L, Serrano M, Kroemer G. Hallmarks of aging: an expanding universe. Cell. 2023;186:243–78.
87. Löwe J, Stock D, Jap B, Zwickl P, Baumeister W, Huber R. Crystal structure of the 20S proteasome from the archaeon T. acidophilum at 3.4 A resolution. Science. 1995;268:533–9.

88. Lualdi M, Alberio T, Fasano M. Proteostasis and proteotoxicity in the network medicine era. Int J Mol Sci. 2020;21:6405.
89. Mahali S, Martinez R, King M, Verbeck A, Harari O, et al. Defective proteostasis in induced pluripotent stem cell models of frontotemporal lobar degeneration. Transl Psychiatry. 2022;12:508.
90. Mayer M, Bukau B. Hsp70 chaperones: cellular functions and molecular mechanism. Cell Mol Life Sci. 2005;62:670–84.
91. McDonald EF, Meiler J, Plate L. CFTR folding: from structure and proteostasis to cystic fibrosis personalized medicine. ACS Chem Biol. 2023;18:2128–43.
92. Mishra R, Upadhyay A, Prajapati VK, Mishra A. Proteasome-mediated proteostasis: novel medicinal and pharmacological strategies for diseases. Med Res Rev. 2018;38:1916–73.
93. Mitra R, Gadkari VV, Meinen BA, van Mierlo CPM, Ruotolo BT, Bardwell JCA. Mechanism of the small ATP-independent chaperone Spy is substrate specific. Nat Commun. 2021;12:851.
94. Mizushima N, Komatsu M. Autophagy: renovation of cells and tissues. Cell. 2011;147:728–41.
95. Monaco A, Fraldi A. Protein aggregation and dysfunction of autophagy-lysosomal pathway: a vicious cycle in lysosomal storage diseases. Front Mol Neurosci. 2020;13:37.
96. Montagut AM, Armengol M, de Pablo GG, Estrada-Tejedor R, Borrell JI, Roué G. Recent advances in the pharmacological targeting of ubiquitin-regulating enzymes in cancer. Semin Cell Dev Biol. 2022;132:213–29.
97. Nakamoto H, Fujita K, Ohtaki A, Watanabe S, Narumi S, et al. Physical interaction between bacterial heat shock protein (Hsp) 90 and Hsp70 chaperones mediates their cooperative action to refold denatured proteins. J Biol Chem. 2014;289:6110–9.
98. Nakamura S, Yoshimori T. New insights into autophagosome-lysosome fusion. J Cell Sci. 2017;130:1209–16.
99. Nandi D, Tahiliani P, Kumar A, Chandu D. The ubiquitin-proteasome system. J Biosci. 2006;31:137–55.
100. Osinalde N, Duarri A, Ramirez J, Barrio R, Perez de Nanclares G, Mayor U. Impaired proteostasis in rare neurological diseases. Semin Cell Dev Biol. 2019;93:164–77.
101. Palmer JE, Wilson N, Son SM, Obrocki P, Wrobel L, et al. Autophagy, aging, and age-related neurodegeneration. Neuron. 2025;113(1):29–48.
102. Papaevgeniou N, Chondrogianni N. UPS activation in the battle against aging and aggregation-related diseases: an extended review. In: Matthiesen R, editor. Proteostasis: methods and protocols. New York: Springer New York; 2016. p. 1–70.
103. Papanikolopoulou K, Skoulakis EM. Altered proteostasis in neurodegenerative tauopathies. In: Proteostasis and disease: from basic mechanisms to clinics. Cham: Springer; 2020. p. 177–94.
104. Powers ET, Balch WE. Diversity in the origins of proteostasis networks –a driver for protein function in evolution. Nat Rev Mol Cell Biol. 2013;14:237–48.
105. Rao NR, Upadhyay A, Savas JN. Derailed protein turnover in the aging mammalian brain. Mol Syst Biol. 2024;20:120–39.
106. Rosenzweig R, Nillegoda NB, Mayer MP, Bukau B. The Hsp70 chaperone network. Nat Rev Mol Cell Biol. 2019;20:665–80.
107. Roth DM, Balch WE. Modeling general proteostasis: proteome balance in health and disease. Curr Opin Cell Biol. 2011;23:126–34.
108. Ruz C, Alcantud JL, Vives Montero F, Duran R, Bandres-Ciga S. Proteotoxicity and neurodegenerative diseases. Int J Mol Sci. 2020;21(16):5646.
109. Saibil H. Chaperone machines for protein folding, unfolding and disaggregation. Nat Rev Mol Cell Biol. 2013;14:630–42.
110. Sala AJ, Bott LC, Morimoto RI. Shaping proteostasis at the cellular, tissue, and organismal level. J Cell Biol. 2017;216:1231–41.
111. Saliba RS, Munro PMG, Luthert PJ, Cheetham ME. The cellular fate of mutant rhodopsin: quality control, degradation and aggresome formation. J Cell Sci. 2002;115:2907–18.
112. Santos PI, Outeiro TF. Protein trapping leads to altered synaptic proteostasis in synucleinopathies. FEBS J. 2020;287:5294–303.

113. Santra M, Dill KA, de Graff AMR. Proteostasis collapse is a driver of cell aging and death. Proc Natl Acad Sci USA. 2019;116:22173–8.
114. Saric T, Graef CI, Goldberg AL. Pathway for degradation of peptides generated by proteasomes: a key role for thimet oligopeptidase and other metallopeptidases*. J Biol Chem. 2004;279:46723–32.
115. Schlieker C, Bukau B, Mogk A. Prevention and reversion of protein aggregation by molecular chaperones in the E. coli cytosol: implications for their applicability in biotechnology. J Biotechnol. 2002;96:13–21.
116. Schopf FH, Biebl MM, Buchner J. The HSP90 chaperone machinery. Nat Rev Mol Cell Biol. 2017;18:345–60.
117. Schreiner P, Chen X, Husnjak K, Randles L, Zhang N, et al. Ubiquitin docking at the proteasome through a novel pleckstrin-homology domain interaction. Nature. 2008;453:548–52.
118. Sharon M, Taverner T, Ambroggio XI, Deshaies RJ, Robinson CV. Structural organization of the 19S proteasome lid: insights from MS of intact complexes. PLoS Biol. 2006;4:e267.
119. Somogyvári M, Khatatneh S, Sőti C. Hsp90: from cellular to organismal proteostasis. Cells. 2022;11(16):2479.
120. Stewart MD, Ritterhoff T, Klevit RE, Brzovic PS. E2 enzymes: more than just middle men. Cell Res. 2016;26:423–40.
121. Suraweera A, Münch C, Hanssum A, Bertolotti A. Failure of amino acid homeostasis causes cell death following proteasome inhibition. Mol Cell. 2012;48:242–53.
122. Talis AL, Huibregtse JM, Howley PM. The role of E6AP in the regulation of p53 protein levels in human papillomavirus (HPV)-positive and HPV-negative cells. J Biol Chem. 1998;273:6439–45.
123. Tanaka K. The proteasome: overview of structure and functions. Proc Jpn Acad Ser B Phys Biol Sci. 2009;85:12–36.
124. Taylor RC, Dillin A. Aging as an event of proteostasis collapse. Cold Spring Harb Perspect Biol. 2011;3(5):a004440.
125. Thellung S, Corsaro A, Dellacasagrande I, Nizzari M, Zambito M, Florio T. Proteostasis unbalance in prion diseases: mechanisms of neurodegeneration and therapeutic targets. Front Neurosci. 2022;16:966019.
126. Tomko RJ Jr, Funakoshi M, Schneider K, Wang J, Hochstrasser M. Heterohexameric ring arrangement of the eukaryotic proteasomal ATPases: implications for proteasome structure and assembly. Mol Cell. 2010;38:393–403.
127. Treweek TM, Meehan S, Ecroyd H, Carver JA. Small heat-shock proteins: important players in regulating cellular proteostasis. Cell Mol Life Sci. 2015;72:429–51.
128. Tsutsumi S, Mollapour M, Graf C, Lee CT, Scroggins BT, et al. Hsp90 charged-linker truncation reverses the functional consequences of weakened hydrophobic contacts in the N domain. Nat Struct Mol Biol. 2009;16:1141–7.
129. Upadhyay A. Structure of proteins: evolution with unsolved mysteries. Prog Biophys Mol Biol. 2019;149:160–72.
130. Upadhyay A. Natural compounds in the regulation of proteostatic pathways: an invincible artillery against stress, ageing, and diseases. Acta Pharm Sin B. 2021;11:2995–3014.
131. Upadhyay A, Joshi V. The ubiquitin tale: current strategies and future challenges. ACS Pharmacol Transl Sci. 2024;7:2573–87.
132. Upadhyay A, Amanullah A, Chhangani D, Mishra R, Mishra A. Selective multifaceted E3 ubiquitin ligases barricade extreme defense: potential therapeutic targets for neurodegeneration and ageing. Ageing Res Rev. 2015;24:138–59.
133. Upadhyay A, Sundaria N, Dhiman R, Prajapati VK, Prasad A, Mishra A. Complex inclusion bodies and defective proteome hubs in neurodegenerative disease: new clues, new challenges. Neuroscientist. 2022;28:271–82.
134. Vosper J, Masuccio A, Kullmann M, Ploner C, Geley S, Hengst L. Statin-induced depletion of geranylgeranyl pyrophosphate inhibits cell proliferation by a novel pathway of Skp2 degradation. Oncotarget. 2015;6:2889–902.

135. Wang Y, Xie Y, Sun B, Guo Y, Song L, et al. The degradation of Rap1GAP via E6AP-mediated ubiquitin-proteasome pathway is associated with HPV16/18-infection in cervical cancer cells. Infect Agent Cancer. 2021;16:71.
136. Wang M, Law ME, Law BK. Chapter 5: Proteotoxicity and endoplasmic reticulum stress-mediated cell death. In: Liao D, editor. Mechanisms of cell death and opportunities for therapeutic development. Academic Press; 2022. p. 119–74.
137. Wang Q, Shao X, Zhang Y, Zhu M, Wang FXC, et al. Role of tumor microenvironment in cancer progression and therapeutic strategy. Cancer Med. 2023;12:11149–65.
138. Wang W, Albadari N, Du Y, Fowler JF, Sang HT, et al. MDM2 inhibitors for cancer therapy: the past, present, and future. Pharmacol Rev. 2024;76:414–53.
139. Wellen KE, Thompson CB. Cellular metabolic stress: considering how cells respond to nutrient excess. Mol Cell. 2010;40:323–32.
140. Whitesell L, Lindquist SL. HSP90 and the chaperoning of cancer. Nat Rev Cancer. 2005;5:761–72.
141. Wiersma M, Henning RH, Brundel BJ. Derailed proteostasis as a determinant of cardiac aging. Can J Cardiol. 2016;32:1166. e11-66. e20
142. Wong MY, Shoulders MD. Targeting defective proteostasis in the collagenopathies. Curr Opin Chem Biol. 2019;50:80–8.
143. Xu W, Lukkarila JL, da Silva SR, Paiva SL, Gunning PT, Schimmer AD. Targeting the ubiquitin E1 as a novel anti-cancer strategy. Curr Pharm Des. 2013;19:3201–9.
144. Yamamoto J, Ito T, Yamaguchi Y, Handa H. Discovery of CRBN as a target of thalidomide: a breakthrough for progress in the development of protein degraders. Chem Soc Rev. 2022;51:6234–50.
145. Yerbury JJ, Ooi L, Dillin A, Saunders DN, Hatters DM, et al. Walking the tightrope: proteostasis and neurodegenerative disease. J Neurochem. 2016;137:489–505.
146. Zeng Q, Zeng S, Dai X, Ding Y, Huang C, et al. MDM2 inhibitors in cancer immunotherapy: current status and perspective. Genes Dis. 2024;11:101279.
147. Zhang Y, Nicholatos J, Dreier JR, Ricoult SJ, Widenmaier SB, et al. Coordinated regulation of protein synthesis and degradation by mTORC1. Nature. 2014;513:440–3.
148. Zhao L, Zhao J, Zhong K, Tong A, Jia D. Targeted protein degradation: mechanisms, strategies and application. Signal Transduct Target Ther. 2022;7:113.

Functional Coupling of the Endoplasmic Reticulum and Ubiquitin Proteasome System

7

Sebabrata Maity and Oishee Chakrabarti

7.1 Introduction

Protein homeostasis, or proteostasis, is a cellular process that maintains protein conformation, concentration, and subcellular localization [5]. It encompasses a diverse array of pathways that regulate protein synthesis, folding, transport, and degradation [64, 74].

Endoplasmic reticulum (ER) is the largest membrane-bound organelle and plays a pivotal role in orchestrating crucial and diverse cellular functions. This primarily includes accurate folding of proteins destined for secretion or localization at membranes; it also actively participates in lipid synthesis and in calcium storage [95, 149]. ER homeostasis is regulated through a highly preserved network of signaling cascades, collectively termed the unfolded protein response (UPR). The mechanisms by which the UPR detects perturbations in ER homeostasis have been a focus of sustained investigation for several decades. The ER is turned over constitutively to maintain its function. It can be more active under certain stress conditions [87, 107, 116, 125, 173]. ER quality control relies mainly on two mechanisms: the ubiquitin-proteasome system (UPS) and autophagy (specifically termed "ER-phagy").

Roughly 30% of the eukaryotic proteome is orchestrated at the ER. Post-synthesis, these proteins get entered into the ER in an unfolded state through the protein translocation complex [154]. To attain active form confirmation and translocate to their native destinations, newly synthesized proteins of the secretory pathway need to be folded into their desired tertiary or quaternary structures. Most of these proteins typically endure post-translational modifications, such as disulfide

S. Maity (✉) · O. Chakrabarti (✉)
Biophysics and Structural Genomics Division, Saha Institute of Nuclear Physics, Kolkata, West Bengal, India

Homi Bhabha National Institute, Mumbai, Maharashtra, India
e-mail: sebabrata.maity@saha.ac.in; oishee.chakrabarti@saha.ac.in

S. Singh (ed.), *Ubiquitin Proteasome System in CNS Neurodegenerative Diseases*, https://doi.org/10.1007/978-981-95-6099-8_7

bond formation, glycosylation, proline isomerization, and lipidation, to ensure proper folding and function [16, 32]. Protein folding and assembly is a complex process [95]. This involves multiple chaperones and enzymes for stabilizing folding intermediates and to minimize energy barriers [9, 95]. However, nearly a third of all translocated proteins fail to fold correctly [95, 165]. Thus, the ER implements stringent measures to maintain cellular protein quality. Misfolded protein intermediates undergo either refolding or are directed toward a process called ER-associated degradation (ERAD) for removal. This process allows for the exclusive transit of correctly folded proteins through the secretory pathway [6, 95, 165]. Constitutive autophagy also plays a crucial role in cellular homeostasis by orchestrating the clearance of cytosolic proteins primarily through basal macroautophagy. This mechanism prevents accumulation of misfolded or aggregated proteins, thereby preventing cellular dysfunction and contributing to disease prevention [57, 66, 102].

The UPS is one of the key cellular disposal machineries, which degrades protein substrates post-translationally modified with ubiquitin through a cascade of enzymatic reactions, in a process called ubiquitylation. UPS is the major system for the degradation of cytosolic, nuclear, and transmembrane proteins [91].

In recent years, studies have increasingly highlighted the significance of the UPS in regulating intracellular protein levels and function. This biological process, primarily involving ubiquitin and its activating and conjugating enzymes, is crucial for maintaining cellular and organellar homeostasis. By targeting proteins for degradation, the UPS ensures homeostasis and functioning of various cellular components [166]. UPS is also known as one of the key regulators of neural development and maintains brain structure and physiology [144]. The terminally differentiated cells, neurons, harbor numerous components of the UPS at their synapses, thereby plausibly influencing synaptic function [181].

7.2 ER Function

Intracellular protein homeostasis is critical for the health of the organism. Therefore, it is essential to control the balance between protein synthesis and its breakdown. Proteins, either intracellular or secreted, need to be synthesized and maintained within a threshold that ensures smooth execution of cellular processes and maintains intra- and intercellular homeostasis. The proper functionality of newly synthesized proteins necessitates accurate folding into their native three-dimensional conformations, frequently involving a series of post-translational modifications, for precise intracellular trafficking to their designated subcellular locations. They also need to be degraded efficiently as and when needed.

The ER comprises a large network of sheets and tubules responsible for the biogenesis and conformational maturation of secreted, transmembrane, and ER-resident proteins. These nascent proteins are modified to fold into their proper structures by ER-resident enzymes and chaperones [8, 20, 24, 25, 161]—enzymatic catalysis by peptidyl prolyl isomerases potentiates the folding process, while glycosidases and mannosidases mediate glycosylation, a post-translational modification. Further, optimal concentrations of calcium ions, glucose, and ATP, along with a suitable redox potential, are crucial for protein maturation.

ER quality control mechanisms rely on the recognition of glycan moieties associated with nascent polypeptides. ER-resident lectin-like chaperones, calnexin and calreticulin, significantly contribute to glycoprotein folding. Uridine diphosphate (UDP)–glucose:glycoprotein glucosyltransferase (UGGT) catalyzes the re-glucosylation of unfolded or misfolded proteins, while ER mannosidases influence the trimming of these modifications to smaller oligosaccharides. The concerted action of calnexin, UGGT, and mannosidases provides misfolded proteins with iterative opportunities to achieve their native conformation. Canonical chaperones, for example, glucose-regulated protein 94 (Grp94), oxygen-regulated protein 150 (Orp150), and BiP (Grp78), along with the lectin-like chaperones and oxidoreductases/isomerases, contribute to the stabilization of folding-competent intermediates [8].

In comparison with the cytosolic proteins, the ones synthesized at the ER are challenged with an oxidative milieu and elevated calcium ion concentrations within the lumen; this makes synthesis and folding at the ER far more complicated [43, 45]. Additionally, the elevated basal rate of translation and the frequent presence of stressors, like elevated glucose levels [33], make protein biogenesis an inherently error-prone process. This could cause the incorporation of incorrect amino acids or disrupt protein folding. This results in the accretion of misfolded or unfolded proteins within the ER lumen and leads to the activation of ER-stress. The misfolded/unfolded proteins are efficiently cleared using the primary protein degradation machine, the 26S proteasome; this helps in the removal of erroneous proteins, mostly those that are ER localized, as well as some from the cytoplasmic pool [48]. Autophagy facilitates the degradation of misfolded proteins and damaged ER components [105]. The excess buildup of unfolded proteins in the lumen causes ER dysfunction. To counteract the deleterious effects of ER stress, cells initiate a coordinated cellular response involving the unfolded UPR and ERAD [195]. Upon resolution of the unfolded protein load, UPR returns to its basal levels. Sustained or aberrant UPR activation is implicated in the pathogenesis of diverse diseases, including obesity, diabetes mellitus, and neurodegenerative disorders [81, 119, 196].

The protein degradation and clearance of unfolded proteins, particularly those originating from the ER, are collectively and precisely designated as ERAD. Dysfunctional ERAD pathways have been increasingly connected in the etiology of various human pathologies [62]. Multiple protein folding-associated ER chaperones have also been associated in the selective targeting of misfolded proteins to the ERAD pathway. This supports the hypothesis that UPR and ERAD are subject to concurrent regulatory control [19, 133] (Table 7.1).

Table 7.1 The key ER-associated pathways of proteostasis

Quality control pathway	Major components	Brief description	References
UPR	PERK, ATF6, IRE1α, GRP78, an ER chaperone	A stress response due to the overaccumulation of misfolded or unfolded proteins inside the ER lumen. It triggers adaptive or destructive cellular activity depending on the protein folding capacity	Read and Schröder [156], Ron and Walter[158]
ERAD	Grp94, calnexin, OS-9, XTP3 (protein recognition); Derlin1, p97, UFD1, NP14 (retrotranslocation); HRD1, SEL1L, GP78 (ubiquitylation); 26S proteasome (degradation)	A quality control system in the ER, targets misfolded or unassembled proteins, extracted through the ER membrane, ubiquitylated, and destroyed by the UPS	Christianson and Ye [35], Wu and Rapoport [197], Christianson and Carvalho [34]
pQC	Derlin, SRP, SR, E3 ligase, p97, BAG6, proteasome	A quality control system in the ER selectively degrades translocationally attenuated ER proteins and reduces protein load	Oyadomari et al. [142], Orsi et al. [141], Kang et al. [94], Kadowaki et al. [89]
Reticulophagy (ER-phagy)	FAM134B, RTN3L, ATL3, TEX264, CALCOCO1, p62, LC3	A selective organellar autophagy that involves the degradation of ER fragments, such as membranes and luminal content, for maintaining ER homeostasis	Khaminets et al. [97], Mookherjee et al. [127], Zhang et al. [211]

7.3 Unfolded Protein Responses

The UPR is a crucial cellular pathway that safeguards proteostasis within the ER. This organelle is dedicated to the folding and maturation of proteins destined for secretion or localization to other cellular compartments. The UPR is activated by a variety of ER stress signals, including the accumulation of unfolded or misfolded proteins, alterations in lipid homeostasis, calcium dysregulation, and glucose deprivation [158, 201].

The UPR is an elaborated, well-orchestrated cellular response system encompassing multiple signaling pathways and three primary branches. It is activated by the accumulation of unfolded or misfolded proteins within the ER, which impairs protein folding capacity and disrupts ER homeostasis. To restore balance, the UPR initiates a coordinated response to reduce protein load, enhance folding capacity, and eliminate misfolded proteins [156].

To mitigate ER stress, global protein translation is generally attenuated, while the translation of UPR-related transcripts is selectively increased to lower protein influx into the ER lumen [68]. Concurrently, in response to increased demand, the ER augments its functional capacity through the transcriptional upregulation of genes encoding proteins involved in ER membrane maintenance, protein folding, ERAD, and protein secretion [1]. However, sustained ER stress can initiate pro-apoptotic signaling cascades, culminating in cellular death [213].

The UPR is regulated by three transmembrane ER stress sensors: activating transcription factor-6 (ATF6), protein kinase RNA-like ER kinase (PERK), and inositol-requiring enzyme 1α (IRE1α). These proteins feature luminal domains capable of detecting unfolded proteins and cytosolic domains that propagate downstream signaling cascades. GRP78, an ER chaperone, binds to unfolded proteins and the luminal domains of these sensors under normal conditions. Upon ER stress, GRP78 dissociates, leading to the activation of the UPR sensors through oligomerization, autophosphorylation, and/or translocation [152, 209, 210].

7.3.1 ATF6-Mediated Signaling in UPR

Following activation, the full-length ATF6 isoform (ATF6p90) is translocated from the ER to the Golgi apparatus, to sequential proteolytic cleavage by site-1 (S1P) and site-2 (S2P) proteases. This cleavage releases the cytosolic domain, ATF6p50, which contains a basic leucine zipper transcription factor domain. ATF6p50 translocates to the nucleus and activates the transcription of UPR target genes [72].

Moreover, ATF6p50 functions in parallel, often overlapping with other pathways, to regulate the transcription of genes responsible for ER chaperones and enzymes. These chaperones and enzymes enable the translocation of ER protein, their folding, maturation, and secretion, along with the degradation of misfolded proteins in response to ER stress [14, 70, 110, 172].

7.3.2 PERK-Mediated Signaling in UPR

In reciprocation of ER stress, the PERK induces the phosphorylation of eIF2α, leading to an overall attenuation of protein synthesis and limiting the entry of newly synthesized proteins into the ER lumen. However, the eIF2α phosphorylation results in the selective enhancement of translation for a subset of specific mRNAs, including ATF4 [72, 188]. ATF4, a transcription factor activated in response to cellular stress, enhances the expression of genes participating in redox homeostasis, autophagy, amino acid metabolism, and apoptosis.

ATF4 also contributes to a negative feedback loop by promoting eIF2α dephosphorylation and restoring protein synthesis by transcriptional upregulation of GADD34 [63, 67, 69]. GADD34 is a directing subunit of protein phosphatase 1 [88, 135]. Further, ATF4 influences downstream signaling molecules, resulting in increased expression of the transcription factor C/EBP homologous protein (CHOP) or modulation of BCL-2 family proteins that regulate apoptosis.

7.3.3 IRE1α-Mediated Signaling in UPR

IRE1α is a transmembrane protein with a cytosolic kinase domain and a luminal endoribonuclease domain. Upon activation, IRE1α oligomerizes and autophosphorylates, leading to the activation of its endoribonuclease activity. This enables IRE1α to cleave XBP-1 mRNA, resulting in the formation of the active transcription factor, spliced X-box binding protein (sXBP)-1 [41, 205]. sXBP-1 enhances ER protein folding capacity by upregulating genes involved in lipid biosynthesis, quality control system of ER, and also ERAD of misfolded proteins, thereby promoting cell survival. Furthermore, IRE1α, via its ribonuclease activity, effectively reduces the overall protein folding burden within the ER lumen by cleaving select mRNAs and miRNA precursors, through a specialized process called regulated IRE1-dependent decay [75, 76, 185, 193]. Moreover, the interaction of IRE1α with the adapter proteins facilitates the diverse stress response pathways, encompassing macroautophagy and the MAPK pathway, underscoring the pleiotropic function of IRE1α in orchestrating cellular responses to ER stress.

The temporal and spatial regulation of ER stress sensor activation is pivotal in dictating cellular fate. An early activation of IRE1α can induce pro-survival pathways, whereas activation of PERK-ATF4 and PERK-CHOP signaling can trigger apoptotic cell death [115, 140]. The intricate interplay among the key mediators within the UPR pathways is reflected in the plethora of pathways that are affected and, in turn, lead to multiple cellular outcomes. A multivariate understanding of the spatial interconnectivity and temporal dynamicity may be achieved by a multiomics approach [118].

7.4 ER-Associated Degradation

When the UPR system fails to reduce the load of misfolded, unfolded, or incorrectly modified proteins, the majority of these substrates are removed from the ER via the ERAD pathway [34, 112, 180]. ERAD is a complex and well-coordinated process that involves the actions of many proteins in the ER, cytoplasm, and the nucleus through UPR [35]. The ERAD pathway is conserved among different mammalian species.

In order for a protein inside or at the ER to be disposed of, it must first be identified and then retrotranslocated into the cytosol. These are ubiquitylated and then extracted from the ER membrane for degradation in the cytosol by the ubiquitin/proteasomal system [197]. ERAD pathway consists of various molecular chaperones such as ER-membrane-localized E3 ubiquitin ligases and adaptor proteins, along with cytoplasmic adaptors and the proteasome [34]. ER chaperones and lectins such as BiP, glucose-regulated protein 170, osteosarcoma amplified 9 (OS-9), and xanthosine 50-triphosphate-B are mainly responsible for the recognition of ERAD substrates [21, 36, 80, 138, 146, 167, 186]. Molecular chaperones that recognize the trimmed glycans of glycoproteins or the exposed hydrophobic domains of non-glycosylated proteins help identify and distinguish between the faulty proteins from those in their native state. The protein disulfide isomerases participate in the recognition and trafficking of defective proteins having disulfide bonds [60, 71, 137, 186].

Client substrates are transported into the cytosol by the retrotranslocation channels located in the ER membrane. In mammalian cells, the well-characterized translocation channel is a multi-subunit protein complex, composed of suppressor/enhancer of lin-12-like (Sel1L) and hydroxyl methyl glutaryl reductase degradation protein 1 (HRD1) [34, 111, 132]. Sel1L, a type-I transmembrane protein, facilitates substrate transfer to HRD1 and also regulates its stability and activity [85, 139, 175]. HRD1, an eight-pass transmembrane protein, is an ER membrane-bound E3 ubiquitin ligase. It can self-associate to form dimers or higher-order oligomeric structures [83, 164, 187]. A variety of adaptor proteins, including degradation in ER protein (Derlin) and homocysteine-inducible ER protein, further contribute to the organization of the HRD1 complex, facilitating retrotranslocation of target substrates [78, 111, 131, 163]. Derlins are ER-membrane-localized multifunctional proteins with six transmembrane domains, and with both of their N- and C-terminal regions facing the cytosol. Notably, human Derlin-1 has been shown to form a tetrameric translocation tunnel with a cavity that accommodates specific proteins [153]. Cytosolic RING finger domain of HRD1 ubiquitylates client substrates in the cytosolic cavity, thereby facilitating the engagement of these faulty proteins with valosin-containing protein (VCP), enabling their final extraction. Besides the HRD1, a plethora of E3 ubiquitin ligases, such as TRC8, RNF170, and MARCH6, are implicated in the ubiquitylation of ERAD substrates [111, 121]; these either cooperate with the HRD1 or function in parallel. GP78 (also known as autocrine motility factor receptor), another such ER E3 ubiquitin ligase, is implicated in the clearance of misfolded proteins; it also exhibits the capacity for self-ubiquitylation, ultimately resulting in its own proteasomal degradation [46]. GP78 functions downstream of HRD1 to target ERAD cargos for proteasomal degradation [10, 171].

The ubiquitylation events catalyzed by these E3 ubiquitin ligases, which rely on the cytoplasmic E2 ubiquitin-conjugating enzymes, UBE2J and UBE2G2, to perform ubiquitylation [35]. Upon ubiquitylation and binding to the cytosolic VCP-UFD1-NPL4 complex, substrates are dislocated from the ER membrane. The ATP hydrolytic activity of the AAA+ ATPase, VCP, provides the energy required for this extraction [34, 124, 147]. The mechanosensing enzyme, VCP or p97, consists of two ATPase domains and is essential for the retrotranslocation of nearly all substrates targeted for ERAD. Moreover, ERAD factors such as UBXD8, Derlin1, and RHBDL4 induce the localization of the VCP-UFD1-NPL4 complex to designated retrotranslocation sites within the ER membrane [100, 111, 128]. Two ATPase domains of VCP are responsible for the protein's assembly into a hexameric ring, which defines a central pore [42, 182]. To allow passage through the VCP central pore, client substrates are subjected to partial or complete unfolding. Additionally, disulfide bonds present in such substrates are disrupted to ensure protein linearity [60, 186]. The ERAD pathway culminates in the delivery of retro-translocated substrates to the cytosole-located 26S proteasome. Cytoplasmic chaperones such as BAG6 and UBL4A help facilitate this process [4, 192, 202]. A

UBL4A-BAG6-TRC35 holdase complex plays a critical role in preserving the cytosolic solubility of retrotranslocated proteins. The complex's interaction with the proteasome optimizes the delivery of client substrates, thereby promoting efficient proteasomal degradation.

7.4.1 UPR-Independent ERAD

It is reported that ERAD, in addition to processing misfolded proteins, plays a role in regulating the abundance of properly folded proteins and also their expression [3, 12]. This comprises the constitutive ERAD pathway, a UPR-independent protein quality control mechanism, involved in regulating ER protein load and maintaining cellular homeostasis. Some examples of endogenous substrates of this pathway are IRE1α, CREBH, CPT2, PGC1B, RMND1, NRF2, and CDKN1B [86, 176, 194, 199, 203, 204]. Notably, impairment of ERAD has been implicated in physiological dysregulation and the onset of numerous disease states [13, 177, 206].

7.5 Pre-emptive Quality Control

In addition to the ERAD system, the selective degradation of ER-targeted proteins by translocation attenuation further limits protein load into the ER [94, 141, 142]. This system is termed pre-emptive quality control (pQC). The mechanisms of ER pQC can be broadly categorized into two sequential steps of substrate rerouting and degradation. Co-translational translocation initiates with the signal recognition particle (SRP) recognizing the hydrophobic signal peptide of nascent polypeptides emerging from the ribosome. This ribosome nascent chain (RNC)–SRP complex is subsequently targeted to the translocon. Notably, the translocation of specific ER proteins is attenuated under conditions of ER stress. These translocation-attenuated nascent peptides are fully translated in the cytosol and targeted thereafter for degradation by the UPS [89]. During translocation into the ER, newly formed polypeptide chains (with failed cleavage of the signal peptide) on the translationally arrested RNC–SRP complex are rerouted to the degradation pathway. This rerouting of client substrates is dependent on the Derlin family proteins, which capture SRP and SRP receptor (SR) α and β during ER stress. The degradation process involves ubiquitylation of the fully translated polypeptides by an E3 ubiquitin ligase. Subsequently, these ubiquitylated polypeptides are efficiently delivered to the proteasome via the concerted action of thc p97 and BAG6 chaperones, ultimately culminating in proteasomal degradation [18, 89].

Within the ERAD pathway, p97 and BAG6 function in the retrograde transport of substrates and, through holdase activity, maintain retrotranslocated proteins in a soluble state. Subsequently, both retrotranslocated and rerouted ER pQC substrates are delivered to the proteasome via p97 and BAG6. While multiple ERAD-associated E3 ubiquitin ligases exist, HRD1 has been established as the principal E3 ubiquitin ligase responsible for ER pQC substrates [90]. While in the case of ERAD, several

E3 ligases are required for different types of substrates, in the case of ER pQC, one or a few E3 ligases may be sufficient, as all the substrates are nascent polypeptides emerging from the ribosome, without signal peptide cleavage and ER-mediated modifications.

ER pQC is understood to play a protective role against cellular stress associated with diminished ER folding capacity. Conversely, disruption of ER pQC can contribute to the development of diseases associated with the presence of altered protein conformations, like those with an abundance of mislocalized protein aggregates, as with various isoforms of the prion protein [26, 94, 141, 142, 151].

7.6 Ubiquitin-Proteasome System

The UPS is the main cellular route for non-lysosomal protein degradation. This system not only clears faulty proteins, but it is also responsible for the removal of proteins with normal functions, to maintain their correct intracellular quantity. Crucially, this mechanism functions to uphold protein homeostasis in physiological and pathological states [104]. Moreover, the UPS participates in a variety of cellular functions, including endocytic trafficking [73], DNA repair [183], and immunological response [11]. Deregulation of the UPS also plays a vital role in multiple neurodegenerative disorders, such as Alzheimer's disease, Parkinson's disease, Huntington's disease, amyotrophic lateral sclerosis, Angelman syndrome, Rett syndrome, and autism, as it is involved in the degradation and removal of aberrant proteins [56, 58, 77, 82, 93, 99, 109, 162, 168, 170].

7.6.1 Ubiquitylation Process

Ubiquitylation is primarily mediated by the UPS. Ubiquitin, ubiquitin-activating enzyme (E1), ubiquitin-conjugating enzyme (E2), ubiquitin protein ligase (E3), deubiquitinase (DUB), and the proteasome are the major functional elements of the UPS [92]. Ubiquitin, a 76-amino acid and 8.5-kDa peptide, is characterized by its highly conserved sequence and abundant expression in eukaryotic cells. It consists of seven lysine residues (K6, K11, K27, K29, K33, K48, and K63) and an N-terminal methionine (M1), which facilitate the formation of covalent ubiquitin chains, a crucial aspect that governs its cellular function [106]. Ubiquitin participates in modifying client substrates by engaging in homotypic or heterotypic conjugation of its molecules [101]. Further, ubiquitin can be modified post-translationally through processes such as phosphorylation and acetylation [136, 207], thus complicating the ubiquitylation process.

Typically, the E1 enzyme, utilizing ATP, catalyzes the formation of a high-energy thioester bond between the C-terminal glycine carboxyl group of ubiquitin and the sulfhydryl side chain of its active site cysteine residue [38]. Serving as an intermediary in the ubiquitylation cascade, E2 enzymes mediate the transfer of ubiquitin between E1 and E3. The E2 enzyme has a highly conserved ubiquitin-conjugating enzyme domain with a critical cysteine residue. This cysteine residue is the site of thioester bond formation with ubiquitin, following activation by E1 [212]. Subsequently, E2 with activated ubiquitin binds to E3. Finally, E2 mediates the conjugation of ubiquitin to the substrate, forming an isopeptide bond between the C-terminus of ubiquitin and a lysine residue within the substrate. E3 specifically recognizes target proteins during this process of ubiquitylation [23].

E3 ubiquitin ligases play a crucial role in determining the specificity and efficiency of ubiquitylation. They can exist as single-subunit proteins or multi-subunit complexes. Single subunit E3s are classified into four major families: HECT domain family, RING domain family, U-box domain family, and N-recognition domain family. Among multi-subunit E3 complexes, Cullin-RING E3s and the Anaphase-Promoting Complex/Cyclosome (APC/C) are the most prevalent [212]. The human genome encodes for over 600 distinct E3 ubiquitin ligases, each capable of catalyzing the formation of specific ubiquitin chain types. The functional outcomes of ubiquitylation are highly dependent on the type of ubiquitin chain formed. Polyubiquitylation chains linked via K48 or K11 residues typically target proteins for proteasomal degradation. Numerous E3 enzymes, including SCF, GP78, and E6AP, can generate K48-linked polyubiquitin chains on substrate proteins, thereby directing them toward proteasomal degradation [51, 179, 191]. The K11 ubiquitin chain was first considered as a regulator of cell fission, and its abundance is linked with APC/C activity. APC/C, as an E3 enzyme, mediates the substrate Cyclin B1 to form a K11-linked ubiquitin chain, which is, in turn, degraded by the proteasome to promote the cellular mitosis [174]. Cytosolic E3 ligase Mahogunin RING finger 1 (MGRN1) interacts with the ERAD-associated ER ubiquitin E3 ligase GP78 and degrades it through non-canonical K11-linked trans-polyubiquitylation [129]. MGRN1 additionally can lead to the partial degradation of the outer mitochondrial membrane protein, Mitofusin1 via proteasome by involving K63-linked polyubiquitin linkages; this, in turn, alters mitochondrial morphology [7, 40, 130]. It has been reported that K11-linked polyubiquitin chains demonstrate limited binding to the 26S proteasome; the formation of hybrid K11/K48 chains, however, significantly improves their susceptibility to proteasomal degradation [59]. It has been previously established that the linear ubiquitin assembly complex mediates NF-κB essential modulator (NF-κB) pathway activation through methionine 1 (M1)-linked ubiquitination of NEMO [150]. This designates that the M1 ubiquitin chain is intricate in NF-κB pathway activation. Itchy E3 ubiquitin protein ligase (ITCH) and smad ubiquitination regulatory factor 1 (SMURF1), members of the HECT family of E3 ubiquitin ligases, are known to catalyze the formation of K29-linked ubiquitin chains. SMURF1, in particular, has been implicated in the K29 ubiquitylation of AXIN, a key component of the WNT signaling pathway. Notably, K29-linked ubiquitylation of AXIN does not target it for proteasomal degradation but rather impairs its capability to interact with the LRP5/6 receptors, thereby modulating WNT signaling [47]. It has been shown that ITCH E3 ubiquitin ligase facilitates the formation of K29-linked ubiquitin chains on Deltex, a key protein in the NOTCH signaling pathway. This ubiquitylation event targets Deltex for degradation via the lysosomal pathway [27]. In contrast, RNF168, another E3 ubiquitin ligase, is known to catalyze the formation of K27-linked polyubiquitin chains on histone proteins, specifically at sites of DNA damage. This modification serves as a platform for the recruitment of DNA repair factors, including 53BP1 and BRCA1, to initiate the DNA damage repair process [22, 52]. In addition to the polyubiquitylation events, multiple proteins can be degraded by monoubiquitylation [17]. Lysine-free nonpolymerizable ubiquitin (K0-Ub) is a modified form of ubiquitin that is generally used to study this ubiquitin-signaling pathway. K0-Ub promotes only monoubiquitylation [117, 159]. It has been reported that the mitochondrial E3 ligase, MITOL, promotes multi-monoubiquitylation and proteasomal degradation of DRP1, a key player in mitochondrial fission; this has implications on mitochondrial dynamics and changes in their morphology [44].

7.6.2 Proteasome

The proteasome, a multi-subunit protease complex, is ubiquitously distributed within the cytoplasm and nucleus of eukaryotic cells. In mammalian cells, the 26S proteasome constitutes the primary ubiquitin-dependent neutral protein hydrolase complex within the UPS. The 26S proteasome is a multi-subunit complex comprising a 20S core particle and 19S regulatory particles [54]. Each 19S sub-complex of the proteasome consists of several ATPase active sites as well as ubiquitin binding sites. These ubiquitin binding sites can recognize the ubiquitylated target proteins, deubiquitinate them, and open up their folded structures prior to their entry into the 20S proteasome [120].

The 20S proteasome core particle has a cylindrical architecture, comprising four stacked rings with two outer α-rings and two inner β-rings. Each ring consists of seven distinct subunits. The α-rings primarily function as gatekeepers, regulating substrate recognition and entry into the proteolytic chamber, while the β-rings house the active sites responsible for protein degradation [50]. The β1, β2, and β5 of the seven β subunits exhibit proteolytic activities—β1 has a CASPASE-like activity, β2 is shown to have trypsin-like activity, and β5 possesses chymotrypsin-like activity [184].

In contrast to ubiquitylation, DUBs antagonize the activity of ubiquitin-conjugating enzymes and ligases by removing ubiquitin modifications from substrates. In humans, approximately 90 DUBs have been identified, exhibiting diverse patterns of expression and abundance across various tissues [122]. Some DUBs recognize and cleave linear ubiquitin chains, whereas other DUBs are only restricted to iso-peptide bonds in side chain ubiquitylation [96, 123] (Clague et al. 2019). DUBs are primarily cysteine proteases or metalloproteases. The cysteine proteases include ubiquitin-specific proteases (USPs), ubiquitin C-terminal hydrolases, motif interacting with Ub-containing novel DUB family, and Zn-finger and UFSP domain protein. The zinc-dependent metalloproteases include JAB1/MPN/MOV34. Beyond their critical role in modulating ubiquitin signals generated by E2 and E3 enzymes, DUBs play an essential role in maintaining cellular ubiquitin homeostasis—like recycling ubiquitin moieties from ubiquitylated substrates and facilitating the processing of newly synthesized ubiquitin fusion proteins [65, 145].

Further, different studies have shown the importance of ubiquitylation in autophagy, whereby it can influence the autophagic flux by regulating autophagy-related protein players or help identify cargo meant for degradation [29–31, 37, 98, 113, 114, 200, 203, 204]. However, the role of ubiquitylation in the context of organellar turnover, like reticulophagy (ER-phagy), is still far from clear. In a very recent study, the DUB, USP20, has been identified as a crucial factor regulating reticulophagy under starvation conditions. USP20 exhibits specificity for the cleavage of K48- and K63-linked ubiquitin chains on the reticulophagy receptor family with sequence similarity 134 member B (FAM134B) or reticulophagy regulator 1 (RETREG1). The deubiquitination of RETREG1 results in its stabilization, leads to reticulophagy, and helps mitigate ER stress [211] (Fig. 7.1).

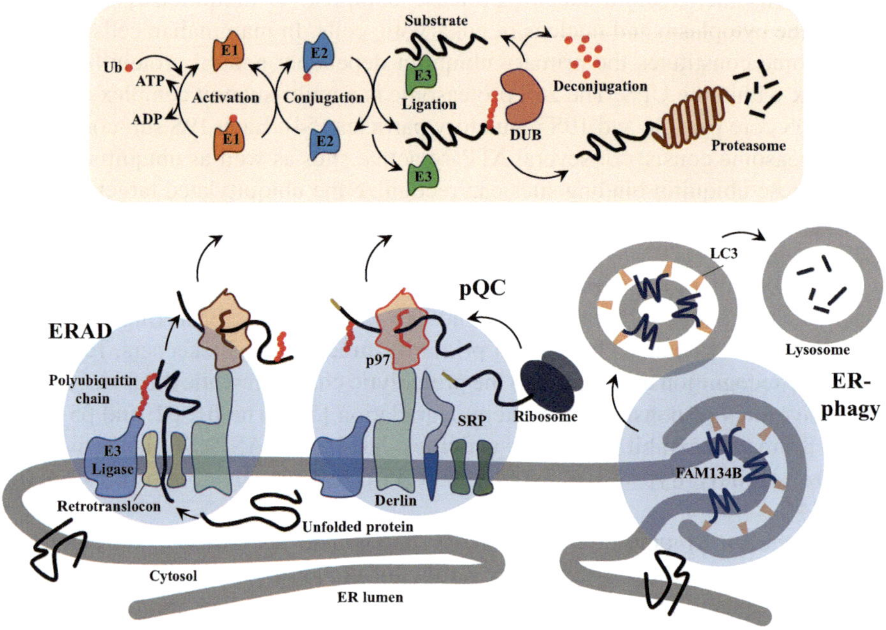

Fig. 7.1 Molecular pathways of proteostasis that couple the ER with the UPS. The figure schematically depicts some of the key molecular pathways that regulate ER protein quality and feed into the UPS. ERAD: linearized unfolded ER proteins are retro-translocated and degraded by the UPS. The ubiquitination of these substrates is mediated by different types of E3 ubiquitin ligases. ERpQC: ER-targeting newly synthesized substrates are rerouted from the translocation pathway to the degradation pathway. The rerouted substrates are degraded by UPS via the activity of p97. ER degradation by autophagy (ERphagy): mammalian ER-resident receptor FAM134B/RETREG1 binds to the autophagy modifiers, like LC3 or GABARAP, and facilitates selective autophagosomal–lysosomal degradation

7.6.3 Functional Association Between ER and UPS

The functionality of the ER is tightly coupled with the UPS to ensure cellular proteostasis [148]. The ER relies on the UPS as a critical component of its quality control machinery [53]. Misfolded proteins within the ER lumen are often retrotranslocated to the cytoplasm, where they are subsequently ubiquitylated by ERAD components, marking them for degradation by the 26S proteasome [79]. This collaborative effort extends beyond ERAD, as the UPS also contributes to the turnover of ER components via ER-phagy. Further, UPR, activated as a consequence of ER stress, modulates the expression of UPS components to enhance cellular capacity for protein degradation [2]. This intricate interplay between ER function and the UPS underscores their essential roles in maintaining protein homeostasis and preventing the accumulation of aberrant proteins.

UPS plays a multifaceted role in regulating ER function, extending beyond simply degrading misfolded proteins. It also fine-tunes UPR by degrading some of its key components, such as IRE1α and processing transcription factors involved in ER stress management [72]. Furthermore, UPS modulates lipid biosynthesis by regulating the activity of lipid synthesis enzymes resident in the ER [126]. Calcium signaling, crucial for various cellular processes, is also influenced by UPS through the degradation of calcium channels and the regulation of calcium pumps within the ER [108]. Additionally, UPS contributes to ER-phagy, a selective form of autophagy whereby ER components are *en masse* targeted for degradation, by ubiquitylating its membrane proteins and regulating ER-phagy receptors [61]. These examples highlight the dynamic interplay between the UPS and ER, demonstrating the former's active role in shaping ER activities and maintaining cellular homeostasis.

7.7 Balance Between the UPS and Autophagosomal–Lysosomal Pathway

The two major degradation mechanisms, involving the proteasome and the autophagosomal–lysosomal pathways, are linked together by one common factor—ubiquitin [39]. PSMD2, PSMD4, and ADRM1, the three principal receptor components of the 19S proteasome complex, enable the proteasome to serve as a versatile platform for the recognition of ubiquitin and ubiquitin-like (Ubl) domains [29, 30, 160]. The endosomal complex required for transport (ESCRT) system, which contains multiple ubiquitin-binding domains within the limiting membrane of endocytic compartments, is responsible for directing cargo to the lysosome. The ESCRT-0 complex, consisting of HRS (also known as HGS) and STAM, provides five potential ubiquitin-binding sites. Further sites are provided by components of the ESCRT-I, consisting of TSG101, UBAP1, MVB12, and ESCRT-II (VPS36) complexes [49]. Protein cargo targeted for autophagy is decorated with ubiquitin chains; these are recognized by adaptor proteins, like p62, which possess both LC3-interacting regions and ubiquitin-binding domains. This interaction facilitates the recruitment of autophagosomes to the targeted structures through the formation of protein-protein complexes between the adaptor proteins, LC3, and ubiquitin [28, 97, 127]. Examples of such adapters include p62 (SQSTM1), optineurin (OPTN), NDP52, NBR1, and TAX1BP1. The degradation of specific substrates, such as mutant huntingtin fragments containing expanded polyglutamine repeats, has been demonstrated to be dependent upon both the UPS and autophagy [103, 155].

Multiple studies have demonstrated a reciprocal regulatory relationship between the UPS and autophagy in mammalian cells and various model organisms, where UPS impairment can lead to the activation of autophagy [84, 143, 157, 178]. A Drosophila model of spinobulbar muscular atrophy, a neurodegenerative condition resulting from polyglutamine expansions within the androgen receptor, revealed that autophagy could serve as a compensatory mechanism for compromised UPS function. Notably, rapamycin-mediated activation of autophagy was shown to rescue neurodegeneration induced by a mutation in the β2-subunit of the proteasome [143]. However, contradictory studies reported that in mice with neuron-specific autophagy inhibition, motor and behavioral deficits resembling neurodegeneration, along with the accumulation of polyubiquitylated proteins within the brain, did not alter proteasomal activity or its component protein levels [66, 102, 103]. Therefore, particularly in neurons, the UPS and autophagy pathways function in parallel to degrade polyubiquitylated substrates, which are harmful. Furthermore, prolonged inhibition of autophagy can directly compromise proteasomal degradation, as evidenced by an increase in substrates specifically targeted for degradation by the UPS [103]. Additional evidence of mutual compensation between UPS and autophagy comes from using MG132, a proteasome inhibitor, in colon cancer cells. MG132 has been shown to not only induce autophagy by increasing the expression of LC3I and LC3II but also arrest cellular proliferation [198]. All these examples clearly show that these two pathways responsible for maintaining cellular proteostasis may get activated in parallel, have a reciprocal relationship in compensating for each other, or may directly affect each other's functionality.

7.8 Conclusion

The ER is the most abundant intracellular membranous structure and consists of an interconnected network of lamellar, tubular, and matrix components [134]. ER is an extremely dynamic organelle, and its domain structures are characterized by the presence of different resident proteins [169, 189]. This organelle is a crucial site for various important intracellular activities, which include lipid and protein synthesis, ion homeostasis, inter-organellar communication, and quality surveillance of newly synthesized proteins [15]. The protein constituents of the ER remain in a constant state of renewal, with a half-life of around 2–5 days [55] to maintain the organelle, structurally and functionally fit. Moreover, as an essential site for protein synthesis as well as the quality control of newly synthesized proteins, the ER is associated with various components that regulate its quality and quantity. External environmental factors, internal genetic mutations, or accumulation of unfolded or misfolded proteins leading to ER stress trigger the UPR, which eases stress by modulating protein folding capacity and clearing misfolded proteins [190, 208]. In response to stress or as a basal quality control mechanism, different molecular pathways such as ERAD, pQC, or reticulophagy (ER-phagy) can get activated to dispose of unwanted cellular garbage and maintain homeostasis. The UPS is one of the major regulators contributing to maintaining ER plasticity, degrading malfunctional ER domains, disposing of preexisting or newly synthesized faulty proteins resulting from ER stress or reciprocally, whose accumulation could lead to ER stress. While the knowledge of the structural and functional connection between the ER and the UPS and a correlation in multiple diseases is not very new, the actual cause-and-effect relationship between these still remains elusive. This becomes even more complicated because the ER maintains structural and functional dynamicity with multiple organelles, such as the mitochondria, nucleus, Golgi body, peroxisomes, plasma membrane, and any alterations in any of these organelles can also directly affect the ER. Thus, diseases classically thought to affect these organelles would affect the ER and vice versa—in effect, deregulation of quality control of any of these would affect proteostasis at the ER and, in turn, cellular homeostasis. The field would benefit from a dynamic and holistic understanding at a multi-organelle, multi-pathway, and multi-omics level.

Acknowledgments We thank OC laboratory members for their help and support. OC is supported by the intramural funding of the Department of Atomic Energy, Government of India. OC is partially funded by SERB, Department of Science and Technology (CRG/2021/000638), Government of India; S.M. acknowledges CSIR (09/489(0109)/2018-EMR-I) for a Ph.D. fellowship.

Author Contributions S.M. and O.C. conceived the idea and wrote the chapter.

Conflict of Interest Statement The authors have declared no conflict of interest.

References

1. Acosta-Alvear D, Zhou Y, Blais A, Tsikitis M, Lents NH, Arias C, Lennon CJ, Kluger Y, Dynlacht BD. XBP1 controls diverse cell type-and condition-specific transcriptional regulatory networks. Mol Cell. 2007;27(1):53–66.
2. Adams CJ, Kopp MC, Larburu N, Nowak PR, Ali MM. Structure and molecular mechanism of ER stress signaling by the unfolded protein response signal activator IRE1. Front Mol Biosci. 2019;6:11.
3. Adle DJ, Wei W, Smith N, Bies JJ, Lee J. Cadmium-mediated rescue from ER-associated degradation induces expression of its exporter. Proc Natl Acad Sci. 2009;106(25):10189–94.
4. Ahner A, Nakatsukasa K, Zhang H, Frizzell RA, Brodsky JL. Small heat-shock proteins select ΔF508-CFTR for endoplasmic reticulum-associated degradation. Mol Biol Cell. 2007;18(3):806–14.
5. Anfinsen CB. Principles that govern the folding of protein chains. Science. 1973;181(4096):223–30.
6. Anken EV, Braakman I. Versatility of the endoplasmic reticulum protein folding factory. Crit Rev Biochem Mol Biol. 2005;40(4):191–228.

7. Anton F, Fres JM, Schauss A, Pinson B, Praefcke GJ, Langer T, Escobar-Henriques M. Ugo1 and Mdm30 act sequentially during Fzo1-mediated mitochondrial outer membrane fusion. J Cell Sci. 2011;124(7):1126–35.
8. Araki K, Nagata K. Protein folding and quality control in the ER. Cold Spring Harb Perspect Biol. 2011;3(11):a007526.
9. Balchin D, Hayer-Hartl M, Hartl FU. In vivo aspects of protein folding and quality control. Science. 2016;353(6294):aac4354.
10. Ballar P, Ors AU, Yang H, Fang S. Differential regulation of CFTRΔF508 degradation by ubiquitin ligases gp78 and Hrd1. Int J Biochem Cell Biol. 2010;42(1):167–73.
11. Bednash JS, Mallampalli RK. Regulation of inflammasomes by ubiquitination. Cell Mol Immunol. 2016;13(6):722–8.
12. Bhattacharya A, Qi L. ER-associated degradation in health and disease–from substrate to organism. J Cell Sci. 2019;132(23):jcs232850.
13. Bhattacharya A, Sun S, Wang H, Liu M, Long Q, Yin L, Kersten S, Zhang K, Qi L. Hepatic Sel1L-Hrd1 ER-associated degradation (ERAD) manages FGF21 levels and systemic metabolism via CREBH. EMBO J. 2018;37(22):e99277.
14. Bommiasamy H, Back SH, Fagone P, Lee K, Meshinchi S, Vink E, Sriburi R, Frank M, Jackowski S, Kaufman RJ, Brewer JW. ATF6α induces XBP1-independent expansion of the endoplasmic reticulum. J Cell Sci. 2009;122(10):1626–36.
15. Borgese N, Francolini M, Snapp E. Endoplasmic reticulum architecture: structures in flux. Curr Opin Cell Biol. 2006;18(4):358–64.
16. Braakman I, Hebert DN. Protein folding in the endoplasmic reticulum. Cold Spring Harb Perspect Biol. 2013;5(5):a013201.
17. Braten O, Livneh I, Ziv T, Admon A, Kehat I, Caspi LH, Gonen H, Bercovich B, Godzik A, Jahandideh S, Jaroszewski L. Numerous proteins with unique characteristics are degraded by the 26S proteasome following monoubiquitination. Proc Natl Acad Sci. 2016;113(32):E4639–47.
18. Braunstein I, Zach L, Allan S, Kalies KU, Stanhill A. Proteasomal degradation of preemptive quality control (pQC) substrates is mediated by an AIRAPL–p97 complex. Mol Biol Cell. 2015;26(21):3719–27.
19. Brodsky JL. The protective and destructive roles played by molecular chaperones during ERAD (endoplasmic-reticulum-associated degradation). Biochem J. 2007;404(3):353–63.
20. Brodsky JL, Skach WR. Protein folding and quality control in the endoplasmic reticulum: Recent lessons from yeast and mammalian cell systems. Curr Opin Cell Biol. 2011;23(4):464–75.
21. Brodsky JL, Werner ED, Dubas ME, Goeckeler JL, Kruse KB, McCracken AA. The requirement for molecular chaperones during endoplasmic reticulum-associated protein degradation demonstrates that protein export and import are mechanistically distinct. J Biol Chem. 1999;274(6):3453–60.
22. Brown JS, Jackson SP. Ubiquitylation, neddylation and the DNA damage response. Open Biol. 2015;5(4):150018.
23. Buetow L, Huang DT. Structural insights into the catalysis and regulation of E3 ubiquitin ligases. Nat Rev Mol Cell Biol. 2016;17(10):626–42.
24. Bukau B, Weissman J, Horwich A. Molecular chaperones and protein quality control. Cell. 2006;125(3):443–51.
25. Caramelo JJ, Parodi AJ. Getting in and out from calnexin/calreticulin cycles. J Biol Chem. 2008;283(16):10221–5.
26. Chakrabarti O, Rane NS, Hegde RS. Cytosolic aggregates perturb the degradation of non-translocated secretory and membrane proteins. Mol Biol Cell. 2011;22(10):1625–37.
27. Chastagner P, Israël A, Brou C. Itch/AIP4 mediates Deltex degradation through the formation of K29-linked polyubiquitin chains. EMBO Rep. 2006;7(11):1147–53.
28. Chen RH, Chen YH, Huang TY. Ubiquitin-mediated regulation of autophagy. J Biomed Sci. 2019;26:1–12.

29. Chen S, Liu Y, Zhou H. Advances in the development ubiquitin-specific peptidase (USP) inhibitors. Int J Mol Sci. 2021a;22(9):4546.
30. Chen X, Yu C, Kang R, Kroemer G, Tang D. Cellular degradation systems in ferroptosis. Cell Death Differ. 2021b;28(4):1135–48.
31. Chen J, Feng H, Wang Y, Bai X, Sheng S, Li H, Huang M, Chu X, Lei Z. The involvement of E3 ubiquitin ligases in the development and progression of colorectal cancer. Cell Death Discov. 2023;9(1):458.
32. Cherepanova N, Shrimal S, Gilmore R. N-linked glycosylation and homeostasis of the endoplasmic reticulum. Curr Opin Cell Biol. 2016;41:57–65.
33. Cheruiyot A, Hollister-Lock J, Sullivan B, Pan H, Dreyfuss JM, Bonner-Weir S, Schaffer JE. Sustained hyperglycemia specifically targets translation of mRNAs for insulin secretion. J Clin Invest. 2024;134(3):e173280.
34. Christianson JC, Carvalho P. Order through destruction: how ER-associated protein degradation contributes to organelle homeostasis. EMBO J. 2022;41(6):e109845.
35. Christianson JC, Ye Y. Cleaning up in the endoplasmic reticulum: ubiquitin in charge. Nat Struct Mol Biol. 2014;21(4):325–35.
36. Christianson JC, Shaler TA, Tyler RE, Kopito RR. OS-9 and GRP94 deliver mutant α1-antitrypsin to the Hrd1–SEL1L ubiquitin ligase complex for ERAD. Nat Cell Biol. 2008;10(3):272–82.
37. Chu Y, Kang Y, Yan C, Yang C, Zhang T, Huo H, Liu Y. LUBAC and OTULIN regulate autophagy initiation and maturation by mediating the linear ubiquitination and the stabilization of ATG13. Autophagy. 2021;17(7):1684–99.
38. Ciechanover A. The unravelling of the ubiquitin system. Nat Rev Mol Cell Biol. 2015;16(5):322–4.
39. Clague MJ, Urbé S. Ubiquitin: same molecule, different degradation pathways. Cell. 2010;143(5):682–5.
40. Cohen MM, Amiott EA, Day AR, Leboucher GP, Pryce EN, Glickman MH, McCaffery JM, Shaw JM, Weissman AM. Sequential requirements for the GTPase domain of the mitofusin Fzo1 and the ubiquitin ligase SCFMdm30 in mitochondrial outer membrane fusion. J Cell Sci. 2011;124(9):1403–10.
41. Concha NO, Smallwood A, Bonnette W, Totoritis R, Zhang G, Federowicz K, Yang J, Qi H, Chen S, Campobasso N, Choudhry AE. Long-range inhibitor-induced conformational regulation of human IRE1α endoribonuclease activity. Mol Pharmacol. 2015;88(6):1011–23.
42. Cooney I, Han H, Stewart MG, Carson RH, Hansen DT, Iwasa JH, Price JC, Hill CP, Shen PS. Structure of the Cdc48 segregase in the act of unfolding an authentic substrate. Science. 2019;365(6452):502–5.
43. Dandekar A, Mendez R, Zhang K. Cross talk between ER stress, oxidative stress, and inflammation in health and disease. In: Stress responses: methods and protocols. Springer; 2015. p. 205–14.
44. Das R, Kamal IM, Das S, Chakrabarti S, Chakrabarti O. MITOL-mediated DRP1 ubiquitylation and degradation promotes mitochondrial hyperfusion in a CMT2A-linked MFN2 mutant. J Cell Sci. 2022;135(2):jcs257808.
45. Dill KA, Ozkan SB, Shell MS, Weikl TR. The protein folding problem. Annu Rev Biophys. 2008;37(1):289–316.
46. Fang S, Ferrone M, Yang C, Jensen JP, Tiwari S, Weissman AM. The tumor autocrine motility factor receptor, gp78, is a ubiquitin protein ligase implicated in degradation from the endoplasmic reticulum. Proc Natl Acad Sci. 2001;98(25):14422–7.
47. Fei C, Li Z, Li C, Chen Y, Chen Z, He X, Mao L, Wang X, Zeng R, Li L. Smurf1-mediated Lys29-linked nonproteolytic polyubiquitination of axin negatively regulates Wnt/β-catenin signaling. Mol Cell Biol. 2013;33(20):4095–105.
48. Finley D. Recognition and processing of ubiquitin-protein conjugates by the proteasome. Annu Rev Biochem. 2009;78(1):477–513.
49. Frankel EB, Audhya A. February. ESCRT-dependent cargo sorting at multivesicular endosomes. In: Seminars in cell & developmental biology, vol. 74. Academic Press; 2018. p. 4–10.

50. Fricker LD. Proteasome inhibitor drugs. Annu Rev Pharmacol Toxicol. 2020;60(1):457–76.
51. Gao Y, Li Y, Zhang C, Zhao M, Deng C, Lan Q, Liu Z, Su N, Wang J, Xu F, Xu Y. Enhanced purification of ubiquitinated proteins by engineered tandem hybrid ubiquitin-binding domains (ThUBDs). Mol Cell Proteomics. 2016;15(4):1381–96.
52. Gatti M, Pinato S, Maiolica A, Rocchio F, Prato MG, Aebersold R, Penengo L. RNF168 promotes noncanonical K27 ubiquitination to signal DNA damage. Cell Rep. 2015;10(2):226–38.
53. Gavilán E, Medina-Guzman R, Bahatyrevich-Kharitonik B, Ruano D. Protein quality control systems and ER stress as key players in SARS-CoV-2-induced neurodegeneration. Cells. 2024;13(2):123.
54. Gillette TG, Kumar B, Thompson D, Slaughter CA, DeMartino GN. Differential roles of the COOH termini of AAA subunits of PA700 (19 S regulator) in asymmetric assembly and activation of the 26 S proteasome. J Biol Chem. 2008;283(46):31813–22.
55. Glass RD, Doyle D. On the measurement of protein turnover in animal cells. J Biol Chem. 1972;247(16):5234–42.
56. Glessner JT, Wang K, Cai G, Korvatska O, Kim CE, Wood S, Zhang H, Estes A, Brune CW, Bradfield JP, Imielinski M. Autism genome-wide copy number variation reveals ubiquitin and neuronal genes. Nature. 2009;459(7246):569–73.
57. Gómez-Virgilio L, Silva-Lucero MDC, Flores-Morelos DS, Gallardo-Nieto J, Lopez-Toledo G, Abarca-Fernandez AM, Zacapala-Gómez AE, Luna-Muñoz J, Montiel-Sosa F, Soto-Rojas LO, Pacheco-Herrero M. Autophagy: a key regulator of homeostasis and disease: an overview of molecular mechanisms and modulators. Cells. 2022;11(15):2262.
58. Graham SH, Liu H. Life and death in the trash heap: the ubiquitin proteasome pathway and UCHL1 in brain aging, neurodegenerative disease and cerebral Ischemia. Ageing Res Rev. 2017;34:30–8.
59. Grice GL, Lobb IT, Weekes MP, Gygi SP, Antrobus R, Nathan JA. The proteasome distinguishes between heterotypic and homotypic lysine-11-linked polyubiquitin chains. Cell Rep. 2015;12(4):545–53.
60. Grubb S, Guo L, Fisher EA, Brodsky JL. Protein disulfide isomerases contribute differentially to the endoplasmic reticulum–associated degradation of apolipoprotein B and other substrates. Mol Biol Cell. 2012;23(4):520–32.
61. Grumati P, Dikic I. Ubiquitin signaling and autophagy. J Biol Chem. 2018;293(15):5404–13.
62. Guerriero CJ, Brodsky JL. The delicate balance between secreted protein folding and endoplasmic reticulum-associated degradation in human physiology. Physiol Rev. 2012;92(2):537–76.
63. Han J, Back SH, Hur J, Lin YH, Gildersleeve R, Shan J, Yuan CL, Krokowski D, Wang S, Hatzoglou M, Kilberg MS. ER-stress-induced transcriptional regulation increases protein synthesis leading to cell death. Nat Cell Biol. 2013;15(5):481–90.
64. Hanley SE, Cooper KF. Sorting nexins in protein homeostasis. Cells. 2020;10(1):17.
65. Hanna J, Meides A, Zhang DP, Finley D. A ubiquitin stress response induces altered proteasome composition. Cell. 2007;129(4):747–59.
66. Hara T, Nakamura K, Matsui M, Yamamoto A, Nakahara Y, Suzuki-Migishima R, Yokoyama M, Mishima K, Saito I, Okano H, Mizushima N. Suppression of basal autophagy in neural cells causes neurodegenerative disease in mice. Nature. 2006;441(7095):885–9.
67. Harding HP, Zhang Y, Ron D. Protein translation and folding are coupled by an endoplasmic-reticulum-resident kinase. Nature. 1999;397(6716):271–4.
68. Harding HP, Calfon M, Urano F, Novoa I, Ron D. Transcriptional and translational control in the mammalian unfolded protein response. Annu Rev Cell Dev Biol. 2002;18(1):575–99.
69. Harding HP, Zhang Y, Zeng H, Novoa I, Lu PD, Calfon M, Sadri N, Yun C, Popko B, Paules R, Stojdl DF. An integrated stress response regulates amino acid metabolism and resistance to oxidative stress. Mol Cell. 2003;11(3):619–33.
70. Hassler JR, Scheuner DL, Wang S, Han J, Kodali VK, Li P, Nguyen J, George JS, Davis C, Wu SP, Bai Y. The IRE1α/XBP1s pathway is essential for the glucose response and protection of β cells. PLoS Biol. 2015;13(10):e1002277.

71. He K, Cunningham CN, Manickam N, Liu M, Arvan P, Tsai B. PDI reductase acts on Akita mutant proinsulin to initiate retrotranslocation along the Hrd1/Sel1L-p97 axis. Mol Biol Cell. 2015;26(19):3413–23.
72. Hetz C, Zhang K, Kaufman RJ. Mechanisms, regulation and functions of the unfolded protein response. Nat Rev Mol Cell Biol. 2020;21(8):421–38.
73. Hicke L. A new ticket for entry into budding vesicles—ubiquitin. Cell. 2001;106(5):527–30.
74. Hipp MS, Hartl FU. Interplay of proteostasis capacity and protein aggregation: implications for cellular function and disease. J Mol Biol. 2024;436(14):168615.
75. Hollien J, Weissman JS. Decay of endoplasmic reticulum-localized mRNAs during the unfolded protein response. Science. 2006;313(5783):104–7.
76. Hollien J, Lin JH, Li H, Stevens N, Walter P, Weissman JS. Regulated Ire1-dependent decay of messenger RNAs in mammalian cells. J Cell Biol. 2009;186(3):323–31.
77. Holmberg CI, Staniszewski KE, Mensah KN, Matouschek A, Morimoto RI. Inefficient degradation of truncated polyglutamine proteins by the proteasome. EMBO J. 2004;23(21):4307–18.
78. Horn SC, Hanna J, Hirsch C, Volkwein C, Schütz A, Heinemann U, Sommer T, Jarosch E. Usa1 functions as a scaffold of the HRD-ubiquitin ligase. Mol Cell. 2009;36(5):782–93.
79. Hoseki J, Ushioda R, Nagata K. Mechanism and components of endoplasmic reticulum-associated degradation. J Biochem. 2010;147(1):19–25.
80. Hosokawa N, Wada I, Nagasawa K, Moriyama T, Okawa K, Nagata K. Human XTP3-B forms an endoplasmic reticulum quality control scaffold with the HRD1-SEL1L ubiquitin ligase complex and BiP. J Biol Chem. 2008;283(30):20914–24.
81. Hotamisligil GS. Endoplasmic reticulum stress and the inflammatory basis of metabolic disease. Cell. 2010;140(6):900–17.
82. Hunter JM, Lesort M, Johnson GV. Ubiquitin-proteasome system alterations in a striatal cell model of Huntington's disease. J Neurosci Res. 2007;85(8):1774–88.
83. Hwang J, Walczak CP, Shaler TA, Olzmann JA, Zhang L, Elias JE, Kopito RR. Characterization of protein complexes of the endoplasmic reticulum-associated degradation E3 ubiquitin ligase Hrd1. J Biol Chem. 2017;292(22):9104–16.
84. Iwata A, Christianson JC, Bucci M, Ellerby LM, Nukina N, Forno LS, Kopito RR. Increased susceptibility of cytoplasmic over nuclear polyglutamine aggregates to autophagic degradation. Proc Natl Acad Sci. 2005;102(37):13135–40.
85. Jeong H, Sim HJ, Song EK, Lee H, Ha SC, Jun Y, Park TJ, Lee C. Crystal structure of SEL1L: insight into the roles of SLR motifs in ERAD pathway. Sci Rep. 2016;6(1):20261.
86. Ji Y, Kim H, Yang L, Sha H, Roman CA, Long Q, Qi L. The Sel1L-Hrd1 endoplasmic reticulum-associated degradation complex manages a key checkpoint in B cell development. Cell Rep. 2016;16(10):2630–40.
87. Jiang X, Wang X, Ding X, Du M, Li B, Weng X, Zhang J, Li L, Tian R, Zhu Q, Chen S. FAM 134B oligomerization drives endoplasmic reticulum membrane scission for ER-phagy. EMBO J. 2020;39(5):e102608.
88. Jousse C, Oyadomari S, Novoa I, Lu P, Zhang Y, Harding HP, Ron D. Inhibition of a constitutive translation initiation factor 2α phosphatase, CReP, promotes survival of stressed cells. J Cell Biol. 2003;163(4):767–75.
89. Kadowaki H, Nagai A, Maruyama T, Takami Y, Satrimafitrah P, Kato H, Honda A, Hatta T, Natsume T, Sato T, Kai H. Pre-emptive quality control protects the ER from protein overload via the proximity of ERAD components and SRP. Cell Rep. 2015;13(5):944–56.
90. Kadowaki H, Satrimafitrah P, Takami Y, Nishitoh H. Molecular mechanism of ER stress-induced pre-emptive quality control involving association of the translocon, Derlin-1, and HRD1. Sci Rep. 2018;8(1):7317.
91. Kandel R, Jung J, Neal S. Proteotoxic stress and the ubiquitin proteasome system. In: Seminars in cell & developmental biology, vol. 156. Academic Press; 2024. p. 107–20.
92. Kane EI, Spratt DE. Structural insights into ankyrin repeat-containing proteins and their influence in ubiquitylation. Int J Mol Sci. 2021;22(2):609.

93. Kaneko M, Koike H, Saito R, Kitamura Y, Okuma Y, Nomura Y. Loss of HRD1-mediated protein degradation causes amyloid precursor protein accumulation and amyloid-β generation. J Neurosci. 2010;30(11):3924–32.
94. Kang SW, Rane NS, Kim SJ, Garrison JL, Taunton J, Hegde RS. Substrate-specific translocational attenuation during ER stress defines a pre-emptive quality control pathway. Cell. 2006;127(5):999–1013.
95. Karagöz GE, Acosta-Alvear D, Walter P. The unfolded protein response: detecting and responding to fluctuations in the protein-folding capacity of the endoplasmic reticulum. Cold Spring Harb Perspect Biol. 2019;11(9):a033886.
96. Keusekotten K, Elliott PR, Glockner L, Fiil BK, Damgaard RB, Kulathu Y, Wauer T, Hospenthal MK, Gyrd-Hansen M, Krappmann D, Hofmann K. OTULIN antagonizes LUBAC signaling by specifically hydrolyzing Met1-linked polyubiquitin. Cell. 2013;153(6):1312–26.
97. Khaminets A, Behl C, Dikic I. Ubiquitin-dependent and independent signals in selective autophagy. Trends Cell Biol. 2016;26(1):6–16.
98. Kim JH, Seo D, Kim SJ, Choi DW, Park JS, Ha J, Choi J, Lee JH, Jung SM, Seo KW, Lee EW. The deubiquitinating enzyme USP20 stabilizes ULK1 and promotes autophagy initiation. EMBO Rep. 2018;19(4):e44378.
99. Kitada T, Asakawa S, Hattori N, Matsumine H, Yamamura Y, Minoshima S, Yokochi M, Mizuno Y, Shimizu N. Mutations in the parkin gene cause autosomal recessive juvenile parkinsonism. Nature. 1998;392(6676):605–8.
100. Klemm EJ, Spooner E, Ploegh HL. Dual role of ancient ubiquitous protein 1 (AUP1) in lipid droplet accumulation and endoplasmic reticulum (ER) protein quality control. J Biol Chem. 2011;286(43):37602–14.
101. Komander D, Rape M. The ubiquitin code. Annu Rev Biochem. 2012;81(1):203–29.
102. Komatsu M, Waguri S, Chiba T, Murata S, Iwata JI, Tanida I, Ueno T, Koike M, Uchiyama Y, Kominami E, Tanaka K. Loss of autophagy in the central nervous system causes neurodegeneration in mice. Nature. 2006;441(7095):880–4.
103. Korolchuk VI, Mansilla A, Menzies FM, Rubinsztein DC. Autophagy inhibition compromises degradation of ubiquitin-proteasome pathway substrates. Mol Cell. 2009;33(4):517–27.
104. Kramer LT, Zhang X. Expanding the landscape of E3 ligases for targeted protein degradation. Curr Res Chem Biol. 2022;2:100020.
105. Kruse KB, Dear A, Kaltenbrun ER, Crum BE, George PM, Brennan SO, McCracken AA. Mutant fibrinogen cleared from the endoplasmic reticulum via endoplasmic reticulum-associated protein degradation and autophagy: an explanation for liver disease. Am J Pathol. 2006;168(4):1299–308.
106. Kwon YT, Ciechanover A. The ubiquitin code in the ubiquitin-proteasome system and autophagy. Trends Biochem Sci. 2017;42(11):873–86.
107. Lahiri V, Klionsky DJ. CCPG1 is a noncanonical autophagy cargo receptor essential for reticulophagy and pancreatic ER proteostasis. Autophagy. 2018;14(7):1107–9.
108. Lee D, Hong JH. Physiological overview of the potential link between the UPS and Ca2+ signaling. Antioxidants (Basel). 2022;11(5):997.
109. Lehman NL. The ubiquitin proteasome system in neuropathology. Acta Neuropathol. 2009;118:329–47.
110. Lei Y, Yu H, Ding S, Liu H, Liu C, Fu R. Molecular mechanism of ATF6 in unfolded protein response and its role in disease. Heliyon. 2024;10(5):e25937.
111. Lemberg MK, Strisovsky K. Maintenance of organellar protein homeostasis by ER-associated degradation and related mechanisms. Mol Cell. 2021;81(12):2507–19.
112. Li H, Sun S. Protein aggregation in the ER: calm behind the storm. Cells. 2021;10(12):3337.
113. Liu J, Xia H, Kim M, Xu L, Li Y, Zhang L, Cai Y, Norberg HV, Zhang T, Furuya T, Jin M. Beclin1 controls the levels of p53 by regulating the deubiquitination activity of USP10 and USP13. Cell. 2011;147(1):223–34.
114. Liu WJ, Ye L, Huang WF, Guo LJ, Xu ZG, Wu HL, Yang C, Liu HF. p62 links the autophagy pathway and the ubiqutin–proteasome system upon ubiquitinated protein degradation. Cell Mol Biol Lett. 2016;21:1–14.

115. Lu M, Lawrence DA, Marsters S, Acosta-Alvear D, Kimmig P, Mendez AS, Paton AW, Paton JC, Walter P, Ashkenazi A. Opposing unfolded-protein-response signals converge on death receptor 5 to control apoptosis. Science. 2014;345(6192):98–101.
116. Ma C, Liu Y, Fu Z. Implications of endoplasmic reticulum stress and autophagy in aging and cardiovascular diseases. Front Pharmacol. 2024;15:1413853.
117. Majumder P, Chakrabarti O. Mahogunin regulates fusion between amphisomes/MVBs and lysosomes via ubiquitination of TSG101. Cell Death Dis. 2015;6(11):e1970.
118. Mallick P, Maity S, Chakrabarti O, Chakrabarti S. Role of systems biology and multi-omics analyses in delineating spatial interconnectivity and temporal dynamicity of ER stress mediated cellular responses. Biochim Biophys Acta Mol Cell Res. 2022;1869(4):119210.
119. Marciniak SJ, Ron D. Endoplasmic reticulum stress signaling in disease. Physiol Rev. 2006;86(4):1133–49.
120. Matyskiela ME, Lander GC, Martin A. Conformational switching of the 26S proteasome enables substrate degradation. Nat Struct Mol Biol. 2013;20(7):781–8.
121. Mehrtash AB, Hochstrasser M. Ubiquitin-dependent protein degradation at the endoplasmic reticulum and nuclear envelope. In: Seminars in cell & developmental biology, vol. 93. Academic Press; 2019. p. 111–24.
122. Michael J, Clague Sylvie, Urbé David, Komander (2019) Breaking the chains: deubiquitylating enzyme specificity begets function Nature Reviews Molecular Cell Biology 20(6) 338–352. https://doi.org/10.1038/s41580-019-0099-1.
123. Mevissen TE, Komander D. Mechanisms of deubiquitinase specificity and regulation. Annu Rev Biochem. 2017;86(1):159–92.
124. Meyer H, Bug M, Bremer S. Emerging functions of the VCP/p97 AAA-ATPase in the ubiquitin system. Nat Cell Biol. 2012;14(2):117–23.
125. Mo ZT, Liao YL, Zheng J, Li WN. Icariin protects neurons from endoplasmic reticulum stress-induced apoptosis after OGD/R injury via suppressing IRE1α-XBP1 signaling pathway. Life Sci. 2020;255:117847.
126. Moncan M, Mnich K, Blomme A, Almanza A, Samali A, Gorman AM. Regulation of lipid metabolism by the unfolded protein response. J Cell Mol Med. 2021;25(3):1359–70.
127. Mookherjee D, Das S, Mukherjee R, Bera M, Jana SC, Chakrabarti S, Chakrabarti O. RETREG1/FAM134B mediated autophagosomal degradation of AMFR/GP78 and OPA1—a dual organellar turnover mechanism. Autophagy. 2021;17(7):1729–52.
128. Mueller B, Klemm EJ, Spooner E, Claessen JH, Ploegh HL. SEL1L nucleates a protein complex required for dislocation of misfolded glycoproteins. Proc Natl Acad Sci. 2008;105(34):12325–30.
129. Mukherjee R, Chakrabarti O. Ubiquitin-mediated regulation of the E3 ligase GP78 by MGRN1 in trans affects mitochondrial homeostasis. J Cell Sci. 2016;129(4):757–73.
130. Mukherjee R, Das A, Chakrabarti S, Chakrabarti O. Calcium dependent regulation of protein ubiquitination–interplay between E3 ligases and calcium binding proteins. Biochim Biophys Acta Mol Cell Res. 2017;1864(7):1227–35.
131. Neal S, Jaeger PA, Duttke SH, Benner C, Glass CK, Ideker T, Hampton RY. The Dfm1 Derlin is required for ERAD retrotranslocation of integral membrane proteins. Mol Cell. 2018;69(2):306–20.
132. Needham PG, Guerriero CJ, Brodsky JL. Chaperoning endoplasmic reticulum–associated degradation (ERAD) and protein conformational diseases. Cold Spring Harb Perspect Biol. 2019;11(8):a033928.
133. Nishikawa SI, Fewell SW, Kato Y, Brodsky JL, Endo T. Molecular chaperones in the yeast endoplasmic reticulum maintain the solubility of proteins for retrotranslocation and degradation. J Cell Biol. 2001;153(5):1061–70.
134. Nixon-Abell J, Obara CJ, Weigel AV, Li D, Legant WR, Xu CS, Pasolli HA, Harvey K, Hess HF, Betzig E, Blackstone C. Increased spatiotemporal resolution reveals highly dynamic dense tubular matrices in the peripheral ER. Science. 2016;354(6311):aaf3928.
135. Novoa I, Zeng H, Harding HP, Ron D. Feedback inhibition of the unfolded protein response by GADD34-mediated dephosphorylation of eIF2α. J Cell Biol. 2001;153(5):1011–22.

136. Ohtake F, Tsuchiya H. The emerging complexity of ubiquitin architecture. J Biochem. 2017;161(2):125–33.
137. Oka OBV, Pringle MA, Schopp IM, Braakman I, Bulleid NJ. ERdj5 is the ER reductase that catalyzes the removal of non-native disulfides and correct folding of the LDL receptor. Mol Cell. 2013;50(6):793–804.
138. Olivari S, Galli C, Alanen H, Ruddock L, Molinari M. A novel stress-induced EDEM variant regulating endoplasmic reticulum-associated glycoprotein degradation. J Biol Chem. 2005;280(4):2424–8.
139. Olzmann JA, Kopito RR, Christianson JC. The mammalian endoplasmic reticulum-associated degradation system. Cold Spring Harb Perspect Biol. 2013;5(9):a013185.
140. Ong G, Ragetli R, Mnich K, Doble BW, Kammouni W, Logue SE. IRE1 signaling increases PERK expression during chronic ER stress. Cell Death Dis. 2024;15(4):276.
141. Orsi A, Fioriti L, Chiesa R, Sitia R. Conditions of endoplasmic reticulum stress favor the accumulation of cytosolic prion protein. J Biol Chem. 2006;281(41):30431–8.
142. Oyadomari S, Yun C, Fisher EA, Kreglinger N, Kreibich G, Oyadomari M, Harding HP, Goodman AG, Harant H, Garrison JL, Taunton J. Cotranslocational degradation protects the stressed endoplasmic reticulum from protein overload. Cell. 2006;126(4):727–39.
143. Pandey UB, Nie Z, Batlevi Y, McCray BA, Ritson GP, Nedelsky NB, Schwartz SL, DiProspero NA, Knight MA, Schuldiner O, Padmanabhan R. HDAC6 rescues neurodegeneration and provides an essential link between autophagy and the UPS. Nature. 2007;447(7146):860–4.
144. Park J, Cho J, Song EJ. Ubiquitin–proteasome system (UPS) as a target for anticancer treatment. Arch Pharm Res. 2020;43(11):1144–61.
145. Pickart CM, Rose IA. Functional heterogeneity of ubiquitin carrier proteins. J Biol Chem. 1985;260(3):1573–81.
146. Plemper RK, Böhmler S, Bordallo J, Sommer T, Wolf DH. Mutant analysis links the translocon and BiP to retrograde protein transport for ER degradation. Nature. 1997;388(6645):891–5.
147. Pye VE, Beuron F, Keetch CA, McKeown C, Robinson CV, Meyer HH, Zhang X, Freemont PS. Structural insights into the p97-Ufd1-Npl4 complex. Proc Natl Acad Sci. 2007;104(2):467–72.
148. Qu J, Zou T, Lin Z. The roles of the ubiquitin–proteasome system in the endoplasmic reticulum stress pathway. Int J Mol Sci. 2021;22(4):1526.
149. Radanović T, Ernst R. The unfolded protein response as a guardian of the secretory pathway. Cells. 2021;10(11):2965.
150. Rahighi S, Ikeda F, Kawasaki M, Akutsu M, Suzuki N, Kato R, Kensche T, Uejima T, Bloor S, Komander D, Randow F. Specific recognition of linear ubiquitin chains by NEMO is important for NF-κB activation. Cell. 2009;136(6):1098–109.
151. Rane NS, Kang SW, Chakrabarti O, Feigenbaum L, Hegde RS. Reduced translocation of nascent prion protein during ER stress contributes to neurodegeneration. Dev Cell. 2008;15(3):359–70.
152. Rao RV, Peel A, Logvinova A, del Rio G, Hermel E, Yokota T, Goldsmith PC, Ellerby LM, Ellerby HM, Bredesen DE. Coupling endoplasmic reticulum stress to the cell death program: role of the ER chaperone GRP78. FEBS Lett. 2002;514(2–3):122–8.
153. Rao B, Li S, Yao D, Wang Q, Xia Y, Jia Y, Shen Y, Cao Y. The cryo-EM structure of an ERAD protein channel formed by tetrameric human Derlin-1. Sci Adv. 2021;7(10):eabe8591.
154. Rapoport TA, Li L, Park E. Structural and mechanistic insights into protein translocation. Annu Rev Cell Dev Biol. 2017;33(1):369–90.
155. Ravikumar B, Sarkar S, Berger Z, Rubinsztein DC. The roles of the ubiquitin-proteasome and autophagy–lysosome pathways in Huntington's disease and related conditions. Clin Neurosci Res. 2003;3(3):141–8.
156. Read A, Schröder M. The unfolded protein response: an overview. Biology. 2021;10(5):384.
157. Rideout HJ, Dietrich P, Wang Q, Dauer WT, Stefanis L. α-synuclein is required for the fibrillar nature of ubiquitinated inclusions induced by proteasomal inhibition in primary neurons. J Biol Chem. 2004;279(45):46915–20.

158. Ron D, Walter P. Signal integration in the endoplasmic reticulum unfolded protein response. Nat Rev Mol Cell Biol. 2007;8(7):519–29.
159. Ronai ZEA. Monoubiquitination in proteasomal degradation. Proc Natl Acad Sci. 2016;113(32):8894–6.
160. Saeki Y. Ubiquitin recognition by the proteasome. J Biochem. 2017;161(2):113–24.
161. Salnikov L. Role of the endoplasmic reticulum in mechanisms of aging and formation of senescent cells. Transl Med Aging. 2024;8:29–32.
162. Schmidt M, Finley D. Regulation of proteasome activity in health and disease. Biochim Biophys Acta Mol Cell Res. 2014;1843(1):13–25.
163. Schmidt CC, Vasic V, Stein A. Doa10 is a membrane protein retrotranslocase in ER-associated protein degradation. elife. 2020;9:e56945.
164. Schoebel S, Mi W, Stein A, Ovchinnikov S, Pavlovicz R, DiMaio F, Baker D, Chambers MG, Su H, Li D, Rapoport TA. Cryo-EM structure of the protein-conducting ERAD channel Hrd1 in complex with Hrd3. Nature. 2017;548(7667):352–5.
165. Schubert U, Anton LC, Gibbs J, Norbury CC, Yewdell JW, Bennink JR. Rapid degradation of a large fraction of newly synthesized proteins by proteasomes. Nature. 2000;404(6779):770–4.
166. Shang F, Taylor A. Ubiquitin–proteasome pathway and cellular responses to oxidative stress. Free Radic Biol Med. 2011;51(1):5–16.
167. Shenkman M, Groisman B, Ron E, Avezov E, Hendershot LM, Lederkremer GZ. A shared endoplasmic reticulum-associated degradation pathway involving the EDEM1 protein for glycosylated and nonglycosylated proteins. J Biol Chem. 2013;288(4):2167–78.
168. Shi Q, Chen LN, Zhang BY, Xiao K, Zhou W, Chen C, Zhang XM, Tian C, Gao C, Wang J, Han J. Proteomics analyses for the global proteins in the brain tissues of different human prion diseases*[S]. Mol Cell Proteomics. 2015;14(4):854–69.
169. Shibata Y, Shemesh T, Prinz WA, Palazzo AF, Kozlov MM, Rapoport TA. Mechanisms determining the morphology of the peripheral ER. Cell. 2010;143(5):774–88.
170. Shimura H, Hattori N, Kubo SI, Mizuno Y, Asakawa S, Minoshima S, Shimizu N, Iwai K, Chiba T, Tanaka K, Suzuki T. Familial Parkinson disease gene product, parkin, is a ubiquitin-protein ligase. Nat Genet. 2000;25(3):302–5.
171. Shmueli A, Tsai YC, Yang M, Braun MA, Weissman AM. Targeting of gp78 for ubiquitin-mediated proteasomal degradation by Hrd1: cross-talk between E3s in the endoplasmic reticulum. Biochem Biophys Res Commun. 2009;390(3):758–62.
172. Shoulders MD, Ryno LM, Genereux JC, Moresco JJ, Tu PG, Wu C, Yates JR, Su AI, Kelly JW, Wiseman RL. Stress-independent activation of XBP1s and/or ATF6 reveals three functionally diverse ER proteostasis environments. Cell Rep. 2013;3(4):1279–92.
173. Smith MD, Harley ME, Kemp AJ, Wills J, Lee M, Arends M, von Kriegsheim A, Behrends C, Wilkinson S. CCPG1 is a non-canonical autophagy cargo receptor essential for ER-phagy and pancreatic ER proteostasis. Dev Cell. 2018;44(2):217–32.
174. Song L, Rape M. Regulated degradation of spindle assembly factors by the anaphase-promoting complex. Mol Cell. 2010;38(3):369–82.
175. Sun S, Shi G, Han X, Francisco AB, Ji Y, Mendonça N, Liu X, Locasale JW, Simpson KW, Duhamel GE, Kersten S. Sel1L is indispensable for mammalian endoplasmic reticulum-associated degradation, endoplasmic reticulum homeostasis, and survival. Proc Natl Acad Sci. 2014;111(5):E582–91.
176. Sun S, Shi G, Sha H, Ji Y, Han X, Shu X, Ma H, Inoue T, Gao B, Kim H, Bu P. IRE1α is an endogenous substrate of endoplasmic-reticulum-associated degradation. Nat Cell Biol. 2015;17(12):1546–55.
177. Sun S, Lourie R, Cohen SB, Ji Y, Goodrich JK, Poole AC, Ley RE, Denkers EY, McGuckin MA, Long Q, Duhamel GE. Epithelial Sel1L is required for the maintenance of intestinal homeostasis. Mol Biol Cell. 2016;27(3):483–90.
178. Suraweera A, Münch C, Hanssum A, Bertolotti A. Failure of amino acid homeostasis causes cell death following proteasome inhibition. Mol Cell. 2012;48(2):242–53.

179. Thacker G, Mishra M, Sharma A, Singh AK, Sanyal S, Trivedi AK. E3 ligase SCFSKP2 ubiquitinates and degrades tumor suppressor C/EBPα in acute myeloid leukemia. Life Sci. 2020;257:118041.
180. Thepsuwan P, Bhattacharya A, Song Z, Hippleheuser S, Feng S, Wei X, Das NK, Sierra M, Wei J, Fang D, Huang YMM. Hepatic SEL1L-HRD1 ER-associated degradation regulates systemic iron homeostasis via ceruloplasmin. Proc Natl Acad Sci. 2023;120(2):e2212644120.
181. Tsakiri EN, Trougakos IP. The amazing ubiquitin-proteasome system: structural components and implication in aging. Int Rev Cell Mol Biol. 2015;314:171–237.
182. Twomey EC, Ji Z, Wales TE, Bodnar NO, Ficarro SB, Marto JA, Engen JR, Rapoport TA. Substrate processing by the Cdc48 ATPase complex is initiated by ubiquitin unfolding. Science. 2019;365(6452):eaax1033.
183. Uckelmann M, Sixma TK. Histone ubiquitination in the DNA damage response. DNA Repair. 2017;56:92–101.
184. Unno M, Mizushima T, Morimoto Y, Tomisugi Y, Tanaka K, Yasuoka N, Tsukihara T. The structure of the mammalian 20S proteasome at 2.75 Å resolution. Structure. 2002;10(5):609–18.
185. Upton JP, Wang L, Han D, Wang ES, Huskey NE, Lim L, Truitt M, McManus MT, Ruggero D, Goga A, Papa FR. IRE1α cleaves select microRNAs during ER stress to derepress translation of proapoptotic Caspase-2. Science. 2012;338(6108):818–22.
186. Ushioda R, Hoseki J, Araki K, Jansen G, Thomas DY, Nagata K. ERdj5 is required as a disulfide reductase for degradation of misfolded proteins in the ER. Science. 2008;321(5888):569–72.
187. Vasic V, Denkert N, Schmidt CC, Riedel D, Stein A, Meinecke M. Hrd1 forms the retrotranslocation pore regulated by auto-ubiquitination and binding of misfolded proteins. Nat Cell Biol. 2020;22(3):274–81.
188. Vattem KM, Wek RC. Reinitiation involving upstream ORFs regulates ATF4 mRNA translation in mammalian cells. Proc Natl Acad Sci. 2004;101(31):11269–74.
189. Voeltz GK, Prinz WA, Shibata Y, Rist JM, Rapoport TA. A class of membrane proteins shaping the tubular endoplasmic reticulum. Cell. 2006;124(3):573–86.
190. Walter P, Ron D. The unfolded protein response: from stress pathway to homeostatic regulation. Science. 2011;334(6059):1081–6.
191. Wang M, Pickart CM. Different HECT domain ubiquitin ligases employ distinct mechanisms of polyubiquitin chain synthesis. EMBO J. 2005;24(24):4324–33.
192. Wang F, Song W, Brancati G, Segatori L. Inhibition of endoplasmic reticulum-associated degradation rescues native folding in loss of function protein misfolding diseases. J Biol Chem. 2011;286(50):43454–64.
193. Wang JM, Qiu Y, Yang ZQ, Li L, Zhang K. Inositol-requiring enzyme 1 facilitates diabetic wound healing through modulating MicroRNAs. Diabetes. 2017;66(1):177–92.
194. Wei J, Chen L, Li F, Yuan Y, Wang Y, Xia W, Zhang Y, Xu Y, Yang Z, Gao B, Jin C. HRD1-ERAD controls production of the hepatokine FGF21 through CREBH polyubiquitination. EMBO J. 2018;37(22):e98942.
195. Wilkinson S. ER-phagy: shaping up and destressing the endoplasmic reticulum. FEBS J. 2019;286(14):2645–63.
196. Wu S, Lin W. The physiological role of the unfolded protein response in the nervous system. Neural Regen Res. 2024;19(11):2411–20.
197. Wu X, Rapoport TA. Mechanistic insights into ER-associated protein degradation. Curr Opin Cell Biol. 2018;53:22–8.
198. Wu WKK, Sung JJY, Wu YC, Li ZJ, Yu L, Cho CH. Bone morphogenetic protein signalling is required for the anti-mitogenic effect of the proteasome inhibitor MG-132 on colon cancer cells. Br J Pharmacol. 2008;154(3):632–8.
199. Wu T, Zhao F, Gao B, Tan C, Yagishita N, Nakajima T, Wong PK, Chapman E, Fang D, Zhang DD. Hrd1 suppresses Nrf2-mediated cellular protection during liver cirrhosis. Genes Dev. 2014;28(7):708–22.
200. Xie W, Jin S, Wu Y, Xian H, Tian S, Liu DA, Guo Z, Cui J. Auto-ubiquitination of NEDD4-1 recruits USP13 to facilitate autophagy through deubiquitinating VPS34. Cell Rep. 2020;30(8):2807–19.

201. Xu F, Wang L. Deciphering ER stress-unfolded protein response relationship by visualizing unfolded proteins in the ER. Cell Rep. 2024;43(6):114358.
202. Xu Y, Cai M, Yang Y, Huang L, Ye Y. SGTA recognizes a noncanonical ubiquitin-like domain in the Bag6-Ubl4A-Trc35 complex to promote endoplasmic reticulum-associated degradation. Cell Rep. 2012;2(6):1633–44.
203. Xu D, Shan B, Sun H, Xiao J, Zhu K, Xie X, Li X, Liang W, Lu X, Qian L, Yuan J. USP14 regulates autophagy by suppressing K63 ubiquitination of Beclin 1. Genes Dev. 2016a;30(15):1718–30.
204. Xu Y, Zhao F, Qiu Q, Chen K, Wei J, Kong Q, Gao B, Melo-Cardenas J, Zhang B, Zhang J, Song J. The ER membrane-anchored ubiquitin ligase Hrd1 is a positive regulator of T-cell immunity. Nat Commun. 2016b;7(1):12073.
205. Yang J, Liu H, Li L, Liu H, Shi W, Yuan X, Wu L. Structural insights into IRE1 functions in the unfolded protein response. Curr Med Chem. 2016;23(41):4706–16.
206. Yang Y, Kong S, Zhang Y, Melo-Cardenas J, Gao B, Zhang Y, Zhang DD, Zhang B, Song J, Thorp E, Zhang K. The endoplasmic reticulum–resident E3 ubiquitin ligase Hrd1 controls a critical checkpoint in B cell development in mice. J Biol Chem. 2018;293(33):12934–44.
207. Yau R, Rape M. The increasing complexity of the ubiquitin code. Nat Cell Biol. 2016;18(6):579–86.
208. Yuan S, She D, Jiang S, Deng N, Peng J, Ma L. Endoplasmic reticulum stress and therapeutic strategies in metabolic, neurodegenerative diseases and cancer. Mol Med. 2024;30(1):40.
209. Zhang LH, Zhang X. Roles of GRP78 in physiology and cancer. J Cell Biochem. 2010;110(6):1299–305.
210. Zhang Y, Liu R, Ni M, Gill P, Lee AS. Cell surface relocalization of the endoplasmic reticulum chaperone and unfolded protein response regulator GRP78/BiP. J Biol Chem. 2010;285(20):15065–75.
211. Zhang M, Wang Z, Zhao Q, Yang Q, Bai J, Yang C, Zhang ZR, Liu Y. USP20 deubiquitinates and stabilizes the reticulophagy receptor RETREG1/FAM134B to drive reticulophagy. Autophagy. 2024;20(8):1780–97.
212. Zheng N, Shabek N. Ubiquitin ligases: structure, function, and regulation. Annu Rev Biochem. 2017;86(1):129–57.
213. Zinszner H, Kuroda M, Wang X, Batchvarova N, Lightfoot RT, Remotti H, Stevens JL, Ron D. CHOP is implicated in programmed cell death in response to impaired function of the endoplasmic reticulum. Genes Dev. 1998;12(7):982–95.

Neuroinflammatory Response and Ubiquitin-Proteasome System

8

Sabiya Khan and Dharmendra Kumar Khatri

8.1 Introduction

Neurodegenerative conditions are marked by a common feature, that is, abnormalities in the maintenance of proteostasis. Cellular homeostasis is strictly maintained in the central nervous system (CNS) by the cell growth, cellular metabolism, and decomposition mechanisms. Cellular decomposition happens by autophagy and ubiquitin-proteasome system (UPS) mediated degradation. UPS is critically important to maintain this homeostasis in the brain by controlling the degradation of misfolded, damaged, or unwanted proteins. Autophagy (including chaperone-mediated autophagy, macroautophagy, and microautophagy) involves lysosomal degradation through intracellular vesicle (autophagosome) formation. The UPS targets 80–90% of the intracellular protein, including damaged, short-lived, or misfolded proteins [23].

Two key components accomplish UPS-mediated degradation; the first is the ubiquitination system, which aids in the damaged (target) protein detection and their ubiquitination, while the second is the proteasome system, which executes the degradation by identifying the ubiquitinated targets. In three steps, the ubiquitin system conjugates ubiquitin (76-amino-acid-long modifiers) to the target protein, identified under strict regulation. First, the ubiquitin-activating enzyme (E1) is linked to ubiquitin by forming a thioester bond between the C-terminal residue Gly76 of ubiquitin and an active site cysteine residue of the E1 enzyme, with the help of ATP. Then, the thioester bond is formed between the activated ubiquitin and the E2 ubiquitin-conjugating enzyme.

S. Khan
Department of Pharmacology & Toxicology, National Institute of Pharmaceutical Education & Research (NIPER), Hyderabad, Telangana, India

D. K. Khatri (✉)
Department of Pharmacology, NIMS Institute of Pharmacy, NIMS University Rajasthan, Jaipur, Rajasthan, India

S. Singh (ed.), *Ubiquitin Proteasome System in CNS Neurodegenerative Diseases*, https://doi.org/10.1007/978-981-95-6099-8_8

Immediately, the E2 conjugating enzyme transfers the activated ubiquitin and attaches it to the specific enzyme E3 ubiquitin ligase, which could bind to the target protein at its Lys residue. Ultimately, the ubiquitin is transferred to the target substrate and is then identified by the proteasome system. There are more than 500 E3 ligases encoded by the human genome, which might be a single polypeptide or made up of subunits of a protein complex (e.g., Skp1-Cullin1-F-box E3 complex). The lysine residue of the target will decide the fate of the process, for example, Lys48 (most widely used) linkage initiates proteasomal degradation, while Lys63 linkage does not lead to proteasomal degradation but signals another type of protein-protein interaction; similarly, Lys11 linkage helps in cell-cycle regulation [14]. The substrate protein is then unfolded, translocated into the proteasomal core, and degraded into smaller peptides. The entire process is strictly regulated and is reversible with the help of deubiquitinating enzymes (DUBs). A chain of four or more Ub at Lys48 can serve as a secondary degron, delivering substrates to the 26S proteasome. The 26S cylindrical machine is composed of 19S regulatory protein and 20S proteolytic core protein. 19S subunit binds to the polyubiquitinated substrate of the protein and feeds forward the unfolded polypeptide into the 20S narrow chamber (a 13 Å diameter), while also deubiquitinating the polyubiquitinated substrate, and, hence, recycles ubiquitin. The substrate particles are cleaved into small peptides while passing through 20S's β1, β2, and β5 subunits, which have caspase-like, trypsin-like, and chymotrypsin-like peptidase activities, respectively [6]. The process is explained in Fig. 8.1. Impairment of these 26S proteasome subunits (i.e., 19S and 20S) is strongly linked to various neurological disorders.

The UPS is an extremely critical component in neurons because these cells are post-mitotic and cannot rely on cell division to remove or dilute harmful protein aggregates [38]. Physiologically, the UPS plays an important role in rendering synaptic plasticity to neurons and neurodevelopment by regulating the cellular turnover of proteins like NMDA (N-methyl-D-aspartate) receptors, which regulate synaptic strength, long-term potentiation, and long-term depression membrane potentials that are important for memory formation. The UPS also helps regulate the quality control of physiological proteins, which, when misfolded or damaged, undergo aggregation and formation of toxic peptide agglomerates like Lewy bodies made up of α-synuclein (α-Syn) protein, causing neurodegenerative Parkinson's pathology. Another important role of UPS is to regulate the inflammatory response by immune cells in the brain upon cellular insult. UPS aids in the clearance of toxic aggregates by a regulated ubiquitination mechanism.

UPS dysfunction leads to inflammation exacerbation in the cellular environment. UPS does not independently cause neuroinflammation, but UPS dysfunction may act as a protagonist in aiding the underlying pathological processes headed by an inflammatory response. Any factor like genetic mutation, aging, or environmental toxicant exposure causes impaired UPS, which leads to abnormal protein aggregation and neurodegenerative pathology. These protein aggregates trigger the activation of CNS microglia, astrocytes, and peripheral macrophages. These cells signal the secretion of pro-inflammatory mediators that further drive neuroinflammation pathology [5]. Considering the importance of the UPS in regulating the

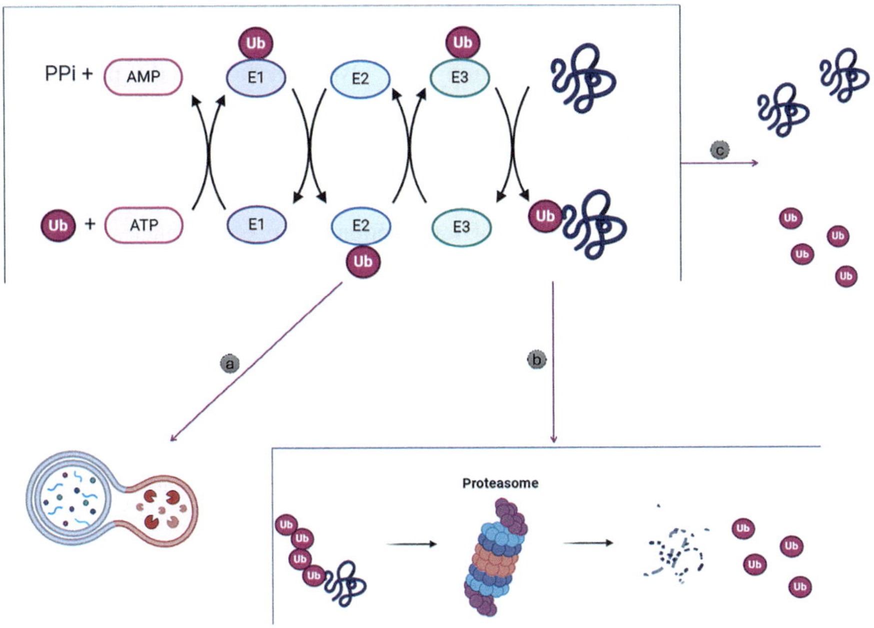

Fig. 8.1 UPS: The UPS is regulated by E1, E2, and E3 subunits with the help of ubiquitin and ATP molecules. E3-ligases play the most crucial role in regulating protein degradation and neuroinflammatory processes. The UPS is controlled by the lysine residue, which is ubiquitinated on the substrate. (**a**) K63 ubiquitination leads to autophagic degradation. (**b**) K48/K11 ubiquitination leads to 26S proteasome-mediated protein processing. (**c**) Deubiquitinates lead to the removal of the ubiquitin chain from the residue, leading to the recycling of ubiquitin and freeing up the intact peptide

neuroinflammatory processes, we ought to descriptively elaborate on the correlation between the two in this work.

8.2 Protein Homeostasis and Immune Activation: UPS at the Interplay

The proteins, due to underlying reasons, are spontaneously misfolded with enriched β-sheet structures, and they form oligomeric aggregates [16]. The endoplasmic reticulum is an important intracellular organelle, facilitating protein folding, translocation, and post-translational modification. Abnormally processed proteins accumulate in the ER lumen, triggering the unfolded protein response (UPR). The main goal of the UPR process is to restore protein homeostasis (proteostasis) in the cell. The malfunctioning proteins undergo proteasomal degradation under physiological conditions, but the ER homeostasis and UPR function are impaired under the influence of different cellular perturbations, leading to a pathological state. ER-associated protein degradation is an important cellular mechanism that prevents the accumulation of toxic aggregates and cell death, which is triggered by excessive ER stress-mediated apoptotic cascade stimulation [17]. Overactivation of the proteasome mediates neuroinflammatory processes in disease conditions [22].

The UPS plays a vital role in cellular immunity in cells of myeloid origin, such as dendritic cells, granulocytes, macrophages, and monocytes. Pattern-recognition receptors (PRRs) recognize the pathogen-associated molecular pattern and induce the innate immune response with the help of transcription factors such as nuclear factor-κB (NF-κB) and interferon regulatory factors (IRF3 and IRF7) that regulate inflammation. These transcription factors translocate to the nucleus to induce the translation of inflammatory cytokines and proteins. UPS is participating in prolonged activation of PRR in response to microbial insult, and E3 ligases, such as inhibitors of apoptosis proteins (IAPs), act downstream of TNF-R (tumor necrosis factor receptor), TLR (toll like receptor), and NOD2 (Nucleotide-Binding Oligomerization Domain-Containing protein 2) signaling pathways to regulate NF-κB and MAPK activity. Production of multiple non-proteolytic ubiquitin chains is required for the assembly of NEMO-TBK1/IKK-ε (NF-κB Essential Modulator/ TANK1 binding kinase/inhibitor of kappa B kinase) through TRAF3 (tumor necrosis factor receptor-associated factor) E3 ligase for linking K63-linked polyubiquitin chains. During viral infection, TRIM27 (tripartite motif-containing 27) (E3 ligase) is recruited through tyrosine phosphatases protein, leading to negative regulation of type 1 interferon induction and degradation of TBK1 [2]. Upon activation by pro-inflammatory mediators like IFN-γ, a proteasome variant, that is, immunoproteasome, is formed. This immunoproteasome has a slightly different proteolytic unit, namely, β1i, β2i, and β5i, making it more efficient in performing antigen presentation as its main function [21]. This immunoproteasome has the ability to regulate innate immune signaling, which is different from that of the constitutive proteasome. Activated glia is reported to show increased activity of immunoproteasome in the AD animal model [32].

UPS dysfunction has been widely reported to cause abnormal protein accumulation, immune response dysregulation, inflammation, defective microglial responsiveness, and increased oxidative stress. These states of cells collectively cause a heightened and prolonged neuroinflammatory state, leading to neurodegenerative states and neurological diseases. Several interlinked events contribute to this exacerbated immune cell response, leading to inflammation where the lead position is borne by impaired protein degradation. Misfolded proteins generated in the cytoplasm, nucleus, and endoplasmic reticulum are effectively removed by quality control components of the cell, that is, UPS, chaperone-mediated autophagy, and macroautophagy. Certain specific abnormal proteins undergo degradation by the KFERQ degradation signal (a part of autophagy), where substrates are selectively recognized by heat-shock cognate 70 chaperone and are directly subjected to lysosomal degradation into amino acids [36]. Misfolded aggregates form the basis of the molecular mechanism behind cytotoxicity and neuronal death-mediated neurodegeneration. Neurons are more prone to protein quality control dysfunction-mediated neuroinflammatory damage because of their cellular structure, which expands based on the axon and dendrite contacts, where aggregate cargo formation for autophagic vacuole becomes challenging. Further, its retrograde transport to the cell body for lysosomal degradation is another factor making protein quality control difficult in neurons. Aging becomes yet another saboteur for complicating this process, which has been considered an important factor for neurodegenerative disease conditions.

Age-dependent decline in proteasomal activity is widely documented for several pathological conditions, including neurodegeneration. A transgenic AD model with proteasomal dysfunction showed tau hyperphosphorylation and ER-stress-related learning and memory impairment. The JNK pathway, a component of ER stress-mediated apoptosis, plays a vital role in neurofibrillary tangle formation, which has been linked to aggravating the proteostatic stress in the AD model [40]. Impairment of the 26S proteasomal component in a 19S proteasome subunit-deficient mouse led to NF-κB-mediated neuroinflammation in the forebrain region of the brain, with pronounced neuroinflammation and glial activation. Other inflammatory proteins such as Stat1 and Trem2 were also increased, leading to behavioral and cognitive deficiencies [18].

8.3 Neuron Versus Glia: UPS Divergence/Bifurcation

Neurodegenerative conditions have been linked to dysfunctional neurons, but the role of glial cells, including astrocytes and microglia, has started to gain attention from the scientific community after the emergence of their roles as immune cells of the brain. Neurons cannot divide and have a low-to-no tendency to regenerate after damage due to their post-mitotic nature. On the other hand, glial cells are proliferative and, hence, remove or dilute misfolded protein toxicity during cell division. Neurons rely on UPS for protein homeostasis and quality control as they have limited autophagy capacity, besides other factors such as aging, proteasomal inhibition, or component loss due to overloaded protein burden in neurons [20, 41]. In neurodegenerative conditions, neurons have shown proteinopathies, but similar pathology was not observed in glial cells, which may be due to cellular differences in protein synthesis and metabolism [42]. These differences may henceforth be extrapolated to the UPS variability among the two cell types because proteasome inhibition has been reported to affect nucleic acid oxidation in neurons more than astrocytes. Hence, making neurons more prone to neurodegenerative conditions such as Alzheimer's disease (AD) and Parkinson's disease (PD) [7].

The glial function is vital for regulating neuronal survival. Hence, disturbances in glial function cause neurodegeneration and excitotoxicity. Glia cells (present in the white matter) are hypothesized to clear the aggregated forms of protein more efficiently than neuronal cells, but proteasomal impairment has been linked to gliosis. 26S proteasomal depletion in the neurons of an experimental animal model caused inclusion body formation and neurodegeneration, besides activating GFAP and vimentin proteins in astrocytes, that is, reactive astrogliosis, which may be extrapolated to heat shock protein HSP25 participation [8, 11]. Proteasome inhibitors (epoxomicin and MG-132) aided in decreasing the mRNA expression levels of astrocyte reactive proteins such as vimentin, GFAP, and nestin, suggesting the regulatory participation of the proteasomal system in glial activities [30]. A similar regulation pattern was observed in oligodendrocytes and microglia cells for proteasomal regulation-mediated gliosis activation [3, 12]. Proteasome inhibition leads to increased nucleic acid oxidation in primary neurons and astrocytes. However, the former was affected with greater intensity. Cytokines such as TNF-α, IFN-γ, IL-2, and G-CSF lead to the upregulation of Ub-like protein ubiquitin-D (FAT10), which targets the long-surviving proteins in the proteasome of neuronal cells. The same effect was not observed in astrocytes or glial cells, suggesting differential regulation of UPS in glia and neurons [24]. In the case of intranuclear inclusion body disease, most of the aggregates are found in glial cells compared to neurons, with slightly more preference for astrocytes than oligodendrocytes. All glial aggregates are positive for ubiquitin, but neurons also showed expression of ubiquitin-like proteins SUMO1, NUB1, and NEDD8, which was not observed in glial cells. The UPS components found in the inclusion bodies differ when compared between neurons and glial cells [31]. Hence, one can say that neurons and glia react differently to proteasome inhibition and cytokines [20].

8.4 UPS-Mediated Neuroinflammatory Pathology in Neurodegenerative Disorders

An important component of cellular homeostasis regulation is the UPS due to its protein quality control process by regulating malfunctioned or overproduced protein degradation and clearance. However, over the period of time, the role of UPS in regulating the neuroinflammatory processes has started to dominate the neurodegenerative disease pathology. Protein misfolding is a major participant in neurodegenerative conditions. These misfolded proteins lead to aggregation and further activation of cytokines like IL-6 and TNF-α that can reduce the expression and activation of proteasomal proteins, leading to exacerbated neuroinflammatory response. Neurodegeneration is accelerated due to such a feedback loop, also impeding the brain's ability to recover from toxic insults. Besides cytokines, inflammatory proteins like NF-κB play a vital role in regulating UPS. It can downregulate the expression of proteasomal subunits, impairing protein degradation and promoting persistent damaged protein aggregation.

8.4.1 Alzheimer's Disease

AD patients have brain atrophies, ventricular enlargement, neuronal shrinkage, synaptic dysfunction, and cognitive impairment due to amyloid β (Aβ) and tau protein aggregates in the form of senile plaques and neurofibrillary tangles, respectively. Over the past decades, the role of astrocyte and microglia-mediated neuroinflammation has evolved as a principal contributor to AD pathology due to Aβ plaques, resulting in increased expression of pro-inflammatory markers [37, 39]. Due to chronic activation, microglia cause increased expression of pro-inflammatory markers such as IL-1β, TNF-α, and IL-6 [43]. Ubiquitin hydrolase increases the cellular level of mono-ubiquitin, participating in increasing the protein turnover by UPS. Aβ accumulation has been linked to low ubiquitin hydrolase levels in AD. Neuronal F-box protein (Fbx2) is able to improve cognitive performance and synaptic plasticity by regulating the β-secretase ubiquitination, thus targeting Aβ generation and its proteasomal degradation. Both these proteins, ubiquitin hydrolase and Fbx2, significantly participate in improving cognitive function and synaptic plasticity in AD mouse models. A frameshift mutation in the ubiquitin gene (UBB + 1) inhibits proteasomal functions and causes an increased neurofibrillary tangle production in disease conditions [13]. The deubiquitinating system (DUB) is also said to play a significant role in disease pathology [35].

8.4.2 Parkinson's Disease

Accumulation of Lewy bodies, comprised of an aggregated form of α-Syn protein, is a major saboteur in PD pathology. Neuroinflammation is caused by impaired clearance of these aggregates, where UPS plays a significant role. These aggregates lead to the activation of microglia in the substantia nigra region of the brain, causing cytokine-mediated neuroinflammation and dopaminergic neurodegeneration. In chronic conditions, peripheral cytokines and the immune system are activated, causing a neuroinflammatory milieu [1]. Mutation of the SNCA gene causes an improper form of the synuclein protein, which the UPS system cannot degrade. Parkin (E3 ubiquitin ligase) is important in dysfunctional protein removal, including the mitochondrial proteins. Mutation in PARK2 (encoding Parkin) leads to impaired ubiquitination and proteasomal degradation of proteins, triggering cellular stress and neuroinflammation. Another protein, PTEN-induced kinase 1, works in tandem with Parkin and is vitally involved in mitophagy regulation. In the disease state, ubiquitin carboxyl-terminal hydrolase L1 (UCHL1) is reported to be impaired, leading to the accumulation of polyubiquitinated proteins and impaired protein clearance. DJ-1 (encoded by the PARK7 gene) is reported to regulate oxidative stress by acting as a chaperone, modulating the activity of UPS and mitochondria. Other important proteins involved in proteasomal stress regulation in PD are Trosin A, 26S proteasome, along with catalytic core PSMA1, PSMB2, and LC3 autophagy regulator [19, 26].

8.4.3 Amyotrophic Lateral Sclerosis

Amyotrophic lateral sclerosis (ALS) is a progressive motor neurodegenerative disease affecting neurons in the brain and the spinal cord, causing muscular weakness, paralysis, and death. Proteins such as SOD1, TAR DNA-binding protein (TDP-43), and fused in sarcoma (FUS) are principally involved in ALS pathology [29]. Hence, UPS regulation plays an important role in regulating these proteins' activity. Ubiquitin-positive inclusions of TDP-43 and FUS in motor neurons are a hallmark of ALS pathology [4]. E3 ligases such as Parkin and Cbl are involved in protein turnover by UPS for TDP-43 protein. SQSTM1/p62, an important autophagic protein, is found upregulated in ALS. The p62-positive inclusions can sequester ubiquitinated substrate and facilitate UPS and autophagic dysfunction, exacerbating ALS pathology. Other important proteasomal components that participate in ALS pathology are UCHL1, PSMA1, PSMB1, and valosin-containing protein/p97 (VCP). Molecular chaperones like heat shock protein (HSP70 and HSP90) are also upregulated in response to cellular stress in ALS condition, exacerbating motor neuron degeneration [27, 34].

8.4.4 Frontotemporal Dementia

Frontotemporal dementia (FTD) is a neurodegenerative condition where the frontal and temporal lobes of the brain are progressively affected, leading to severe behavioral, lingual, and personality changes, causing functional and cognitive impairment [25]. TDP-43, primarily located in the nucleus and involved in regulating mRNA splicing and gene expression, is mislocalized to the cytoplasm after improper post-translational modification, leading to toxic accumulation in neurons in FTD conditions. FUS (RNA-binding protein regulating transcription, splicing, and RNA transport) is also found localized in the cytoplasm instead of the nucleus and becomes resistant to proteasomal degradation. Similarly, UCHL1, Tau, Parkin, VCP, PSMA1, PSMB1, HSP70, and HSP90 are found targets for proteasomal dysregulation in FTD [9].

8.4.5 Huntington's Disease

Huntington's disease's characteristic feature is the mutant form of the huntingtin protein (encoded by the HTT gene) with polyglutamine (CAG) repeats, causing motor dysfunction. These extra polyglutamine repeats cause an altered degradation of the peptide by UPS, causing toxic aggregate accumulation and cellular function disruption, further causing microglial and neuroinflammatory cascade activation in the striatal region of the brain. Like other neurodegenerative conditions, proteasomal 26S (20S and 19S), E3 ubiquitin ligase (Parkin, Cbl), TDP-43, HSP70, HSP90, SQSTM1/p62, and VCP are directly or indirectly involved in regulating the mutant form of HTT protein, causing proteotoxic stress and UPS dysfunction [15, 33].

8.5 Therapeutic Targeting of UPS-Mediated Neuroinflammation

Based on the widespread participation of UPS in neurodegenerative disease pathology, UPS-mediated neuroinflammation is an active area of research for therapeutic targeting and disease management. Small molecule proteasome activators or proteasome degradation inhibitors (e.g., bortezomib) can regulate proteasomal function in disease conditions. Alternatively, targeted protein degraders are also being explored for the same purpose. Two E1 enzymes, nearly 40 E2 enzymes, around 600 E3-ligases, and almost 100 DUBs, and 32 proteasome subunits are theoretically considered potential druggable targets. Targeting E1 or E2 enzymes is comparatively less efficient, while targeting specific E3 ligase or DUBs will offer a more efficient druggable UPS node as they have the ability to bind to the substrate with minimum-to-no side effects directly [28]. FBXO3 and βTrCP are previously reported to be suitable targets for inflammatory conditions using inhibitors BC-1215 and GS143, respectively. Similarly, IAPs are explored in cancer immunotherapy. Neo-substrates targeting using E3s repurposing has been explored for protein degradation, for example, PROTAC (protein-targeting chimeric molecule). PROTACs comprises of three subunits, a substrate binding moiety, a linker region, and an E3-binding unit. PROTAC1 was designed using IKBα phosphopeptide, which binds to $SCF^{TRCP\text{-}\beta}$, and ovacilin, which ligates MetAp-2, getting it closer to E3-ligases, aiding ubiquitylation and degradation. Another group used P300/CBP-associated factor and general control nonderepressible 5 targeting PROTAC strategy for promoting its degradation, hence reducing immune system-mediated inflammation in LPS-stimulated dendritic cells and macrophages. Inhibition of IRAK4 PROTAC helps in inhibiting cytokine production in peripheral blood mononuclear cells. The major limitation for such advanced peptide therapy is its diffusion through the cell membrane due to its larger size. To overcome this challenge, molecular glues were discovered, which are simple small molecules promoting protein–protein interactions. This kind of molecular glue binds to both E3-ligase and the substrate, aiding substrate degradation. Auxin is one such glue that aids the interaction between the F-box protein Arabidopsis auxin receptor TIR1 and its substrate. Lenalidomide, acting as a glue, helps in the multiple myeloma condition by targeting ubiquitin-mediated degradation of IKZF1 and IKZF3 (B-cell transcription factors) by CRBN-CRL4 ubiquitin ligase [10]. Consequently, targeting microglial activation and other immune system components might aid in targeting neuroinflammatory responses in neurodegenerative conditions.

8.6 Future Prospective and Conclusion

Neuroinflammation has evolved as an important contributor to neurodegeneration, and the interplay between neuroinflammation and UPS dysfunction is inevitable. Protein degradation impairment causes the accumulation of toxic aggregates, leading to proteotoxic stress in neurons and exacerbating neuronal damage. These two processes are critically interlinked, and understanding the molecular mechanism in the interplay might help develop therapeutic strategies and better refine the disease pathology. E3 targeted therapy offers a promising window for drug development, as nearly 600 E3 ligases are coded by the human genome. Although the E3-ligase/substrate pairs repertoire limits this aspect due to limited exploration in this direction. Understanding the molecular mechanism for substrate degradation is crucial currently. PROTACs and glue-mediated targeting have evolved as promising approaches in such situations. Exploring CRISPR-Cas9 gene editing technology for an easy E3-ligase identification warrants further investigations in suitable disease settings. As the research in this field advances, targeted therapies addressing the underlying mechanism of UPS dysfunction and neuroinflammation might offer promising avenues for mitigating neurodegenerative disease progression.

References

1. Behl T, et al. Exploring the role of ubiquitin-proteasome system in Parkinson's disease. Mol Neurobiol. 2022;59(7):4257–73. Available at: https://doi.org/10.1007/S12035-022-02851-1.
2. Bhat SA, et al. Ubiquitin proteasome system in immune regulation and therapeutics. Curr Opin Pharmacol. 2022;67:102310. Available at: https://doi.org/10.1016/J.COPH.2022.102310.
3. Bi W, et al. Inhibition of 26S protease regulatory subunit 7 (MSS1) suppresses neuroinflammation. PLoS One. 2012;7(5):e36142. Available at: https://doi.org/10.1371/JOURNAL.PONE.0036142.
4. Chisholm C, et al. Ubiquitin homeostasis disruption, a common cause of proteostasis collapse in amyotrophic lateral sclerosis? Neural Regen Res. 2022;17(10):2218–20. Available at: https://doi.org/10.4103/1673-5374.335786.
5. Ciechanover A, Kwon YT. Degradation of misfolded proteins in neurodegenerative diseases: therapeutic targets and strategies. Exp Mol Med. 2015;47(3):e147. Available at: https://doi.org/10.1038/emm.2014.117.
6. Clague MJ, Urbé S. Ubiquitin: same molecule, different degradation pathways. Cell. 2010;143(5):682–5. Available at: https://doi.org/10.1016/J.CELL.2010.11.012.
7. Ding Q, et al. Proteasome inhibition increases DNA and RNA oxidation in astrocyte and neuron cultures. J Neurochem. 2004;91(5):1211–8. Available at: https://doi.org/10.1111/J.1471-4159.2004.02802.X.
8. Elkharaz J, et al. Implications for oxidative stress and astrocytes following 26S proteasomal depletion in mouse forebrain neurones. Biochim Biophys Acta. 2013;1832(12):1930–8. Available at: https://doi.org/10.1016/J.BBADIS.2013.07.002.
9. Ferrari R, et al. Frontotemporal dementia: insights into the biological underpinnings of disease through gene co-expression network analysis. Mol Neurodegener. 2016;11(1):1–19. Available at: https://doi.org/10.1186/S13024-016-0085-4/FIGURES/7.
10. Fink EC, Ebert BL. The novel mechanism of lenalidomide activity. Blood. 2015;126(21):2366–9. Available at: https://doi.org/10.1182/BLOOD-2015-07-567958.
11. Goldbaum O, et al. The small heat shock protein HSP25 protects astrocytes against stress induced by proteasomal inhibition. Glia. 2009;57(14):1566–77. Available at: https://doi.org/10.1002/GLIA.20870.
12. Goldbaum O, Vollmer G, Richter-Landsberg C. Proteasome inhibition by MG-132 induces apoptotic cell death and mitochondrial dysfunction in cultured rat brain oligodendrocytes but not in astrocytes. Glia. 2006;53(8):891–901. Available at: https://doi.org/10.1002/GLIA.20348.
13. Gong B, et al. The ubiquitin-proteasome system: potential therapeutic targets for alzheimer's disease and spinal cord injury. Front Mol Neurosci. 2016;9(Jan):168119. Available at: https://doi.org/10.3389/FNMOL.2016.00004/BIBTEX.
14. Hadian K, et al. NF-κB essential modulator (NEMO) interaction with linear and lys-63 ubiquitin chains contributes to NF-κB activation. J Biol Chem. 2011;286(29):26107–17. Available at: https://doi.org/10.1074/JBC.M111.233163.
15. Harding RJ, Tong YF. Proteostasis in Huntington's disease: disease mechanisms and therapeutic opportunities. Acta Pharmacol Sin. 2018;39(5):754–69. Available at: https://doi.org/10.1038/aps.2018.11.

16. Hardy J, Selkoe DJ. The amyloid hypothesis of Alzheimer's disease: progress and problems on the road to therapeutics. Science (New York, NY). 2002;297(5580):353–6. Available at: https://doi.org/10.1126/SCIENCE.1072994.
17. Hetz C, Mollereau B. Disturbance of endoplasmic reticulum proteostasis in neurodegenerative diseases. Nat Rev Neurosci. 2014;15(4):233–49. Available at: https://doi.org/10.1038/NRN3689.
18. Huber CC, et al. Impaired 26S proteasome causes learning and memory deficiency and induces neuroinflammation mediated by NF-κB in mice. bioRxiv. 2024:2024.02.09.579699. Available at: https://doi.org/10.1101/2024.02.09.579699.
19. Jang HJ, Chung KC. The ubiquitin–proteasome system and autophagy mutually interact in neurotoxin-induced dopaminergic cell death models of Parkinson's disease. FEBS Lett. 2022;596(22):2898–913. Available at: https://doi.org/10.1002/1873-3468.14479.
20. Jansen AHP, Reits EAJ, Hol EM. The ubiquitin proteasome system in glia and its role in neurodegenerative diseases. Front Mol Neurosci. 2014;7(Aug):103096. Available at: https://doi.org/10.3389/FNMOL.2014.00073/BIBTEX.
21. Kincaid EZ, et al. Mice completely lacking immunoproteasomes show major changes in antigen presentation. Nat Immunol. 2011;13(2):129–35. Available at: https://doi.org/10.1038/NI.2203.
22. Liew HK, et al. Over-activated proteasome mediates neuroinflammation on acute intracerebral hemorrhage in rats. Cells. 2019;8(11):1326. Available at: https://doi.org/10.3390/CELLS8111326.
23. Lilienbaum A. Relationship between the proteasomal system and autophagy. Int J Biochem Mol Biol. 2013;4(1):1. Available at: https://pmc.ncbi.nlm.nih.gov/articles/PMC3627065/. Accessed 14 Jan 2025.
24. Lisak RP, et al. Cytokines regulate neuronal gene expression: differential effects of Th1, Th2 and monocyte/macrophage cytokines. J Neuroimmunol. 2011;238(1–2):19–33. Available at: https://doi.org/10.1016/J.JNEUROIM.2011.06.010.
25. Mackenzie IRA, Neumann M. Molecular neuropathology of frontotemporal dementia: insights into disease mechanisms from postmortem studies. J Neurochem. 2016;138:54–70. Available at: https://doi.org/10.1111/JNC.13588.
26. McKinnon C, et al. Early-onset impairment of the ubiquitin-proteasome system in dopaminergic neurons caused by α-synuclein. Acta Neuropathol Commun. 2020;8(1):1–16. Available at: https://doi.org/10.1186/S40478-020-0894-0/FIGURES/7.
27. Medinas DB, Valenzuela V, Hetz C. Proteostasis disturbance in amyotrophic lateral sclerosis. Hum Mol Genet. 2017;26(R2):R91–R104. Available at: https://doi.org/10.1093/HMG/DDX274.
28. Meiners S, Evankovich J, Mallampalli RK. The ubiquitin proteasome system as a potential therapeutic target for systemic sclerosis. Transl Res. 2018;198:17–28. Available at: https://doi.org/10.1016/J.TRSL.2018.03.003.
29. Merjane J, et al. Molecular mechanisms of amyotrophic lateral sclerosis as broad therapeutic targets for gene therapy applications utilizing adeno-associated viral vectors. Med Res Rev. 2023;43(4):829–54. Available at: https://doi.org/10.1002/MED.21937.
30. Middeldorp J, et al. Intermediate filament transcription in astrocytes is repressed by proteasome inhibition. FASEB J. 2009;23(8):2710–26. Available at: https://doi.org/10.1096/FJ.08-127696.
31. Mori F, et al. Ubiquitin-related proteins in neuronal and glial intranuclear inclusions in intranuclear inclusion body disease. Pathol Int. 2012;62(6):407–11. Available at: https://doi.org/10.1111/J.1440-1827.2012.02812.X.
32. Orre M, et al. Reactive glia show increased immunoproteasome activity in Alzheimer's disease. Brain. 2013;136(Pt 5):1415–31. Available at: https://doi.org/10.1093/BRAIN/AWT083.
33. Ortega Z, Lucas JJ. Ubiquitin-proteasome system involvement in Huntington's disease. Front Mol Neurosci. 2014;7(Sep):99006. Available at: https://doi.org/10.3389/FNMOL.2014.00077/BIBTEX.

34. Paez-Colasante X, et al. Amyotrophic lateral sclerosis: mechanisms and therapeutics in the epigenomic era. Nat Rev Neurol. 2015;11(5):266–79. Available at: https://doi.org/10.1038/nrneurol.2015.57.
35. Qin B, et al. DUBs in Alzheimer's disease: mechanisms and therapeutic implications. Cell Death Discov. 2024;10(1):1–15. Available at: https://doi.org/10.1038/s41420-024-02237-3.
36. Rothenberg C, et al. Ubiquilin functions in autophagy and is degraded by chaperone-mediated autophagy. Hum Mol Genet. 2010;19(16):3219. Available at: https://doi.org/10.1093/HMG/DDQ231.
37. Rozemuller JM, Eikelenboom P, Stam FC. Role of microglia in plaque formation in senile dementia of the Alzheimer type - an immunohistochemical study. Virchows Arch B Cell Pathol Incl Mol Pathol. 1986;51(1):247–54. Available at: https://doi.org/10.1007/BF02899034/METRICS.
38. Schmidt MF, et al. Ubiquitin signalling in neurodegeneration: mechanisms and therapeutic opportunities. Cell Death Differ. 2021;28(2):570–90. Available at: https://doi.org/10.1038/s41418-020-00706-7.
39. Sobue A, Komine O, Yamanaka K. Neuroinflammation in Alzheimer's disease: microglial signature and their relevance to disease. Inflamm Regen. 2023;43(1):1–6. Available at: https://doi.org/10.1186/S41232-023-00277-3/FIGURES/1.
40. Tomaru U, et al. Decreased proteasomal function induces neuronal loss and memory impairment. Am J Pathol. 2021;191(1):144–56. Available at: https://doi.org/10.1016/J.AJPATH.2020.10.004/ATTACHMENT/9EA0FBE0-D82E-4763-AE5A-6AA4CDCA78F7/MMC4.PDF.
41. Tydlacka S, et al. Differential activities of the ubiquitin–proteasome system in neurons versus glia may account for the preferential accumulation of misfolded proteins in neurons. J Neurosci. 2008;28(49):13285–95. Available at: https://doi.org/10.1523/JNEUROSCI.4393-08.2008.
42. Vilchez D, et al. Increased proteasome activity in human embryonic stem cells is regulated by PSMD11. Nature. 2012;489(7415):304–8. Available at: https://doi.org/10.1038/nature11468.
43. Weng FL, He L. Disrupted ubiquitin proteasome system underlying tau accumulation in Alzheimer's disease. Neurobiol Aging. 2021;99:79–85. Available at: https://doi.org/10.1016/J.NEUROBIOLAGING.2020.11.015.

Ubiquitin Signaling in CNS Neurodegenerative Diseases

9

Amritpal Kaur, Shareen Singh, and Thakur Gurjeet Singh

Abbreviations

AD	Alzheimer's disease
PD	Parkinson's disease
HD	Huntington's disease
ALS	Amyotrophic lateral sclerosis
UPS	Ubiquitin-proteasome system
E1	Ubiquitin-activating enzyme
E2	Ubiquitin-conjugating enzyme
E3	Ubiquitin ligase
DUBs	Deubiquitinating enzymes

9.1 Introduction

Ubiquitination is a crucial posttranslational modification that plays a key role in regulating protein turnover, signaling, and maintaining cellular balance by attaching ubiquitin molecules to target proteins. This process involves a series of enzymatic activities carried out by E1 ubiquitin-activating enzymes, E2 ubiquitin-conjugating

A. Kaur · S. Singh (✉)
Chitkara College of Pharmacy, Chitkara University, Rajpura, Punjab, India
e-mail: k.amritpal@chitkara.edu.in; shareen.singh@chitkara.edu.in

T. G. Singh
Chitkara College of Pharmacy, Chitkara University, Rajpura, Punjab, India

School of Public Health, Faculty of Health, University of Technology Sydney, Ultimo, NSW, Australia
e-mail: gurjeet.singh@chitkara.edu.in

S. Singh (ed.), *Ubiquitin Proteasome System in CNS Neurodegenerative Diseases*, https://doi.org/10.1007/978-981-95-6099-8_9

enzymes, and E3 ubiquitin ligases, which provide specificity for the substrates [65, 68]. Depending on the type of ubiquitin chain that is added, ubiquitination can either signal for proteasomal degradation, facilitate autophagic clearance, or influence protein activity and localization [82]. Disruption of this complex system has been linked to the development of various neurodegenerative diseases, such as Alzheimer's disease (AD), Parkinson's disease (PD), Huntington's disease (HD), and amyotrophic lateral sclerosis (ALS). These conditions are marked by the buildup of misfolded and aggregated proteins, including beta-amyloid and tau in AD, alpha-synuclein in PD, and mutant huntingtin in HD, which overwhelm the cell's ability to maintain protein balance and lead to neuronal dysfunction and death. In neurodegenerative diseases, the ubiquitin-proteasome system (UPS) and autophagy-lysosome pathways, which both depend heavily on ubiquitination, are crucial for protein quality control [77]. Abnormal ubiquitination has been noted in protein aggregates specific to these diseases, indicating issues with the recognition, tagging, or degradation of misfolded proteins. For example, in Parkinson's disease (PD), mutations in the E3 ligase Parkin and the ubiquitin kinase PINK1 hinder mitophagy, leading to mitochondrial dysfunction and neuronal degeneration [60]. Likewise, in Huntington's disease (HD), the mutant huntingtin protein evades effective proteasomal degradation despite being ubiquitinated, resulting in aggregate formation and cellular toxicity [7]. These findings highlight the importance of ubiquitination in preserving neuronal health and its potential as a target for therapy. Recent progress in our understanding of the ubiquitination machinery and its involvement in neurodegeneration has paved the way for new therapeutic options. Researchers are investigating strategies to modulate the activity of E3 ligases, boost proteasomal degradation, or enhance the autophagic clearance of protein aggregates [74]. Furthermore, targeting deubiquitinating enzymes (DUBs), which reverse the process of ubiquitination and influence substrate stability, offers an additional method for restoring proteostasis [39, 64]. However, despite these encouraging advancements, challenges persist, such as ensuring the specificity of therapies that target ubiquitination and navigating the complexity of ubiquitin's various roles in cellular functions. This chapter explores the molecular mechanisms behind ubiquitination, how its dysregulation contributes to neurodegenerative diseases, and the new therapeutic strategies being developed. It emphasizes the promise of targeting this pathway to meet the pressing needs in the treatment of these challenging conditions.

9.2 Ubiquitination Pathway and Its Components

9.2.1 The Ubiquitin-Proteasome System and Ubiquitination Process

The ubiquitin-proteasome system (UPS) serves as the principal intracellular mechanism for protein breakdown. Ubiquitination is a posttranslational modification process that entails the covalent binding of ubiquitin peptides to designated target proteins [69]. This pathway is essential for numerous physiological functions, including protein degradation, trafficking, DNA repair, and cell signaling. This process occurs in a cascade, with the cessation of ubiquitination assisted by E3 (ubiquitin ligase), E1 (ubiquitin-activating enzyme), and E2 (ubiquitin-conjugating enzyme) [61]. The ubiquitin-activating enzyme commences ubiquitin activation by transferring it to the ubiquitin-conjugating enzyme. This procedure is predominantly accomplished through the cooperation of E2 and E3 ligases, which enable the transfer of ubiquitin to the target protein [30, 38]. Ubiquitination can occur through multiple methods, leading to diverse results. The mono-ubiquitination process involves the addition of a single ubiquitin moiety, which may influence the localization or activity of proteins. The assembly of polyubiquitin chains comprises many chains, each possessing distinct activities that dictate the fate of the changed protein. Ubiquitination not only fulfils its basic job in protein degradation but also performs crucial non-degradative functions. K63-linked polyubiquitination is implicated in multiple physiological activities, such as DNA repair, protein sorting, and cell signaling [62]. It is essential for the formation of protein complexes and the establishment of molecular linkages. Deubiquitinating enzymes (DUBs) cleave ubiquitin moieties from proteins, thereby modulating ubiquitin-mediated signaling pathways [3].

9.3 Mechanisms of Ubiquitination in Neurodegeneration

9.3.1 Ubiquitination and Protein Degradation

The ubiquitination process has several stages: activation, conjugation, ligation, and polyubiquitination. In the activation phase, an E1 enzyme activates ubiquitin by establishing a thioester bond in an ATP-dependent way [6]. During conjugation, the active ubiquitin is conveyed to an E2 enzyme. In the ligation phase, the E2 enzyme collaborates with an E3 ligase to conjugate ubiquitin to the lysine residue of the target protein [79]. The successive addition of numerous ubiquitin moieties in polyubiquitination can produce chains with diverse biological effects. Upon being tagged with polyubiquitin chains, the target protein is recognized and destroyed by the 26S proteasome complex, a substantial multi-subunit protease [53].

9.4 Ubiquitination in Specific Neurodegenerative Diseases

9.4.1 Ubiquitination Process in Alzheimer's Disease

Ubiquitination is an ATP-dependent process, facilitated by enzymes, that involves the covalent attachment of a 76-amino acid ubiquitin to a target protein. This attachment transpires via one of the seven Lys side chains or the C-terminus of the target protein. The outlined method, which includes the activation, conjugation, and ligation of proteins by enzymes E1, E2, and E3, is essential for maintaining proteostasis. The principal role is to tag ubiquitin-conjugated proteins for degradation by the

26S proteasomal complex [20]. The ubiquitin code demonstrates considerable plasticity, facilitating the elongation of ubiquitin chains and their subsequent degradation by deubiquitinases (DUBs). Ubiquitin is subject to posttranslational modifications (PTMs), such as phosphorylation and acetylation, which enhance its regulatory functions [42]. The deregulation of the ubiquitin-proteasome degradation pathway contributes to the accumulation of misfolded proteins and aggregates, ultimately resulting in the pathogenesis of Alzheimer's disease [67]. The interference obstructs the removal of aggregation species and the proteasomal degradation process, hence exacerbating disease progression indicate that proteins associated with the etiology of Alzheimer's disease, including amyloid precursor protein (APP), beta-secretase 1 (BACE1), and tau, undergo aberrant ubiquitination (Fig. 9.1). The utilization of an antibody that specifically targets anti-ubiquitin conjugates has demonstrated efficacy in identifying neuritic plaques and neurofibrillary tangles (NFTs) in cerebral cortex samples impacted by Alzheimer's disease (AD) [45]. This conclusion indicates that the ubiquitination process may significantly contribute to the formation and progression of Alzheimer's disease.

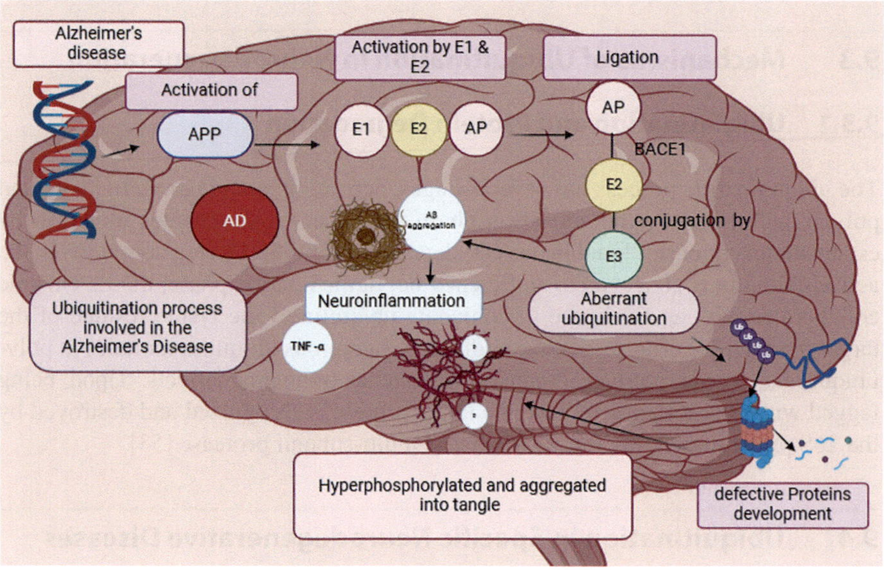

Fig. 9.1 The ubiquitination process begins with the E1 enzyme (Activation), which uses ATP to activate ubiquitin, transferring it to the E2 enzyme (Conjugation). The E3 enzyme (Ligation) facilitates the transfer of ubiquitin from E2 to the target protein, tagging it for degradation. Ubiquitin, a small 76-amino acid protein, attaches to target proteins, signaling their path. The proteasome, a large protein complex, degrades ubiquitin-tagged proteins. In Alzheimer's disease, misfolded proteins like APP, BACE1, and tau undergo aberrant ubiquitination, contributing to the formation of amyloid plaques and neurofibrillary tangles (NFTs), leading to the AD

9.4.2 Ubiquitination of Tau Protein in Alzheimer's Disease

Focusing on the ubiquitination of tau protein has become a promising treatment approach for Alzheimer's disease. Tau pathology is a hallmark of Alzheimer's disease and may lead to neuronal dysfunction and cognitive decline. Proteins undergo degradation by the proteasome following ubiquitination [29]. In Alzheimer's disease, tau protein undergoes ubiquitination, signifying that the ubiquitin-proteasome system eliminates dysfunctional tau protein. Tau is a protein that generally preserves neuronal architecture, but in Alzheimer's disease, it undergoes hyperphosphorylation, resulting in the formation of toxic neurofibrillary tangles (NFTs). [32]. Tau ubiquitination is impaired in Alzheimer's disease. The proteasome system destroys proteins via ubiquitination. Ubiquitination of tau protein is compromised in Alzheimer's disease. Tau protein is a microtubule-associated protein that facilitates the assembly and stabilization of microtubules in neuronal cells. Tau hyperphosphorylation induces misfolding and the formation of neurofibrillary tangles (NFTs) in Alzheimer's disease. Tau protein is subject to ubiquitination at many lysine residues, notably K48, which facilitates proteasomal breakdown, and K63, which may lead to the formation of neurofibrillary tangles, a hallmark of Alzheimer's disease (AD) [19]. Their role in promoting or inhibiting tau aggregation remains ambiguous; nevertheless, their capacity to regulate tau degradation and localization suggests they may indirectly influence it in the brain. E3 ligases modulate the clearance of tau protein in the brain to prevent Alzheimer's disease-associated neurofibrillary tangles (NFTs) [37]. The E3 ligases CHIP, TRAF6, and auxotrophic/MARCH 7 conjugate ubiquitin moieties to tau protein, facilitating its degradation via the proteasome or lysosome-autophagy pathways. Ubiquitin is involved in the aggregation of brain tau; however, the role of E3 ligases in either facilitating or inhibiting this process remains ambiguous. Research indicates that the ubiquitination by E3 ligases may alter the solubility and localization of tau protein, hence indirectly influencing tau aggregation [32]. CHIP governs the destruction of both soluble and aggregated tau through the proteasome and lysosomal-autophagy pathways, potentially diminishing tau protein buildup and inhibiting tau aggregation, which may offer therapeutic benefits for tauopathies in Alzheimer's disease (AD) [56]. It seems improbable that TRAF6-mediated tau ubiquitination influences its localization to autophagosomes and lysosomes, indicating a potential regulatory role in tau protein degradation via both pathways [46].

9.4.3 Ubiquitination of Amyloid Protein Precursor (APP) in Alzheimer's Disease (AD)

In Alzheimer's disease (AD), the amyloid protein precursor (APP) undergoes ubiquitination, influencing protein stability and degradation. APP engages with E3 ubiquitin-protein ligase Stub1, a modulator of E3 ubiquitin-protein ligases, and PPAR [11]. The lysines ACR 649, 650, 651, 676, and 688 are ubiquitinated by E3 ligases, predominantly Stub1. APP is directed into endosomes through the ubiquitination of lysine in the ACR domain, where pSEN2 subsequently converts it into amyloid peptides. The etiology of Alzheimer's disease may entail the ubiquitination of amyloid precursor protein, which impedes secretase processing and reduces amyloid peptide synthesis [49]. CBD treatment diminishes APP expression and Aβ production by activating PPAR, resulting in APP degradation via ubiquitination. Ubiquitination modulates the amyloid peptide-generating enzyme BACE1. The proteasome destroys BACE1 upon ubiquitination at lysine residues 203 and 382 [47]. The E3 ligase Fbx2 associates with and ubiquitinates BACE1, resulting in reduced Aβ synthesis and enhanced degradation. PGC-1 regulates Fbx2 expression, which facilitates the ubiquitin-mediated degradation of BACE1 in Alzheimer's disease. Consequently, the ubiquitination of APP and BACE1 governs their stability and degradation, resulting in the production of amyloid peptides that contribute to Alzheimer's disease pathogenesis [83]. Consequently, the disruption of ubiquitination stabilizes beta-secretase (BACE1) inside the cell membrane and early and late endosomes, whereas the di-leucine motif facilitates endocytosis and BACE1 accumulation, thereby enhancing APP processing and amyloid production. It was discovered that USP8 deletion enhances BACE1 ubiquitination and degradation via the lysosomal pathway, hence diminishing APP processing and amyloidogenesis in H4 human neuroglioma cells [81]. The results indicate that focusing on the ubiquitination degradation pathway, namely USP8 activity, could be a potential treatment for Alzheimer's disease. Consequently, regulating the ubiquitin-proteasome system may provide a therapeutic approach for Alzheimer's disease.

9.4.4 SUMOylation in Alzheimer's Disease (AD)

Posttranslational SUMOylation is reversible like ubiquitination. Normal physiological activities include cell cycle regulation, nuclear-cytosolic transport, gene transcription, protein stability, stress response, and neuronal death, which depend on SUMOylation, unlike ubiquitin, which signals protein breakdown [22, 44]. The compact 12 kDa protein SUMO (small ubiquitin-like modifier) covalently binds to target protein lysine residues to change their activity. This posttranslational modification (PTM) controls many cell biology processes and is linked to certain neuron-specific functions [78]. In Alzheimer's disease (AD), SUMOylation process changes damage brain amyloid (Aβ) and tau protein aggregation, leading to neurodegeneration. NF-kB, PKC, c-Jun N-terminal kinase (JNK), p38 MAPK, PI3-K, ERK, and STAT, which are crucial to Alzheimer's disease's neuroinflammation, are modulated by SUMOylation [33]. The SUMOylation process regulates the caspase-3 pathway, which is linked to neurodegeneration caused by tau and Aβ proteins in Alzheimer's disease. AD is linked to ubc9 alteration [31]. Dysregulation of ubiquitin-conjugating enzyme 9 (Ubc9) causes amyloid beta protein aggregation and neuronal death by endoplasmic reticulum stress or mitochondrial oxidative stress in Alzheimer's disease. In Alzheimer's disease, a changed SUMOylation substrate (ubiquitin-conjugating enzyme 9) is linked to neuronal death signals via ER stress and mitochondrial dysfunction via aggregated proteins (Tau or Aβ). The cytotoxic effects of amyloid beta and tau protein cause brain neuroinflammation, which is linked to SUMOylation and glial cell hyperactivation. SUMOylation is facilitated by Ubc9, which modulates proteins like IκBα (inhibitor of kB), cytosolic nuclear factor kappa-B, and Rel proteins. Ubc9 binds to IκBα and Rel proteins, degrading them and activating NF-kB, which then transcribes pro-inflammatory mediators in the nucleus, worsening neuroinflammation in Alzheimer's disease. In Alzheimer's disease, astrocyte SUMO pathway dysregulation worsens neuroinflammation-induced oxidative stress in neural tissue [1]. Alzheimer's disease patients' brains have increased inflammatory mediators and reactive oxygen species, defined by nitric oxide synthase two enzyme-mediated neurodegeneration. By suppressing neuroinflammation-induced neurodegeneration, targeting ubiquitin-conjugating enzyme 9 (Ubc9), an enzyme in the SUMOylation pathway, may treat Alzheimer's disease. Ubiquitin-conjugating enzyme 9 (ubc9) is conjugated with amyloid-beta and tau proteins, suggesting it contributes to their aggregation in Alzheimer's disease [72]. In Alzheimer's disease, lipopolysaccharides target Ubiquitin-conjugating enzyme 9 (ubc9), reducing amyloid beta and tau protein aggregation, oxidative stress, and pro-inflammatory mediators that cause neurodegeneration [4]. The SUMO system regulates the neuroimmune system by modifying astrocyte activity, reducing neuroinflammation-causing pro-inflammatory mediators. Furthermore, the treatment of curcumin demonstrated efficient downregulation of Ubiquitin-conjugating enzyme 9 (ubc9) and inactivated the STAT1-dependent inflammatory pathway, thereby reducing neuroinflammation in Alzheimer's disease [57]. Consequently, the SUMOylation process may serve as a therapeutic target for neuroinflammation-induced neurodegeneration in Alzheimer's disease.

9.4.5 Parkinson's Disease

Parkinson's disease is a prevalent neurological disorder characterized by bradykinesia, resting tremor, rigidity, gait disturbances, and postural instability. The reduction of dopaminergic neurons in the substantia nigra pars compacta and the accumulation of Lewy bodies, mostly alpha-synuclein protein, promote their degeneration [25]. Though extensive, Parkinson's disease's processes of dopaminergic neuronal death and alpha-synuclein aggregation remain unclear. Genetic research has connected UCHL1, alpha-synuclein, DJ1, Parkin, and PINK1 to early-onset Parkinson's disease, revealing pathogenic circumstances. Parkinson's disease patients' substantia nigra pars compacta post-mortem shows Lewy bodies and dopaminergic neuron loss, supporting the diagnosis [43]. Decreased lysosomal activity, endoplasmic reticulum stress, and inflammatory and autophagic-apoptotic cell death cause alpha-synuclein fibrillar clumps and dopaminergic neurodegeneration in Parkinson's disease. Alpha-synuclein accumulates, segregates, and recycles neurotransmitters in presynaptic terminals, affecting neurophysiology. Alpha-synuclein abnormalities cause neurodegeneration in Parkinson's disease by affecting synaptic integrity, neurotransmitter release, and vesicle recycling [66]. Autosomal recessive juvenile parkinsonism (ARJP) often causes mitochondrial dysfunction due to Parkin and PINK1 gene mutations [54]. Alpha-synuclein increases cytotoxicity, inflammation, neuronal apoptosis, mitochondrial dysfunction, and endoplasmic reticulum stress-induced dopaminergic neurodegeneration in Parkinson's disease. Aggregated alpha-synuclein has been linked to mitochondrial dysfunction, biogenesis problems, and lysosomal diseases-mediated neurodegeneration in Parkinson's disease [63]. Alpha-synuclein monomers and oligomers inhibit mitochondrial protein import receptor TOM20 translocation. Alpha-synuclein blocks the outer mitochondrial membrane voltage-dependent anion channel (VDAC), affecting metabolite outflow and influx. Alpha-synuclein oligomers cause mitochondrial damage, calcium-mediated inflammation, depolarization, and cytochrome C release, causing dopaminergic neuron cell death [12]. Parkin mutations lead to mitochondrial malfunction and ubiquitination defects in Parkinson's disease. E3 ubiquitin ligase Parkin identifies damaged proteins for ubiquitin-proteasome destruction and protects mitochondrial function. DJ1, Parkin, and PINK1 genetic variants suggest that posttranslational modifications (PTMs) control Parkinson's disease proteins [58]. Cellular equilibrium requires DJ-1's reversible oxidative changes, Parkin's ubiquitination and phosphorylation, and PINK1's phosphorylation. PD pathophysiology is revealed by PTM dysregulation, causing cellular dysfunction. PD symptoms are linked to alpha-synuclein aggregation caused by posttranslational modifications (PTMs), especially in mutant genes like Parkin. LRRK2 phosphorylation dysregulation is essential to aberrant kinase activity and neurodegeneration in Parkinson's disease [48]. In familial and sporadic Parkinson's disease (PD), posttranslational modifications (PTMs) of alpha-synuclein and LRRK2 in Lewy bodies are often linked to neurological symptoms. Alpha-synuclein and LRRK2 pathogenic pathways are regulated by ubiquitination, phosphorylation, and sumoylation. Mutations in the LRRK2 gene cause familial and sporadic Parkinson's disease [51]. Several studies have examined how phosphorylation affects LRRK2, a multidomain protein that acts as a kinase and GTPase. Aberrant LRRK2 phosphorylation increases kinase activity and may cause Parkinson's [27]. Understanding how posttranslational modifications, especially phosphorylation, affect LRRK2 is crucial to understanding Parkinson's disease's complex molecular pathways and developing treatments that target these modifications (Fig. 9.2). This study reveals the complex relationship between posttranslational modifications, genetic changes, and dopaminergic neurodegeneration in Parkinson's disease, suggesting treatment options [51].

9.4.6 Parkinson's Disease SUMOylation

SUMOylation, a reversible posttranslational modification that covalently links the 12 kDa small ubiquitin-like modifier (SUMO) to lysine residues of target proteins, is essential for a variety of physiological functions [28, 44]. These include cell cycle control, nuclear-cytosolic transport, gene transcription, protein stability, stress response, and apoptosis. SUMOylation, despite its participation in a variety of biological processes, is most commonly used to facilitate protein breakdown. This posttranslational alteration has a substantial impact on protein structure, stability,

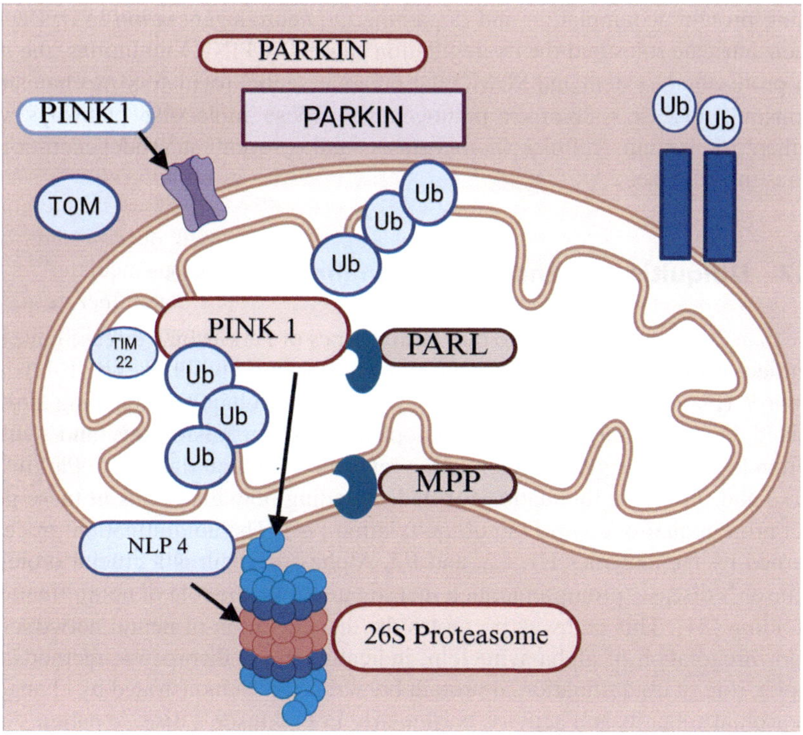

Fig. 9.2 The PINK1 and PARKIN pathway in mitochondria involved in PD

localization, and interactions. SUMOylation has an effect on neurodegenerative proteins like tau and alpha-synuclein. Parkinson's disease (PD) is distinguished by alpha-synuclein buildup and aggregation, although the underlying processes have yet to be uncovered. Posttranslational changes, particularly SUMOylation, have an effect on alpha-synuclein breakdown, protein-protein interactions, and subcellular localization. Stress increases protein SUMOylation, particularly of alpha-synuclein, perhaps leading to neurodegeneration in Parkinson's disease. Dopaminergic neurodegeneration in Parkinson's disease is linked to alpha-synuclein, which is encoded by the soluble NSF attachment protein receptor (SNARE). In alpha-synucleinopathies, SNARE mutations impair mitochondrial and lysosomal function, as well as DJ-1 function. DJ-1, an antioxidant, inhibits p53 and BAX SUMOylation, which modulates Nrf2 and prevents neuronal death [84]. DJ-1 mutations, such as L166P, have been linked to aberrant SUMOylation, suggesting a link between mutant proteins and Parkinson's disease-related neurodegeneration. Parkin is an RBR family (RING) E3 ligase that destroys misfolded proteins to safeguard cells [14]. Subcellular expression controls gene expression, mitochondrial quality control, oxidative stress reduction, and endoplasmic reticulum stress relief. Parkin activates ubiquitin, allowing protein breakdown by monoubiquitination and polyubiquitination of substrates. The mutant PTEN-induced putative kinase 1 (PINK1) gene disrupts the ubiquitin-proteasome pathway in Parkinson's disease (PD), causing protein accumulation and dopaminergic neurodegeneration [55]. Because of their intricate roles and the dysregulation caused by PINK1 mutations, the ubiquitin-proteasome system and SUMOylation are essential regulatory mechanisms in Parkinson's disease's complex pathogenesis. These molecular pathways work together to sustain cellular homeostasis and prevent neurodegeneration in Parkinson's disease.

9.4.7 Ubiquitin-Dependent PD Signaling Pathways

More than 20 genes connected to rare family types of Parkinson's disease have been identified, providing insight into the disease's molecular basis. Parkin, F-box only protein 7 (Fbxo7), PINK1, and alpha-synuclein are ubiquitin signaling proteins involved in membrane trafficking, autophagy, protein misfolding, and synaptic function [5, 8]. These genes highlight the relevance of ubiquitination in Parkinson's disease and provide a foundation for understanding how mutations in these genes affect protein quality control via ubiquitylation [76]. The ubiquitination process is governed by the enzymes E1, E2, and E3. Alpha-synuclein is a crucial protein in Parkinson's disease, prompting much investigation into the role of ubiquitination in misfolding [21]. This pathway promotes the disintegration of neural networks and the downregulation of alpha-synuclein, indicating novel therapeutic methods. The complex role of ubiquitination in protein breakdown is demonstrated by changes in proteasomal subunits and activity, particularly in Parkinson's disease patients' substantia nigra [75]. Chronic systemic proteasome inhibitor delivery has been linked to a Parkinsonian phenotype marked by dopaminergic neuron degradation and the appearance of alpha-synuclein-positive inclusions, emphasizing the complex role of ubiquitination in Parkinson's disease. Understanding ubiquitin-mediated pathways is critical for developing targeted Parkinson's disease therapeutics [13].

9.4.8 Huntington's Disease

9.4.8.1 Huntingtin Ubiquitination and Protein Aggregates

Huntington's disease (HD) is a genetic neurological condition caused by a CAG repeat expansion in exon 1 of the Huntingtin (HTT) gene. The CAG repeat expansion causes a polyglutamine (polyQ) expansion in the HTT protein. The N-terminal regions of mutant HTT (mHTT) proteins containing polyQ repeats are prone to aggregation and formation of intracellular inclusion bodies (IBs), as observed in human Huntington's disease post-mortem brain tissue and various animal or cellular models [73]. Huntington's disease is caused by 40 or more CAG repeats, and the age of onset is inversely related to repeat length. Huntington's disease symptoms include physical and cognitive disability, as well as psychosocial issues, which deteriorate with time and usually result in death around 15 years after the disease begins [9]. There is no treatment for Huntington's disease. Reducing mHTT through increased degradation before aggregation and toxicity is a therapeutic strategy for preventing, delaying, or mitigating disease onset and progression. The principal intracellular protein breakdown mechanisms are the ubiquitin-proteasome system (UPS) and autophagy [36]. Both approaches contribute to the removal of mHTT. The UPS works in both the nucleus and the cytoplasm. However, it can only break down unfolded monomeric HTT proteins. The autophagic pathway is a cytoplasmic degradation system that transports soluble and aggregated HTT proteins to the lysosome for destruction. The proteasome is not hindered by mHTT fragments, contrary to original assumptions, but can completely digest mHTT when effectively directed to proteasomes through ubiquitination, and the proteasome is capable of degrading the polyQ sequence itself [10, 15]. Furthermore, both the proteasome and ubiquitin are not sequestered within mHTT aggregates but are instead dynamically and reversibly recruited. This suggests that the UPS remains functional in cells expressing mHTT, and that effective targeting of mHTT to proteasomes leads to its full destruction [18]. What accounts for the inefficient degradation of mHTT? Recent studies indicate that polyQ expansion influences the ubiquitination of mHTT, as demonstrated in various HD mice and rat models, revealing discrepancies in ubiquitination locations between wtHTT and mHTT. HTT breakdown and aggregation are linked to several ubiquitin-modifying enzymes, and we will delineate the different ubiquitinating and deubiquitinating enzymes that may contribute to mHTT degradation [16]. In Huntington's disease (HD), ubiquitination plays a critical role in the handling of mutant huntingtin (mHTT) protein, characterized by an expanded polyglutamine (polyQ) tract. Under normal conditions, ubiquitination involves the sequential action of E1 ubiquitin-activating enzymes, E2 ubiquitin-conjugating enzymes, and E3 ubiquitin ligases, tagging proteins for proteasomal degradation or other regulatory processes [26]. In HD, mHTT undergoes aberrant ubiquitination, but its misfolded conformation and aggregation-prone nature hinder efficient proteasomal clearance. Instead, ubiquitinated mHTT forms insoluble aggregates, which accumulate in neurons as inclusion bodies. These aggregates impair proteasome activity, sequester key cellular components, and disrupt protein homeostasis, triggering compensatory mechanisms like autophagy. Despite attempts to clear the aggregates via autophagic pathways, these processes are often overwhelmed, leading to persistent aggregate accumulation. This disruption of cellular proteostasis contributes to neuronal dysfunction and degeneration, hallmark features of HD pathology [70]. In Huntington's disease (HD), the process of ubiquitination is crucial for managing the mutant huntingtin (mHTT) protein, which has an expanded polyglutamine (polyQ) tract. Normally, ubiquitination involves a series of steps carried out by E1 ubiquitin-activating enzymes, E2 ubiquitin-conjugating enzymes, and E3 ubiquitin ligases, which tag proteins for degradation by the proteasome or for other regulatory functions [38]. However, in HD, mHTT experiences abnormal ubiquitination, and its misfolded structure and tendency to aggregate make it difficult for the proteasome to clear it effectively. Instead, the ubiquitinated mHTT forms insoluble aggregates that build up in neurons as inclusion bodies [23]. These aggregates not only hinder proteasome function but also sequester essential cellular components and disrupt protein balance, prompting compensatory responses like autophagy. Unfortunately, the efforts to eliminate these aggregates through autophagic pathways are frequently insufficient, resulting in the continued accumulation of aggregates. This breakdown of cellular protein management contributes to the neuronal dysfunction and degeneration that are characteristic of HD pathology [17].

9.4.8.2 Htt Ubiquitination and SUMOylation

Htt is a protein with high ubiquitination potential. The protein has 124 lysine residues, a significant number of which are located on the surface, as demonstrated by cryo-EM structural analysis, and may be ubiquitinated or conjugated with other Ub-like (Ubl) proteins, such as SUMO [44]. Ubiquitination and SUMOylation processes have been shown to compete for lysine residues K6, K9, and K15 in the N-terminal tail, which are not reflected in the resolved cryo-EM structure, and they may also compete for other lysine residues [16]. Ubiquitination reduces mHtt toxicity, most likely due to increased proteasome degradation. In contrast, SUMOylation has been shown to stabilize mHtt, reduce mHtt aggregation, aggravate mHtt-induced transcriptional dysregulation, and increase mHtt toxicity in a Drosophila model [35, 41]. The involvement of SUMOylation in mHtt toxicity may be aided by mHtt targeting the nucleus, which prevents ubiquitination and subsequent elimination. Many E3 ligases facilitate Htt ubiquitination; however, RHES (Ras homolog enriched in striatum) is the only protein that mediates mHtt SUMOylation. RHES has a higher affinity for mHtt than wild-type Htt, and its preferential expression in the striatum strongly suggests that this protein is involved in the pathophysiology of HD. As a result, RHES has been identified as a compelling target for HD therapy. The presence of seven lysine residues allows ubiquitin to communicate a variety of signals to target proteins. As a result, Htt polyubiquitination can occur by a variety of mechanisms that are still poorly understood. Two of the 40 E2 enzymes in the human genome have been shown to interact with Htt: UBE2W targets the N-termini of disordered proteins such as Tau, RBPB8, ATXN3, and mHtt. UBE2K, also known as Htt-interacting protein (Hip2) or E2-25k, is a ubiquitin-conjugating enzyme that interacts directly with Htt, potentially facilitating its ubiquitination [59]. Numerous E3 Ub ligases have been shown to target polyQ-containing proteins, including mHtt, resulting in lower mHtt cellular levels, aggregation, and cytotoxicity. Certain E3 ubiquitin ligases can polyubiquitinate Htt directly via ubiquitin's K48 attachment to Htt, allowing it to be degraded via the ubiquitin-proteasome system. These include UBE3A/E6AP, CHIP, HRD1, Parkin, and the SCF complex. Additional E3 ligases, such as HACE1, reduce the negative impact of mHtt on redox metabolism. Additional E3 ligases, such as WWP1 and TRAF6, which are elevated in the post-mortem brains of Huntington's disease patients, enable atypical Htt ubiquitination (at K6, K27, and K29), preventing mHtt breakdown and increasing the formation of ubiquitinated aggregates. Huntington's disease (HD) is associated with broad changes in the ubiquitin-proteasome system (UPS), including the accumulation of K48-, K63-, and K11-linked polyubiquitin chains. The hypothesized loss in proteasome activity by mutant huntingtin (mHtt) may contribute to neurotoxicity, although the molecular mechanisms underlying this phenomenon are unknown (Fig. 9.3). The autophagic system is involved in the clearance of Htt aggregates, and activating

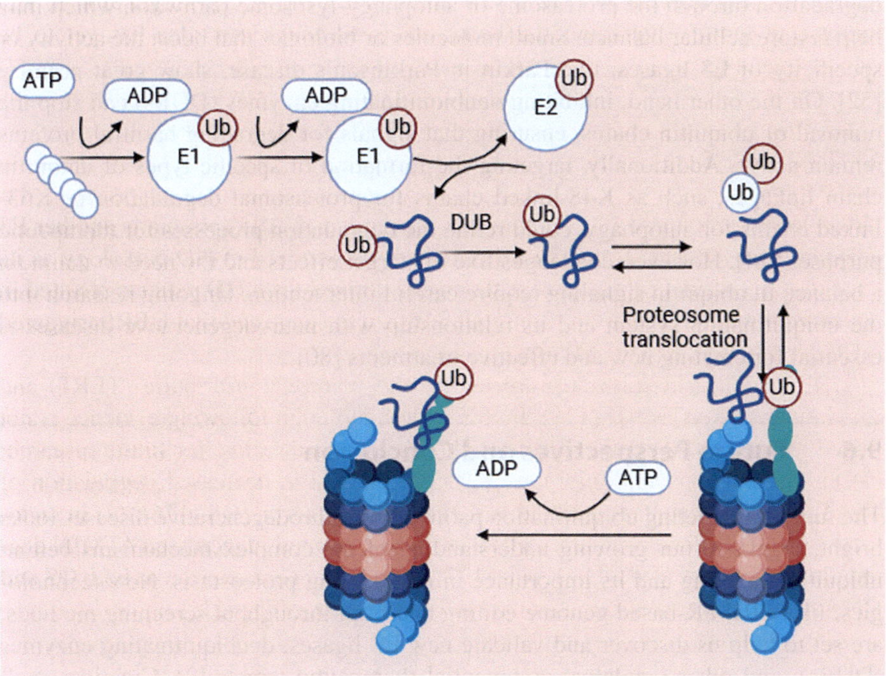

Fig. 9.3 The E1, E2, E3, and E4 enzymes are responsible for transferring Ub molecules (pink circles) to substrates (blue). This process requires energy in the form of ATP molecules. The proteasome is the protease that breaks down the substrates into tiny peptides. The proteasome does not destroy the polyubiquitin chain; instead, DUB enzymes detach it from the substrate and divide it into monomers that are ready to be reused

autophagy has been shown to reduce polyQ expansion toxicity in mouse models of Huntington's disease. Only a few deubiquitinating enzymes (DUBs) have been shown to deubiquitinate Htt. Apart from proteasome-associated deubiquitinating enzymes, the only deubiquitinating enzymes that target Htt are ATXN3, which has a polyQ region, and USP19.

9.5 Therapeutic Implications of Ubiquitination in Neurodegenerative Diseases

The ubiquitination pathway presents a promising therapeutic avenue for neurodegenerative diseases, where the misfolding and aggregation of proteins play a key role in their progression. By modulating the activity of enzymes involved in the ubiquitination process specifically E1, E2, and especially E3 ubiquitin ligases, we can enhance the removal of harmful protein aggregates, such as mutant huntingtin in Huntington's disease or alpha-synuclein in Parkinson's disease [40]. Potential strategies include promoting the ubiquitination of misfolded proteins to aid their degradation through the proteasome or autophagy-lysosome pathways, which may help restore cellular balance. Small molecules or biologics that boost the activity or specificity of E3 ligases, like Parkin in Parkinson's disease, show great promise [52]. On the other hand, inhibiting deubiquitinating enzymes (DUBs) can stop the removal of ubiquitin chains, ensuring that signals for degrading harmful proteins remain active. Additionally, targeting the formation of specific types of ubiquitin chain linkages, such as K48-linked chains for proteasomal degradation or K63-linked chains for autophagy, could refine the degradation processes for therapeutic purposes [24]. However, challenges like off-target effects and the need to maintain a balance in ubiquitin signaling require careful intervention. Ongoing research into the ubiquitination system and its relationship with neurodegenerative diseases is essential for creating new and effective treatments [80].

9.6 Future Perspectives and Conclusion

The future of targeting ubiquitination pathways in neurodegenerative diseases looks bright, thanks to our growing understanding of the complex mechanisms behind ubiquitin signaling and its importance in maintaining proteostasis. New technologies, like CRISPR-based genome editing and high-throughput screening methods, are set to help us discover and validate new E3 ligases, deubiquitinating enzymes (DUBs), and other regulators as potential therapeutic targets [2]. Creating small molecules or PROTACs (proteolysis-targeting chimeras) aimed at boosting the ubiquitination and removal of specific harmful proteins presents a customized approach to treating conditions such as Huntington's and Parkinson's. Furthermore, developing synthetic ubiquitin chain linkages or adjusting specific autophagy-ubiquitin interactions could allow for more precise therapeutic control. The use of artificial intelligence and computational modeling can speed up drug discovery and help predict off-target effects, improving safety profiles [50, 71]. Future studies will also aim to tackle challenges like delivering therapeutics to specific tissues, reducing unintended disruptions in ubiquitin signaling, and managing the complexity of ubiquitin's various roles in cellular functions. Merging ubiquitination-targeting strategies with other treatment methods, such as gene therapy or anti-aggregation agents, shows promise for creating synergistic effects. These developments highlight a transformative future for ubiquitination-based therapies in the fight against neurodegenerative diseases [34, 74].

Focusing on the ubiquitination pathway presents a promising therapeutic strategy for neurodegenerative diseases by addressing the root causes of protein aggregation and imbalances in proteostasis. Recent advancements in our molecular understanding and drug development techniques are enabling more targeted modulation of ubiquitin signaling, which can improve the removal of causing proteins. As research and innovation progress, therapies based on ubiquitination could greatly enhance the quality of life for patients suffering from these challenging conditions.

References

1. Agostinho P, A Cunha R, Oliveira C. Neuroinflammation, oxidative stress and the pathogenesis of Alzheimer's disease. Curr Pharm Des. 2010;16(25):2766–78.
2. Aljoudi S, Ayotunde PI, Chang-Gonzalez A, Hoegnifioh C, Durie C, Dayna P, Deol H, Deossa RA, Dey S, Esposto J, Fuente G. 2021 Protein society diversity award winners.
3. Amerik AY, Hochstrasser M. Mechanism and function of deubiquitinating enzymes. Biochimica et Biophysica Acta (BBA)-Mol Cell Res. 2004;1695(1–3):189–207.
4. Bedford L, Lowe J, Dick LR, Mayer RJ, Brownell JE. Ubiquitin-like protein conjugation and the ubiquitin–proteasome system as drug targets. Nat Rev Drug Discov. 2011;10(1):29–46.
5. Burchell VS, Nelson DE, Sanchez-Martinez A, Delgado-Camprubi M, Ivatt RM, Pogson JH, Randle SJ, Wray S, Lewis PA, Houlden H, Abramov AY. The Parkinson's disease–linked proteins Fbxo7 and Parkin interact to mediate mitophagy. Nat Neurosci. 2013;16(9):1257–65.
6. Callis J. The ubiquitination machinery of the ubiquitin system. Arabidopsis Book. 2014;12:e0174.
7. Chow WN, Luk HW, Chan HY, Lau KF. Degradation of mutant huntingtin via the ubiquitin/proteasome system is modulated by FE65. Biochem J. 2012;443(3):681–9.
8. Cook C, Petrucelli L. A critical evaluation of the ubiquitin–proteasome system in Parkinson's disease. Biochimica et Biophysica Acta (BBA)-Mol Basis Dis. 2009;1792(7):664–75.
9. Craufurd D, Snowden J. Neuropsychological and neuropsychiatric aspects of Huntington's disease. Oxf Monogr Med Genet. 2002;45:62–94.
10. Dahlmann B, Overkleeft H, Hartmann-Petersen R, Reits EA. Modulating proteasome activity in Huntington's disease. The role of the proteasome in Huntington's disease. 2013; 163.
11. Del Prete D, Rice RC, Rajadhyaksha AM, D'Adamio L. Amyloid precursor protein (APP) may act as a substrate and a recognition unit for CRL4CRBN and Stub1 E3 ligases facilitating ubiquitination of proteins involved in presynaptic functions and neurodegeneration. J Biol Chem. 2016;291(33):17209–27.
12. Di Maio R, Barrett PJ, Hoffman EK, Barrett CW, Zharikov A, Borah A, Hu X, McCoy J, Chu CT, Burton EA, Hastings TG. α-Synuclein binds to TOM20 and inhibits mitochondrial protein import in Parkinson's disease. Sci Transl Med. 2016;8(342):342ra78.
13. Ebrahimi-Fakhari D, Wahlster L, McLean PJ. Protein degradation pathways in Parkinson's disease. Acta Neuropathol. 2012;124(2):153–72.
14. Eckermann K. SUMO and Parkinson's disease. NeuroMolecular Med. 2013;15:737–59.
15. Ehrnhoefer DE, Martin DD, Schmidt ME, Qiu X, Ladha S, Caron NS, Skotte NH, Nguyen YT, Vaid K, Southwell AL, Engemann S. Preventing mutant huntingtin proteolysis and intermittent fasting promote autophagy in models of Huntington disease. Acta Neuropathol Commun. 2018;6:1–6.
16. Fiorillo A, Morea V, Colotti G, Ilari A. Huntingtin ubiquitination mechanisms and novel possible therapies to decrease the toxic effects of mutated huntingtin. J Pers Med. 2021;11(12):1309.
17. Gandhi J, Antonelli AC, Afridi A, Vatsia S, Joshi G, Romanov V, Murray IV, Khan SA. Protein misfolding and aggregation in neurodegenerative diseases: a review of pathogeneses, novel detection strategies, and potential therapeutics. Rev Neurosci. 2019;30(4):339–58.
18. Harding RJ, Tong YF. Proteostasis in Huntington's disease: disease mechanisms and therapeutic opportunities. Acta Pharmacol Sin. 2018;39(5):754–69.
19. Harris LD, Jasem S, Licchesi JD. The ubiquitin system in Alzheimer's disease. In: Proteostasis and Disease: From Basic Mechanisms to Clinics; 2020. p. 195–221.
20. Hillert EK. Mechanistic Studies of a Novel Inhibitor of the Ubiquitin-Proteasome System. Karolinska Institutet; 2019.
21. Ho HH, Wing SS. α-Synuclein ubiquitination–functions in proteostasis and development of Lewy bodies. Front Mol Neurosci. 2024;17:1498459.
22. Hoppe JB, Salbego CG, Cimarosti H. SUMOylation: novel neuroprotective approach for Alzheimer's disease? Aging Dis. 2015 Oct;6(5):322.

23. Hyrskyluoto A. Endoplasmic reticulum stress, regulation of autophagy and ubiquitin-proteasome system in cellular models of Huntington's disease. University of Helsinki; 2014.
24. Jacomin AC, Taillebourg E, Fauvarque MO. Deubiquitinating enzymes related to autophagy: new therapeutic opportunities? Cells. 2018;7(8):112.
25. Jankovic J. Parkinson's disease: clinical features and diagnosis. J Neurol Neurosurg Psychiatry. 2008;79(4):368–76.
26. Jarosińska OD, Rüdiger SG. Molecular strategies to target protein aggregation in Huntington's disease. Front Mol Biosci. 2021;8:769184.
27. Jeong GR, Lee BD. Pathological functions of LRRK2 in Parkinson's disease. Cells. 2020;9(12):2565.
28. Joshi G, Arya P, Mahajan V, Chaturvedi A, Mishra RK. SUMO and SUMOylation pathway at the forefront of host immune response. Front Cell Dev Biol. 2021;9:681057.
29. Kammula SV, Tripathi S, Wang N, Dawson VL, Dawson TM, Mao X. Unraveling the tau puzzle: a review of mechanistic targets and therapeutic interventions to prevent tau pathology in Alzheimer's disease. Ageing Neurodegener Dis. 2023;3(4):22.
30. Kaushik A, Parashar S, Ambasta RK, Kumar P. Ubiquitin E3 ligases assisted technologies in protein degradation: sharing pathways in neurodegenerative disorders and cancer. Ageing Res Rev. 2024;96:102279.
31. Krumova P, Weishaupt JH. Sumoylation in neurodegenerative diseases. Cell Mol Life Sci. 2013;70:2123–38.
32. Kumar P, Jha NK, Jha SK, Ramani K, Ambasta RK. Tau phosphorylation, molecular chaperones, and ubiquitin E3 ligase: clinical relevance in Alzheimer's disease. J Alzheimer's Dis. 2015;43(2):341–61.
33. Kumar D, Ambasta RK, Kumar P. Ubiquitin biology in neurodegenerative disorders: from impairment to therapeutic strategies. Ageing Res Rev. 2020;61:101078.
34. Le Guerroué F, Youle RJ. Ubiquitin signaling in neurodegenerative diseases: an autophagy and proteasome perspective. Cell Death Differ. 2021;28(2):439–54.
35. Lee JH. Altered mTOR signaling in Huntington's disease. University of Iowa; 2015.
36. Li XJ, Li S. Proteasomal dysfunction in aging and Huntington disease. Neurobiol Dis. 2011;43(1):4–8.
37. Lin N, Gao XY, Li X, Chu WM. Involvement of ubiquitination in Alzheimer's disease. Front Neurol. 2024;15:1459678.
38. Liu W, Tang X, Qi X, Fu X, Ghimire S, Ma R, Li S, Zhang N, Si H. The ubiquitin conjugating enzyme: an important ubiquitin transfer platform in ubiquitin-proteasome system. Int J Mol Sci. 2020;21(8):2894.
39. Liu B, Ruan J, Chen M, Li Z, Manjengwa G, Schlüter D, Song W, Wang X. Deubiquitinating enzymes (DUBs): decipher underlying basis of neurodegenerative diseases. Mol Psychiatry. 2022 Jan;27(1):259–68.
40. Liu N, Lin MM, Wang Y. The emerging roles of E3 ligases and DUBs in neurodegenerative diseases. Mol Neurobiol. 2023 Jan;60(1):247–63.
41. Lontay B, Kiss A, Virág L, Tar K. How do post-translational modifications influence the pathomechanistic landscape of Huntington's disease? A comprehensive review. Int J Mol Sci. 2020;21(12):4282.
42. Lu L, Jifu C, Xia J, Wang J. E3 ligases and DUBs target ferroptosis: a potential therapeutic strategy for neurodegenerative diseases. Biomed Pharmacother. 2024;175:116753.
43. Maiti P, Manna J, Dunbar GL. Current understanding of the molecular mechanisms in Parkinson's disease: targets for potential treatments. Transl Neurodegener. 2017;6:1–35.
44. Mandel N, Agarwal N. Role of SUMOylation in neurodegenerative diseases. Cells. 2022;11(21):3395.
45. Maniv I, Sarji M, Bdarneh A, Feldman A, Ankawa R, Koren E, Magid-Gold I, Reis N, Soteriou D, Salomon-Zimri S, Lavy T. Altered ubiquitin signaling induces Alzheimer's disease-like hallmarks in a three-dimensional human neural cell culture model. Nat Commun. 2023;14(1):5922.

46. Masperone L. TRAF6 Involvement In Prion Disease: A Possible Crosstalk Among Neurodegenerative Disorders. SISSA; 2016.
47. Mayer J. The role of autophagy in Aβ and tau metabolism in Alzheimer's disease. Inst för neurobiologi, vårdvetenskap och samhälle/Department of Neurobiology, Care Sciences and Society; 2024.
48. Mishra T, Singh S, Singh TG. Therapeutic implications and regulations of protein post-translational modifications in Parkinsons disease. Cell Mol Neurobiol. 2024;44(1):53.
49. O'brien RJ, Wong PC. Amyloid precursor protein processing and Alzheimer's disease. Annu Rev Neurosci. 2011;34(1):185–204.
50. O'Keefe S. Small molecule inhibitors targeting protein biogenesis at the endoplasmic reticulum. The University of Manchester (United Kingdom); 2019.
51. Pajarillo E, Rizor A, Lee J, Aschner M, Lee E. The role of posttranslational modifications of α-synuclein and LRRK2 in Parkinson's disease: potential contributions of environmental factors. Biochimica et Biophysica Acta (BBA) Mol Basis Dis. 2019;1865(8):1992–2000.
52. Pao KC, Stanley M, Han C, Lai YC, Murphy P, Balk K, Wood NT, Corti O, Corvol JC, Muqit MM, Virdee S. Probes of ubiquitin E3 ligases enable systematic dissection of parkin activation. Nat Chem Biol. 2016 May;12(5):324–31.
53. Peters JM, Harris JR, Finley D, editors. Ubiquitin and the biology of the cell. Springer Science & Business Media; 2013.
54. Quinn PM, Moreira PI, Ambrósio AF, Alves CH. PINK1/PARKIN signalling in neurodegeneration and neuroinflammation. Acta Neuropathol Commun. 2020;8(1):189.
55. Rakovic A, Shurkewitsch K, Seibler P, Grünewald A, Zanon A, Hagenah J, Krainc D, Klein C. Phosphatase and tensin homolog (PTEN)-induced putative kinase 1 (PINK1)-dependent ubiquitination of endogenous Parkin attenuates mitophagy: study in human primary fibroblasts and induced pluripotent stem cell-derived neurons. J Biol Chem. 2013;288(4):2223–37.
56. Rawat P, Sehar U, Bisht J, Selman A, Culberson J, Reddy PH. Phosphorylated tau in Alzheimer's disease and other tauopathies. Int J Mol Sci. 2022;23(21):12841.
57. Rodrigues B, Ventura E, Moreira P, Resende R, Bicker J, Santos AE, Pereira CF, Cruz MT, Robalo MP, Silva A, Silva S. New low-dose curcumin derivative with therapeutic potential in Alzheimer's disease: results from an in vitro and in vivo study in mice. Neurobiol Aging. 2024;147:105–23.
58. Safadi SS, Barber KR, Shaw GS. Impact of autosomal recessive juvenile Parkinson's disease mutations on the structure and interactions of the parkin ubiquitin-like domain. Biochemistry. 2011;50(13):2603–10.
59. Sap KA, Geijtenbeek KW, Schipper-Krom S, Guler AT, Reits EA. Ubiquitin-modifying enzymes in Huntington's disease. Front Mol Biosci. 2023;10:1107323.
60. Schmidt MF, Gan ZY, Komander D, Dewson G. Ubiquitin signalling in neurodegeneration: mechanisms and therapeutic opportunities. Cell Death Differ. 2021;28(2):570–90.
61. Sharma A, Trivedi AK. Regulation of apoptosis by E3 ubiquitin ligases in ubiquitin proteasome system. Cell Biol Int. 2020;44(3):721–34.
62. Silva GM, Finley D, Vogel C. K63 polyubiquitination is a new modulator of the oxidative stress response. Nat Struct Mol Biol. 2015;22(2):116–23.
63. Siwecka N, Saramowicz K, Galita G, Rozpędek-Kamińska W, Majsterek I. Inhibition of protein aggregation and endoplasmic reticulum stress as a targeted therapy for α-synucleinopathy. Pharmaceutics. 2023;15(8):2051.
64. Snyder NA, Silva GM. Deubiquitinating enzymes (DUBs): regulation, homeostasis, and oxidative stress response. J Biol Chem. 2021;297(3):101077.
65. Song L, Luo ZQ. Post-translational regulation of ubiquitin signaling. J Cell Biol. 2019;218(6):1776–86.
66. Stefanis L. A-Synuclein in Parkinson's disease. Cold Spring Harb Perspect Med. 2012;2(2):a009399.
67. Sulistio YA, Heese K. The ubiquitin-proteasome system and molecular chaperone deregulation in Alzheimer's disease. Mol Neurobiol. 2016;53:905–31.

68. Suresh B, Lee J, Kim KS, Ramakrishna S. The importance of ubiquitination and deubiquitination in cellular reprogramming. Stem Cells Int. 2016;2016(1):6705927.
69. Swatek KN, Komander D. Ubiquitin modifications. Cell Res. 2016;26(4):399–422.
70. Sweeney P, Park H, Baumann M, Dunlop J, Frydman J, Kopito R, McCampbell A, Leblanc G, Venkateswaran A, Nurmi A, Hodgson R. Protein misfolding in neurodegenerative diseases: implications and strategies. Transl Neurodegener. 2017;6:1–3.
71. Tenchov R, Sasso JM, Zhou QA. Polyglutamine (PolyQ) diseases: navigating the landscape of neurodegeneration. ACS Chem Neurosci. 2024;15(15):2665–94.
72. Tomasi ML, Ramani K, Ryoo M. Ubiquitin-conjugating enzyme 9 phosphorylation as a novel mechanism for potentiation of the inflammatory response. Am J Pathol. 2016;186(9):2326–36.
73. Tong H, Yang T, Xu S, Li X, Liu L, Zhou G, Yang S, Yin S, Li XJ, Li S. Huntington's disease: complex pathogenesis and therapeutic strategies. Int J Mol Sci. 2024;25(7):3845.
74. Upadhyay A, Joshi V, Amanullah A, Mishra R, Arora N, Prasad A, Mishra A. E3 ubiquitin ligases neurobiological mechanisms: development to degeneration. Front Mol Neurosci. 2017;10:151.
75. Walden H, Muqit MM. Ubiquitin and Parkinson's disease through the looking glass of genetics. Biochem J. 2017;474(9):1439–51.
76. Wanka V, Fottner M, Cigler M, Lang K. Genetic code expansion approaches to decipher the ubiquitin code. Chem Rev. 2024;124(20):11544–84.
77. Watanabe Y, Taguchi K, Tanaka M. Ubiquitin, autophagy and neurodegenerative diseases. Cells. 2020;9(9):2022.
78. Wilson VG, editor. SUMOylation and ubiquitination: current and emerging concepts. Caister Academic Press; 2019.
79. Winborn BJ, Travis SM, Todi SV, Scaglione KM, Xu P, Williams AJ, Cohen RE, Peng J, Paulson HL. The deubiquitinating enzyme ataxin-3, a polyglutamine disease protein, edits Lys63 linkages in mixed linkage ubiquitin chains. J Biol Chem. 2008;283(39):26436–43.
80. Xian Y, Ye J, Tang Y, Zhang N, Peng C, Huang W, He G. Deubiquitinases as novel therapeutic targets for diseases. MedComm. 2024 Dec;5(12):e70036.
81. Yeates EF, Tesco G. The endosome-associated deubiquitinating enzyme USP8 regulates BACE1 enzyme ubiquitination and degradation. J Biol Chem. 2016;291(30):15753–66.
82. Yin Z, Popelka H, Lei Y, Yang Y, Klionsky DJ. The roles of ubiquitin in mediating autophagy. Cells. 2020;9(9):2025.
83. Zhang X, Song W. The role of APP and BACE1 trafficking in APP processing and amyloid-β generation. Alzheimer's Res Ther. 2013 Oct;5:1–8.
84. Zhang J, Li X, Li JD. The roles of post-translational modifications on α-Synuclein in the pathogenesis of Parkinson's diseases. Front Neurosci. 2019;13:381.

E3 Ubiquitin Ligases in Alzheimer's Disease

10

Elena Gavilán

Abbreviations

ACH	Amyloid cascade hypothesis
AD	Alzheimer's disease
AICD	Amyloid precursor protein intracellular domain
AMPARs	AMPA receptors
APC/C	Anaphase promoting complex/cyclosome
APOE ε4	Apolipoprotein Epsilon 4 allele
APP	Amyloid precursor protein
ATG	Autophagy-related genes
ATN	Amyloid deposition (A), tau pathology (T), and neurodegeneration (N)
Aβ	Amyloid-beta
BACE1	β-Secretase 1
BRCA1	Breast cancer type 1 susceptibility protein
C/EBPβ	CCAAT/enhancer-binding protein beta
Cav1.2	L-type voltage-gated calcium channel 1.2
CHIP	C-terminus Hsp70 interacting protein (also STUB1)
CNS	Central nervous system
COP1	Constitutive photomorphogenic 1
CRL	Cullin-RING ligase
CRL4(CRBN)	Cullin-RING ligase 4 (cereblon)
CSF	Cerebrospinal fluid
CUL4B	Cullin-4B

E. Gavilán (✉)
University of Seville, Department of Biochemistry and Molecular Biology, Faculty of Pharmacy, Seville, Spain
e-mail: egavilan@us.es

S. Singh (ed.), *Ubiquitin Proteasome System in CNS Neurodegenerative Diseases*, https://doi.org/10.1007/978-981-95-6099-8_10

DDB1	DNA damage-binding protein 1
DMT1	Divalent metal transporter 1
DNA	Deoxyribonucleic acid
DSBs	DNA double-strand breaks
DUBs	Deubiquitinating enzymes
ER	Endoplasmic reticulum
ERAD	Endoplasmic reticulum-associated degradation
ESCRT	Endosomal sorting complex required for transport
FBL2	F-box/LRR-repeat protein 2
FBXO2	F-box only protein 2
HECT	Homologous to E6-AP carboxyl terminus
HECTD1	HECT domain E3 ubiquitin protein ligase 1
HOIP	HOIL-1-interacting protein
HRD1	Homocysteine-responsive endoplasmic reticulum-resident ubiquitin-like domain member 1 protein
HRS	Hepatocyte growth factor-regulated tyrosine kinase substrate
HSPs	Heat shock proteins
IL-1β	Interleukin-1 beta
IL-6	Interleukin-6
ITCH	ITCHy E3 ubiquitin ligase
JNK	c-Jun N-terminal kinase
LUBAC	Linear ubiquitin chain assembly complex
MARCH	Membrane-associated RING-CH proteins, including MARCH7, MARCH8, and MARCH5 (also known as MITOL)
MCI	Mild cognitive impairment
MDM2	Murine double minute 2
MITOL	Mitochondrial outer membrane E3 ubiquitin ligase (MARCH5)
MLC2	Myosin light chain 2
MVB	Multivesicular bodies
NDFIP1	NEDD4 family interacting protein 1
NEDD4	Neural precursor cell expressed developmentally downregulated 4
NFTs	Neurofibrillary tangles
NF-κB	Nuclear factor kappa-light-chain-enhancer of activated B cells
NLRP3	NOD-like receptor family pyrin domain containing 3
NMDARs	N-methyl-D-aspartate receptors
NRBP1	Nuclear receptor binding protein 1
PAR-1	Partitioning-defective protein 1
PET	Positron emission tomography
PIAS1	Protein inhibitor of activated STAT 1
PINK1	PTEN-induced kinase 1
PSEN1	Presenilin 1
PSEN2	Presenilin 2
PTMs	Post-translational modifications
RAD18	RAD18 E3 ubiquitin protein ligase
RIPK1/3	Receptor-interacting serine/threonine-protein kinases 1 and 3

RNF	Ring finger proteins, including, RNF8, RNF167, RFN168, RNF182, and RNF220
SCF	Skp1-Cullin1-F-box complex
SK2	Small conductance calcium-activated potassium channel 2
SSTR4	Somatostatin receptor 4
TAp73	Transcriptionally active isoform of p73
TDP-43	TAR DNA binding protein 43
TLRs	Toll-like receptors
TNF-α	Tumor necrosis factor alpha
TPR	Tetratricopeptide repeat
TRAF6	Tumor necrosis factor receptor-associated factor 6
TRIM	Tripartite motif proteins, including TRIM21, TRIM28, TRIM32, TRIM37
UBE3A	Ubiquitin protein ligase E3A
UBE4B	Ubiquitination factor E4B
UBR5	Ubiquitin protein ligase E3 component N-recognin 5
UPS	Ubiquitin-proteasome system
USP46	Ubiquitin-specific peptidase 46
XIAP	X-linked inhibitor of apoptosis

10.1 Introduction

Neurodegenerative diseases, particularly dementia, represent a growing and urgent challenge in today's healthcare landscape. Among these, Alzheimer's disease (AD) stands out as the leading cause of dementia, making it one of the most critical public health issues of our time. As populations continue to age, the prevalence of AD is rising sharply, creating significant concerns for healthcare systems, caregivers, and families worldwide [1]. For the majority of patients, effective treatments are still unavailable, and the underlying causes of the disease remain largely unidentified except in a limited number of inherited cases linked to genetic mutations [2].

Protein homeostasis, or proteostasis, is crucial for neuronal health [3, 4]. This involves maintaining optimal levels of functional proteins at all times by balancing their production and degradation when they become damaged, accumulated, or are no longer needed. The two main systems that sustain intracellular protein degradation are the ubiquitin-proteasome system (UPS) and autophagy. The UPS plays a central role by tagging misfolded or damaged proteins with ubiquitin, targeting them for degradation by the proteasome [5, 6]. Autophagy complements this process by degrading larger protein aggregates and damaged organelles, while also selectively degrading certain ubiquitinated proteins [7, 8]. Postmortem studies of AD brains heightened levels of ubiquitin and increased ubiquitination associated with pathological protein aggregates, emphasizing the UPS dysfunction as a key factor in AD pathogenesis [9].

The ubiquitination process involves the covalent linkage of ubiquitin molecules (mono- or polyubiquitination) to lysine residues on target proteins. This modification is carried out through a sequential enzymatic cascade that includes E1 (ubiquitin-activating), E2 (ubiquitin-conjugating), and E3 (ubiquitin ligase) enzymes. Additionally, deubiquitinating enzymes (DUBs) help regulate this process by removing ubiquitin from target proteins [10].

The determination of ubiquitination specificity relies on the type of bonds formed between ubiquitin molecules [11]. As a result, E3 ubiquitin ligases are essential for selecting specific proteins for degradation [12, 13]. Bioinformatics analysis has revealed more than 600 E3 ubiquitin ligases encoded in the human genome. These ligases are classified into four main types: HECT, RING, U-box, and RBR, each distinguished by its unique structural features and mechanisms of ubiquitin transfer. For instance, HECT E3s engage in a two-step reaction involving a catalytic cysteine, while RING E3s facilitate direct transfer from E2 ligases to substrates. The diversity among these ligases underscores their distinct roles in numerous cellular processes [14].

As key components of the UPS, E3 ligases' activity, expression levels, and turnover are tightly controlled to ensure proper ubiquitination and maintain cellular integrity [15]. Dysregulation of E3 ubiquitin ligases is markedly implicated in AD progression [16–18]. Their impairment can lead to the buildup of misfolded proteins such as amyloid-beta (Aβ) and tau, which disrupts cellular homeostasis and induces chronic endoplasmic reticulum (ER) stress. ER stress activates the unfolded protein response (UPR), a protective mechanism aimed at restoring protein folding capacity; however, persistent ER stress in AD can lead to the activation of pro-apoptotic pathways of the UPR, contributing to neuronal toxicity and cell death [19]. Moreover, in addition to Aβ and tau accumulation and the consequent ER stress, dysregulated E3 ligases impact multiple cellular processes implicated in AD pathology. These processes include the maintenance of synaptic integrity, mitochondrial function, neuroinflammatory responses, and the control of the cell cycle and cell stress responses. Additionally, E3 ligases have been implicated in mechanisms of DNA repair, calcium homeostasis, and broader cell survival pathways [17, 18, 20–22]. Disruption in these interconnected systems, in which E3 ligases are crucial players, may exacerbate neurodegenerative processes in AD. Understanding the multifunctional roles of E3 ligases in these pathways offers valuable insights into the molecular underpinnings of AD and highlights potential therapeutic targets aimed at preserving neuronal health and function [16, 18].

This unit of the chapter will explore the role of E3 ubiquitin ligases in AD, examining their influence on key pathological processes, including Aβ and tau metabolism, and broader aspects of neuronal function. Understanding the multifaceted roles of E3 ubiquitin ligases in AD could provide novel insights into disease mechanisms and identify potential therapeutic targets.

10.2 Brief Overview of AD Pathology

AD is a complex neurodegenerative disorder that leads to memory loss, cognitive decline and the eventual inability to perform everyday tasks. This degeneration is often associated with the shrinkage of key brain regions, including the hippocampus and cortex [1, 23]. It is characterized by loss of proteostasis, accumulation of pathological protein aggregates, and significant disruptions in neuronal networks. Of all pathological features, the major hallmarks of AD include the accumulation of extracellular Aβ plaques and the formation of intracellular neurofibrillary tangles (NFTs) composed of hyperphosphorylated tau protein. These protein aggregates disrupt neuronal function and lead to synaptic loss and neuronal death [2, 24]. While these proteinopathies are hallmark features of AD, other key contributors to cell death in AD include chronic neuroinflammation, metabolic and vascular impairments, mitochondrial and synaptic dysfunction, and various cellular stresses [1, 2, 24, 25].

Recent scientific advances in AD research have increasingly shifted from the traditional emphasis on Aβ plaques and NFTs to explore a wider array of mechanisms that may contribute to neurodegeneration [26]. This evolving perspective recognizes that AD and similar conditions are complex, involving multiple interacting factors, each offering potential new pathways for therapeutic intervention. New advances in neuroimaging, genetic research, biomarker analysis, and omics technologies are significantly enhancing our understanding of the factors that may contribute to the late onset of AD [27].

The need for a biological definition of AD, based on biomarkers rather than purely clinical symptoms, has become widely recognized in recent years. The ATN classification system, which focuses on amyloid deposition (A), tau pathology (T), and neurodegeneration (N), has revolutionized the ability to diagnose and stage AD, even in its preclinical phases [28]. By identifying and tracking the disease through biomarkers such as cerebrospinal fluid (CSF) levels of Aβ and tau, or via neuroimaging techniques like positron emission tomography (PET), AD pathology can be detected years before cognitive symptoms appear. This shift is vital for the development of early interventions, as the disease is most likely to respond to treatments in its earlier stages [28–31].

Currently, the global burden of AD is immense, with approximately 32 million people diagnosed worldwide. Estimates suggest that this number will rise significantly due to the aging population, particularly in regions with longer life expectancies [32]. In addition to those diagnosed, around 69 million individuals are in the mild cognitive impairment (MCI) phase, a condition that often precedes AD. Even more remarkably, over 300 million individuals are in the preclinical phase, harboring AD pathology without yet exhibiting cognitive symptoms [33]. This enormous population of at-risk individuals points to the need for both preventive strategies and more effective therapies. By 2050, the number of AD cases is projected to triple, further escalating the economic and societal impact of the disease [34].

10.2.1 Pathological Models of AD

AD can manifest in two forms: familial and sporadic. Despite the identification of genetic markers for familial AD through genome-wide association studies, such as Apolipoprotein Epsilon 4 (APOE ε4) allele, the underlying molecular mechanisms that initiate and drive the progression of both forms of the disease remain largely unclear [22]. As detailed by Granzotto and Sensi in 2024, several key models have been proposed to explain AD pathogenesis [26]:

(i) Amyloid cascade hypothesis (ACH): The ACH remains the most well-established model for AD. It posits that the overproduction or impaired clearance of Aβ peptides results in their accumulation in the brain, where they form plaques. These plaques are thought to trigger a cascade of downstream events, including tau hyperphosphorylation and NFTs formation, synaptic dysfunction, and neuronal death, ultimately leading to cognitive decline [26, 35]. Genetic mutations in the amyloid precursor protein (APP) and presenilin (PSEN1, PSEN2) genes, which increase Aβ production or reduce its clearance, are associated with familial AD, providing strong support for this hypothesis [36, 37].

(ii) Multifactorial model: This model recognizes that AD is likely driven by a combination of factors beyond just Aβ and tau, indicating a non-linear progression of the disease. These include other proteinopathies (e.g., α-synuclein and TDP-43 accumulations), vascular contributions, metabolic dysregulation, oxidative stress, and chronic neuroinflammation [26, 38]. Genetic risk factors, like the APOE ε4 allele, interact with lifestyle factors (e.g., physical activity, diet, and education) to influence disease susceptibility and progression [39, 40]. In this line, AD would be a heterogeneous disorder with different pathways contributing to its pathology in different individuals, making it more difficult to target with single therapeutic strategies.

(iii) Misfolded protein accumulation-independent model: This model challenges the central role of Aβ and tau accumulation in AD, suggesting that other molecular disruptions, such as mitochondrial dysfunction, inflammation, or epigenetic changes, may initiate the disease. In this framework, Aβ and tau accumulation could be secondary effects rather than primary drivers of AD, contributing to but not initiating the disease process [26].

All three models underscore the importance of cellular mechanisms that maintain proteostasis and prevent the appearance of pathological protein aggregates.

10.2.2 Neuroinflammation and AD

Chronic neuroinflammation is a critical driver of AD pathology [25, 41]. At the center of this inflammatory response are microglia, the resident immune cells of the central nervous system (CNS), which play vital roles in maintaining brain homeostasis.

Under normal conditions, microglia are responsible for clearing cellular debris, removing apoptotic cells, and regulating synaptic pruning. However, in the context of AD, their function becomes dysregulated, contributing to disease progression [42].

One of the primary triggers for microglial activation in AD is the accumulation of Aβ plaques. Upon recognizing these pathological protein aggregates, microglia shift into an activated state and begin releasing pro-inflammatory cytokines, such as interleukin-1β (IL-1β), tumor necrosis factor-alpha (TNF-α), and interleukin-6 (IL-6), along with chemokines that recruit other immune cells to the site of inflammation [43]. While this response is initially aimed at clearing Aβ and protecting neurons, prolonged activation leads to a maladaptive inflammatory environment. Chronic neuroinflammation results in the release of neurotoxic factors, further exacerbating neuronal injury and contributing to synaptic dysfunction, ultimately accelerating neurodegeneration. The persistence of this inflammatory state can also promote the spread of tau pathology, which, like Aβ, contributes to the progression of cognitive decline. Studies show that pro-inflammatory cytokines from activated microglia can enhance tau hyperphosphorylation and facilitate its aggregation, creating a vicious cycle of inflammation and proteinopathies within the AD brain [25, 41–44].

In addition to the direct impact of Aβ and tau, genetic factors play a significant role in modulating neuroinflammation. The APOE ε4 allele, a major genetic risk factor for late-onset AD, has been shown to influence microglial function. Individuals carrying the APOE ε4 allele tend to have a more pronounced pro-inflammatory response. APOE ε4 is less efficient at facilitating the clearance of Aβ compared to the APOE ε3 and APOE ε2 alleles. This impaired clearance capacity results in higher Aβ burdens and greater microglial activation, perpetuating a harmful inflammatory state that accelerates disease progression [45, 46].

E3 ubiquitin ligases regulate inflammation by controlling the degradation of key proteins involved in immune responses [47], such as NF-κB and NLRP3, a process explained in more detail later. Normally, E3 ligases help maintain immune balance by degrading proteins, preventing excessive inflammation. When disrupted, they increase cytokine production, such as IL-1β and TNF-α by microglia [48].

In summary, chronic neuroinflammation in AD is driven by a multifaceted interplay of factors, including impaired proteostasis, microglial dysfunction, genetic predispositions and disruptions in critical signaling pathways. The regulation of immune responses by E3 ubiquitin ligases represents a promising therapeutic target, offering potential strategies to mitigate neuroinflammation.

10.3 E3 Ubiquitin Ligases and Protein Homeostasis in AD

10.3.1 Influence of E3 Ubiquitin Ligases on UPS Dysfunction and ER Stress in AD

Dysfunctions in the UPS, involving E3 ubiquitin ligases, have been implicated in AD pathology [16–18]. The malfunction of the E3 ubiquitin ligases can lead to the accumulation and aggregation of toxic proteins such as Aβ and tau [21].

Maintaining a precise balance of protein synthesis, folding, and breakdown is crucial for the health of brain cells, particularly neurons, which have limited regenerative abilities and high metabolic demands [4]. Aging is a major risk factor for AD, and one of the primary reasons is the decline in the ability to maintain protein quality control. Both of the key systems responsible for degrading proteins, the UPS and autophagy, become less efficient as we age [49]. The UPS is effective in degrading ubiquitinated soluble proteins, but in AD, it becomes overwhelmed when proteins begin to aggregate [50]. This places an increased reliance on autophagy, which is capable of degrading larger protein aggregates and damaged organelles, including mitochondria. However, autophagy is also affected by age-related decline, leading to reduced efficiency in eliminating harmful proteins and accumulation of damaged organelles [51, 52]. As a result of the protein degradation impairment, the buildup of Aβ and hyperphosphorylated tau worsens, further driving the progression of the disease [53, 54].

E3 ubiquitin ligases are essential components of the UPS. When misfolded or damaged proteins accumulate, they play a crucial role in resolving ER stress by tagging these proteins for degradation, helping to restore proteostasis and prevent cell death induced by prolonged ER stress [55]. It is plausible to consider that impairment in certain E3 ubiquitin ligases, particularly those involved in the degradation of APP/Aβ and tau, significantly contributes to ER stress in AD [56–58]. HRD1 (HMG-CoA Reductase Degradation Protein 1), an E3 ubiquitin ligase located in the ER membrane, facilitates the ER-associated degradation (ERAD) by targeting misfolded proteins like APP for degradation, thereby reducing protein buildup and mitigating stress. Similarly, CHIP (Carboxy Terminus of Hsc70-Interacting Protein), another critical E3 ligase, is involved in targeting tau for proteasomal degradation. Dysfunction in HRD1 or CHIP allows toxic proteins to accumulate and exacerbate ER stress, favoring AD pathology [56–58]. As well, E3 ligases serve as regulatory agents within autophagy pathways, particularly in the selective targeting of protein aggregates and organelles that the UPS cannot handle alone [59].

These enzymes confer specificity to the ubiquitination process, ensuring that the target substrates are accurately identified for degradation. When the E3 ubiquitin ligase activity is disrupted, however, this specificity is compromised, which can severely impair protein degradation systems. Consequently, toxic proteins, short-lived proteins that are no longer needed, and dysfunctional organelles accumulate, producing chronic ER stress, disturbing proteostasis and altering various critical cellular processes. Such disruptions can accelerate AD progression [12, 13, 60].

As discussed in the following sections, while the role of E3 ubiquitin ligases in UPS dysfunction in AD is well-documented, their role in autophagy regulation in AD is primarily associated with the impairment in the selective targeting and clearance of damaged mitochondria [18, 59]. These alterations can present as upregulation/overactivation, such as with NEDD4-1 (neuronal precursor cell-expressed developmentally down-regulated 4-1), MARCH8 (membrane-associated RING-CH 8), and RNF182 (Ring Finger Protein 182), or downregulation/inhibition, as in the case of PARKIN.

Besides ubiquitination, there are other post-translational modifications (PTMs) that influence the metabolism of toxic proteins in AD such as phosphorylation, neddylation, acetylation, and SUMOylation, among others [10, 61]. For example, PIAS1 (Protein Inhibitor of Activated STAT 1) is an E3 SUMO (small ubiquitin-like modifier) ligase which promotes Aβ degradation and reduces amyloid pathology by SUMOylating the intracellular domain of APP (AICD), enhancing its nuclear translocation and activating the transcription of Aβ-degrading enzymes, neprilysin and transthyretin [61]. Moreover, these PTMs can become impaired with aging and AD [20] and can interact with each other, forming complex regulatory networks that influence the metabolism of toxic proteins. For instance, tau SUMOylation can block its ubiquitination, which reduces tau degradation and promotes its aggregation [62]. Similarly, the SUMOylation of β-secretase (BACE1) interacts with and affects its phosphorylation and ubiquitination, impacting the buildup of Aβ in AD [63]. Such cross-talk between PTMs adds an additional layer of complexity to the regulation of protein processing and aggregation in AD.

Due to their ability to target a broad range of substrates, E3 ubiquitin ligases have a significant impact on diverse cellular pathways, making them integral to various aspects of AD and offering potential avenues for therapeutic intervention [16–18, 22].

10.3.2 E3 Ubiquitin Ligases in APP and Aβ Metabolism

In AD, one of the primary pathogenic events is the accumulation of Aβ peptides, which are derived from the sequential cleavage of APP by the enzymes BACE1 and γ-secretase. APP is involved in important cellular functions like synaptic regulation and neuronal growth [64]. However, APP can undergo cleavage via the amyloidogenic pathway. The initial cleavage is performed by BACE1, generating a soluble APP fragment (sAPPβ) and a membrane-bound C-terminal fragment. This C-terminal fragment is further cleaved by γ-secretase, producing Aβ peptides of varying lengths, with Aβ42 being the most aggregation-prone and neurotoxic form [65, 66].

APP has been found to be ubiquitinated at multiple lysine residues, specifically Lys-649 to Lys-651 and Lys-688. This ubiquitination plays a crucial role in regulating the production of Aβ peptides and influences the maturation and degradation processes of APP [67, 68]. E3 ubiquitin ligases play a significant role in modulating the processing of APP, thus influencing Aβ production and clearance [69–78].

For example, protein levels of the human HRD1 were significantly reduced in the cerebral cortex of AD patients. Reduced HRD1 levels impair APP degradation, leading to its accumulation, increased Aβ production, and the onset of ER stress, contributing to neuronal dysfunction and apoptosis [69].

Likewise, a component of the E3 ubiquitin ligase complex Skp1-Cullin1-F-box (SCF), known as FBL2, regulates APP by promoting its ubiquitination and inhibition of APP endocytosis, which decreases Aβ production and plaque formation. Its overexpression lowered Aβ levels, while knockdown increased them. FBL2 also enhanced APP degradation via the proteasome and reduced its presence in lipid rafts [70]. In fact, in the brains of AD patients, the levels of FBL2 are reduced [71].

Another key E3 ubiquitin ligase is CHIP (C-terminus Hsp70 interacting protein), also known as STUB1, which serves dual functions as a co-chaperone and ubiquitin ligase in the metabolism of Aβ. It stabilizes normal holo-βAPP while assisting in the ubiquitination and proteasomal degradation of misfolded or excess βAPP molecules. Additionally, it enhances the degradation of Aβ42 by forming complexes with heat shock proteins (HSPs) and promoting the ubiquitination and degradation of BACE1. By accelerating the clearance of Aβ, CHIP reduces its cellular accumulation, offering a protective mechanism against AD [79, 80]. It has been reported that Aβ42 accumulation reduces CHIP levels, but also affects tau turnover, creating a cycle that exacerbates tau pathology. Restoring CHIP levels in the presence of Aβ42 can reverse tau degradation impairment, underscoring CHIP's potential as a therapeutic target in AD [81].

A study identified ubiquitination factor E4B (UBE4B) as an E3 ubiquitin ligase that binds to ESCRT components and interacts with APP [72]. UBE4B regulates the ubiquitination of APP, which is crucial for its internalization into multivesicular bodies (MVBs). When UBE4B was depleted in neurons, APP ubiquitination was impaired, leading to increased levels of Aβ42 in endosomes and elevated secretion of Aβ42 from neurons. Moreover, analysis of brain samples from AD patients revealed a significant reduction in the levels of hepatocyte growth factor-regulated tyrosine kinase substrate (HRS), an ESCRT component that interacts with UBE4B. This reduction points to alterations in the ESCRT machinery, which plays a vital role in membrane protein sorting in the endocytic pathway, as part of AD pathology [72].

The E3 ubiquitin ligase IDOL seems to control the microglial uptake and removal of Aβ [73]. The loss of IDOL expression in a transgenic mouse model of Aβ amyloidosis led to a decrease in both soluble and insoluble Aβ, as well as a reduction in amyloid plaque burden. This suggests that IDOL deficiency promotes the clearance of Aβ in the brain [73].

It has also been reported that deficiency in ubiquitin protein ligase E3A (UBE3A) inhibits Aβ plaque formation in a APPswe/PS1δE9 mouse model of AD, by enhancing α-secretase activity. However, its deficiency also led to worsened cognitive and motor deficits, obesity, and shortened lifespan [74].

Besides, the age-related downregulation of E3 ubiquitin ligase mahogunin (MGRN1) promotes amyloidogenic APP processing, leading to increased production of Aβ, which may contribute to AD progression [75].

Furthermore, the F-box only protein 2 (FBXO2), a neuron-enriched ubiquitin ligase substrate adaptor, regulates APP levels and processing in the brain and may play a role in modulating AD pathogenesis. The loss of FBXO2 led to increased neuronal APP levels and processing in a brain region-specific manner [76].

Cullin-RING ubiquitin ligase (CRL) is another E3 ubiquitin ligase recently identified as a regulator of APP and Aβ metabolism, which has been suggested to have a role in AD. The CRL2/CRL4A complex, interacting with NRBP1, regulates Aβ production through targeted degradation of BRI2 and BRI3 proteins. The knockdown in neuronal cells of the nuclear receptor binding protein 1 (NRBP1), a substrate receptor of this ligase, inhibits APP processing and Aβ oligomerization [77].

In the AD-like five familial AD (5 × FAD) mouse model, which mimics Aβ pathology, the E3 ubiquitin ligase PELLINO-1 (PELI1) mediates the ubiquitination-induced degradation of C/EBPβ, a key transcription factor that regulates the expression of the scavenger receptor CD36, crucial for Aβ phagocytosis. By promoting C/EBPβ degradation, PELI1 impairs microglial Aβ clearance, contributing to Aβ accumulation [78].

With this evidence, it is likely that the effect on APP metabolism and Aβ plaque formation extends beyond proteostasis imbalance, impacting a wider array of dysregulated cellular functions involved in the pathogenesis of AD. Continued analysis of the interplay between E3 ligases, APP processing, and Aβ clearance is crucial for developing therapeutic strategies that target these ligases, with the potential to reduce Aβ accumulation and slow the progression of AD.

10.3.3 Influence of E3 Ubiquitin Ligases on Tau Metabolism

Tau is a microtubule-associated protein essential for maintaining neuronal structure and function. In AD, tau becomes abnormally hyperphosphorylated, leading to its detachment from microtubules and aggregation into NFTs, one of the hallmark pathologies of AD [82]. Ubiquitination of tau proteins, which occurs in both mono- and polyubiquitinated forms, plays a crucial role in tau metabolism and its degradation [83]. Twenty-eight tau ubiquitination sites have been detected in human AD brain samples. The most frequently reported sites are Lys-254, Lys-257, Lys-311, Lys-317, and Lys-353 [84, 85]. Interestingly, ubiquitination seems to play a dual role, being essential for tau clearance but also potentially promoting tau aggregation, leading to a pathological cycle where aggregation impedes proteasome function, thus increasing tau levels and further aggregation [83]. Under physiological conditions, tau proteins, whether in monomeric or oligomeric form, are ubiquitinated and directed to the proteasome for degradation. However, during the progression of AD, proteasome impairment leads to the accumulation of ubiquitinated tau and other proteins. This dysfunction in the protein degradation pathway may contribute to the formation of NFTs [85].

The E3 ubiquitin ligases are largely involved in tau metabolism, and their alterations can lead to the accumulation of pathological tau in AD brains [10, 22, 80, 81, 83, 86–95].

The E3 ubiquitin ligase CHIP/STUB1 is crucial in tau metabolism, particularly in the context of AD. In human AD brains and in tauopathy mouse models, CHIP expression is elevated. Conversely, a deficiency in CHIP can lead to increased levels of insoluble tau, indicating its protective role against tau aggregation. It can directly ubiquitinate and degrade phosphorylated tau (with a ten-fold greater affinity compared to unmodified tau) and interacts more strongly with caspase-cleaved tau than with full-length tau, promoting its rapid degradation [81, 86]. Additionally, CHIP's binding to tau requires both its U-box and TPR domains, with Hsp70 interaction being essential for reducing tau levels. Overexpression of CHIP has been shown to mitigate tau phosphorylation and protect against neurofibrillary tangle formation, highlighting its potential as a therapeutic target in tau-related pathologies [80, 87]. However, CHIP can also enhance the ubiquitination of tau, increasing tau aggregation, which probably depends on its interaction with Hsp70 and the efficiency of the proteasome activity [88].

Tau can also be targeted by the E3 ubiquitin ligase known as tumor necrosis factor receptor-associated factor 6 (TRAF6). This ligase localizes to tau inclusions in the brains of AD [89]. This ligase, in conjunction with the E2 enzyme UbcH7, mediates Lys-63-linked polyubiquitination of tau, which can alter tau's degradation pathway and promote aggregation [90].

The ligase TRIM21 plays a protective role in tau metabolism by recognizing and neutralizing misfolded tau assemblies that enter the cell, similar to how it responds to viral infections. TRIM21 is recruited by tau-associated antibodies, which accompany tau seeds into the cell. Once bound to tau seeds, TRIM21 mediates their degradation via the proteasome, involving the ATPase p97/VCP. This process helps to prevent the spread of tau misfolding, highlighting a mechanism where intracellular immunity targets misfolded tau to counter AD progression [91].

TRIM28 is a key regulator in tau metabolism, but its role does not involve typical ubiquitin ligase activity. Instead, it stabilizes tau by promoting its nuclear accumulation, which contributes to its toxic effects. Reduction of TRIM28 levels has been shown to decrease tau toxicity in vivo, suggesting that TRIM28 is involved in maintaining tau's pathogenic form [92].

HRD1, expressed in neurons and reactive astrocytes, could be associated with the degradation of hyperphosphorylated tau in the hippocampus, as evidenced by its colocalization with Alz-50-labeled pretangles [96]. However, its function seems to be inhibited by the accumulation of soluble tau [93].

Axotrophin/MARCH7, a RING-variant protein resembling E3 ubiquitin ligases, interacts with tau, preferentially modifying the 4-repeat isoform through monoubiquitination, which decreases tau's microtubule binding and may contribute to its accumulation and aggregation [94].

SLIMB, a key component of the SCF ubiquitin ligase complex, plays a vital role in tau metabolism by regulating the degradation of phosphorylated PAR-1, a kinase that controls tau activity. When the SLIMB function is impaired, phosphorylated PAR-1 accumulates, leading to excessive tau phosphorylation. This disrupts tau homeostasis, contributing to the formation of toxic tau aggregates, promoting neurodegeneration through tau-mediated neuronal damage [95].

These E3 ubiquitin ligases play crucial roles in maintaining tau homeostasis by targeting misfolded or hyperphosphorylated tau for degradation. Impairments in their function can disrupt this degradation process, leading to the accumulation of pathogenic tau species, which contributes to the development and progression of neurodegenerative diseases such as AD.

Table 10.1 summarizes the roles of E3 ubiquitin ligases specifically involved in the metabolism of Aβ and tau proteins in AD. These ligases play crucial roles in

Table 10.1 Summary of E3 ubiquitin ligases involved in the metabolism of Aβ and tau proteins, highlighting their roles and alterations in AD pathology

Targeted proteins	E3 ubiquitin ligase	E3 ubiquitin ligase's role in AD	References
APP and Aβ metabolism	HRD1	Facilitates APP degradation through the ERAD pathway. Reduced levels in AD impair APP degradation, leading to increased Aβ production and accumulation, inducing ER stress	[69]
	FBL2	Promotes APP ubiquitination and degradation via the proteasome and reduces Aβ production by inhibiting APP endocytosis. Levels are reduced in AD brains, together with increased Aβ production.	[70, 71]
	CHIP	Assists in the degradation of misfolded APP and enhances BACE1 degradation, facilitating Aβ clearance. Aβ42 accumulation may decrease CHIP levels in AD.	[79–81]
	UBE4B	Regulates APP ubiquitination, influencing Aβ accumulation and secretion within endosomes. Depletion leads to increased Aβ42 secretion and accumulation in endosomes, contributing to AD pathology.	[72]
	IDOL	Controls microglial uptake and clearance of Aβ. Loss of IDOL expression leads to reduced Aβ clearance, contributing to plaque burden in the brain.	[73]
	UBE3A	Facilitates α-secretase activity, inhibiting Aβ plaque formation. Deficiency reduces Aβ plaques but worsens cognitive deficits in a mouse model of AD.	[74]
	MGRN1	Age-related downregulation promotes amyloidogenic APP processing, leading to increased Aβ production.	[75]
	FBXO2	Modulates APP processing in neurons. Loss leads to increased APP levels, enhancing amyloid pathology in a brain region-specific manner.	[76]
	CRL	CRL2/CRL4A regulates APP and Aβ production in neurons by targeting BRI2 and BRI3 for degradation.	[77]
	PELI1	By promoting C/EBPβ degradation, PELI1 impairs microglial Aβ clearance in a familial AD mouse model.	[78]

(continued)

Table 10.1 (continued)

Targeted proteins	E3 ubiquitin ligase	E3 ubiquitin ligase's role in AD	References
Tau metabolism	CHIP	Degrades phosphorylated tau, preventing NFT formation. Increased CHIP levels mitigate tau aggregation; impairment contributes to NFT formation in AD.	[80, 81, 86–88]
	TRAF6	Promotes Lys-63-linked polyubiquitination of tau, which influences degradation pathways and potentially leads to tau aggregation. TRAF6 localizes to tau inclusions in the brains of AD.	[89, 90]
	TRIM21	Degrades misfolded tau, acting as a protective factor against tau pathology. Potential role in AD.	[91]
	TRIM28	Stabilizes tau, promoting its nuclear accumulation, possibly contributing to tau toxicity in AD.	[92]
	HRD1	Degrades hyperphosphorylated tau, but its function is inhibited by accumulated tau in AD.	[93, 96]
	MARCH7	Modifies tau by mono-ubiquitination, reducing microtubule binding, which may lead to tau accumulation and aggregation in AD.	[94]
	SLIMB	Impaired SLIMB leads to excessive tau phosphorylation and the formation of toxic tau aggregates. Potential role in AD.	[95]

processing and regulating Aβ and tau, with their dysfunction contributing significantly to AD pathology.

10.4 Multifaceted Roles of E3 Ubiquitin Ligases in AD Beyond Proteostasis

E3 ubiquitin ligases extend their functions beyond proteostasis by introducing ubiquitination as a regulatory modification in various cellular processes, not limited to protein degradation. They are essential for cell functions such as signal transduction, cell cycle control, autophagy (including mitophagy), and immune responses [16], all of which are interconnected processes. In AD, dysregulated E3 ligases can impair synaptic plasticity, mitochondrial function, neuroinflammation, cell cycle regulation, DNA damage response, calcium homeostasis, and induce neuronal death, which accelerates the disease progression [17, 18, 20–22]. Thus, E3 ligases play a complex role in both protein homeostasis and other key cellular functions in AD.

10.4.1 Synaptic Plasticity and Function

Synaptic dysfunction and loss are early pathological events in AD, closely associated with cognitive deterioration [97]. E3 ubiquitin ligases play a crucial role in maintaining synaptic plasticity, fundamental to memory and learning, by regulating the endocytosis and degradation of various synaptic receptors and proteins [98]. In AD, dysfunction of these ligases significantly contributes to synaptic pathology through disrupted trafficking of essential receptors, such as AMPA receptors (AMPARs), NMDA receptors (NMDARs), and neurotrophic receptors [22].

One of the key E3 ligases implicated in AD-related synaptic dysfunction is UBE3A/E6-AP, which regulates proteins like Arc and Ephexin-5. In AD models, such as the Tg2576 mouse model, which overexpresses human APP carrying the Swedish mutation, UBE3A levels are reduced, leading to synaptic abnormalities. This loss results in the accumulation of Arc and Ephexin-5, triggering the internalization of GluR1 (an AMPAR subunit) and activating RhoA, a small GTPase, both of which contribute to synapse loss and dysfunction. Restoration of UBE3A expression in these models has been shown to mitigate these deleterious effects, positioning UBE3A as a potential therapeutic target in AD [99]. Further exploring UBE3A's functions, another study found that UBE3A regulates the endocytosis of SK2 channels, which are vital for synaptic plasticity and cognitive functions. In UBE3A-deficient mice, elevated SK2 levels impair NMDARs activity and long-term synaptic plasticity, leading to memory deficits. Inhibiting SK2 channels in these mice alleviated these impairments, underscoring the importance of UBE3A in cognitive processes [100]. Additionally, a recent review highlights the importance of UBE3A/E6-AP in maintaining neuronal and synaptic health. While the review covers a range of disorders, the principles discussed offer valuable insights into how E6AP dysfunction may contribute to the synaptic impairments and cognitive deficits characteristic of AD. Targeting E6AP-mediated pathways emerges as a promising strategy for developing treatments aimed at preserving synaptic function and slowing AD progression [101].

Also, the activity of NEDD4-1 has been implicated in the homeostatic scaling of synaptic strength, where it modulates AMPAR surface expression in response to neuronal activity changes [102]. The activity-dependent ubiquitination of the GluA1 subunit by NEDD4-1 triggers a specific pathway for AMPAR endocytosis and sorting. This process occurs in response to AMPAR activation but not NMDAR activation, highlighting specificity in AMPAR trafficking regulation [103]. In the context of AD, the pathogenic peptide Aβ has been shown to exacerbate NEDD4-1-mediated ubiquitination of AMPARs. Aβ promotes the internalization and degradation of AMPARs by enhancing their ubiquitination, a process dependent on NEDD4-1 [104]. This ubiquitination leads to a decrease in surface AMPAR expression, resulting in synaptic weakening and reduced synaptic strength. Studies using cultured neurons have demonstrated that Aβ treatment increases the recruitment of NEDD4-1 to synapses, correlating with enhanced ubiquitination of AMPARs and a reduction in dendritic spine density [104, 105]. Similarly, analysis of human AD brain tissues has revealed increased levels of NEDD4-1, probably favored by oxidative stress [106] and decreased levels of the deubiquitinase USP46, further supporting the role of ubiquitination in AMPAR downregulation in AD [107]. Moreover, attenuation of NEDD4-1 expression has been shown to rescue synaptic function and memory deficits in AD models. In 3xTg-AD mice, reducing the expression of NEDD4-1 improved synaptic plasticity and enhanced excitatory postsynaptic potentials [108].

The Ring Finger Proteins RNF182, RNF220, and RNF167 are three distinct E3 ubiquitin ligases that influence AMPAR stability and degradation, contributing to the synaptic dysfunction seen in AD [17, 109]. The RNF182 E3 ubiquitin ligase serves as an important regulator of synaptic function by controlling the degradation of ATP6V0C, a vital protein in neurotransmitter release. Its increased expression in AD suggests it may contribute to the synaptic impairments observed in the condition [109]. RNF220 binds to AMPARs and triggers the ubiquitination of the GluA1 subunit, leading to its internalization and degradation. This process ultimately results in the inhibition of excitatory synaptic transmission by reducing the number of functional AMPARs at the synapse. On the other hand, RNF167 interacts with the E2 coupling enzyme UBE2N, and through its RING structural domain, facilitates the ubiquitination of the GluA2 subunit. This ubiquitination enhances the degradation of GluA2, which paradoxically improves synaptic transmission efficacy by altering the balance of AMPAR subunit composition at the synapse. These regulatory mechanisms highlight the critical role of E3 ligases in modulating synaptic function through precise control of AMPAR turnover [17, 109].

E3 ubiquitin ligases CHIP/STUB1 and CRL4(CRBN) interact specifically with APP at its cytosolic region, ubiquitinating multiple lysine residues. This ubiquitination facilitates the modification of presynaptic proteins involved in neurotransmitter release. Dysfunction in CHIP and CRL4(CRBN) impairs APP-mediated ubiquitination, leading to synaptic deficits and the accumulation of dementia-related proteins like tau and ApoE. These disruptions contribute to the synaptic dysfunction and protein aggregation observed in AD, also positioning CHIP/STUB1 and CRL4(CRBN) as critical targets for therapeutic intervention [110].

Another E3 ubiquitin ligase potentially involved in the synaptic plasticity impairment in AD could be the Anaphase Promoting Complex/Cyclosome (APC/C), which regulates synaptic plasticity, and its dysfunction is associated with deficits in learning and memory in mice [111]. Thus, it is reasonable to hypothesize that alterations in APC/C function may contribute to synaptic dysfunction in AD.

As detailed below, synaptic plasticity is indirectly affected by alterations of E3 ubiquitin ligases, which regulates the synaptic distribution of mitochondria, like the broad activation of the PARKIN pathway [112].

10.4.2 Mitochondrial Function

Mitochondrial dysfunction, a characteristic feature of AD, is intricately linked to impaired mitophagy, the selective autophagic degradation of damaged mitochondria [113]. Among the E3 ubiquitin ligases, PARKIN is the primary one documented to play a pivotal role in mitochondrial alterations observed in AD. PARKIN is an E3 ubiquitin ligase that is more commonly associated with Parkinson's disease (PD) but plays a role in the pathogenesis of AD and other neurodegenerative diseases [114]. This ligase mediates mitophagy and UPS-associated protein degradation, which regulate mitochondrial distribution and function, and these processes are disrupted in AD [115].

PARKIN-mediated mitophagy, regulated through the PINK1-PARKIN pathway, is crucial for maintaining mitochondrial quality control, and its dysfunction exacerbates the accumulation of defective mitochondria, contributing to AD progression [115, 116]. Studies have shown that both Aβ and tau can interfere with PARKIN function. Pathologically phosphorylated tau inhibits PARKIN translocation to damaged mitochondria, blocking their clearance and leading to synaptic dysfunction and neurodegeneration [112, 117]. Similarly, Aβ oligomers trigger mitochondrial fragmentation and reduce Miro1, an essential downstream component of the mitochondrial quality control pathway regulated by PARKIN. This impairs mitochondrial motility and further aggravates mitophagy deficits [118]. Restoration of PARKIN activity has been shown to reverse cognitive deficits and reduce Aβ and tau toxicity by enhancing mitophagy in various AD models [119, 120]. Moreover, pharmacological approaches that increase PARKIN expression or activity have demonstrated neuroprotective effects, highlighting the therapeutic potential of targeting E3 ligases to mitigate mitochondrial dysfunction in AD [10, 121]. These findings emphasize the importance of E3 ligases, particularly PARKIN, in the maintenance of mitochondrial health and in counteracting the neurodegenerative processes underlying AD.

The mitochondrial ubiquitin ligase MITOL (MARCH5), which is located on the outer mitochondrial membrane, plays a dual role in AD by regulating both mitochondrial homeostasis and neuroinflammation [17]. In terms of mitochondrial regulation, MITOL maintains mitochondrial integrity by promoting the clearance of protein aggregates that accumulate on the mitochondrial surface, preserving proper mitochondrial dynamics and function. However, its involvement in AD is more prominently linked to neuroinflammation [122].

10.4.3 Neuroinflammation

In AD, several E3 ubiquitin ligases are crucial for regulating neuroinflammation and immune responses, contributing to the disease's progression:

TRAF6 is a pivotal mediator of neuroinflammation, overactivated by toll-like receptors (TLRs) on microglia in response to Aβ. TRAF6 activates the NF-κB and MAPK pathways, leading to the production of pro-inflammatory cytokines like IL-1β, TNF-α, and IL-6. These cytokines drive chronic inflammation in the brain, which exacerbates AD pathology [18, 123].

Another significant E3 ubiquitin ligase involved in AD is COP1 (also known as RFWD2). This ligase plays a protective role in suppressing neuroinflammation by regulating the degradation of the transcription factor CCAAT/enhancer-binding protein beta (c/EBPβ), which controls pro-inflammatory gene expression in microglia [124]. In AD, the absence of COP1 leads to the accumulation of c/EBPβ, which drives a robust pro-inflammatory gene expression program. This increase in neuroinflammation, primarily driven by microglia, exacerbates neurodegeneration. Experiments have shown that COP1-deficient microglia significantly worsen tau-related neurodegeneration in mouse models of AD. Notably, the neurotoxic effects of this process are largely mediated by complement system activation, further implicating COP1 as a critical suppressor of c/EBPβ-dependent inflammatory pathways in AD [124, 125].

Furthermore, the E3 ubiquitin ligase MARCH5 is implicated in regulating the NLRP3 inflammasome, which is involved in neuroinflammatory responses [122]. MARCH5 mediates K27-linked polyubiquitination of NLRP3, which is essential for the formation of the NLRP3-NEK7 complex, a crucial step in inflammasome activation. This activation leads to the production of pro-inflammatory cytokines like IL-1β and IL-18, which drive neuroinflammation. In models lacking MARCH5, the inability to ubiquitinate NLRP3 results in reduced inflammasome activation and diminished secretion of these cytokines, highlighting MARCH5 as a key regulator of neuroinflammation that may play a role in AD progression. Although the direct link between MARCH5 and AD has not been fully established, its role in controlling NLRP3 inflammasome activity suggests that it could contribute to the neuroinflammatory environment in AD [122].

In summary, these E3 ubiquitin ligases are integral to the regulation of neuroinflammation and immune responses in AD. They influence critical signaling pathways that control the production of pro-inflammatory cytokines and the activation of inflammasomes, thereby contributing to the progression of neurodegeneration.

10.4.4 Cell Cycle Regulation

In AD, neurons aberrantly attempt to re-enter the cell cycle, leading to neuronal stress and death [126]. Several studies highlight the critical roles of these ligases in aberrant cell cycle re-entry and neuronal dysfunction in AD.

One of the key E3 ubiquitin ligases implicated in AD-related cell cycle dysfunction is the APC/C. This ligase, regulated by its co-activator Cdh1, is essential for maintaining post-mitotic neurons in their non-dividing state. Cdh1-APC/C normally promotes the degradation of cell cycle proteins, such as cyclin B1, to prevent unscheduled cell cycle progression in neurons. However, in AD, APC/C-Cdh1 is inactivated, often due to the phosphorylation of Cdh1 by cyclin-dependent kinase-5 (Cdk5) [127]. This inactivation leads to the stabilization and accumulation of cyclin B1, a hallmark of aberrant cell cycle re-entry in neurons, which contributes to neuronal apoptosis rather than mitotic division [10, 128]. The presence of Aβ leads to increased glutamate release and excitotoxic stress, which exacerbates the downregulation of APC/C-Cdh1. This further drives the accumulation of cyclin B1 and the cell cycle re-entry, leading to neuronal degeneration [129].

Another E3 ubiquitin ligase involved in cell cycle dysregulation in AD is ITCH, a member of the NEDD4 family of E3 ligases. ITCH plays a role in regulating neuronal apoptosis through its interaction with the transcription factor TAp73, which inhibits cell cycle progression. In AD, Aβ exposure leads to the activation of the JNK signaling pathway, causing hyperphosphorylation and autoubiquitination of ITCH. This activation of ITCH promotes the degradation of TAp73, contributing to aberrant cell cycle re-entry and apoptosis in neurons [10, 130].

10.4.5 DNA Damage Response

The involvement of E3 ubiquitin ligases in DNA damage response in AD has gained increasing attention due to their crucial role in maintaining genomic stability and modulating cellular responses to oxidative stress and other pathological conditions.

BRCA1 (Breast Cancer Type 1 susceptibility protein) functions as an E3 ubiquitin ligase by forming a heterodimer with BARD1 (BRCA1-associated RING domain protein 1), together facilitating crucial DNA repair processes [131]. In the context of AD, BRCA1 is found to be depleted in the brains of affected individuals, and this depletion impairs cognitive function in mouse models [132]. This highlights the essential role of BRCA1-mediated DNA repair in maintaining neuronal integrity and cognitive functions, suggesting that its loss contributes to neurodegeneration observed in AD.

A recent study highlights PIRH2 as an E3 ubiquitin ligase that modulates the mitochondrial function and DNA damage response in AD [133]. PIRH2 upregulation in experimental AD models is associated with increased oxidative stress and DNA damage, including DNA double-strand breaks (DSBs). Silencing PIRH2 mitigates DNA damage, suggesting that its aberrant activity exacerbates mitochondrial dysfunction and contributes to the pathology of AD. PIRH2 regulation of cytochrome c-mediated neuronal apoptosis further emphasizes its role in disease progression through impaired DNA repair and mitochondrial integrity [133].

Additionally, LUBEL (the Drosophila ortholog of human HOIP), a key component of the linear ubiquitin chain assembly complex (LUBAC), may play a role in DNA repair in AD [134]. LUBEL knockdown in a Drosophila AD model increased neuronal DNA damage, particularly DSBs, and aggravated neurodegeneration. This suggests that LUBEL plays a protective role against DNA damage and Aβ-induced neurotoxicity. Although the direct human equivalent of LUBEL is HOIP, this study proposes that linear ubiquitination, facilitated by LUBAC, might be a crucial mechanism protecting neurons from DNA damage in AD, highlighting the potential role of HOIP in similar processes in humans [134].

Furthermore, the Cullin4 E3 ubiquitin ligase complex, particularly through DDB1 and HECTD1, is involved in base excision repair, addressing oxidative DNA damage, which is common in AD. Although not directly linked to AD, their role in DNA repair suggests a potential connection to AD-related neurodegeneration. Besides, RNF8, RNF168, and RAD18 are key in repairing double-strand breaks through histone ubiquitination, and dysfunction in these ligases can impair DNA repair, contributing to neurodegeneration and potentially accelerating AD progression [12].

10.4.6 Calcium Homeostasis

AD is characterized by early disruptions in calcium homeostasis, which precede clinical symptoms and contribute to neuronal dysfunction and degeneration [135]. Recent studies have highlighted the pivotal roles of E3 ubiquitin ligases, particularly MDM2 and NEDD4-1, in modulating calcium channels and receptors implicated in AD [10].

One key player in calcium regulation is the L-type voltage-gated calcium channel Cav1.2. Abnormal activity of Cav1.2 leads to intracellular calcium overload, a condition observed in AD that promotes neurotoxicity [136]. The E3 ubiquitin ligase MDM2 has been shown to facilitate the ubiquitination and subsequent proteasomal degradation of Cav1.2. In APP/PS1 double-mutant mice, overexpression of MDM2 reduced Cav1.2 levels, ameliorated calcium overload, and improved cognitive functions [10, 137]. Activation of estrogen receptor α (ERα) by agonists like 17β-estradiol (E2) or propylpyrazoletriol (PPT) enhances MDM2 expression, thereby promoting Cav1.2 degradation and restoring calcium balance in neuronal cells [137, 138]. The mechanism involves MDM2 recognizing specific sequences in the Cav1.2 protein, tagging it with ubiquitin chains linked through Lysine-29. This post-translational modification marks Cav1.2 for degradation via the ubiquitin-proteasome system, thus reducing excessive calcium influx into neurons [137]. The protective effects of ERα activation on calcium homeostasis and cognitive function emphasize the therapeutic potential of targeting this pathway in AD.

Another E3 ubiquitin ligase, NEDD4-1, significantly influences neuronal calcium homeostasis by modulating AMPARs. As previously mentioned, Aβ peptides enhance NEDD4-1 activity, leading to increased ubiquitination and internalization of AMPARs from the neuronal surface [104]. While AMPARs are not primary conduits for calcium ions, their activation affects neuronal depolarization, which regulates calcium influx through voltage-gated calcium channels and NMDARs. Thus, NEDD4-1-mediated reduction of AMPAR surface expression indirectly impacts calcium homeostasis by altering neuronal excitability and downstream calcium signaling pathways. Inhibiting or knocking down NEDD4-1 prevents Aβ-induced AMPAR internalization and associated disturbances in calcium signaling, highlighting its crucial role in Aβ-related calcium dysregulation in AD [10, 104].

These findings suggest that E3 ubiquitin ligases MDM2 and NEDD4-1 are central to the regulation of calcium homeostasis in AD. By controlling the degradation of calcium channels and receptors, they modulate intracellular calcium levels and neuronal function.

10.4.7 Neuronal Death

Neuronal cell death often results from cumulative stressors that disrupt cellular homeostasis, with dysfunction of E3 ubiquitin ligases being a key factor. E3 ligases contribute to cell death through different mechanisms. When these ligases are impaired, the inadequate clearance of toxic proteins leads to their accumulation, causing proteotoxic stress that can trigger cell death. In addition to disrupting proteostasis, impaired E3 ubiquitin ligases indirectly affect cell death by altering the previously described cellular processes, while directly influencing signaling pathways involved in AD, including apoptosis, necroptosis, and ferroptosis [12, 18].

10.4.7.1 Apoptosis

Apoptosis, a form of programmed cell death, plays a crucial role in AD [139]. It is activated through extrinsic and intrinsic pathways, involving Bcl-2 family proteins (e.g., Bax, Bad, Bcl-XL) and caspases, which trigger a cascade of events that ultimately result in cell disintegration [139].

XIAP (X-linked Inhibitor of Apoptosis) is an E3 ubiquitin ligase that suppresses apoptosis by ubiquitinating caspase-3, -7, and -9. In AD, XIAP is inactivated by S-nitrosylation, leading to increased caspase activity and excessive apoptosis in neurons [12, 140].

MARCH8 regulates TNF-α-induced apoptosis by ubiquitinating myosin light chain 2 (MLC2). Elevated TNF-α in AD brains enhances MARCH8 activity, which accelerates neuronal apoptosis by degrading MLC2. Blocking MARCH8 can inhibit caspase-3 activation, reducing apoptosis and offering potential therapeutic value [141].

CUL4B, another E3 ligase, is involved in neuronal apoptosis by targeting somatostatin receptor 4 (SSTR4) for degradation. Upregulation of CUL4B in AD leads to the enhanced degradation of SSTR4, which triggers apoptotic pathways in neurons. Inhibiting CUL4B has been shown to reduce neuronal apoptosis and improve cognitive function in AD models [142].

RNF182 is an E3 ubiquitin ligase specifically enriched in the brain, which becomes upregulated in AD. By promoting ATP6V0C degradation, RNF182 disrupts synaptic function and induces apoptosis in neurons, contributing to the overall neurodegeneration seen in AD [109].

TRAF6 plays protective roles by preventing Aβ-induced apoptosis. This ligase, in collaboration with p62, regulates neuronal survival by promoting the ubiquitination of the p75 neurotrophin receptor (p75NTR). In AD, Aβ disrupts this ubiquitination process, leading to reduced p75NTR signaling and increased neuronal death. TRAF6 restores p75NTR ubiquitination and prevents Aβ-induced apoptosis by enhancing NF-κB-mediated survival pathways [143].

MDM2, an E3 ubiquitin ligase that targets p53 for ubiquitination, interacts with tau in AD brains. Tau inhibits MDM2's ability to ubiquitinate p53, resulting in p53 accumulation, which promotes apoptotic and other cell death pathways. The disruption of MDM2 function by tau contributes to neurodegeneration in AD [144].

10.4.7.2 Necroptosis

Necroptosis is an inflammatory form of regulated cell death that occurs when apoptosis is blocked, often driven by receptor-interacting kinases (RIPK1 and RIPK3) [145, 146]. Necroptosis has been observed to be activated in the brains of both human and mouse models of AD, playing a significant role in disease progression and contributing to various pathological processes [145, 146].

PARKIN (PARK2) regulates necroptosis by ubiquitinating RIPK1 and RIPK3. In AD, a deficiency of PARKIN leads to increased RIPK1-RIPK3 complex formation, which drives necroptosis and inflammation-induced neuronal death. PARKIN's role in preventing necroptosis makes it a potential therapeutic target for limiting neuroinflammation in AD [147, 148]. In addition to regulating necroptosis, PARKIN also influences NF-κB signaling by ubiquitinating RIPK1, promoting cell survival pathways while inhibiting necrosome formation. This balance between cell survival and necroptosis is crucial in AD pathology [148].

10.4.7.3 Ferroptosis

Ferroptosis is a regulated form of cell death characterized by the accumulation of iron-dependent lipid peroxides, leading to oxidative damage of cell membranes. Ferroptosis has been identified as a key mechanism of neurodegeneration in AD, as increased iron accumulation in the neocortex is strongly associated with the progression of AD [149]. The E3 ubiquitin ligase NEDD4 regulates ferroptosis by ubiquitinating divalent metal transporter 1 (DMT1), which controls iron uptake. In AD, elevated DMT1 levels lead to excessive iron accumulation and trigger ferroptosis. NEDD4-mediated ubiquitination and degradation of DMT1 help prevent iron overload and protect neurons from ferroptosis. Dysfunction in this regulatory pathway exacerbates iron-driven neuronal loss in AD [17]. Additionally, reduced expression of NEDD4 family interacting protein 1 (NDFIP1) may contribute to AD by impairing DMT1 degradation, leading to iron accumulation and increased Aβ production. Overexpression of NDFIP1 in cell models decreased DMT1 levels, reduced iron influx, and lowered Aβ secretion, highlighting the potential role of NDFIP1 in moderating iron-induced neurotoxicity in AD [150].

This evidence highlights the key roles of E3 ubiquitin ligases in regulating different forms of cell death in AD. Their modulation of apoptosis, necroptosis, ferroptosis, and Aβ- and tau-associated toxicity makes them key players in neuronal survival.

In Table 10.2, the roles of E3 ubiquitin ligases involved in distinct cellular processes disrupted in AD are summarized. These ligases influence pathways beyond Aβ and tau metabolism and ER stress management, including mitochondrial function, neuroinflammation, cell cycle regulation, DNA damage response, calcium homeostasis, and neuronal death:

Table 10.2 Summary of E3 ubiquitin ligases involved in various cellular processes beyond Aβ and tau metabolism, highlighting their implications in AD pathology

Cell process	E3 ubiquitin ligase	E3 ubiquitin ligase's role in AD	References
Synaptic plasticity	UBE3A	Regulates Arc and Ephexin-5 to maintain synaptic function. Impairment leads to GluR1 internalization and synapse loss in a mouse model of AD.	[99–101]
	NEDD4-1	Regulates AMPAR ubiquitination and expression on the surface, reducing synaptic strength. In AD, Aβ enhances this activity, leading to synaptic weakening.	[102–108]
	RNF182, RNF220, RNF167	Regulate AMPAR stability and degradation. Increased expression in AD impairs neurotransmission and contributes to synaptic dysfunction.	[17, 109]
	CHIP and CRL4	Interact with APP at the cytosolic region, promoting ubiquitination of presynaptic proteins. Dysfunction leads to synaptic deficits and accumulation of APOE	[110]
	APC/C	Regulates synaptic plasticity; impairment in AD models leads to learning and memory deficits, suggesting its role in AD-associated synaptic dysfunction.	[111]
Mitochondrial function	PARKIN	Facilitates mitophagy via the PINK1-PARKIN pathway, aiding in the clearance of defective mitochondria. Impairment in AD leads to defective mitochondrial clearance, exacerbating synaptic dysfunction and neurodegeneration.	[10, 112, 114–121]
	MITOL (MARCH5)	Maintains mitochondrial integrity by clearing surface protein aggregates. Linked to neuroinflammation in AD, which disrupts mitochondrial dynamics.	[17]

(continued)

Table 10.2 (continued)

Cell process	E3 ubiquitin ligase	E3 ubiquitin ligase's role in AD	References
Neuroinflammation	TRAF6	Activated by TLRs in response to Aβ, triggering chronic inflammation through NF-κB signaling in AD brains.	[18, 123]
	COP1	Degrades C/EBPβ, suppressing neuroinflammation. In its absence, C/EBPβ accumulates, promoting pro-inflammatory gene expression and worsening neurodegeneration in AD.	[124, 125]
	MARCH5	Regulates NLRP3 inflammasome activation, influencing cytokine release and contributing to neuroinflammation. Potential role in AD.	[122]
Cell cycle regulation	APC/C	Maintains neurons in a non-dividing state. Inactivation in AD leads to aberrant cell cycle re-entry and neuronal apoptosis, exacerbating neurodegeneration.	[10, 127–129]
	ITCH	Prevents apoptosis in neurons by interacting with TAp73. Hyperactivation in AD leads to degradation of TAp73, contributing to cell cycle re-entry and neuronal apoptosis.	[10, 130]
DNA damage response	BRCA1	Essential for DNA repair. Depletion in AD brains impairs cognitive function, promoting neurodegeneration.	[132]
	PIRH2	Induces oxidative stress, DNA damage, and mitochondrial dysfunction, exacerbating AD pathology.	[133]
	LUBEL	Protects neurons against DNA damage. Knockdown in Drosophila AD models leads to increased neurodegeneration.	[134]
Calcium homeostasis	MDM2	Degrades Cav1.2 calcium channels, reducing calcium overload in neurons. Enhancing its activity can ameliorate calcium dysregulation and cognitive deficits in AD.	[10, 137, 138]
	NEDD4-1	Modulates AMPAR surface expression, indirectly affecting calcium influx. In AD, Aβ enhances NEDD4-1 activity, leading to calcium dysregulation.	[10, 104]

(continued)

Table 10.2 (continued)

Cell process	E3 ubiquitin ligase	E3 ubiquitin ligase's role in AD	References
Neuronal death	XIAP	Suppresses apoptosis by ubiquitinating caspases. Inactivation in AD leads to excessive neuronal apoptosis.	[12, 140]
	MARCH8	Regulates TNF-α-induced apoptosis. Increased activity in AD accelerates apoptosis through degradation of MLC2.	[141]
	CUL4B	Degrades SSTR4, triggering apoptotic pathways. Upregulated in AD, contributing to neuronal death.	[142]
	RNF182	Degrades ATP6V0C, impairing synaptic function and inducing apoptosis in neurons, exacerbating AD pathology.	[109]
	TRAF6	Prevents Aβ-induced apoptosis by promoting p75NTR ubiquitination. In AD, reduced p75NTR signaling increases neuronal death.	[143]
	MDM2	Ubiquitinates p53, promoting cell survival. In AD, tau disrupts MDM2 function, leading to p53 accumulation and apoptosis.	[144]
	PARKIN	Prevents necroptosis by ubiquitinating RIPK1 and RIPK3, maintaining cell survival. Deficiency in AD increases necroptosis and neuroinflammation.	[147, 148]
	NEDD4	Prevents ferroptosis by regulating iron uptake via DMT1 degradation. Dysfunction in AD leads to iron accumulation and increased neurotoxicity.	[17, 150]

10.5 Future Directions and Conclusion

This work explores the dysfunctions of E3 ubiquitin ligases and their impact on cellular processes central to AD pathology. These ligases play a critical role in AD by influencing interconnected processes such as proteostasis, synaptic stability, mitochondrial function, neuroinflammation, DNA repair, cell cycle regulation, calcium balance, and overall cell survival. The analysis highlights how dysfunctions in E3 ubiquitin ligases disrupt protein quality control and drive mechanisms linked to AD progression, revealing an intricate network of molecular disturbances underlying the disease. Figure 10.1 recopilates the documented E3 ubiquitin ligases linked to dysfunctions in cellular processes associated with AD.

The dual roles of E3 ligases in AD pathogenesis are particularly significant; while they can facilitate neuronal protection through targeted degradation of

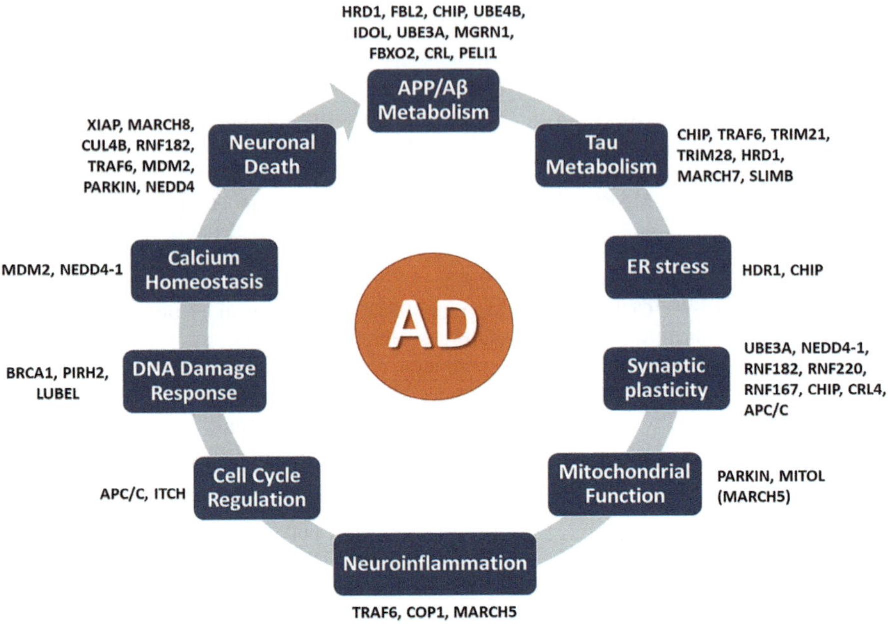

Fig. 10.1 E3 ubiquitin ligases implicated in impaired cellular processes associated with AD. This diagram highlights key interconnected cellular processes affected in AD, including APP/Aβ metabolism and ER stress management, tau metabolism, mitochondrial function, neuroinflammation, cell cycle regulation, DNA damage response, calcium homeostasis, and neuronal death mechanisms. Each process lists specific E3 ubiquitin ligases reported to contribute to the dysfunction of these pathways, underscoring their roles in the molecular pathology of AD

pathogenic proteins, their dysregulation may instead exacerbate the protein accumulation, cell stress, and neurodegeneration. This complexity needs an advanced understanding of E3 ubiquitin ligase activities across disease stages and within specific cellular environments in the brain.

Intriguingly, altered expression of several E3 ubiquitin ligases has been observed in AD brains, yet the affected molecular pathways remain underexplored. For instance, Dactylidin (RNF146), a RING finger protein with probable E3 ubiquitin ligase activity, is upregulated in neurons vulnerable to early AD stages, hinting at its possible role in initiating neurodegenerative processes [151]. Additionally, UBR5, another E3 ubiquitin ligase upregulated in AD, suggests further investigation is needed to clarify its function in AD biology [152]. Altered expression of tripartite motif-containing ligases in AD brains, such as TRIM32 and TRIM37 (which exhibit anti-apoptotic and ubiquitin-protein ligase activities), also suggests that E3 ubiquitin ligase dysregulation may contribute to neurodegenerative mechanisms, warranting further study [153].

Therapeutic strategies aimed at restoring or fine-tuning E3 ubiquitin ligase function could thus yield considerable benefits, especially if adapted to the disease's temporal progression and the unique characteristics of affected cellular populations.

Future research must further delineate the intricate interplay between ubiquitination and other PTMs, such as phosphorylation and SUMOylation, and their cumulative impact on AD progression.

Given the integrative and multifaceted roles of the E3 ubiquitin ligases, they represent compelling therapeutic targets for interventions focused on restoring cellular homeostasis and reducing pathological protein burdens. Therapeutic approaches targeting E3 ligases in AD encompass the development of small-molecule inhibitors to attenuate hyperactive ligases implicated in Aβ and tau pathologies, as well as activators to enhance the degradation of misfolded proteins and mitigate neuroinflammation [12, 18]. Additionally, PROTACs (Proteolysis Targeting Chimeras) offer an innovative strategy, harnessing E3 ligases to selectively degrade disease-associated proteins, thereby offering potential to modify neurodegenerative progression in AD [17, 154].

In conclusion, E3 ubiquitin ligases represent both a rich source of mechanistic insight and a promising avenue for therapeutic innovation in AD. A deeper, comprehensive understanding of these enzymes has the potential to transform current approaches to AD treatment, fostering interventions that address central aspects of disease pathology, including Aβ and tau accumulation, neuroinflammation, mitochondrial and synaptic dysfunctions, and neuronal death. By modulating E3 ubiquitin ligase activity with precision, it could be possible to significantly alter or slow the progression of AD, addressing the fundamental molecular disruptions that underlie this disease.

References

1. Scheltens P, De Strooper B, Kivipelto M, Holstege H, Chetelat G, Teunissen CE, et al. Alzheimer's disease. Lancet. 2021;397(10284):1577–90.
2. DeTure MA, Dickson DW. The neuropathological diagnosis of Alzheimer's disease. Mol Neurodegener. 2019;14(1):32.
3. Thapa R, Ahmad Bhat A, Shahwan M, Ali H, PadmaPriya G, Bansal P, et al. Proteostasis disruption and senescence in Alzheimer's disease pathways to neurodegeneration. Brain Res. 2024;1845:149202.
4. Lautrup S, Sinclair DA, Mattson MP, Fang EF. NAD(+) in brain aging and neurodegenerative disorders. Cell Metab. 2019;30(4):630–55.
5. Gong B, Radulovic M, Figueiredo-Pereira ME, Cardozo C. The ubiquitin-proteasome system: potential therapeutic targets for Alzheimer's disease and spinal cord injury. Front Mol Neurosci. 2016;9:4.
6. Ciechanover A. The unravelling of the ubiquitin system. Nat Rev Mol Cell Biol. 2015;16(5):322–4.
7. Pohl C, Dikic I. Cellular quality control by the ubiquitin-proteasome system and autophagy. Science. 2019;366(6467):818–22.
8. Chen RH, Chen YH, Huang TY. Ubiquitin-mediated regulation of autophagy. J Biomed Sci. 2019;26(1):80.
9. Abreha MH, Dammer EB, Ping L, Zhang T, Duong DM, Gearing M, et al. Quantitative analysis of the brain ubiquitylome in Alzheimer's disease. Proteomics. 2018;18(20):e1800108.
10. Liu N, Lin MM, Wang Y. The emerging roles of E3 ligases and DUBs in neurodegenerative diseases. Mol Neurobiol. 2023;60(1):247–63.

11. Lopata A, Kniss A, Lohr F, Rogov VV, Dotsch V. Ubiquitination in the ERAD process. Int J Mol Sci. 2020;21(15):5369.
12. Schmidt MF, Gan ZY, Komander D, Dewson G. Ubiquitin signalling in neurodegeneration: mechanisms and therapeutic opportunities. Cell Death Differ. 2021;28(2):570–90.
13. Komander D, Rape M. The ubiquitin code. Annu Rev Biochem. 2012;81:203–29.
14. Yang Q, Zhao J, Chen D, Wang Y. E3 ubiquitin ligases: styles, structures and functions. Mol Biomed. 2021;2(1):23.
15. Vembar SS, Brodsky JL. One step at a time: endoplasmic reticulum-associated degradation. Nat Rev Mol Cell Biol. 2008;9(12):944–57.
16. Kaushik A, Parashar S, Ambasta RK, Kumar P. Ubiquitin E3 ligases assisted technologies in protein degradation: sharing pathways in neurodegenerative disorders and cancer. Ageing Res Rev. 2024;96:102279.
17. Lin N, Gao XY, Li X, Chu WM. Involvement of ubiquitination in Alzheimer's disease. Front Neurol. 2024;15:1459678.
18. Potjewyd FM, Axtman AD. Exploration of aberrant E3 ligases implicated in Alzheimer's disease and development of chemical tools to modulate their function. Front Cell Neurosci. 2021;15:768655.
19. Ajoolabady A, Lindholm D, Ren J, Pratico D. ER stress and UPR in Alzheimer's disease: mechanisms, pathogenesis, treatments. Cell Death Dis. 2022;13(8):706.
20. Ramazi S, Dadzadi M, Darvazi M, Seddigh N, Allahverdi A. Protein modification in neurodegenerative diseases. MedComm. 2024;5(8):e674.
21. Li W, Li HL, Wang JZ, Liu R, Wang X. Abnormal protein post-translational modifications induces aggregation and abnormal deposition of protein, mediating neurodegenerative diseases. Cell Biosci. 2024;14(1):22.
22. Harris LD, Jasem S, Licchesi JDF. The ubiquitin system in Alzheimer's disease. Adv Exp Med Biol. 2020;1233:195–221.
23. Breijyeh Z, Karaman R. Comprehensive review on Alzheimer's disease: causes and treatment. Molecules. 2020;25(24):5789.
24. Serrano-Pozo A, Frosch MP, Masliah E, Hyman BT. Neuropathological alterations in Alzheimer disease. Cold Spring Harb Perspect Med. 2011;1(1):a006189.
25. Kinney JW, Bemiller SM, Murtishaw AS, Leisgang AM, Salazar AM, Lamb BT. Inflammation as a central mechanism in Alzheimer's disease. Alzheimers Dement (NY). 2018;4:575–90.
26. Granzotto A, Sensi SL. Once upon a time, the amyloid cascade hypothesis. Ageing Res Rev. 2024;93:102161.
27. Graff-Radford J, Yong KXX, Apostolova LG, Bouwman FH, Carrillo M, Dickerson BC, et al. New insights into atypical Alzheimer's disease in the era of biomarkers. Lancet Neurol. 2021;20(3):222–34.
28. Jack CR Jr, Bennett DA, Blennow K, Carrillo MC, Dunn B, Haeberlein SB, et al. NIA-AA research framework: toward a biological definition of Alzheimer's disease. Alzheimers Dement. 2018;14(4):535–62.
29. Zou K, Abdullah M, Michikawa M. Current biomarkers for Alzheimer's disease: from CSF to blood. J Pers Med. 2020;10(3):85.
30. Auso E, Gomez-Vicente V, Esquiva G. Biomarkers for Alzheimer's disease early diagnosis. J Pers Med. 2020;10(3):114.
31. Barthelemy NR, Salvado G, Schindler SE, He Y, Janelidze S, Collij LE, et al. Highly accurate blood test for Alzheimer's disease is similar or superior to clinical cerebrospinal fluid tests. Nat Med. 2024;30(4):1085–95.
32. Gustavsson A, Norton N, Fast T, Frolich L, Georges J, Holzapfel D, et al. Global estimates on the number of persons across the Alzheimer's disease continuum. Alzheimers Dement. 2023;19(2):658–70.
33. van der Flier WM, de Vugt ME, Smets EMA, Blom M, Teunissen CE. Towards a future where Alzheimer's disease pathology is stopped before the onset of dementia. Nat Aging. 2023;3(5):494–505.

34. Jaisa-Aad M, Munoz-Castro C, Serrano-Pozo A. Update on modifiable risk factors for Alzheimer's disease and related dementias. Curr Opin Neurol. 2024;37(2):166–81.
35. Frisoni GB, Altomare D, Thal DR, Ribaldi F, van der Kant R, Ossenkoppele R, et al. The probabilistic model of Alzheimer disease: the amyloid hypothesis revised. Nat Rev Neurosci. 2022;23(1):53–66.
36. Weggen S, Beher D. Molecular consequences of amyloid precursor protein and presenilin mutations causing autosomal-dominant Alzheimer's disease. Alzheimer's Res Ther. 2012;4(2):9.
37. Petit D, Fernandez SG, Zoltowska KM, Enzlein T, Ryan NS, O'Connor A, et al. Abeta profiles generated by Alzheimer's disease causing PSEN1 variants determine the pathogenicity of the mutation and predict age at disease onset. Mol Psychiatry. 2022;27(6):2821–32.
38. Dallora AL, Minku L, Mendes E, Rennemark M, Anderberg P, Sanmartin Berglund J. Multifactorial 10-year prior diagnosis prediction model of dementia. Int J Environ Res Public Health. 2020;17(18):6674.
39. Lyall DM, Celis-Morales C, Lyall LM, Graham C, Graham N, Mackay DF, et al. Assessing for interaction between APOE epsilon4, sex, and lifestyle on cognitive abilities. Neurology. 2019;92(23):e2691–8.
40. Angelopoulou E, Paudel YN, Papageorgiou SG, Piperi C. APOE genotype and Alzheimer's disease: the influence of lifestyle and environmental factors. ACS Chem Neurosci. 2021;12(15):2749–64.
41. Leng F, Edison P. Neuroinflammation and microglial activation in Alzheimer disease: where do we go from here? Nat Rev Neurol. 2021;17(3):157–72.
42. Heneka MT, Carson MJ, El Khoury J, Landreth GE, Brosseron F, Feinstein DL, et al. Neuroinflammation in Alzheimer's disease. Lancet Neurol. 2015;14(4):388–405.
43. Heneka MT, Golenbock DT, Latz E. Innate immunity in Alzheimer's disease. Nat Immunol. 2015;16(3):229–36.
44. Salter MW, Stevens B. Microglia emerge as central players in brain disease. Nat Med. 2017;23(9):1018–27.
45. Lee S, Devanney NA, Golden LR, Smith CT, Schwartz JL, Walsh AE, et al. APOE modulates microglial immunometabolism in response to age, amyloid pathology, and inflammatory challenge. Cell Rep. 2023;42(3):112196.
46. Ferrari-Souza JP, Lussier FZ, Leffa DT, Therriault J, Tissot C, Bellaver B, et al. APOEepsilon4 associates with microglial activation independently of Abeta plaques and tau tangles. Sci Adv. 2023;9(14):eade1474.
47. Dong L, Li YZ, An HT, Wang YL, Chen SH, Qian YJ, et al. The E3 ubiquitin ligase c-Cbl inhibits microglia-mediated CNS inflammation by regulating PI3K/Akt/NF-kappaB pathway. CNS Neurosci Ther. 2016;22(8):661–9.
48. Humphries F, Bergin R, Jackson R, Delagic N, Wang B, Yang S, et al. The E3 ubiquitin ligase Pellino2 mediates priming of the NLRP3 inflammasome. Nat Commun. 2018;9(1):1560.
49. Ruano D. Proteostasis dysfunction in aged mammalian cells. The stressful role of inflammation. Front Mol Biosci. 2021;8:658742.
50. Rao G, Croft B, Teng C, Awasthi V. Ubiquitin-proteasome system in neurodegenerative disorders. J Drug Metab Toxicol. 2015;6(4):187.
51. Watanabe Y, Taguchi K, Tanaka M. Ubiquitin, autophagy and neurodegenerative diseases. Cells. 2020;9(9):2022.
52. Gavilan E, Pintado C, Gavilan MP, Daza P, Sanchez-Aguayo I, Castano A, et al. Age-related dysfunctions of the autophagy lysosomal pathway in hippocampal pyramidal neurons under proteasome stress. Neurobiol Aging. 2015;36(5):1953–63.
53. Sun-Wang JL, Ivanova S, Zorzano A. The dialogue between the ubiquitin-proteasome system and autophagy: implications in ageing. Ageing Res Rev. 2020;64:101203.
54. Wong E, Cuervo AM. Autophagy gone awry in neurodegenerative diseases. Nat Neurosci. 2010;13(7):805–11.

55. Salminen A, Kauppinen A, Suuronen T, Kaarniranta K, Ojala J. ER stress in Alzheimer's disease: a novel neuronal trigger for inflammation and Alzheimer's pathology. J Neuroinflammation. 2009;6:41.
56. Sakagami Y, Kudo T, Tanimukai H, Kanayama D, Omi T, Horiguchi K, et al. Involvement of endoplasmic reticulum stress in tauopathy. Biochem Biophys Res Commun. 2013;430(2):500–4.
57. de Vrij FM, Fischer DF, van Leeuwen FW, Hol EM. Protein quality control in Alzheimer's disease by the ubiquitin proteasome system. Prog Neurobiol. 2004;74(5):249–70.
58. Nomura J, Hosoi T, Kaneko M, Ozawa K, Nishi A, Nomura Y. Neuroprotection by endoplasmic reticulum stress-induced HRD1 and chaperones: possible therapeutic targets for Alzheimer's and Parkinson's disease. Med Sci (Basel). 2016;4(3):14.
59. Kuang E, Qi J, Ronai Z. Emerging roles of E3 ubiquitin ligases in autophagy. Trends Biochem Sci. 2013;38(9):453–60.
60. David Y, Ternette N, Edelmann MJ, Ziv T, Gayer B, Sertchook R, et al. E3 ligases determine ubiquitination site and conjugate type by enforcing specificity on E2 enzymes. J Biol Chem. 2011;286(51):44104–15.
61. Liu YC, Hsu WL, Ma YL, Lee EHY. Melatonin induction of APP intracellular domain 50 SUMOylation alleviates AD through enhanced transcriptional activation and Abeta degradation. Mol Ther. 2021;29(1):376–95.
62. Luo HB, Xia YY, Shu XJ, Liu ZC, Feng Y, Liu XH, et al. SUMOylation at K340 inhibits tau degradation through deregulating its phosphorylation and ubiquitination. Proc Natl Acad Sci USA. 2014;111(46):16586–91.
63. Zhao Y, Zhou H, Zhao Y, Liang Z, Gong X, Yu J, et al. BACE1 SUMOylation deregulates phosphorylation and ubiquitination in Alzheimer's disease pathology. J Neurochem. 2023;166(2):318–27.
64. Ludewig S, Korte M. Novel insights into the physiological function of the APP (gene) family and its proteolytic fragments in synaptic plasticity. Front Mol Neurosci. 2016;9:161.
65. Orobets KS, Karamyshev AL. Amyloid precursor protein and Alzheimer's disease. Int J Mol Sci. 2023;24(19):14794.
66. Tamagno E, Guglielmotto M, Monteleone D, Manassero G, Vasciaveo V, Tabaton M. The unexpected role of Abeta1-42 monomers in the pathogenesis of Alzheimer's disease. J Alzheimers Dis. 2018;62(3):1241–5.
67. Morel E, Chamoun Z, Lasiecka ZM, Chan RB, Williamson RL, Vetanovetz C, et al. Phosphatidylinositol-3-phosphate regulates sorting and processing of amyloid precursor protein through the endosomal system. Nat Commun. 2013;4:2250.
68. El Ayadi A, Stieren ES, Barral JM, Boehning D. Ubiquilin-1 regulates amyloid precursor protein maturation and degradation by stimulating K63-linked polyubiquitination of lysine 688. Proc Natl Acad Sci USA. 2012;109(33):13416–21.
69. Kaneko M, Koike H, Saito R, Kitamura Y, Okuma Y, Nomura Y. Loss of HRD1-mediated protein degradation causes amyloid precursor protein accumulation and amyloid-beta generation. J Neurosci. 2010;30(11):3924–32.
70. Watanabe T, Hikichi Y, Willuweit A, Shintani Y, Horiguchi T. FBL2 regulates amyloid precursor protein (APP) metabolism by promoting ubiquitination-dependent APP degradation and inhibition of APP endocytosis. J Neurosci. 2012;32(10):3352–65.
71. Blalock EM, Geddes JW, Chen KC, Porter NM, Markesbery WR, Landfield PW. Incipient Alzheimer's disease: microarray correlation analyses reveal major transcriptional and tumor suppressor responses. Proc Natl Acad Sci USA. 2004;101(7):2173–8.
72. Gireud-Goss M, Reyes S, Tewari R, Patrizz A, Howe MD, Kofler J, et al. The ubiquitin ligase UBE4B regulates amyloid precursor protein ubiquitination, endosomal trafficking, and amyloid beta42 generation and secretion. Mol Cell Neurosci. 2020;108:103542.
73. Choi J, Gao J, Kim J, Hong C, Kim J, Tontonoz P. The E3 ubiquitin ligase Idol controls brain LDL receptor expression, ApoE clearance, and Abeta amyloidosis. Sci Transl Med. 2015;7(314):314ra184.

74. Singh BK, Vatsa N, Kumar V, Shekhar S, Sharma A, Jana NR. Ube3a deficiency inhibits amyloid plaque formation in APPswe/PS1deltaE9 mouse model of Alzheimer's disease. Hum Mol Genet. 2017;26(20):4042–54.
75. Benvegnu S, Wahle T, Dotti CG. E3 ligase mahogunin (MGRN1) influences amyloid precursor protein maturation and secretion. Oncotarget. 2017;8(52):89439–50.
76. Atkin G, Hunt J, Minakawa E, Sharkey L, Tipper N, Tennant W, et al. F-box only protein 2 (Fbxo2) regulates amyloid precursor protein levels and processing. J Biol Chem. 2014;289(10):7038–48.
77. Yasukawa T, Tsutsui A, Tomomori-Sato C, Sato S, Saraf A, Washburn MP, et al. NRBP1-containing CRL2/CRL4A regulates amyloid beta production by targeting BRI2 and BRI3 for degradation. Cell Rep. 2020;30(10):3478–91 e6.
78. Xu J, Yu T, Pietronigro EC, Yuan J, Arioli J, Pei Y, et al. Peli1 impairs microglial Abeta phagocytosis through promoting C/EBPbeta degradation. PLoS Biol. 2020;18(10):e3000837.
79. Kumar P, Ambasta RK, Veereshwarayya V, Rosen KM, Kosik KS, Band H, et al. CHIP and HSPs interact with beta-APP in a proteasome-dependent manner and influence Abeta metabolism. Hum Mol Genet. 2007;16(7):848–64.
80. Zhang S, Hu ZW, Mao CY, Shi CH, Xu YM. CHIP as a therapeutic target for neurological diseases. Cell Death Dis. 2020;11(9):727.
81. Oddo S, Caccamo A, Tseng B, Cheng D, Vasilevko V, Cribbs DH, et al. Blocking Abeta42 accumulation delays the onset and progression of tau pathology via the C terminus of heat shock protein70-interacting protein: a mechanistic link between Abeta and tau pathology. J Neurosci. 2008;28(47):12163–75.
82. Rawat P, Sehar U, Bisht J, Selman A, Culberson J, Reddy PH. Phosphorylated tau in Alzheimer's disease and other tauopathies. Int J Mol Sci. 2022;23(21):12841.
83. Papanikolopoulou K, Skoulakis EMC. Altered proteostasis in neurodegenerative tauopathies. Adv Exp Med Biol. 2020;1233:177–94.
84. Alquezar C, Arya S, Kao AW. Tau post-translational modifications: dynamic transformers of tau function, degradation, and aggregation. Front Neurol. 2020;11:595532.
85. Li L, Jiang Y, Wang JZ, Liu R, Wang X. Tau ubiquitination in Alzheimer's disease. Front Neurol. 2021;12:786353.
86. Dolan PJ, Johnson GV. A caspase cleaved form of tau is preferentially degraded through the autophagy pathway. J Biol Chem. 2010;285(29):21978–87.
87. Nadel CM, Thwin AC, Callahan M, Lee K, Connelly E, Craik CS, et al. The E3 ubiquitin ligase, CHIP/STUB1, inhibits aggregation of phosphorylated proteoforms of microtubule-associated protein tau (MAPT). J Mol Biol. 2023;435(11):168026.
88. Petrucelli L, Dickson D, Kehoe K, Taylor J, Snyder H, Grover A, et al. CHIP and Hsp70 regulate tau ubiquitination, degradation and aggregation. Hum Mol Genet. 2004;13(7):703–14.
89. Babu JR, Geetha T, Wooten MW. Sequestosome 1/p62 shuttles polyubiquitinated tau for proteasomal degradation. J Neurochem. 2005;94(1):192–203.
90. Seibenhener ML, Babu JR, Geetha T, Wong HC, Krishna NR, Wooten MW. Sequestosome 1/p62 is a polyubiquitin chain binding protein involved in ubiquitin proteasome degradation. Mol Cell Biol. 2004;24(18):8055–68.
91. McEwan WA, Falcon B, Vaysburd M, Clift D, Oblak AL, Ghetti B, et al. Cytosolic Fc receptor TRIM21 inhibits seeded tau aggregation. Proc Natl Acad Sci USA. 2017;114(3):574–9.
92. Rousseaux MW, Revelli JP, Vazquez-Velez GE, Kim JY, Craigen E, Gonzales K, et al. Depleting Trim28 in adult mice is well tolerated and reduces levels of alpha-synuclein and tau. Elife. 2018;7:e36768.
93. Abisambra JF, Jinwal UK, Blair LJ, O'Leary JC 3rd, Li Q, Brady S, et al. Tau accumulation activates the unfolded protein response by impairing endoplasmic reticulum-associated degradation. J Neurosci. 2013;33(22):9498–507.
94. Flach K, Ramminger E, Hilbrich I, Arsalan-Werner A, Albrecht F, Herrmann L, et al. Axotrophin/MARCH7 acts as an E3 ubiquitin ligase and ubiquitinates tau protein in vitro impairing microtubule binding. Biochim Biophys Acta. 2014;1842(9):1527–38.

95. Lee S, Wang JW, Yu W, Lu B. Phospho-dependent ubiquitination and degradation of PAR-1 regulates synaptic morphology and tau-mediated Abeta toxicity in Drosophila. Nat Commun. 2012;3:1312.
96. Hou HL, Shen YX, Zhu HY, Sun H, Yan XB, Fang H, et al. Alterations of hHrd1 expression are related to hyperphosphorylated tau in the hippocampus in Alzheimer's disease. J Neurosci Res. 2006;84(8):1862–70.
97. Lapresa R, Agulla J, Bolanos JP, Almeida A. APC/C-Cdh1-targeted substrates as potential therapies for Alzheimer's disease. Front Pharmacol. 2022;13:1086540.
98. Kawabe H, Stegmuller J. The role of E3 ubiquitin ligases in synapse function in the healthy and diseased brain. Mol Cell Neurosci. 2021;112:103602.
99. Olabarria M, Pasini S, Corona C, Robador P, Song C, Patel H, et al. Dysfunction of the ubiquitin ligase E3A Ube3A/E6-AP contributes to synaptic pathology in Alzheimer's disease. Commun Biol. 2019;2:111.
100. Sun J, Zhu G, Liu Y, Standley S, Ji A, Tunuguntla R, et al. UBE3A regulates synaptic plasticity and learning and memory by controlling SK2 channel endocytosis. Cell Rep. 2015;12(3):449–61.
101. Owais A, Mishra RK, Kiyokawa H. The HECT E3 ligase E6AP/UBE3A as a therapeutic target in cancer and neurological disorders. Cancers (Basel). 2020;12(8):2108.
102. Hou Q, Gilbert J, Man HY. Homeostatic regulation of AMPA receptor trafficking and degradation by light-controlled single-synaptic activation. Neuron. 2011;72(5):806–18.
103. Schwarz LA, Hall BJ, Patrick GN. Activity-dependent ubiquitination of GluA1 mediates a distinct AMPA receptor endocytosis and sorting pathway. J Neurosci. 2010;30(49):16718–29.
104. Rodrigues EM, Scudder SL, Goo MS, Patrick GN. Abeta-induced synaptic alterations require the E3 ubiquitin ligase Nedd4-1. J Neurosci. 2016;36(5):1590–5.
105. Lin A, Hou Q, Jarzylo L, Amato S, Gilbert J, Shang F, et al. Nedd4-mediated AMPA receptor ubiquitination regulates receptor turnover and trafficking. J Neurochem. 2011;119(1):27–39.
106. Kwak YD, Wang B, Li JJ, Wang R, Deng Q, Diao S, et al. Upregulation of the E3 ligase NEDD4-1 by oxidative stress degrades IGF-1 receptor protein in neurodegeneration. J Neurosci. 2012;32(32):10971–81.
107. Zhang Y, Guo O, Huo Y, Wang G, Man HY. Amyloid-beta induces AMPA receptor ubiquitination and degradation in primary neurons and human brains of Alzheimer's disease. J Alzheimer's Dis. 2018;62(4):1789–801.
108. Suresh P, Jasmin S, Yen Y, Hsu HJ, Varinthra P, Pairojana T, et al. Attenuation of HECT-E3 ligase expression rescued memory deficits in 3xTg-AD mice. Front Aging Neurosci. 2022;14:916904.
109. Liu QY, Lei JX, Sikorska M, Liu R. A novel brain-enriched E3 ubiquitin ligase RNF182 is up regulated in the brains of Alzheimer's patients and targets ATP6V0C for degradation. Mol Neurodegener. 2008;3:4.
110. Del Prete D, Rice RC, Rajadhyaksha AM, D'Adamio L. Amyloid precursor protein (APP) may act as a substrate and a recognition unit for CRL4CRBN and Stub1 E3 ligases facilitating ubiquitination of proteins involved in presynaptic functions and neurodegeneration. J Biol Chem. 2016;291(33):17209–27.
111. Kuczera T, Stilling RM, Hsia HE, Bahari-Javan S, Irniger S, Nasmyth K, et al. The anaphase promoting complex is required for memory function in mice. Learn Mem. 2011;18(1):49–57.
112. Jeong YY, Han S, Jia N, Zhang M, Sheshadri P, Tammineni P, et al. Broad activation of the Parkin pathway induces synaptic mitochondrial deficits in early tauopathy. Brain. 2022;145(1):305–23.
113. Mary A, Eysert F, Checler F, Chami M. Mitophagy in Alzheimer's disease: molecular defects and therapeutic approaches. Mol Psychiatry. 2023;28(1):202–16.
114. Quinn PMJ, Moreira PI, Ambrosio AF, Alves CH. PINK1/PARKIN signalling in neurodegeneration and neuroinflammation. Acta Neuropathol Commun. 2020;8(1):189.
115. Fang EF, Hou Y, Palikaras K, Adriaanse BA, Kerr JS, Yang B, et al. Mitophagy inhibits amyloid-beta and tau pathology and reverses cognitive deficits in models of Alzheimer's disease. Nat Neurosci. 2019;22(3):401–12.

116. Hou X, Watzlawik JO, Cook C, Liu CC, Kang SS, Lin WL, et al. Mitophagy alterations in Alzheimer's disease are associated with granulovacuolar degeneration and early tau pathology. Alzheimers Dement. 2020;17(3):417–30.
117. Cummins N, Tweedie A, Zuryn S, Bertran-Gonzalez J, Gotz J. Disease-associated tau impairs mitophagy by inhibiting Parkin translocation to mitochondria. EMBO J. 2019;38(3):e99360.
118. Kam MK, Lee DG, Kim B, Huh JW, Lee HJ, Park YH, et al. Amyloid-beta oligomers induce Parkin-mediated mitophagy by reducing Miro1. Biochem J. 2020;477(23):4581–97.
119. Checler F, Alves da Costa C. Parkin as a molecular bridge linking Alzheimer's and Parkinson's diseases? Biomolecules. 2022;12(4):559.
120. Monzio Compagnoni G, Di Fonzo A, Corti S, Comi GP, Bresolin N, Masliah E. The role of mitochondria in neurodegenerative diseases: the lesson from Alzheimer's disease and Parkinson's disease. Mol Neurobiol. 2020;57(7):2959–80.
121. Haakonsen DL, Heider M, Ingersoll AJ, Vodehnal K, Witus SR, Uenaka T, et al. Stress response silencing by an E3 ligase mutated in neurodegeneration. Nature. 2024;626(8000):874–80.
122. Park YJ, Dodantenna N, Kim Y, Kim TH, Lee HS, Yoo YS, et al. MARCH5-dependent NLRP3 ubiquitination is required for mitochondrial NLRP3-NEK7 complex formation and NLRP3 inflammasome activation. EMBO J. 2023;42(19):e113481.
123. Dou Y, Tian X, Zhang J, Wang Z, Chen G. Roles of TRAF6 in central nervous system. Curr Neuropharmacol. 2018;16(9):1306–13.
124. Ndoja A, Reja R, Lee SH, Webster JD, Ngu H, Rose CM, et al. Ubiquitin ligase COP1 suppresses neuroinflammation by degrading c/EBPbeta in microglia. Cell. 2020;182(5):1156–69 e12.
125. Xiong J, Zhang Z, Ye K. C/EBPbeta/AEP signaling drives Alzheimer's disease pathogenesis. Neurosci Bull. 2023;39(7):1173–85.
126. Moh C, Kubiak JZ, Bajic VP, Zhu X, Smith MA, Lee HG. Cell cycle deregulation in the neurons of Alzheimer's disease. Results Probl Cell Differ. 2011;53:565–76.
127. Maestre C, Delgado-Esteban M, Gomez-Sanchez JC, Bolanos JP, Almeida A. Cdk5 phosphorylates Cdh1 and modulates cyclin B1 stability in excitotoxicity. EMBO J. 2008;27(20):2736–45.
128. Fuchsberger T, Lloret A, Vina J. New functions of APC/C ubiquitin ligase in the nervous system and its role in Alzheimer's disease. Int J Mol Sci. 2017;18(5):1057.
129. Vina J, Fuchsberger T, Giraldo E, Lloret A. APC/Cdh E3 ubiquitin ligase in the pathophysiology of Alzheimer's disease. Free Radic Biol Med. 2014;75(Suppl 1):S4.
130. Chauhan M, Modi PK, Sharma P. Aberrant activation of neuronal cell cycle caused by dysregulation of ubiquitin ligase Itch results in neurodegeneration. Cell Death Dis. 2020;11(6):441.
131. Wang M, Li W, Tomimatsu N, Yu CH, Ji JH, Alejo S, et al. Crucial roles of the BRCA1-BARD1 E3 ubiquitin ligase activity in homology-directed DNA repair. Mol Cell. 2023;83(20):3679–91 e8.
132. Suberbielle E, Djukic B, Evans M, Kim DH, Taneja P, Wang X, et al. DNA repair factor BRCA1 depletion occurs in Alzheimer brains and impairs cognitive function in mice. Nat Commun. 2015;6:8897.
133. Singh A, Tiwari S, Singh S. Pirh2 modulates the mitochondrial function and cytochrome c-mediated neuronal death during Alzheimer's disease. Cell Death Dis. 2024;15(5):331.
134. Choi B, Lim C, Lee H, Lee JE, Kim J, Chung C, et al. Neuroprotective effects of linear ubiquitin E3 ligase against aging-induced DNA damage and amyloid beta neurotoxicity in the brain of Drosophila melanogaster. Biochem Biophys Res Commun. 2022;637:196–202.
135. Hadi F, Mortaja M, Hadi Z. Calcium (Ca(2+)) hemostasis, mitochondria, autophagy, and mitophagy contribute to Alzheimer's disease as early moderators. Cell Biochem Funct. 2024;42(5):e4085.
136. Crossley CA, Rajani V, Yuan Q. Modulation of L-type calcium channels in Alzheimer's disease: a potential therapeutic target. Comput Struct Biotechnol J. 2023;21:11–20.

137. Lai YJ, Zhu BL, Sun F, Luo D, Ma YL, Luo B, et al. Estrogen receptor alpha promotes Cav1.2 ubiquitination and degradation in neuronal cells and in APP/PS1 mice. Aging Cell. 2019;18(4):e12961.
138. Li Y, Yang H, He T, Zhang L, Liu C. Post-translational modification of Cav1.2 and its role in neurodegenerative diseases. Front Pharmacol. 2021;12:775087.
139. Kumari S, Dhapola R, Reddy DH. Apoptosis in Alzheimer's disease: insight into the signaling pathways and therapeutic avenues. Apoptosis. 2023;28(7–8):943–57.
140. Nakamura T, Wang L, Wong CC, Scott FL, Eckelman BP, Han X, et al. Transnitrosylation of XIAP regulates caspase-dependent neuronal cell death. Mol Cell. 2010;39(2):184–95.
141. Guo S, Zhang Y, Wei C, Shi L, Feng Y. The E3 ubiquitin ligase MARCH8 regulates TNF-alpha-induced apoptosis in hippocampal neurons by targeting myosin light chain 2 for degradation. Anat Rec (Hoboken). 2019;302(12):2271–8.
142. Weng G, Zhou B, Liu T, Huang Z, Huang S. Tetramethylpyrazine improves cognitive function of Alzheimer's disease mice by regulating SSTR4 ubiquitination. Drug Des Devel Ther. 2021;15:2385–99.
143. Geetha T, Zheng C, McGregor WC, Douglas White B, Diaz-Meco MT, Moscat J, et al. TRAF6 and p62 inhibit amyloid beta-induced neuronal death through p75 neurotrophin receptor. Neurochem Int. 2012;61(8):1289–93.
144. Sola M, Rendon-Angel A, Rojo Martinez V, Sgrignani J, Magrin C, Piovesana E, et al. Tau protein binds to the P53 E3 ubiquitin ligase MDM2. Sci Rep. 2023;13(1):10208.
145. Caccamo A, Branca C, Piras IS, Ferreira E, Huentelman MJ, Liang WS, et al. Necroptosis activation in Alzheimer's disease. Nat Neurosci. 2017;20(9):1236–46.
146. Zhang R, Song Y, Su X. Necroptosis and Alzheimer's disease: pathogenic mechanisms and therapeutic opportunities. J Alzheimer's Dis. 2023;94(s1):S367–S86.
147. Lee SB, Kim JJ, Han SA, Fan Y, Guo LS, Aziz K, et al. The AMPK-Parkin axis negatively regulates necroptosis and tumorigenesis by inhibiting the necrosome. Nat Cell Biol. 2019;21(8):940–51.
148. Wang Y, Shan B, Liang Y, Wei H, Yuan J. Parkin regulates NF-kappaB by mediating site-specific ubiquitination of RIPK1. Cell Death Dis. 2018;9(7):732.
149. Ma H, Dong Y, Chu Y, Guo Y, Li L. The mechanisms of ferroptosis and its role in Alzheimer's disease. Front Mol Biosci. 2022;9:965064.
150. Tian J, Zheng W, Li XL, Cui YH, Wang ZY. Lower expression of Ndfip1 is associated with Alzheimer disease pathogenesis through decreasing DMT1 degradation and increasing iron influx. Front Aging Neurosci. 2018;10:165.
151. von Rotz RC, Kins S, Hipfel R, von der Kammer H, Nitsch RM. The novel cytosolic RING finger protein dactylidin is up-regulated in brains of patients with Alzheimer's disease. Eur J Neurosci. 2005;21(5):1289–98.
152. Hu X, Pickering EH, Hall SK, Naik S, Liu YC, Soares H, et al. Genome-wide association study identifies multiple novel loci associated with disease progression in subjects with mild cognitive impairment. Transl Psychiatry. 2011;1(11):e54.
153. Yokota T, Mishra M, Akatsu H, Tani Y, Miyauchi T, Yamamoto T, et al. Brain site-specific gene expression analysis in Alzheimer's disease patients. Eur J Clin Investig. 2006;36(11):820–30.
154. Inuzuka H, Liu J, Wei W, Rezaeian AH. PROTACs technology for treatment of Alzheimer's disease: advances and perspectives. Acta Mater Med. 2022;1(1):24–41.

E3 Ubiquitin Ligases in Huntington's Disease

11

Luis Angel Monsalvo-Maraver, Sonia Galván-Arzate, Anahí Chavarría, Tessy López-Goerne, Isaac Túnez, Alexey A. Tinkov, Michael Aschner, and Abel Santamaría

Abbreviations

BDNF	brain-derived neurotrophic factor
HD	Huntington's disease
Htt	huntingtin
NPCs	nuclear pore complexes
NF-kB	Nuclear factor-kB
Nrf2	Nuclear factor erythroid 2-related factor 2
OS	oxidative stress
UHRF2	ubiquitin-like with PHD and ring finger domains protein 2
Poly-Q	Polyglutamine
Ran	Ras-related nuclear protein
ROS	reactive oxygen species

L. A. Monsalvo-Maraver (✉)
Postgraduate Program in Biological Sciences, National Autonomous University of Mexico, Mexico City, Mexico

Neurochemistry Department, National Institute of Neurology and Neurosurgery, Mexico City, Mexico
e-mail: luismonsalvo@ciencias.unam.mx

S. Galván-Arzate
Neurochemistry Department, National Institute of Neurology and Neurosurgery, Mexico City, Mexico

A. Chavarría
Unit of Experimental Medicine, Faculty of Medicine, National Autonomous University of Mexico, Mexico City, Mexico

T. López-Goerne
Laboratory of Nanotechnology and Nanomedicine, Department of Health Care, Metropolitan Autonomous University-Xochimilco, Mexico City, Mexico

S. Singh (ed.), *Ubiquitin Proteasome System in CNS Neurodegenerative Diseases*, https://doi.org/10.1007/978-981-95-6099-8_11

SOD1	superoxide dismutase 1
SOD2	superoxide dismutase 2
TRAF6	tumor necrosis factor receptor-associated factor 6
LRSAM1	leucine-rich repeat and sterile alpha motif-containing protein 1
Nqo1	NAD(P)H quinone oxidoreductase 1
Drp1	dynamin-related protein 1
Mfn1	mitofusin 1
Mfn2	mitofusin 2
Opa1	mitochondrial dynamin like GTPase 1
SP1	specificity protein 1
NMDAR	n-methyl-D-aspartate receptor
QA	quinolinic acid
GLT-1	glutamate transporter 1
3-NP	3-nitropropionic acid
PGC-1α	peroxisome proliferator–activated receptor-γ coactivator-1α
ASOs	antisense oligonucleotides
PPARγ	Peroxisome proliferator-activated receptor-γ
CREB	cAMP response element-binding protein
FOXO3	Forkhead box O3
UPS	ubiquitin-proteasome system
E3	E3 ubiquitin-ligase enzymes
HECT	homologous to the E6AP carboxyl terminus
RING	really interesting new gene
RBR	RING-IBR-RING
Mdm2	mouse double minute 2 homolog
HSF1	heat shock factor 1

I. Túnez
Maimonides Institute for Research in Biomedicine of Cordoba (IMIBIC), Cordoba, Spain

Department of Biochemistry and Molecular Biology, Faculty of Medicine and Nursing, University of Cordoba, Córdoba, Spain

A. A. Tinkov
Center of Bioelementology and Human Ecology, IM Sechenov First Moscow State Medical University (Sechenov University), Moscow, Russia

Department of Medical Elementology, Peoples' Friendship University of Russia (RUDN University), Moscow, Russia

Laboratory of Molecular Ecobiomonitoring and Quality Control, Yaroslavl State University, Yaroslavl, Russia

M. Aschner
Department of Molecular Pharmacology, Albert Einstein College of Medicine, Bronx, NY, USA

A. Santamaría
Laboratory of Nanotechnology and Nanomedicine, Department of Health Care, Metropolitan Autonomous University-Xochimilco, Mexico City, Mexico

Faculty of Science, National Autonomous University of Mexico, Mexico City, Mexico

PJA1	Praja Ring Finger Ubiquitin Ligase 1
gp78	glycoprotein 78
CHOP	C/EBP homologous protein
BiP	binding immunoglobulin protein
HRD1	Hydroxymethylglutaryl-CoA reductase degradation protein 1
CUL1	cullin-1
CUL3	cullin-3
CUL4	cullin-4
RBX1	RING-box protein 1
SKP1	S-phase kinase-associated protein 1
SCF	SKP1/CUL1/F-box protein
CRL3	Cullin-RING Ubiquitin Ligase 3
Keap1	Kelch-like ECH-associated protein 1
FipoQ	F-box involved in Poly-Q pathogenesis
FBXO33	F-box only protein 33
FBXW7	F-box and WD repeat domain containing protein 7
BTB	Bric-a-brac
SFN	sulforaphane
DMF	dimethyl fumarate
HACE1	HECT domain and ankyrin repeat containing E3 ubiquitin protein ligase 1
RAC1	Ras-related C3 botulinum toxin substrate 1
HECTD1	HECT Domain E3 Ubiquitin Protein Ligase 1
UBE3A	Ubiquitin protein ligase E3A
UBR5	Ubiquitin protein ligase E3 component n-recognin 5
ATMIN	ATM interactor
TERT	telomerase reverse-transcriptase
WWP1	WW containing-domain E3 ubiquitin protein ligase 1
WWP2	WW containing-domain E3 ubiquitin protein ligase 2
NDP52	nuclear dot protein 52 kDa
OPTN	optineurin
SQSTM1	sequestosome 1
NEDD4	neuronally expressed developmentally downregulated protein 4
IGF-1Rβ	insulin-like growth factor receptor β
CHIP	C-terminus of Hsc70-interacting protein
HOIP	HOIL-1 L-interacting protein
LUBAC	linear ubiquitin chain assembly complex
PINK1	PTEN-induced putative kinase 1
ITCH	Itchy E3 ubiquitin ligase
Siah-1	siah E3 ubiquitin protein ligase 1
TRIM37	Tripartite motif-containing 37
Herp	homocysteine-induced endoplasmic reticulum protein
SOCS2	Suppressor of cytokine signaling 2
MGRN1	mahogunin ring finger-1
Znf179	Zinc finger protein 179

SMURF1	smad ubiquitination regulatory factor 1
COP1	constitutive photomorphogenic 1
RNF20	Ring Finger Protein 20
mHtt	mutant huntingtin
AMPAR	α-amino-3-hydroxy-5-methyl-4-isoxazolepropionic acid
LPS	lipopolysaccharide
PSDP5	postsynaptic density protein 95

11.1 Huntington's Disease

11.1.1 Overview

Huntington's disease (HD) is a neurodegenerative disorder clinically associated with motor abnormalities such as choreiform movements and, to a lesser extent, with oculomotor impairments, ataxia and dystonia [1, 2]. HD is also related to neuropsychiatric symptoms, including anxiety, depression, obsessive compulsive disorder and psychosis, some of which are thought to occur in over half of HD patients [3, 4]. Motor deficits in HD patients are also strongly related to cognitive decline symptoms such as emotion recognition and processing speed [5, 6].

Epidemiologically, HD has a wide range of variation. Studies on Western populations have reported a prevalence ranging from 10.6 to 13.7 per 100,000 individuals; however, it has been suggested that Asian populations have lower prevalence rates when compared to European, Australian or North American populations, in which an increasing prevalence per decade has been observed [6–8].

Unlike other neurodegenerative disorders such as Alzheimer's disease or Parkinson's disease, HD has a strong genetic basis underlying its etiology. An expansion of the CAG triplet in the *HTT* gene causes the production of a mutant huntingtin (mHtt) protein with an abnormal polyglutamine (poly-Q) expansion. mHtt is thought to exert neurotoxic effects by impairing a number of cell functions, ultimately leading to neuronal dysfunction and death [9]. Toxicity of mHtt mainly targets striatal GABAergic medium spiny neurons, although HD also affects a wide range of brain regions, including the globus pallidus, *substantia nigra pars reticulata*, hypothalamus, subthalamic nucleus, hippocampus, cerebellum and brain cortex (affecting both interneurons and pyramidal neurons) [10–15].

11.1.2 Neuropathological Mechanisms Related to HD

The *HTT* gene encodes for the 348 kDa wild-type protein huntingtin (Htt), which is formed by multiple α-helix structures organized in 3 main domains [16]. While the neurotoxic role of mHtt in HD is undeniable, Htt plays a pivotal role in a number of biological processes in neurons [17]: (1) the protein is thought to participate in cell trafficking as it is expressed on clathrin-containing vesicles and the trans-Golgi network [18, 19]; (2) further evidence reported that overexpression of Htt can suppress cell death in primary cultured neurons, suggesting a pro-survival signaling role for the protein [20]; (3) autophagy in both mice and *Drosophila melanogaster* is disrupted when Htt expression is experimentally abolished, suggesting that Htt algo regulates autophagic degradation within cells [21]; (4) Htt plays an important role in ciliogenesis, as it is expressed in sensory cilia and cilia of photoreceptors, whereas mHtt disturbs photoreceptor-cilia function [22, 23]; (5) Htt is thought to participate in early neurodevelopment as it promotes neural stem cell proliferation/differentiation, and it is necessary for normal cell migration/proliferation during corticogenesis [24, 25].

Even though Htt expression is necessary for several physiological functions in the brain, including survival and synaptogenesis of striatal neurons in mice [26], it has been shown to produce several neurotoxic effects associated with HD pathogenesis.

HD has an autosomal dominant inheritance pattern, meaning that the presence of one mutant allele of the *HTT* gene is enough to cause the disease in an individual. Whether HD results from either the reduction of physiological functions of Htt in the heterozygous condition, a gain of function of mHtt that enhances its neurotoxic properties, or a combination of both, the pathogenic nature of the disease still remains unclear [9, 17, 27].

The *HTT* gene can either produce a full-length Htt or a shorter protein derived from the exon 1 of the mRNA [9]. Both forms contain the poly-Q sequence and can contribute to HD pathogenesis. Furthermore, Htt undergoes post-translational modifications such as phosphorylation, palmitoylation, acetylation, SUMOylation, ubiquitination and proteolytic cleavage [28, 29]. Some of these modifications have been shown to decrease mHtt-driven toxicity, whereas others, like SUMOylation, can increase mHtt aggregation and neurotoxicity [30–32].

As observed in other neurodegenerative proteinopathies, mHtt can either remain as an isolated monomer, or assemble into diverse species of oligomers and fibrils [33]. Evidence reported that mHtt oligomerization is positively correlated to Poly-Q sequence length [34]. Moreover, it has been proven that oligomers are more toxic to cells than monomers [35], and that oligomers can spread from one cell to another, potentially being able to diffuse pathology across the brain tissue in HD [36]. Finally, studies on STHDH Q7/111 cells also revealed that diverse mHtt-derived fibrils are cytotoxic [37]. Further evidence suggests that several aggregation forms might have a different impact on cell neurotoxicity, and consequently, on the development of HD.

mHtt disturbs many physiological functions of the cell nucleus. Firstly, mHtt exhibits an enhanced interaction with DNA compared to its wild-type counterpart, obstructing the DNA binding sequences of several transcription factors, thus impairing the expression of a number of genes [38]. In addition, mHtt reduces polynucleotide kinase-3′-phosphatase function, interfering with DNA repair [39]. Nuclear pore complexes (NPCs) are also affected in HD patients, and it has been revealed that mHtt also colocalizes with NPCs and disrupts Ras-related nuclear protein (Ran)-dependent protein/RNA nucleocytoplasmic transport through the NPCs [40].

Finally, Htt forms aberrant perinuclear inclusions that alter the nuclear membrane integrity, eventually provoking cell death [41].

HD is also characterized by an imbalance between the production of reactive oxygen species (ROS) and endogenous antioxidants, a phenomenon known as oxidative stress (OS), which is a hallmark of a number of other neurodegenerative diseases [42]. mHtt is thought to act like a causal factor of OS as it disrupts nuclear factor-kB (NF-kB) and Nuclear factor erythroid 2-related factor 2 (Nrf2) signaling pathways, consequently altering the expression of several endogenous antioxidant genes such as superoxide dismutase 1/2 (SOD1/SOD2), catalase, thioredoxin-2, heme oxygenase-1 and NAD(P)H quinone oxidoreductase 1 (Nqo1) [43, 44]. Conversely, OS can also raise mHtt-toxicity as evidence shows that OS directly modifies Htt structure, enhancing its nuclear deposition while promoting Htt aggregation and Htt-driven cell death [45, 46]. The involvement of OS in HD pathology is clear as early studies have revealed increased levels of OS-related markers (DNA-double strand breaks, lipofuscin, 8-hydroxy-deoxyguanosine, malondialdehyde and 3-nitrotyrosine) in HD-brain samples [47].

Furthermore, HD pathology is closely related to overall mitochondrial dysfunction in affected neurons. In the transgenic HD mice model R6/2, the striatal activity of mitochondrial respiratory complexes II, III and IV was diminished [48]. In addition, in *post-mortem* samples from HD patients, the levels of the mitochondrial fission-related protein Dynamin-related protein 1 (Drp1) and mitochondrial fission protein 1 were augmented, while the levels of mitochondrial fusion-related proteins mitofusin 1/2 (Mfn1/Mfn2) and mitochondrial dynamin-like GTPase 1 (Opa1) were significantly reduced [49]. Moreover, it has been suggested that mHtt pathology is interconnected with mitochondrial impairment, as alterations observed in mitochondrial membrane potential, mitochondrial gene expression and respiration could be consequences of Htt toxicity [50, 51]. Mitochondrial dysfunction has also been used to model HD in rodents by treating them with 3-nitropropionic acid (3-NP), a mitochondrial respiratory complex II (succinate dehydrogenase) blocker; this toxin mimics many of the motor alterations observed in HD patients, and provokes a specific degeneration of striatal medium spiny neurons [52].

Excitotoxic damage to striatal neurons results from a N-methyl-D-aspartate receptor (NMDAR)-dependent cytosolic calcium overload, which leads to the activation of proapoptotic signaling pathways and consequent neuronal death [53]. Early studies have used quinolinic acid (QA) as an NMDAR agonist to produce striatal neuron loss and motor impairments in vivo [54–56]. In addition, overexpressing certain NMDAR subunits leads to an enhanced mHtt-driven toxicity in transgenic HD mice [57]; however, some studies support the idea that only extrasynaptic NMDAR causes excitotoxic damage: preferential blockage of extrasynaptic NMDAR leads to a partial recovery of motor deficits in HD mice, rescues pro-survival signaling of cAMP response element-binding protein (CREB) and peroxisome proliferator-activated receptor-γ coactivator-1α (PGC-1α), and inhibits the p38 Mitogen-Activated Protein Kinase proapoptotic signaling [58–60].

Glial cells play an important role in the pathogenesis of HD. Both astrocytes and microglia produce proinflammatory enzymes and cytokines that provoke neuroinflammation [61]. Regarding microglia: (1) the innate immune complement system recognizes specific molecules within the synapses between cortical and striatal neurons; consequently, these complement molecules bind to microglia, leading to synapse-elimination, thereby contributing to cognitive deficits in HD mice [62]; (2) microglia metabolism include the kynurenine pathway, which produces cytotoxic metabolites like QA and 3-hydroxykynurenine. It has been reported that mHtt can enhance the kynurenine pathway and the production of QA (which activates NMDAR, potentially causing excitotoxicity) and 3-hydroxykynurenine (which causes OS) [63].

Regarding astrocytes: (1) these cells also accumulate mHtt, disrupting several astrocytic glutamate transporters, potentially expanding excitotoxic damage [64]; (2) mHtt expression in astrocytes impairs the secretion of brain-derived neurotrophic factor (BDNF), a protein that promotes neuronal survival [65]; (3) striatal astrocytes predominantly oxidize fatty acids to obtain energy in HD brains due to low glucose availability, and in turn fatty acid metabolism increases the production of ROS potentially vulnerating nearby neurons to OS [66].

HD pathology has also been related to many other factors such as synaptic dysfunction [67], endoplasmic reticulum stress [68], axonal transport deficits [69], adult neurogenesis impairments [70], and dysregulation of calcium storage and signaling [71]. Finally, it has been proposed that HD might emerge from neurodevelopmental alterations derived from *HTT* gene mutations, which only begin to show the neurodegenerative phenotype as affected individuals age [72–74].

11.1.3 Therapeutic Approaches to HD

Even though HD currently represents an incurable condition, preclinical and clinical research have made huge efforts to identify potential therapeutic targets and develop neuroprotective strategies that might help to treat or cure HD in the future.

Clinical research has focused on different key components of HD pathology: (1) Antisense oligonucleotides (ASOs) that target both wild-type Htt and mHtt (nonselective ASOs), or allele-specific ASOs that target mHtt, have been developed [75]; (2) mHtt-targeted siRNA-based gene therapy (AMT-130) has also being tested for clinical tolerability and safety in phase-I/II trials, although some adverse effects were observed, coupled with the fact that the intracerebral administration of this drug is immensely invasive for HD patients [76]; (3) the Sirtulin 1 inhibitor Selisistat was also tested, but serious adverse effects were observed in a double blind phase-II study [77, 78]; (4) in a phase-II trial, HD patients were treated with laquinimod (an neuro-immunomodulator), resulting in limited caudate volume loss and inefficiency to improve the motor pathology [79]; (5) Coenzyme Q10 advanced to phase-III trials, though the aim to prove efficacy to diminish the functional decline of the HD patients enrolled in the trial was unsuccessful [80]; (6) finally, phase-II trials of creatine as a potential disease modifying treatment for HD actually proved to reduce oxidative damage to DNA in phase-II trials, though it failed to delay or stop the overall functional decline of patients in phase-III trials [81, 82].

Ongoing preclinical research has continued to explore potential therapeutic targets and experimental drugs. Stimulation of neuroprotective signaling pathways and transcription factors has become an extensively explored approach to HD: (1) Firstly, enhancement of Nrf2-dependent transcription has diminished OS in both in vivo and in vitro excitotoxicity HD models [83]; Nrf2 signaling also lowered neuroinflammatory and apoptotic markers while rescuing motor capacity and protein levels of BDNF/Nrf2 in rats [84], and proved to stimulate proteasomal degradation of mHtt in mice [85]. (2) Activation of the peroxisome proliferator-activated receptor-γ (PPARγ) pathway has resulted in improved motor performance through the stimulation of mitochondrial biogenesis and BDNF signaling [86]; PPARγ signaling also boosted proteasome and mitochondrial function while diminishing mHtt aggregation and endoplasmic reticulum stress in N2a neuroblastoma cells [87] and augmented mitochondrial biogenesis and survival of R6/2 mice [88]. (3) Endocannabinoid system stimulation has been used to reduce motor learning deficits and striatal plasticity alterations in mHtt-expressing mice [89], and activation of cannabinoid receptor 2 in microglia from R6/2 mice lowered striatal neurodegeneration and neuroinflammation [90]. (4) CREB indirect stimulation has limited oxidative damage, neuroinflammation and motor deficits in 3-NP-induced HD-like phenotype in rats [91]. (5) Forkhead box O3 (FOXO3) is another transcription factor that, upon stimulation by genistein, raised autophagic Htt degradation in HD mice while lessening neuroinflammation, OS-markers and motor impairments [92]. (6) Finally, it is noteworthy that blockage of the NF-kB proinflammatory signaling pathway rescued antioxidant protein levels while abolishing neuroinflammation and oxidative damage in a signaling pathway that involved CREB upregulation [93].

Regarding Htt degradation, growing preclinical evidence identifies the ubiquitin-proteasome system (UPS) as a major pathway for Htt clearance. Research suggests that either mHtt provokes impairments in the UPS or UPS inhibition allows Htt deposition as well [46, 94, 95]. Despite the above, UPS enhancement has proven to be an effective approach to target neurotoxicity in several HD models [96–98].

E3 ubiquitin-ligase enzymes (E3) are enzymes that participate in ubiquitin binding to proteasomal degradation-targeted proteins, thereby playing an important role in the whole function and physiology of the UPS [99]. Herein, the role of different E3 enzymes in the pathogenesis mechanisms related to HD will be discussed, highlighting the potential relevance of E3 ubiquitin ligases to serve as new therapeutic approaches to HD.

11.2 E3 Ligases in Pathogenesis and Potential Treatment of HD

E3 ligases are structurally diverse proteins, with approximately 600 of them being coded in the human genome [100]. Structurally, E3 ligases can be classified into four major groups: (1) homologous to the E6AP carboxyl terminus (HECT); (2) new gene (RING), (3) U-box, and (4) RING-IBR-RING (RBR) [101]. As will be seen in the sections below, members of all four families have been involved in HD pathogenesis.

11.2.1 Mdm2

Mouse double minute 2 homolog (Mdm2) is a RING E3 enzyme that participates in the ubiquitination and consequent proteasomal degradation of p53, a protein that promotes cell cycle arrest and apoptosis [102]. It has been reported that expanded Poly-Q-coding mRNAs are able to interact with nucleolin, affecting its capacity to promote rRNA transcription [103]. When rRNA levels are low, free ribosomal proteins are not able to interact with rRNA and start binding with Mdm2, hence reducing p53 ubiquitination and triggering p53-driven cell death in a process known as nucleolar stress [104]. This phenomenon has been described in both *D. melanogaster* and R6/2 mice models of HD, suggesting that Poly-Q-coding mRNAs trigger nucleolar stress and apoptosis via Mdm2 dysfunction [103].

Further evidence in zQ175 HD mice suggests that mHtt directly alters the interaction between p53 and Mdm2, preventing p53 degradation, thereby downregulating heat shock factor 1 (HSF1) [105]. HSF1 was found to reduce motor deficits and striatal Htt aggregation. Notably, zQ175 HD mice lacking p53 protein presented lower striatal levels of Htt and a better performance at motor tests [105]. Nonetheless, some studies propose that even in the presence of mHtt, p53 and Mdm2 are still capable of interacting and that differences in the ubiquitination pattern of p53 might be responsible for the cell response to mHtt downstream [106].

On the other hand, studies in cell culture models have revealed that Mdm2 augments mitophagy via activating Parkin-dependent mitophagy [107]. This report is relevant in the HD context as it shows that mitophagy is severely compromised by mHtt [108]. Thereby, targeting Mdm2 might boost mitophagy and consequently minimize Htt-driven toxicity, although this hypothesis is yet to be tested directly in an HD experimental model.

11.2.2 PJA1

Praja Ring Finger Ubiquitin Ligase 1 (PJA1) is another RING-type E3 enzyme that is widely expressed in the human brain [109]. In different cell lines transfected with mHtt, diminished levels of PJA1 were observed and conversely, upregulation of PJA1 suppressed the aggregation of Htt, suggesting that mHtt might be a direct target of PJA1 ubiquitin ligase activity [110]. It is noteworthy that PJA1 also inhibits the aggregation of proteins involved in the pathogenesis of other neurodegenerative disorders such as Parkinson's disease, spinocerebellar ataxia 3 and amyotrophic lateral sclerosis [111].

11.2.3 gp78

Glycoprotein 78 (gp78) is an endoplasmic reticulum-associated RING-type E3 enzyme. gp78 is part of the proteic machinery that is employed to maintain proteostasis and reduce misfolded protein aggregation by triggering misfolded protein response [112]. In the HD context, studies on mHtt-transfected HEK293 cells have reported that mHtt directly binds to gp78, and that overexpression of mHtt leads to an increase of gp78 protein levels [113]. Further evidence suggested that mHtt-transfection conducted to a marked rise of endoplasmic reticulum stress markers such as C/EBP homologous protein (CHOP) and binding immunoglobulin protein (BiP), while diminishing the degradation of known gp78 substrates, suggesting that mHtt stimulates the accumulation of non-functional gp78, induces endoplasmic reticulum stress and disrupts proteostasis [113]. In turn, experimental overexpression of gp78 in HEK293 cells lowered the levels of mHtt [113], suggesting that enhancing endoplasmic reticulum-dependent protein degradation might counteract the endoplasmic reticulum stress (and consequent cell death) that has been observed in several HD preclinical models [114].

As previously described, mitophagy appears to be aberrantly down-regulated by mHtt [108]. Mitophagy has been enhanced by gp78 in several cultured cell lines as transfection with the gene-encoding gp78 induced mitophagy in a Parkin-independent manner [115]. Moreover, gp78 silencing in a fibrosarcoma cell line model led to mitochondrial depolarization, accompanied by decreased mitophagy and increased ROS production [116]. Nevertheless, it is yet to be proved if gp78 activity can reduce neurotoxicity in an HD model via mitophagy augmentation and ROS production abolishment.

11.2.4 HRD1

Hydroxymethylglutaryl-CoA reductase degradation protein 1 (HRD1) is yet another RING-type E3 enzyme that plays a key role in proteostasis within the endoplasmic reticulum [117]. It has been reported that HRD1 protects cells against endoplasmic reticulum stress-driven apoptosis and aids in unfolded protein degradation via UPS [118]. In SH-SY5Y cell cultures transfected with the N-terminal fragment of mHtt (containing the cytotoxic Poly-Q sequence), it was revealed that HRD1 is able to use the N-terminal Htt fragment as a substrate for ubiquitination and subsequent degradation. In addition, it was noticed that HRD1 has a preference for Htt N-terminal fragments with a high number of glutamine residues in their Poly-Q sequences as HRD1 ubiquitinates Htt fragments with 130 glutamine residues, however it is not able to use Htt fragments with 25 glutamine residues (similar to the ones observed in wild-type Htt), hence suggesting that HRD1 acts specifically on mHtt [119]. Despite the above, HRD1 enhanced cell viability while decreasing Htt aggregation in SH-SY5Y cell cultures [119]. Special attention should be paid to HRD1 as a potential neuroprotective factor in the HD context, as immunohistochemical assays revealed that HRD1 is present in the striatum [120].

11.2.5 UHRF2

Ubiquitin-like with PHD and ring finger domains protein 2 (UHRF2) is a nuclear E3 ligase that degrades several substrates related to cell cycle control and proliferation [121]. Studies conducted on HeLa cells transfected with nuclear-targeted mHtt demonstrated that UHRF2 ubiquitinates and provokes the degradation of nuclear [122]. The same report also suggested that UHRF2 expression exerts neuroprotection in mHtt-expressing cells [122]. In this regard, early studies have reported that mHtt can enter the cell nucleus, where it increases neuronal cell death and inhibits cell transcription [123, 124]. Together, these results suggest that nuclear Htt is highly neurotoxic and might play a central role in the pathogenesis of HD. Additionally, UHRF2 upregulation can be considered as a viable therapeutic approach to target nuclear-Htt-derived toxicity.

11.2.6 LRSAM1

Another example of E3-ligase potentially involved in pathogenesis is leucine-rich repeat and sterile alpha motif-containing protein 1 (LRSAM1). LRSAM1 is a RING-type E3 ligase that is thought to be primarily expressed in motor neurons and spinal cord tissue although it is also present in a wide range of different cell types as well [125, 126]. LRSAM1 has been related to several neurodegenerative disorders, including amyotrophic lateral sclerosis and Charcot-Marie-Tooth disease [127, 128]. Recent studies validated that LRSAM1 upregulation can enhance the degradation of abnormally misfolded proteins via UPS [129], which makes it relevant in the HD context. In addition, primary striatal neuron cultures expressing a fragment of mHtt provided in vitro evidence that LRSAM1 overexpression was able to rescue the cell viability while decreasing the number of nuclear Htt aggregates [130].

In addition, recent findings demonstrated that LRSAM1 expression is augmented in response to OS, autophagy suppression, endoplasmic reticulum stress and proteasome inhibition (all of these being potential HD-related pathogenic factors), and promotes cell survival and resilience against these cytotoxic stressors [131]. Overexpression of LRSAM1 also limited the aggregation of misfolded luciferase (experimentally introduced to cells) via its E3 ligase activity [131]. This evidence also validates the overexpression of E3 ligases like LRSAM1 as a potential strategy in HD therapeutics, although it has not been directly tested in an HD model.

11.2.7 TRAF6

Up to this point, we have reviewed several RING-type E3 enzymes whose activity is related to the clearance of neurotoxic forms of Htt in different HD models; however, some E3 enzymes appear to have an opposite effect on Htt-driven toxicity. Tumor necrosis factor receptor-associated factor 6 (TRAF6) is a RING-type E3 enzyme involved in the signaling pathways of toll-like receptors, several proinflammatory cytokines, tumor necrosis factor receptors and T-cell receptors, and even though it has multiple functions in the immune system, it has been proposed that TRAF6 might play a role in neurodegeneration [132, 133].

Regarding Htt-driven toxicity in HD, TRAF6 was found to be upregulated in post-mortem brain samples from HD patients while in cell cultures TRAF6 was reported to interact with N-terminal Htt aggregates within cells. Furthermore, it was reported that TRAF6 ubiquitinates in a specific pattern whereby instead of targeting N-terminal Htt aggregates for proteasomal degradation, it enhances Htt aggregation and potential toxicity [134].

Even though the following evidence is not sustained directly by an HD model, TRAF6 has been relevant in multiple neuroinflammation studies that positively correlate TRAF6 expression with multiple neuroinflammatory markers potentially relevant for HD pathogenesis, and therefore will be discussed below: (1) in a neuralgia model, TRAF6 signaling showed to increase microglial reactivity production of proinflammatory cytokines that have also been found to be upregulated in several HD models [61, 135]; (2) in lipopolysaccharide (LPS)-stimulated microglia and microglial-like BV-2 cell line, TRAF6 signaling resulted in incremented microglial reactivity and neuroinflammatory response-induction, while compounds such as schizandrin A, tetrathiomolybdate and a *Gingko biloba* leaf extract were able to reduce neuroinflammation via disrupting TRAF6-axis signaling [136–139]. These results highlight TRAF6 modulation as a potential target for neuroprotection in HD models.

11.2.8 Cullin-RING Ubiquitin E3 Ligase Multi-subunit Complexes

Cullin-RING complexes are multi-subunit proteins that regulate UPS-mediated degradation of several substrates within cells. These proteins are composed by 3 main components: (1) a Cullin-family scaffold protein, e.g., cullin-1 (CUL1), cullin-3 (CUL3) or cullin-4 (CUL4); (2) a RING-type E3 ligase like RING-box protein 1 (RBX1); and (3) variable substrate receptors that interact with the ubiquitination targeted protein [140]. The two main multimeric E3 ligase complexes are SKP1/CUL1/F-box protein (SCF) complex and Cullin-RING Ubiquitin Ligase 3 (CRL3) complex. SCF complexes contain F-box proteins as substrate receptors, while CRL3 complexes contain proteins with Bric-a-brac (BTB) domains as substrate receptors [141].

11.2.8.1 SCF Complex Subunits

Firstly, SCF subunits CUL1 and S-phase kinase-associated protein 1 (SKP1), which acts like an adaptor between CUL1 and the substrate receptor with a F-box domain, are downregulated in the R6/2 HD mice model. Additionally, the expression of N-terminal fragments of mHtt in HD-150Q cells significantly reduced the protein levels of CUL1 [142]. In a *D. melanogaster* HD model, CUL1 silencing was worsened Poly-Q toxicity [143].

On the other hand, in a *D. melanogaster* model, silencing of a F-box-type substrate receptor named F-box, which is involved in Poly-Q pathogenesis (FipoQ), enhanced Poly-Q toxicity in the external eyes of the flies [143]. In the same report, the F-box only protein 33 (FBXO33), a human F-box protein analog to FipoQ in *D. melanogaster*, was reported to mediate the ubiquitination of insoluble aggregates of Poly-Q proteins (via UPS degradation), though it failed to lower the soluble isoforms of Poly-Q proteins [143]. This report supports the idea that FBXO33 might be a target for the insoluble mHtt aggregates in the HD context.

Other studies have shown that β-catenin is abnormally accumulated within striatal-neuron cultures from mice overexpressing mHtt. Besides, the experimental transfection of striatal cells with a vector encoding another F-box protein of the SCF complex (β-transducin repeat-containing protein) resulted in decreased cell death through β-catenin ubiquitination (and consequent degradation), providing further evidence in support of the notion that augmenting SCF-dependent degradation can be an effective neuroprotective approach to HD [144].

F-box and WD repeat domain containing protein 7 (FBXW7) is yet another F-box protein potentially involved in HD pathogenesis. It was previously reviewed that mHtt stabilizes p53 within cells by impairing its proteasomal degradation via Mdm2. Consequently, it has been reported that p53 upregulates the expression of FBXW7, whose substrate is HSF1 [107]. Reduced protein levels of HSF1 have been observed in HD brain samples and in several HD cell and mouse models. In line with the evidence above, experimental rescuing of HSF1 levels restricted synaptic alterations in medium spiny neurons from mice expressing mHtt [145]. Conversely, in a non-HD model based on cultured cortical neurons exposed to glutamate excitotoxicity, it was shown that FBXW7 levels were decreased in response to excitotoxicity, and FBXW7 reduction led to increased levels of the transcription factor c-jun, a proapoptotic factor [146]. Combined, current evidence suggests that FBXW7 upregulation might have dual effects in HD, and future research will be needed to elucidate the potential role of FBXW7 regulation as a neuroprotective approach.

11.2.8.2 Keap1 as a CRL3 Complex Forming Subunit

Kelch-like ECH-associated protein 1 (Keap1) is a substrate receptor protein that interacts with CUL3 and RBX1 to form a CRL3 complex. In basal conditions, Keap1 binds to its substrate Nrf2 and targets it to proteasomal degradation; however, when ROS are overproduced, certain cysteine residues within Keap1 become oxidized and Keap1 suffers a conformational modification that prevents Nrf2 degradation, allowing Nrf2-dependent transcription in OS conditions. Nrf2-transcriptional activity boosts the expression of several endogenous antioxidants that counteract ROS overproduction and OS [147].

Sulforaphane (SFN) is a widely used compound that activates Nrf2-dependent transcription by inducing analogous conformational changes to Keap1 similar to those induced by ROS [148]. SFN has shown promising results in several HD models: (1) in the 3-NP model of HD, SFN minimized striatal neuron loss and the expression of multiple proinflammatory factors [149]; (2) SFN decreased the aggregation of mHtt in human fibroblasts [150]; (3) SFN diminished the aggregation and toxicity of mHtt via UPS degradation [85]; (4) SFN induced the Nrf2-dependent antioxidant response in HD mice primary astrocytes and protected them from oxidative damage [151].

Dimethyl fumarate (DMF) is another Nrf2 activator that works via modifying Keap1 cysteine residues, preventing Keap1-driven Nrf2 degradation [152]. DMF treatment also has potential neuroprotective effects in several HD models: (1) in 3-NP-induced striatal damage in rats, DMF treatment reduced motor deficits, striatal neuron loss, OS markers and endoplasmic reticulum stress-related proteins, while boosting BDNF production [153]; (2) DMF treatment diminished motor alterations and striatal neurodegeneration in YAC128 HD mice [154]; (3) DMF treatment rescued the motor function and prevented neuronal loss in R6/2 HD mice, although it was unable to block mHtt aggregation [154].

Multiple studies converge on the fact that targeting the Keap1/Nrf2 signaling pathway with compounds like SFN or DMF results in positive outcomes regarding neuroprotection in many HD models, and it could be a potential approach in the development of therapeutic advances in the HD context.

11.2.9 HACE1

HECT domain and ankyrin repeat containing E3 ubiquitin protein ligase 1 (HACE1) is a HECT-type E3 ligase [155]. HACE1 is another potential therapeutic factor in HD pathology due to its capacity to activate Nrf2 by suppressing its Keap1-driven degradation in response to ROS overproduction.

HACE1 protein and mRNA levels were reduced in the striatum of HD brain samples. In addition, in HD-cell lines, HACE1 expression proved to enhance the cell tolerance to mutant-Htt toxicity through Nrf2 pathway activation [156].

HACE1 is also involved in Ras-related C3 botulinum toxin substrate 1 (RAC1) ubiquitination and degradation via UPS. RAC1 activates NADPH oxidase, an enzyme that plays a crucial role in ROS production and neurotoxicity in HD. Despite the above, HACE1 inactivation caused NADPH oxidase overactivation and ROS production [157, 158]. This evidence further supports HACE1 as a master regulator of the redox balance within cells, and HACE1 upregulation might be a potential approach to abolish neurotoxicity in HD. Nonetheless, this idea has not been tested yet in an HD model.

Moreover, HACE1 silencing in YAC128 HD mice managed to accelerate the onset of motor pathology in vivo, while reducing mitochondrial respiration and proliferation of astrocytes. Astrogliosis was also observed in HACE1-deficient wild-type mice [159]. Hence, HACE1 is potentially involved in the astroglial pathology observed in HD.

11.2.10 HECTD1

HECT Domain E3 Ubiquitin Protein Ligase 1 (HECTD1) is a HECT-type E3 enzyme with a particular role in glial physiology. In LPS-stimulated BV-2 cells, circRNA (Circ-Bnc2) was able to increase the expression of HECTD1 and, in response, HECTD1 downregulated the secretion of several proinflammatory enzymes and cytokines produced by BV-2 cells to the culture media. Consequently, when the culture media from BV-2 cells was used in the HT-22 hippocampal cell line, the culture medium from the circRNA-treated BV-2 cells induced a lower level of apoptosis in HT-22 cells compared to the medium derived from LPS-treatment alone [160].

Conversely, HECTD1 upregulation enhanced reactivity in primary astrocyte cultures exposed to LPS. Experimental silencing of HECTD1 expression resulted in reduced astroglial reactivity in the striatum of LPS-treated rodents. Furthermore, HECTD1 appeared to be a fundamental part of the signaling pathway needed by LPS to induce astroglial reactivity [161]. In addition, further evidence suggests that astrocytic HECTD1 is necessary for the degradation of the glutamate transporter 1 (GLT-1) [162]. These results are relevant in the context because GLT-1 activity is thought to be hampered in HD models, leading to impaired glutamate uptake from the synaptic clefts and augmented excitotoxicity [163].

This evidence highlights the role of HECTD1 in different pathogenic mechanisms related to excitotoxicity and neuroinflammation in HD; however, none of these studies were strictly tested on HD models, and direct evidence is still needed in this regard.

11.2.11 UBE3A

Ubiquitin protein ligase E3A (UBE3A) is a HECT-type E3 enzyme with a wide range of effects regarding neurodegeneration [164]. UBE3A has a more potent ubiquitination capacity over mHtt than its wild-type counterpart, making UBE3A potentially selective to target the neurotoxic form of Htt. In the same report, overexpression of UBE3A augmented the degradation of mHtt in HD mice [165], and, as expected, in the inverse situation, UBE3A silencing from HD mice's brain resulted in increased Htt aggregation and toxicity [166].

Synaptopathy in HD also appears to be related to UBE3A activity. In R6/2 HD mice, UBE3A levels were diminished in the striatum (and in other brain regions), while the expression of several synaptic markers such as synapsin-1, postsynaptic density protein 95 (PSD95), Arc protein and two subunits of the α-amino-3-hydroxy-5-methyl-4-isoxazolepropionic acid (AMPAR) receptors was also decreased in R6/2 mice [167]. This evidence is in line with reports that revealed a clear role of UBE3A in synapse development, as UBE3A dysregulation is related to synaptic deficits in other neurologic disorders like autism spectrum disorder and Angelman syndrome [168–171].

Transcriptomic analysis of the UBE3A deletion from embryonic fibroblasts revealed that UBE3A knockout resulted in transcriptomic alterations of genes related to mitochondrial glutathione metabolism and homeostasis by regulating the forkhead box O pathway, which is related to mitochondrial biogenesis, fission/fusion rate and mitophagy [172, 173]. Therefore, UBE3A downregulation in HD might also modify mitochondrial homeostasis, although this idea has yet to be proven in an HD model.

11.2.12 UBR5

Ubiquitin protein ligase E3 component n-recognin 5 (UBR5) protein is another E3 enzyme from the HECT-type family. UBR5 is also related to HD pathology as its knockout from induced pluripotent stem cells derived from HD patients resulted in augmented mHtt aggregation. Consequently, it was suggested that UBR5 triggered mHtt degradation via UPS [174].

URB5 additional substrates are telomerase reverse transcriptase (TERT) and ATM interactor (ATMIN) [175]. In this regard, although following evidence is not linked directly to HD pathology, it might represent a directive for future research: (1) in vascular models, inhibition of URB5 resulted in decreased degradation of TERT and consequently protected vascular organoids from OS [176]; (2) in endothelial cells, PPARγ-dependent UBR5 activation resulted in increased ATMIN degradation and augmented DNA repair [177]. This evidence is relevant as both OS and DNA repair impairments are involved in HD pathology [39, 42].

11.2.13 WWP1

WW-containing domain E3 ubiquitin protein ligases 1/2 (WWP1/2) are two HECT-type protein ligases potentially related to several neurodegenerative diseases [178]. Both in the brain of R6/2 mice and in N2a neuroblastoma cells transfected with mHtt, WWP1 was colocalized with mHtt aggregates, as illustrated with confocal fluorescence microscopy. Furthermore, WWP1 overexpression resulted in enhanced mHtt aggregation by promoting its ubiquitination, but blocking its degradation via UPS [179]. Contrasting results were observed in *Caenorhabditis elegans* models, where the overexpression of the ortholog of human WWP1 resulted in decreased aggregation of Poly-Q-containing proteins [180].

WWP2 has multiple targets related to autophagy/mitophagy such as nuclear dot protein 52 kDa (NDP52), optineurin (OPTN), and sequestosome 1 (SQSTM1), as shown by microarray studies. In corroboration, WWP2 knockout results in deficient autophagic degradation [181]. Research has demonstrated that the Poly-Q sequence in mHtt interferes with OPTN and SQSTM1-related mitophagy pathways [108]. Complementary evidence suggests that striatal medium spiny neurons have lower OPTN levels compared to other neuronal populations within the striatum [182], and that OPTN-lacking cells are more sensitive to mHtt toxicity [183]. Therefore, WWP2 regulation might play a role as a cytoprotective factor by regulating OPTN or SQSTM1 levels although this hypothesis still needs to be tested in an HD model.

11.2.14 NEDD4

Neuronally expressed developmentally downregulated protein 4 (NEDD4) is a HECT-type E3 ligase that is widely expressed in neurons and has multiple roles on neurodevelopment, neurodegeneration and other critical physiological functions within the nervous system [184].

In HEK293 cultures transfected with a vector expressing mHtt-constructs, it was revealed that NEDD4 knockdown resulted in an increase in the number of the small (<50 μM^2) Htt aggregates [180].

In addition, brain samples from HD patients show augmented protein levels of a NEDD4 subtype, NEDD4-1. The same study revealed that the OS-inducer H_2O_2 and the apoptosis-inducer camptothecin increased the protein levels of NEDD4-1, thus explaining the NEDD4-1 upregulation in HD patients. In addition, it was revealed that ROS acted through a transcription factor that binds NEDD4-1 promoter, and consequently NEDD4-1 enhanced insulin-like growth factor 1 receptor β (IGF-1Rβ) degradation [185]. The signaling pathway mediated by IGF-1Rβ has been considered as neuroprotective by several studies, especially regarding the mitochondrial impairments observed in HD [186, 187]. Therefore, it is reasonable to hypothesize that NEDD4-1 can reduce IGF-1Rβ signaling in HD, thereby enhancing neurotoxicity.

11.2.15 Chip

The C-terminus of Hsc70-interacting protein (CHIP) is a U-BOX-type E3 ligase with a widely recognized role in unfolded protein degradation as it was able to cooperate with cell chaperones to degrade unfolded luciferase in vitro [188]. Recent studies revealed that CHIP deficiency inhibited the aggregation of ataxin-3, another protein with Poly-Q-related pathologic mechanisms [189]. Despite the above, CHIP reduced mHtt aggregation and apoptotic cell death in different cell culture models. Finally, CHIP deficiency was able to accelerate the motor deficits observed in HD mice [190].

Additional evidence suggested that in HT22 cells, H_2O_2-induced OS and apoptosis were reduced by CHIP overexpression. In the same report, diminished microglial and astrocytic activation was observed in an ischemia model due to CHIP expression [191], further suggesting a neuroprotective role of CHIP in several HD-related pathological mechanisms.

11.2.16 HOIP

HOIL-1 L-interacting protein (HOIP) is an RBR-type E3 enzyme that, together with other proteins, forms a multi-subunit protein complex that can catalyze the addition of linear polyubiquitin chains to target proteins. This complex is named linear ubiquitin chain assembly complex (LUBAC) [192].

HOIP and other LUBAC-forming proteins are colocalized with mHtt aggregates within transfected SH-SY5Y cells, and HOIP specifically is also colocalized with the aggregates in HD-brain samples. This colocalization results in the linear ubiquitination of mHtt aggregates. Linear ubiquitin chains reduce mHtt-driven apoptosis, improve Htt degradation via UPS, and prevent the aberrant sequestration of specificity protein 1 (SP1) into the mHtt aggregates [193]. This evidence helps to explain previous results showing that SP1-dependent transcription is downregulated in HD [194].

Moreover, HOIP/LUBAC-dependent linear ubiquitination of glutathione peroxidase 4 results in increased stability of this protein, reduced lipid peroxidation and minimized ferroptosis in a non-HD model [195]. However, pyroptotic cell death has an important emerging role in HD pathogenesis indeed [196].

HOIP plays a pivotal role in neuroinflammation, as, in an ischemic model, it was suggested that HOIP/LUBAC-dependent ubiquitination boosted NF-kB signaling, microglial activation and consequent production of proinflammatory factors [197]. Combined, this evidence points to a crucial role of HOIP E3 ligase in pathological mechanisms related to HD.

11.2.17 Parkin

Parkin is another RBR-type E3 ligase better known for its pivotal role in Parkinson's disease pathogenesis [198]. In the HD context, increased mRNA and protein levels of parkin were found in HD fibroblasts, which was accompanied by an abnormal reduction of Mfn1, a Parkin-substrate that is involved in mitochondrial fusion [199]. These results are in line with evidence of diminished Mfn1 levels in HD patients and altered mitochondrial dynamics (involving fusion/fission rate) [49]. Moreover, in a *D. melanogaster* model of HD it was shown that the overexpression of PTEN-induced putative kinase 1 (PINK1), a protein that controls Parkin-dependent mitophagy, reduced mitochondrial abnormalities and mHtt-driven neurodegeneration. However, PINK1-driven neuroprotection depends on parkin expression in the fly model [200].

Parkin ligase might also play a role in Htt degradation by itself. Parkin has been shown to interact with Poly-Q sequence-containing proteins in YAC72 HD mice. In agreement, colocalization of Htt aggregates and parkin has also been observed in HD brain samples. The same report suggested that parkin was able to hamper Htt aggregation, Htt-driven reduction of cell viability and Htt-driven proteasomal dysfunction, potentially alleviating endoplasmic reticulum stress within cells as well [201].

Parkin is a crucial molecule for mitophagy induction and it has been suggested that it promotes the mitophagic degradation of impaired mitochondria. Additionally, ROS-induced mitochondrial depolarization is also sensed by parkin to mediate the degradation of depolarized mitochondria [202, 203]. Although this was discovered separately, research has shown that Poly-Q proteins can potentiate mitochondrial depolarization in vitro [204]. Thus, parkin-dependent mitophagy might be a viable experimental alternative to exert neuroprotection in HD.

Finally, lack of microglial parkin resulted in overproduction of proinflammatory factors, which is in line with a Parkinson's disease model in which activation of the PINK1/parkin pathway induced a reduction in the NLRP3-inflammasome-dependent response and microglial activation [205, 206]. Even though this was not proved in an HD model, both the NLRP3-inflammasome pathway and microglial activation phenomena have been observed in the 3-NP HD model [207], highlighting the potential neuroprotective role of parkin.

Further studies identified parkin as an anti-excitotoxic agent as parkin overexpression limited the cell damage exerted by the excitotoxin kainite, and conversely, parkin deficiency potentiated kainate-derived toxicity [208]. This study is relevant in the HD context as excitotoxic damage to striatal neurons might play a key role in HD pathogenesis [57].

11.2.18 Other E3 Ubiquitin Ligases

Apart from the E3 ligases already mentioned in the sections above, little evidence also supports the interplay between HD pathogenesis and several other E3 ligases. Though evidence is limited, it should be considered as a potential research focus in the future: (1) Itchy E3 ubiquitin ligase (ITCH) is a HECT-type E3 ligase that colocalizes with mHtt aggregates, interacting with them and diminishing the Htt-driven cytotoxicity in cell culture models [209]; (2) in YAC128 HD mice it was shown that Siah E3 ubiquitin protein ligase 1 (Siah-1) interacts with mHtt aggregates. In addition, cell culture models revealed that Siah-1 prevented Htt aggregation via UPS degradation while promoting autophagy [210]; (3) Tripartite motif-containing 37 (TRIM37) is a RING-type E3 ligase that also interacts with mHtt, promoting its ubiquitination and degradation, as shown in HEK293 cells experimentally expressing mHtt [211]; (4) homocysteine-induced endoplasmic reticulum protein (Herp) is another E3 ligase that binds to the N-terminal end of mHtt, promoting its degradation via UPS [212]; (5) suppressor of cytokine signaling 2 is a complex protein that contains an E3 Ligase domain. SOCS2 activity was enhanced in striatal cells overexpressing Poly-Q-containing proteins. Moreover, SOCS inhibition resulted in increased autophagy and decreased striatal cell death [213]; (6) mahogunin ring finger-1 (MGRN1) interacts with the N-terminal end of mHtt fragments, ubiquitinating them and preventing aggregation and cell death [214]; (7) Zinc finger protein 179 (Znf179), also named RING finger protein 112, is a RING-type ubiquitin E3 ligase with a pivotal role in neuronal differentiation and brain development [215]; (8) Znf179 has shown neuroprotective effects against H_2O_2-induced OS, and

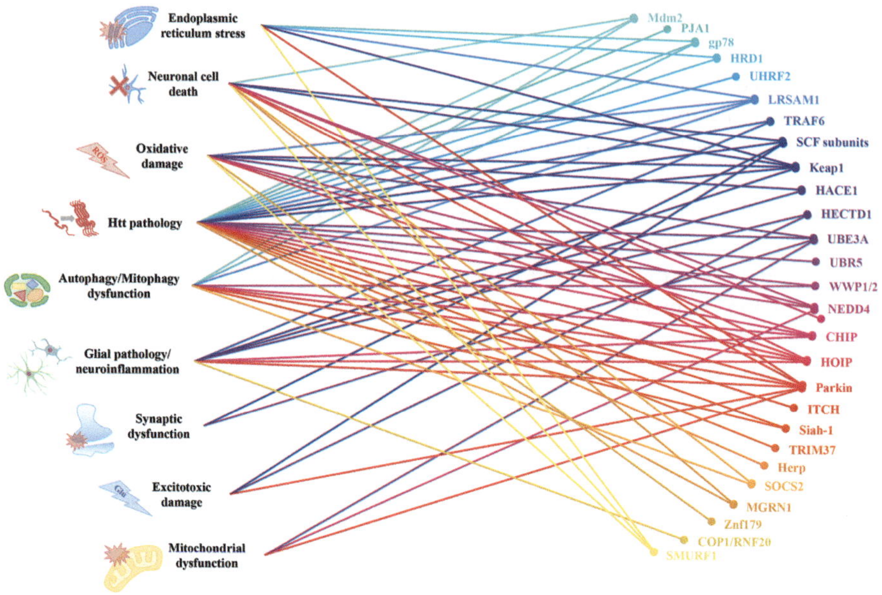

Fig. 11.1 Potential interaction network of the E3 ligases reviewed herein and different pathological hallmarks of HD. Linear connectors show that a specific E3 ubiquitin ligase has beneficial or deleterious effects on a specific HD hallmark. It should be considered that both evidence from HD and non-HD models was considered to design this figure

particularly Znf acetylation has been shown to promote protection [216, 217]; (9) currently no reports provide a direct link between HD and smad ubiquitination regulatory factor 1 (SMURF1); however, in glioblastoma cells it was shown that SMURF1 is able to induce Keap1 degradation and consequent induction of Nrf2-dependent transcription, reduction of ROS production, decreased cell apoptosis and diminished endoplasmic reticulum stress [218]; (10) in several non-HD models, microglia-derived neuroinflammation has been related to E3 ligases like Constitutive photomorphogenic 1 (COP1) and Ring Finger Protein 20 (RNF20) [219, 220]. Figure 11.1 summarizes the interactions of E3 ligases and several hallmarks of toxicity in HD.

11.3 Conclusions and Future Perspectives

While the genetic basis of HD is unequivocal, the development of neurodegeneration in HD patients arises from a multifaceted interplay of pathogenic mechanisms. Similar to Alzheimer's disease, Parkinson's disease, amyotrophic lateral sclerosis, and other neurodegenerative disorders, HD pathology is intricately linked to a widespread disruption of proteostasis in affected cells and tissues [221]. Although the UPS has recently emerged as a central proteostasis regulator, conflicting evidence has been reported, with some reports suggesting that UPS enhancement is beneficial for neurodegeneration, while other reports proposed that UPS activation has deleterious consequences and triggers pathologic processes [222]. The diverse effects observed in the role of UPS in HD and other neurodegenerative disorders may, at least in part, be attributed to the multifaceted functions of the enzymes regulating proteasomal degradation, particularly E3 ubiquitin ligases.

This work proposes that the activation/inhibition of different E3 ubiquitin ligases can lead to various cellular outcomes, depending on the protein targets of these ligases and the cellular pathways influenced by their enzymatic activity. Consequently, while E3 ubiquitin ligases hold significant promise as therapeutic targets for HD, it is crucial to acknowledge that the upregulation of every E3 ligase cannot universally be assumed to have neuroprotective effects. For future perspectives, a deeper understanding of the roles of E3 ligases in HD pathology is essential. Many of the studies cited herein provide valuable insights into the potential neuroprotective properties of particular E3 ligases but do not focus specifically on experimental models or conclusions tailored to HD. To fully evaluate the neuroprotective potential of these ligases in the context of HD, comprehensive experimental validation is needed. Such research will help to clarify E3 ligases' mechanisms of action and their capacity to mitigate the complex pathology of HD.

Acknowledgments LAM-M was supported by CONAHCYT scholarship (program ID No. 000289).

AAT was supported by the Peoples´ Friendship University of Russia (RUDN University) Strategic Academic Leadership Program (award no. 202717–0–000).

References

1. McAllister B, Gusella JF, Landwehrmeyer GB, Lee JM, MacDonald ME, Orth M, et al. Investigators of the European Huntington's disease network. Timing and impact of psychiatric, cognitive, and motor abnormalities in Huntington disease. Neurology. 2021;96(19):e2395–406.
2. Stoker TB, Mason SL, Greenland JC, Holden ST, Santini H, Barker RA. Huntington's disease: diagnosis and management. Pract Neurol. 2022;22(1):32–41.
3. Jellinger KA. Pathomechanisms of behavioral abnormalities in Huntington disease: an update. J Neural Transm. 2024;131(9):999–1012.
4. Craufurd D, Thompson JC, Snowden JS. Behavioral changes in Huntington disease. Cogn Behav Neurol. 2001;14(4):219–26.
5. Garcia-Gorro C, Llera A, Martinez-Horta S, Perez-Perez J, Kulisevsky J, Rodriguez-Dechicha N, et al. Specific patterns of brain alterations underlie distinct clinical profiles in Huntington's disease. NeuroImage: Clinical. 2019;23:101900.
6. McColgan P, Tabrizi SJ. Huntington's disease: a clinical review. Eur J Neurol. 2018;25(1):24–34.
7. Pringsheim T, Wiltshire K, Day L, Dykeman J, Steeves T, Jette N. The incidence and prevalence of Huntington's disease: a systematic review and meta-analysis. Mov Disord. 2012;27(9):1083–91.
8. Rawlins MD, Wexler NS, Wexler AR, Tabrizi SJ, Douglas I, Evans SJ, Smeeth L. The prevalence of Huntington's disease. Neuroepidemiology. 2016;46(2):144–53.

9. Bates GP, Dorsey R, Gusella JF, Hayden MR, Kay C, Leavitt BR, Nance M, Ross CA, Scahill RI, Wetzel R, Wild EJ, Tabrizi SJ. Huntington disease. Nat Rev Dis Primers. 2015;1:15005.
10. Waldvogel HJ, Kim EH, Tippett LJ, Vonsattel JP, Faull RL. The Neuropathology of Huntington's Disease. Curr Top Behav Neurosci. 2015;22:33–80.
11. Atherton JF, McIver EL, Mullen MR, Wokosin DL, Surmeier DJ, Bevan MD. Early dysfunction and progressive degeneration of the subthalamic nucleus in mouse models of Huntington's disease. Elife. 2016;5:e21616.
12. Petersén Å, Gabery S. Hypothalamic and limbic system changes in Huntington's disease. J Huntington's Dis. 2012;1(1):5–16.
13. L'episcopo F, Drouin-Ouellet J, Tirolo C, Pulvirenti A, Giugno R, Testa N, et al. GSK-3β-induced Tau pathology drives hippocampal neuronal cell death in Huntington's disease: involvement of astrocyte–neuron interactions. Cell Death Dis. 2016;7(4):e2206.
14. Padron-Rivera G, Diaz R, Vaca-Palomares I, Ochoa A, Hernandez-Castillo CR, Fernandez-Ruiz J. Cerebellar degeneration signature in Huntington's disease. Cerebellum. 2021;20(6):942–5.
15. Mehrabi NF, Waldvogel HJ, Tippett LJ, Hogg VM, Synek BJ, Faull RL. Symptom heterogeneity in Huntington's disease correlates with neuronal degeneration in the cerebral cortex. Neurobiol Dis. 2016;96:67–74.
16. Guo Q, Bin Huang, Cheng J, Seefelder M, Engler T, Pfeifer G, Oeckl P, Otto M, Moser F, Maurer M, Pautsch A, Baumeister W, Fernández-Busnadiego R, Kochanek S. The cryo-electron microscopy structure of huntingtin. Nature. 2018;555(7694):117–120.
17. Saudou F, Humbert S. The biology of huntingtin. Neuron. 2016;89(5):910–26.
18. Strehlow AN, Li JZ, Myers RM. Wild-type huntingtin participates in protein trafficking between the Golgi and the extracellular space. Hum Mol Genet. 2007;16(4):391–409.
19. Trushina E, Singh RD, Dyer RB, Cao S, Shah VH, Parton RG, et al. Mutant huntingtin inhibits clathrin-independent endocytosis and causes accumulation of cholesterol in vitro and in vivo. Hum Mol Genet. 2006;15(24):3578–91.
20. Buren C, Wang L, Smith-Dijak A, Raymond LA. Region-specific pro-survival signaling and global neuronal protection by wild-type huntingtin. J Huntington's Dis. 2014;3(4):365–76.
21. Ochaba J, Lukacsovich T, Csikos G, Zheng S, Margulis J, Salazar L, et al. Potential function for the huntingtin protein as a scaffold for selective autophagy. Proc Natl Acad Sci. 2014;111(47):16889–94.
22. Maiuri T, Woloshansky T, Xia J, Truant R. The huntingtin N17 domain is a multifunctional CRM1 and ran-dependent nuclear and cilial export signal. Hum Mol Genet. 2013;22(7):1383–94.
23. Karam A, Tebbe L, Weber C, Messaddeq N, Morlé L, Kessler P, et al. A novel function of huntingtin in the cilium and retinal ciliopathy in Huntington's disease mice. Neurobiol Dis. 2015;80:15–28.
24. Tong Y, Ha TJ, Liu L, Nishimoto A, Reiner A, Goldowitz D. Spatial and temporal requirements for huntingtin (htt) in neuronal migration and survival during brain development. J Neurosci. 2011;31(41):14794–9.
25. Nguyen GD, Gokhan S, Molero AE, Mehler MF. Selective roles of normal and mutant huntingtin in neural induction and early neurogenesis. PLoS One. 2013;8(5):e64368.
26. Burrus CJ, McKinstry SU, Kim N, Ozlu MI, Santoki AV, Fang FY, et al. Striatal projection neurons require huntingtin for synaptic connectivity and survival. Cell Rep. 2020;30(3):642–57.
27. Jacobsen JC, Gregory GC, Woda JM, Thompson MN, Coser KR, Murthy V, et al. HD CAG-correlated gene expression changes support a simple dominant gain of function. Hum Mol Genet. 2011;20(14):2846–60.
28. Wang Y, Lin F, Qin ZH. The role of post-translational modifications of huntingtin in the pathogenesis of Huntington's disease. Neurosci Bull. 2010;26(2):153–162.
29. Juenemann K, Wiemhoefer A, Reits EA. Detection of ubiquitinated huntingtin species in intracellular aggregates. Front Mol Neurosci. 2015;8:1.
30. Arbez N, Ratovitski T, Roby E, Chighladze E, Stewart JC, Ren M, et al. Post-translational modifications clustering within proteolytic domains decrease mutant huntingtin toxicity. J Biol Chem. 2017;292(47):19238–49.

31. Gottlieb L, Guo L, Shorter J, Marmorstein R. N-alpha-acetylation of huntingtin protein increases its propensity to aggregate. J Biol Chem. 2021;297(6):101363.
32. Lontay B, Kiss A, Virág L, Tar K. How do post-translational modifications influence the pathomechanistic landscape of Huntington's disease? A comprehensive review. Int J Mol Sci. 2020;21(12):4282.
33. Olshina MA, Angley LM, Ramdzan YM, Tang J, Bailey MF, Hill AF, Hatters DM. Tracking mutant huntingtin aggregation kinetics in cells reveals three major populations that include an invariant oligomer pool. J Biol Chem. 2010;285(28):21807–16.
34. Legleiter J, Mitchell E, Lotz GP, Sapp E, Ng C, DiFiglia M, et al. Mutant huntingtin fragments form oligomers in a polyglutamine length-dependent manner in vitro and in vivo. J Biol Chem. 2010;285(19):14777–90.
35. Lajoie P, Snapp EL. Formation and toxicity of soluble polyglutamine oligomers in living cells. PLoS One. 2010;5(12):e15245.
36. Herrera F, Tenreiro S, Miller-Fleming L, Outeiro TF. Visualization of cell-to-cell transmission of mutant huntingtin oligomers. PLoS Curr. 2011;3:RRN1210.
37. Mario Isas J, Pandey NK, Xu H, Teranishi K, Okada AK, Fultz EK, et al. Huntingtin fibrils with different toxicity, structure, and seeding potential can be interconverted. Nat Commun. 2021;12(1):4272.
38. Benn CL, Sun T, Sadri-Vakili G, McFarland KN, DiRocco DP, Yohrling GJ, et al. Huntingtin modulates transcription, occupies gene promoters in vivo, and binds directly to DNA in a polyglutamine-dependent manner. J Neurosci. 2008;28(42):10720–33.
39. Gao R, Chakraborty A, Geater C, Pradhan S, Gordon KL, Snowden J, et al. Mutant huntingtin impairs PNKP and ATXN3, disrupting DNA repair and transcription. Elife. 2019;8:e42988.
40. Grima JC, Daigle JG, Arbez N, Cunningham KC, Zhang K, Ochaba J, et al. Mutant huntingtin disrupts the nuclear pore complex. Neuron. 2017;94(1):93–107.
41. Liu KY, Shyu YC, Barbaro BA, Lin YT, Chern Y, Thompson LM, et al. Disruption of the nuclear membrane by perinuclear inclusions of mutant huntingtin causes cell-cycle re-entry and striatal cell death in mouse and cell models of Huntington's disease. Hum Mol Genet. 2015;24(6):1602–16.
42. Andersen JK. Oxidative stress in neurodegeneration: cause or consequence? Nat Med. 2004;10(Suppl):S18–25.
43. Jin YN, Yu YV, Gundemir S, Jo C, Cui M, Tieu K, Johnson GV. Impaired mitochondrial dynamics and Nrf2 signaling contribute to compromised responses to oxidative stress in striatal cells expressing full-length mutant huntingtin. PLoS One. 2013;8(3):e57932.
44. Reijonen S, Kukkonen JP, Hyrskyluoto A, Kivinen J, Kairisalo M, Takei N, et al. Downregulation of NF-κB signaling by mutant huntingtin proteins induces oxidative stress and cell death. Cell Mol Life Sci. 2010;67(11):1929–41.
45. DiGiovanni LF, Mocle AJ, Xia J, Truant R. Huntingtin N17 domain is a reactive oxygen species sensor regulating huntingtin phosphorylation and localization. Hum Mol Genet. 2016;25(18):3937–45.
46. Goswami A, Dikshit P, Mishra A, Mulherkar S, Nukina N, Jana NR. Oxidative stress promotes mutant huntingtin aggregation and mutant huntingtin-dependent cell death by mimicking proteasomal malfunction. Biochem Biophys Res Commun. 2006;342(1):184–90.
47. Kumar A, Ratan RR. Oxidative stress and Huntington's disease: the good, the bad, and the ugly. J Huntington's Dis. 2016;5(3):217–37.
48. Aidt FH, Nielsen SMB, Kanters J, Pesta D, Nielsen TT, Nørremølle A, et al. Dysfunctional mitochondrial respiration in the striatum of the Huntington's disease transgenic R6/2 mouse model. PLoS Curr. 2013;5
49. Shirendeb U, Reddy AP, Manczak M, Calkins MJ, Mao P, Tagle DA, Hemachandra Reddy PH. Abnormal mitochondrial dynamics, mitochondrial loss and mutant huntingtin oligomers in Huntington's disease: implications for selective neuronal damage. Hum Mol Genet. 2011;20(7):1438–55.
50. Bossy-Wetzel E, Petrilli A, Knott AB. Mutant huntingtin and mitochondrial dysfunction. Trends Neurosci. 2008;31(12):609–16.

51. Guo X, Disatnik MH, Monbureau M, Shamloo M, Mochly-Rosen D, Qi X. Inhibition of mitochondrial fragmentation diminishes Huntington's disease–associated neurodegeneration. J Clin Invest. 2013;123(12):5371–88.
52. Borlongan CV, Koutouzis TK, Sanberg PR. 3-Nitropropionic acid animal model and Huntington's disease. Neurosci Biobehav Rev. 1997;21(3):289–93.
53. Fan MM, Raymond LA. N-methyl-D-aspartate (NMDA) receptor function and excitotoxicity in Huntington's disease. Prog Neurobiol. 2007;81(5–6):272–93.
54. Beal MF, Kowall NW, Ellison DW, Mazurek MF, Swartz KJ, Martin JB. Replication of the neurochemical characteristics of Huntington's disease by quinolinic acid. Nature. 1986;321(6066):168–71.
55. Sanberg PR, Calderon SF, Giordano M, Tew JM, Norman AB. The quinolinic acid model of Huntington's disease: locomotor abnormalities. Exp Neurol. 1989;105(1):45–53.
56. Beal MF, Ferrante RJ, Swartz KJ, Kowall NW. Chronic quinolinic acid lesions in rats closely resemble Huntington's disease. J Neurosci. 1991;11(6):1649–59.
57. Heng MY, Detloff PJ, Wang PL, Tsien JZ, Albin RL. In vivo evidence for NMDA receptor-mediated excitotoxicity in a murine genetic model of Huntington disease. J Neurosci. 2009;29(10):3200–5.
58. Milnerwood AJ, Gladding CM, Pouladi MA, Kaufman AM, Hines RM, Boyd JD, et al. Early increase in extrasynaptic NMDA receptor signaling and expression contributes to phenotype onset in Huntington's disease mice. Neuron. 2010;65(2):178–90.
59. Dau A, Gladding CM, Sepers MD, Raymond LA. Chronic blockade of extrasynaptic NMDA receptors ameliorates synaptic dysfunction and pro-death signaling in Huntington disease transgenic mice. Neurobiol Dis. 2014;62:533–42.
60. Okamoto S, Pouladi MA, Talantova M, Yao D, Xia P, Ehrnhoefer DE, et al. Balance between synaptic versus extrasynaptic NMDA receptor activity influences inclusions and neurotoxicity of mutant huntingtin. Nat Med. 2009;15(12):1407–13.
61. Palpagama TH, Waldvogel HJ, Faull RL, Kwakowsky A. The role of microglia and astrocytes in Huntington's disease. Front Mol Neurosci. 2019;12:258.
62. Wilton DK, Mastro K, Heller MD, Gergits FW, Willing CR, Fahey JB, et al. Microglia and complement mediate early corticostriatal synapse loss and cognitive dysfunction in Huntington's disease. Nat Med. 2023;29(11):2866–84.
63. Giorgini F, Möller T, Kwan W, Zwilling D, Wacker JL, Hong S, et al. Histone deacetylase inhibition modulates kynurenine pathway activation in yeast, microglia, and mice expressing a mutant huntingtin fragment. J Biol Chem. 2008;283(12):7390–400.
64. Faideau M, Kim J, Cormier K, Gilmore R, Welch M, Auregan G, et al. In vivo expression of polyglutamine-expanded huntingtin by mouse striatal astrocytes impairs glutamate transport: a correlation with Huntington's disease subjects. Hum Mol Genet. 2010;19(15):3053–67.
65. Wang L, Lin F, Wang J, Wu J, Han R, Zhu L, et al. Expression of mutant N-terminal huntingtin fragment (Htt552-100Q) in astrocytes suppresses the secretion of BDNF. Brain Res. 2012;1449:69–82.
66. Polyzos AA, Lee DY, Datta R, Hauser M, Budworth H, Holt A, et al. Metabolic reprogramming in astrocytes distinguishes region-specific neuronal susceptibility in Huntington mice. Cell Metab. 2019;29(6):1258–73.
67. Smith-Dijak AI, Sepers MD, Raymond LA. Alterations in synaptic function and plasticity in Huntington disease. J Neurochem. 2019;150(4):346–65.
68. Leitman J, Ulrich Hartl F, Lederkremer GZ. Soluble forms of polyQ-expanded huntingtin rather than large aggregates cause endoplasmic reticulum stress. Nat Commun. 2013;4(1):2753.
69. Lee WC, Yoshihara M, Littleton JT. Cytoplasmic aggregates trap polyglutamine-containing proteins and block axonal transport in a drosophila model of Huntington's disease. Proc Natl Acad Sci. 2004;101(9):3224–9.
70. Simpson JM, Gil-Mohapel J, Pouladi MA, Ghilan M, Xie Y, Hayden MR, Christie BR. Altered adult hippocampal neurogenesis in the YAC128 transgenic mouse model of Huntington disease. Neurobiol Dis. 2011;41(2):249–60.

71. Czeredys M, Gruszczynska-Biegala J, Schacht T, Methner A, Kuznicki J. Expression of genes encoding the calcium signalosome in cellular and transgenic models of Huntington's disease. Front Mol Neurosci. 2013;6:42.
72. van der Plas E, Schultz JL, Nopoulos PC. The neurodevelopmental hypothesis of Huntington's disease. J Huntington's Dis. 2020;9(3):217–29.
73. van der Plas E, Langbehn DR, Conrad AL, Koscik TR, Tereshchenko A, Epping EA, et al. Abnormal brain development in child and adolescent carriers of mutant huntingtin. Neurology. 2019;93(10):e1021–30.
74. Lisowski P, Lickfett S, Rybak-Wolf A, Menacho C, Le S, Pentimalli TM, et al. Mutant huntingtin impairs neurodevelopment in human brain organoids through CHCHD2-mediated neurometabolic failure. Nat Commun. 2024;15(1):7027.
75. Rook ME, Southwell AL. Antisense oligonucleotide therapy: from design to the Huntington disease clinic. BioDrugs. 2022;36(2):105–19.
76. Stimming EF, Sung V, Testa C, Kostyk S, Ross CA, Samii A, et al. J05 interim results from cohort 1 of the double-blind, dose-escalation phase I/II clinical trial of amt-130 (HD-genetrx-1) for early-stage Huntington's disease (HD). J Neurol Neurosurg Psychiatry. 2022;93(Suppl 1):A95.
77. Süssmuth SD, Haider S, Landwehrmeyer GB, Farmer R, Frost C, Tripepi G, et al. An exploratory double-blind, randomized clinical trial with selisistat, a SirT1 inhibitor, in patients with Huntington's disease. Br J Clin Pharmacol. 2015;79(3):465–76.
78. Reilmann R, Squitieri F, Priller J, Saft C, Mariotti C, Süessmuth SD, et al. N02 safety and tolerability of selisistat for the treatment of huntington's disease: results from a randomised, double-blind, placebo-controlled phase II trial. J Neurol Neurosurg Psychiatry. 2014;85(1):A102.
79. Reilmann R, Gordon MF, Anderson KE, Feigin A, Tabrizi SJ, Leavitt BR, et al. The efficacy and safety results of laquinimod as a treatment for Huntington disease (LEGATO-HD)(S16. 007). Neurology. 2019;92(15_supplement):S16-007.
80. McGarry A, McDermott M, Kieburtz K, de Blieck EA, Beal F, Marder K, et al. A randomized, double-blind, placebo-controlled trial of coenzyme Q10 in Huntington disease. Neurology. 2017;88(2):152–9.
81. Hersch SM, Gevorkian S, Marder K, Moskowitz C, Feigin A, Cox M, et al. Creatine in Huntington disease is safe, tolerable, bioavailable in brain and reduces serum 8OH2′ dG. Neurology. 2006;66(2):250–2.
82. Hersch SM, Schifitto G, Oakes D, Bredlau AL, Meyers CM, Nahin R, et al. The CREST-E study of creatine for Huntington disease: a randomized controlled trial. Neurology. 2017;89(6):594–601.
83. Jiang Z, Liu D, Li T, Gai C, Xin D, Zhao Y, et al. Hydrogen sulfide reduces oxidative stress in Huntington's disease via Nrf2. Neural Regen Res. 2025;20(6):1776–88.
84. Ibrahim WW, Abdel Rasheed NO. Diapocynin neuroprotective effects in 3-nitropropionic acid Huntington's disease model in rats: emphasis on Sirt1/Nrf2 signaling pathway. Inflammopharmacology. 2022;30(5):1745–58.
85. Liu Y, Hettinger CL, Zhang D, Rezvani K, Wang X, Wang H. Sulforaphane enhances proteasomal and autophagic activities in mice and is a potential therapeutic reagent for Huntington's disease. J Neurochem. 2014;129(3):539–47.
86. Jin J, Albertz J, Guo Z, Peng Q, Rudow G, Troncoso JC, et al. Neuroprotective effects of PPAR-γ agonist rosiglitazone in N171-82Q mouse model of Huntington's disease. J Neurochem. 2013;125(3):410–9.
87. Chiang MC, Cheng YC, Nicol CJ, Lin KH, Yen CH, Chen SJ, Huang RN. Rosiglitazone activation of PPARγ-dependent signaling is neuroprotective in mutant huntingtin expressing cells. Exp Cell Res. 2015;338(2):183–93.
88. Chandra A, Sharma A, Calingasan NY, White JM, Shurubor Y, Yang XW, et al. Enhanced mitochondrial biogenesis ameliorates disease phenotype in a full-length mouse model of Huntington's disease. Hum Mol Genet. 2016;25(11):2269–82.

89. Sepers MD, Woodard CL, Ramandi D, Vecchiarelli HA, Hill MN, Raymond LA. Effect of chronic upregulation of endocannabinoid signaling in vivo with JZL184 on striatal synaptic plasticity and motor learning in YAC128 Huntington disease mice. bioRxiv. 2024:2024-09.
90. Palazuelos J, Aguado T, Pazos MR, Julien B, Carrasco C, Resel E, et al. Microglial CB2 cannabinoid receptors are neuroprotective in Huntington's disease excitotoxicity. Brain. 2009;132(11):3152–64.
91. El-Shamarka MES, El-Sahar AE, Saad MA, Assaf N, Sayed RH. Inosine attenuates 3-nitropropionic acid-induced Huntington's disease-like symptoms in rats via the activation of the A2AR/BDNF/TrKB/ERK/CREB signaling pathway. Life Sci. 2022;300:120569.
92. Pierzynowska K, Podlacha M, Gaffke L, Rintz E, Wiśniewska K, Cyske Z, Węgrzyn G. Correction of symptoms of Huntington disease by genistein through FOXO3-mediated autophagy stimulation. Autophagy. 2024;20(5):1159–82.
93. Saroj P, Bansal Y, Singh R, Akhtar A, Sodhi RK, Bishnoi M, et al. Neuroprotective effects of roflumilast against quinolinic acid-induced rat model of Huntington's disease through inhibition of NF-κB mediated neuroinflammatory markers and activation of cAMP/CREB/BDNF signaling pathway. Inflammopharmacology. 2021;29(2):499–511.
94. Li X, Wang CE, Huang S, Xu X, Li XJ, Li H, Li S. Inhibiting the ubiquitin–proteasome system leads to preferential accumulation of toxic N-terminal mutant huntingtin fragments. Hum Mol Genet. 2010;19(12):2445–55.
95. Díaz-Hernández M, Valera AG, Morán MA, Gómez-Ramos P, Alvarez-Castelao B, Castaño JG, et al. Inhibition of 26S proteasome activity by huntingtin filaments but not inclusion bodies isolated from mouse and human brain. J Neurochem. 2006;98(5):1585–96.
96. Seo H, Sonntag KC, Kim W, Cattaneo E, Isacson O. Proteasome activator enhances survival of Huntington's disease neuronal model cells. PLoS One. 2007;2(2):e238.
97. Jeon J, Kim W, Jang J, Isacson O, Seo H. Gene therapy by proteasome activator, PA28γ, improves motor coordination and proteasome function in Huntington's disease YAC128 mice. Neuroscience. 2016;324:20–8.
98. Bauer PO, Nukina N. Enhanced degradation of mutant huntingtin by rho kinase inhibition is mediated through activation of proteasome and macroautophagy. Autophagy. 2009;5(5):747–8.
99. Kleiger G, Mayor T. Perilous journey: a tour of the ubiquitin–proteasome system. Trends Cell Biol. 2014;24(6):352–9.
100. Toma-Fukai S, Shimizu T. Structural diversity of ubiquitin E3 ligase. Molecules. 2021;26(21):6682.
101. Yang Q, Zhao J, Chen D, Wang Y. E3 ubiquitin ligases: styles, structures and functions. Mol Biomed. 2021;2:23.
102. Honda R, Yasuda H. Activity of MDM2, a ubiquitin ligase, toward p53 or itself is dependent on the RING finger domain of the ligase. Oncogene. 2000;19(11):1473–6.
103. Tsoi H, Lau TC, Tsang SY, Lau KF, Chan HY. CAG expansion induces nucleolar stress in polyglutamine diseases. Proc Natl Acad Sci. 2012;109(33):13428–33.
104. Tsoi H, Chan HY. Expression of expanded CAG transcripts triggers nucleolar stress in Huntington's disease. Cerebellum. 2013;12(3):310–312.
105. Mansky RH, Greguske EA, Yu D, Zarate N, Intihar TA, Tsai W, et al. Tumor suppressor p53 regulates heat shock factor 1 protein degradation in Huntington's disease. Cell Rep. 2023;42(3):112198.
106. Illuzzi JL, Vickers CA, Kmiec EB. Modifications of p53 and the DNA damage response in cells expressing mutant form of the protein huntingtin. J Mol Neurosci. 2011;45(2):256–68.
107. Kook S, Zhan X, Thibeault K, Ahmed MR, Gurevich VV, Gurevich EV. Mdm2 enhances ligase activity of parkin and facilitates mitophagy. Sci Rep. 2020;10(1):5028.
108. Franco-Iborra S, Plaza-Zabala A, Montpeyo M, Sebastian D, Vila M, Martinez-Vicente M. Mutant HTT (huntingtin) impairs mitophagy in a cellular model of Huntington disease. Autophagy. 2021;17(3):672–89.
109. Yu P, Chen Y, Tagle DA, Cai T. PJA1, encoding a RING-H2 finger ubiquitin ligase, is a novel human X chromosome gene abundantly expressed in brain. Genomics. 2002;79(6):869–74.

110. Ghosh B, Karmakar S, Prasad M, Mandal AK. Praja1 ubiquitin ligase facilitates degradation of polyglutamine proteins and suppresses polyglutamine-mediated toxicity. Mol Biol Cell. 2021;32(17):1579–93.
111. Watabe K, Niida-Kawaguchi M, Tada M, Kato Y, Murata M, Tanji K, et al. Praja1 RING-finger E3 ubiquitin ligase is a common suppressor of neurodegenerative disease-associated protein aggregation. Neuropathology. 2022;42(6):488–504.
112. Watabe K, Kato Y, Sakuma M, Murata M, Niida-Kawaguchi M, Takemura T, et al. Praja1 RING-finger E3 ubiquitin ligase suppresses neuronal cytoplasmic TDP-43 aggregate formation. Neuropathology. 2020;40(6):570–86.
113. Yang H, Liu C, Zhong Y, Luo S, Monteiro MJ, Fang S. Huntingtin interacts with the cue domain of gp78 and inhibits gp78 binding to ubiquitin and p97/VCP. PLoS One. 2010;5(1):e8905.
114. Maity S, Komal P, Kumar V, Saxena A, Tungekar A, Chandrasekar V. Impact of ER stress and ER-mitochondrial crosstalk in Huntington's disease. Int J Mol Sci. 2022;23(2):780.
115. Fu M, St-Pierre P, Shankar J, Wang PT, Joshi B, Nabi IR. Regulation of mitophagy by the Gp78 E3 ubiquitin ligase. Mol Biol Cell. 2013;24(8):1153–62.
116. Alan P, Vandevoorde KR, Joshi B, Cardoen B, Gao G, Mohammadzadeh Y, et al. Basal Gp78-dependent mitophagy promotes mitochondrial health and limits mitochondrial ROS. Cell Mol Life Sci. 2022;79(11):565.
117. Kikkert M, Doolman R, Dai M, Avner R, Hassink G, Van Voorden S, et al. Human HRD1 is an E3 ubiquitin ligase involved in degradation of proteins from the endoplasmic reticulum. J Biol Chem. 2004;279(5):3525–34.
118. Kaneko M, Ishiguro M, Niinuma Y, Uesugi M, Nomura Y. Human HRD1 protects against ER stress-induced apoptosis through ER-associated degradation. FEBS Lett. 2002;532(1–2):147–52.
119. Yang H, Zhong X, Ballar P, Luo S, Shen Y, Rubinsztein DC, et al. Ubiquitin ligase Hrd1 enhances the degradation and suppresses the toxicity of polyglutamine-expanded huntingtin. Exp Cell Res. 2007;313(3):538–50.
120. Omura T, Kaneko M, Tabei N, Okuma Y, Nomura Y. Immunohistochemical localization of a ubiquitin ligase HRD1 in murine brain. J Neurosci Res. 2008;86(7):1577–87.
121. Lu H, Hallstrom TC. The nuclear protein UHRF2 is a direct target of the transcription factor E2F1 in the induction of apoptosis. J Biol Chem. 2013;288(33):23833–43.
122. Iwata A, Nagashima Y, Matsumoto L, Suzuki T, Yamanaka T, Date H, et al. Intranuclear degradation of polyglutamine aggregates by the ubiquitin-proteasome system. J Biol Chem. 2009;284(15):9796–803.
123. Saudou F, Finkbeiner S, Devys D, Greenberg ME. Huntingtin acts in the nucleus to induce apoptosis but death does not correlate with the formation of intranuclear inclusions. Cell. 1998;95(1):55–66.
124. Kegel KB, Meloni AR, Yi Y, Kim YJ, Doyle E, Cuiffo BG, et al. Huntingtin is present in the nucleus, interacts with the transcriptional corepressor C-terminal binding protein, and represses transcription. J Biol Chem. 2002;277(9):7466–76.
125. Engeholm M, Sekler J, Schöndorf DC, Arora V, Schittenhelm J, Biskup S, et al. A novel mutation in LRSAM1 causes axonal Charcot-Marie-tooth disease with dominant inheritance. BMC Neurol. 2014;14:118.
126. Mishra R, Upadhyay A, Prajapati VK, Dhiman R, Poluri KM, Jana NR, Mishra A. LRSAM1 E3 ubiquitin ligase: molecular neurobiological perspectives linked with brain diseases. Cell Mol Life Sci. 2019;76(11):2093–110.
127. Qin JY, Zhou ZL, Zhou YQ, Chen Y, Yang XY, Zhang HQ, Xu ZC. Double gene mutations of LRSAM1 and REEP1 and a new REEP1 mutation site found in a patient with amyotrophic lateral sclerosis with subjective paresthesia: a case report. Ibrain. 2023;11(3):393–399.
128. Guernsey DL, Jiang H, Bedard K, Evans SC, Ferguson M, Matsuoka M, et al. Mutation in the gene encoding ubiquitin ligase LRSAM1 in patients with Charcot-Marie-tooth disease. PLoS Genet. 2010;6(8):e1001081.

129. Dubey AR, Mishra R, Sundaria N, Jagtap YA, Kumar P, Kinger S, et al. Resveratrol promotes LRSAM1 E3 ubiquitin ligase-dependent degradation of misfolded proteins linked with neurodegeneration. Cell Physiol Biochem. 2022;56(5):530–45.
130. Tang B, Seredenina T, Coppola G, Kuhn A, Geschwind DH, Luthi-Carter R, Thomas EA. Gene expression profiling of R6/2 transgenic mice with different CAG repeat lengths reveals genes associated with disease onset and progression in Huntington's disease. Neurobiol Dis. 2011;42(3):459–67.
131. Mishra R, Amanullah A, Upadhyay A, Dhiman R, Dubey AR, Singh S, et al. Ubiquitin ligase LRSAM1 suppresses neurodegenerative diseases linked aberrant proteins induced cell death. Int J Biochem Cell Biol. 2020;120:105697.
132. Walsh MC, Lee J, Choi Y. Tumor necrosis factor receptor-associated factor 6 (TRAF 6) regulation of development, function, and homeostasis of the immune system. Immunol Rev. 2015;266(1):72–92.
133. Dou Y, Tian X, Zhang J, Wang Z, Chen G. Roles of TRAF6 in central nervous system. Curr Neuropharmacol. 2018;16(9):1306–13.
134. Zucchelli S, Marcuzzi F, Codrich M, Agostoni E, Vilotti S, Biagioli M, et al. Tumor necrosis factor receptor-associated factor 6 (TRAF6) associates with huntingtin protein and promotes its atypical ubiquitination to enhance aggregate formation. J Biol Chem. 2011;286(28):25108–17.
135. Zhao Y, Li T, Zhang L, Yang J, Zhao F, Wang Y, et al. TRAF6 promotes spinal microglial M1 polarization to aggravate neuropathic pain by activating the c-JUN/NF-kB signaling pathway. Cell Biol Toxicol. 2024;40(1):54.
136. Li M, Zhang D, Ge X, Zhu X, Zhou Y, Zhang Y, et al. TRAF6-p38/JNK-ATF2 axis promotes microglial inflammatory activation. Exp Cell Res. 2019;376(2):133–48.
137. Song F, Zeng K, Liao L, Yu Q, Tu P, Wang X. Schizandrin A inhibits microglia-mediated neuroninflammation through inhibiting TRAF6-NF-κB and Jak2-Stat3 signaling pathways. PLoS One. 2016;11(2):e0149991.
138. Liu M, Peng Y, Che Y, Zhou M, Bai Y, Tang W, et al. MiR-146b-5p/TRAF6 axis is essential for Ginkgo biloba L. extract GBE to attenuate LPS-induced neuroinflammation. Front Pharmacol. 2022;13:978587.
139. Wang Z, Zhang YH, Guo C, Gao HL, Zhong ML, Huang TT, et al. Tetrathiomolybdate treatment leads to the suppression of inflammatory responses through the TRAF6/NFκB pathway in LPS-stimulated BV-2 microglia. Front Aging Neurosci. 2018;10:9.
140. Duda DM, Scott DC, Calabrese MF, Zimmerman ES, Zheng N, Schulman BA. Structural regulation of cullin-RING ubiquitin ligase complexes. Curr Opin Struct Biol. 2011;21(2):257–64.
141. Genschik P, Sumara I, Lechner E. The emerging family of CULLIN3-RING ubiquitin ligases (CRL3s): cellular functions and disease implications. EMBO J. 2013;32(17):2307–20.
142. Bhutani S, Das A, Maheshwari M, Lakhotia SC, Jana NR. Dysregulation of core components of SCF complex in poly-glutamine disorders. Cell Death Dis. 2012;3(11):e428.
143. Chen ZS, Wong AKY, Cheng TC, Koon AC, Chan HYE. FipoQ/FBXO33, a Cullin-1-based ubiquitin ligase complex component modulates ubiquitination and solubility of polyglutamine disease protein. J Neurochem. 2019;149(6):781–98.
144. Godin JD, Poizat G, Hickey MA, Maschat F, Humbert S. Mutant huntingtin-impaired degradation of β-catenin causes neurotoxicity in Huntington's disease. EMBO J. 2010;29(14):2433–45.
145. Gomez Pastor R, Burchfiel ET, Neef DW, Jaeger AM, Cabiscol E, McKinstry SU, et al. Abnormal degradation of the neuronal stress-protective transcription factor HSF1 in Huntington's disease. Nat Commun. 2017;8(1):14405.
146. Ko YU, Kim C, Lee J, Kim D, Kim Y, Yun N, Oh YJ. Site-specific phosphorylation of Fbxw7 by Cdk5/p25 and its resulting decreased stability are linked to glutamate-induced excitotoxicity. Cell Death Dis. 2019;10(8):579.
147. Canning P, Sorrell FJ, Bullock AN. Structural basis of Keap1 interactions with Nrf2. Free Radic Biol Med. 2015;88:101–7.

148. Silva-Islas C, Santana RA, Colín-González AL, Maldonado PD. Nrf2 activation, an innovative therapeutic alternative in cerebral ischemia. In: Advances in the preclinical study of ischemic stroke; 2012. p. 347–78.
149. Jang M, Cho IH. Sulforaphane ameliorates 3-nitropropionic acid-induced striatal toxicity by activating the Keap1-Nrf2-ARE pathway and inhibiting the MAPKs and NF-κB pathways. Mol Neurobiol. 2016;53(4):2619–35.
150. Brokowska J, Hać A, Węgrzyn G, Herman-Antosiewicz A. L12 Sulforaphane reduces the level of exogenous mutated huntingtin protein in normal human fibroblasts. J Neurol Neurosurg Psychiatry. 2016;87:A94.1.
151. Moretti D, Tambone S, Cerretani M, Fezzardi P, Missineo A, Sherman LT, et al. NRF2 activation by reversible KEAP1 binding induces the antioxidant response in primary neurons and astrocytes of a Huntington's disease mouse model. Free Radic Biol Med. 2021;162:243–54.
152. Unni S, Deshmukh P, Krishnappa G, Kommu P, Padmanabhan B. Structural insights into the multiple binding modes of Dimethyl Fumarate (DMF) and its analogs to the Kelch domain of Keap1. FEBS J. 2021;288(5):1599–613.
153. Hassab LY, Abbas SS, Mohammed RA, Abdallah DM. Dimethyl fumarate abrogates striatal endoplasmic reticulum stress in experimentally induced late-stage Huntington's disease: focus on the IRE1α/JNK and PERK/CHOP trajectories. Front Pharmacol. 2023;14:1133863.
154. Ellrichmann G, Petrasch-Parwez E, Lee DH, Reick C, Arning L, Saft C, et al. Efficacy of fumaric acid esters in the R6/2 and YAC128 models of Huntington's disease. PLoS One. 2011;6(1):e16172.
155. Düring J, Wolter M, Toplak JJ, Torres C, Dybkov O, Fokkens TJ, et al. Structural mechanisms of autoinhibition and substrate recognition by the ubiquitin ligase HACE1. Nat Struct Mol Biol. 2024;31(2):364–77.
156. Rotblat B, Southwell AL, Ehrnhoefer DE, Skotte NH, Metzler M, Franciosi S, Leprivier G, Somasekharan SP, Barokas A, Deng Y, Tang T, Mathers J, Cetinbas N, Daugaard M, Kwok B, Li L, Carnie CJ, Fink D, Nitsch R, Galpin JD, Ahern CA, Melino G, Penninger JM, Hayden MR, Sorensen PH. HACE1 reduces oxidative stress and mutant Huntingtin toxicity by promoting the NRF2 response. Proc Natl Acad Sci USA. 2014;111(8):3032–7.
157. Daugaard M, Nitsch R, Razaghi B, McDonald L, Jarrar A, Torrino S, et al. Hace1 controls ROS generation of vertebrate Rac1-dependent NADPH oxidase complexes. Nat Commun. 2013;4(1):2180.
158. Maldonado PD, Molina-Jijón E, Villeda-Hernández J, Galván-Arzate S, Santamaría A, Pedraza-Chaverrí J. NAD (P) H oxidase contributes to neurotoxicity in an excitotoxic/prooxidant model of Huntington's disease in rats: protective role of apocynin. J Neurosci Res. 2010;88(3):620–9.
159. Ehrnhoefer DE, Southwell AL, Sivasubramanian M, Qiu X, Villanueva EB, Xie Y, et al. HACE1 is essential for astrocyte mitochondrial function and influences Huntington disease phenotypes in vivo. Hum Mol Genet. 2018;27(2):239–53.
160. Chen Y, Cao P. Circ-Bnc2 alleviates neuroinflammation in LPS-stimulated microglial cells to inhibit neuron cell apoptosis through regulating miR-497a-5p/HECTD1 axis. Brain Behav. 2023;13(5):e2935.
161. Tang Y, Zhou M, Huang R, Shen L, Yang L, Zhou Z, et al. Involvement of HECTD1 in LPS-induced astrocyte activation via σ-1R-JNK/p38-FOXJ2 axis. Cell Biosci. 2021;11(1):62.
162. Zeng X, Dong X, Xiao Q, Yao J. Vitamin C inhibits ubiquitination of glutamate transporter 1 (GLT-1) in astrocytes by downregulating HECTD1. ACS Chem Neurosci. 2022;13(5):676–87.
163. Estrada-Sánchez AM, Rebec GV. Corticostriatal dysfunction and glutamate transporter 1 (GLT1) in Huntington's disease: interactions between neurons and astrocytes. Basal Ganglia. 2012;2(2):57–66.
164. Nash KR, Jinwal UK, Bhat KM. UBE3A: bridging the gap between neurodevelopment, neural function, and neurodegenerative woes. J Alzheimer's Dis. 2024;102(1):3–10.
165. Bhat KP, Yan S, Wang CE, Li S, Li XJ. Differential ubiquitination and degradation of huntingtin fragments modulated by ubiquitin-protein ligase E3A. Proc Natl Acad Sci USA. 2014;111(15):5706–11.

166. Maheshwari M, Shekhar S, Singh BK, Jamal I, Vatsa N, Kumar V, et al. Deficiency of Ube3a in Huntington's disease mice brain increases aggregate load and accelerates disease pathology. Hum Mol Genet. 2014;23(23):6235–45.
167. Maheshwari M, Samanta A, Godavarthi SK, Mukherjee R, Jana NR. Dysfunction of the ubiquitin ligase Ube3a may be associated with synaptic pathophysiology in a mouse model of Huntington disease. J Biol Chem. 2012;287(35):29949–57.
168. Greer PL, Hanayama R, Bloodgood BL, Mardinly AR, Lipton DM, Flavell SW, et al. The Angelman syndrome protein Ube3A regulates synapse development by ubiquitinating arc. Cell. 2010;140(5):704–16.
169. Furusawa K, Ishii K, Tsuji M, Tokumitsu N, Hasegawa E, Emoto K. Presynaptic Ube3a E3 ligase promotes synapse elimination through down-regulation of BMP signaling. Science. 2023;381(6663):1197–205.
170. Lopez SJ, Laufer BI, Beitnere U, Berg EL, Silverman JL, O'Geen H, et al. Imprinting effects of UBE3A loss on synaptic gene networks and Wnt signaling pathways. Hum Mol Genet. 2019;28(22):3842–52.
171. Wallace ML, Burette AC, Weinberg RJ, Philpot BD. Maternal loss of Ube3a produces an excitatory/inhibitory imbalance through neuron type-specific synaptic defects. Neuron. 2012;74(5):793–800.
172. Cheng Z. FoxO transcription factors in mitochondrial homeostasis. Biochem J. 2022;479(4):525–36.
173. Panov J, Simchi L, Feuermann Y, Kaphzan H. Bioinformatics analyses of the transcriptome reveal Ube3a-dependent effects on mitochondrial-related pathways. Int J Mol Sci. 2020;21(11):4156.
174. Koyuncu S, Saez I, Lee HJ, Gutierrez-Garcia R, Pokrzywa W, Fatima A, et al. The ubiquitin ligase UBR5 suppresses proteostasis collapse in pluripotent stem cells from Huntington's disease patients. Nat Commun. 2018;9(1):2886.
175. Wang F, He Q, Zhan W, Yu Z, Finkin-Groner E, Ma X, et al. Structure of the human UBR5 E3 ubiquitin ligase. Structure. 2023;31(5):541–52.
176. Zhao H, Cao N, Liu Q, Zhang Y, Jin R, Lai H, et al. Inhibition of the E3 ligase UBR5 stabilizes TERT and protects vascular organoids from oxidative stress. J Transl Med. 2024;22(1):1080.
177. Li CG, Mahon C, Sweeney NM, Verschueren E, Kantamani V, Li D, Hennigs JK, Marciano DP, Diebold I, Abu-Halawa O, Elliott M, Sa S, Guo F, Wang L, Cao A, Guignabert C, Sollier J, Nickel NP, Kaschwich M, Cimprich KA, Rabinovitch M. PPARγ Interaction with UBR5/ATMIN Promotes DNA Repair to Maintain Endothelial Homeostasis. Cell Rep. 2019;26(5):1333–1343.e7.
178. Haouari S, Vourc'h P, Jeanne M, Marouillat S, Veyrat-Durebex C, Lanznaster D, et al. The roles of NEDD4 subfamily of HECT E3 ubiquitin ligases in neurodevelopment and neurodegeneration. Int J Mol Sci. 2022;23(7):3882.
179. Lin L, Jin Z, Tan H, Xu Q, Peng T, Li H. Atypical ubiquitination by E3 ligase WWP1 inhibits the proteasome-mediated degradation of mutant huntingtin. Brain Res. 2016;1643:103–12.
180. Peters TW, Nelson CS, Gerencser AA, Dumas KJ, Tavshanjian B, Chang KC, Lithgow GJ, Hughes RE. Natural Genetic Variation in Yeast Reveals That NEDD4 Is a Conserved Modifier of Mutant Polyglutamine Aggregation. G3 (Bethesda). 2018;8(11):3421–3431.
181. Jiang H, Chiang CY, Chen Z, Nathan S, D'Agostino G, Paulo JA, et al. Enzymatic analysis of WWP2 E3 ubiquitin ligase using protein microarrays identifies autophagy-related substrates. J Biol Chem. 2022;298(5):101854.
182. Okita S, Morigaki R, Koizumi H, Kaji R, Nagahiro S, Goto S. Cell type-specific localization of optineurin in the striatal neurons of mice: implications for neuronal vulnerability in Huntington's disease. Neuroscience. 2012;202:363–70.
183. Moharir SC, Raghawan AK, Ramaswamy R, Swarup G. Autophagy-independent cytoprotection by optineurin from toxicity of aggregates formed by mutant huntingtin and mutant ataxin-3. J Biochem. 2022;171(5):555–65.
184. Donovan P, Poronnik P. Nedd4 and Nedd 4-2: ubiquitin ligases at work in the neuron. Int J Biochem Cell Biol. 2013;45(3):706–10.

185. Kwak YD, Wang B, Li JJ, Wang R, Deng Q, Diao S, et al. Upregulation of the E3 ligase NEDD4-1 by oxidative stress degrades IGF-1 receptor protein in neurodegeneration. J Neurosci. 2012;32(32):10971–81.
186. Naia L, Ribeiro M, Rodrigues J, Duarte AI, Lopes C, Rosenstock TR, et al. Insulin and IGF-1 regularize energy metabolites in neural cells expressing full-length mutant huntingtin. Neuropeptides. 2016;58:73–81.
187. Naia L, Ferreira IL, Cunha-Oliveira T, Duarte AI, Ribeiro M, Rosenstock TR, et al. Activation of IGF-1 and insulin signaling pathways ameliorate mitochondrial function and energy metabolism in Huntington's disease human lymphoblasts. Mol Neurobiol. 2015;51(1):331–48.
188. Murata S, Minami Y, Minami M, Chiba T, Tanaka K. CHIP is a chaperone-dependent E3 ligase that ubiquitylates unfolded protein. EMBO Rep. 2001;2(12):1133.
189. Williams AJ, Knutson TM, Gould VFC, Paulson HL. In vivo suppression of polyglutamine neurotoxicity by C-terminus of Hsp70-interacting protein (CHIP) supports an aggregation model of pathogenesis. Neurobiol Dis. 2009;33(3):342–53.
190. Miller VM, Nelson RF, Gouvion CM, Williams A, Rodriguez-Lebron E, Harper SQ, et al. CHIP suppresses polyglutamine aggregation and toxicity in vitro and in vivo. J Neurosci. 2005;25(40):9152–61.
191. Hahn KR, Kwon HJ, Yoon YS, Kim DW, Hwang IK. CHIP ameliorates neuronal damage in H2O2-induced oxidative stress in HT22 cells and gerbil ischemia. Sci Rep. 2022;12(1):20659.
192. Oikawa D, Sato Y, Ito H, Tokunaga F. Linear ubiquitin code: its writer, erasers, decoders, inhibitors, and implications in disorders. Int J Mol Sci. 2020;21(9):3381.
193. van Well EM, Bader V, Patra M, Sánchez-Vicente A, Meschede J, Furthmann N, et al. A protein quality control pathway regulated by linear ubiquitination. EMBO J. 2019;38(9):e100730.
194. Li SH, Cheng AL, Zhou H, Lam S, Rao M, Li H, Li XJ. Interaction of Huntington disease protein with transcriptional activator Sp1. Mol Cell Biol. 2002;22(5):1277–87.
195. Dong K, Wei R, Jin T, Zhang M, Shen J, Xiang H, et al. HOIP modulates the stability of GPx4 by linear ubiquitination. Proc Natl Acad Sci U S A. 2022;119(44):e2214227119.
196. Dar NJ, John U, Bano N, Khan S, Bhat SA. Oxytosis/ferroptosis in neurodegeneration: the underlying role of master regulator glutathione peroxidase 4 (GPX4). Mol Neurobiol. 2024;61(3):1507–26.
197. Xu H, Qin W, Hu X, Mu S, Zhu J, Lu W, Luo Y. Lentivirus-mediated overexpression of OTULIN ameliorates microglia activation and neuroinflammation by depressing the activation of the NF-κB signaling pathway in cerebral ischemia/reperfusion rats. J Neuroinflammation. 2018;15(1):3.
198. Arkinson C, Walden H. Parkin function in Parkinson's disease. Science. 2018;360(6386):267–8.
199. Aladdin A, Király R, Boto P, Regdon Z, Tar K. Juvenile Huntington's disease skin fibroblasts respond with elevated parkin level and increased proteasome activity as a potential mechanism to counterbalance the pathological consequences of mutant huntingtin protein. Int J Mol Sci. 2019;20(21):5338.
200. Khalil B, El Fissi N, Aouane A, Cabirol-Pol MJ, Rival T, Liévens JC. PINK1-induced mitophagy promotes neuroprotection in Huntington's disease. Cell Death Dis. 2015;6(1):e1617.
201. Tsai YC, Fishman PS, Thakor NV, Oyler GA. Parkin facilitates the elimination of expanded polyglutamine proteins and leads to preservation of proteasome function. J Biol Chem. 2003;278(24):22044–55.
202. Narendra D, Tanaka A, Suen DF, Youle RJ. Parkin is recruited selectively to impaired mitochondria and promotes their autophagy. J Cell Biol. 2008;183(5):795–803.
203. Wang Y, Nartiss Y, Steipe B, McQuibban GA, Kim PK. ROS-induced mitochondrial depolarization initiates PARK2/PARKIN-dependent mitochondrial degradation by autophagy. Autophagy. 2012;8(10):1462–76.
204. Panov AV, Burke JR, Strittmatter WJ, Greenamyre JT. In vitro effects of polyglutamine tracts on Ca2+-dependent depolarization of rat and human mitochondria: relevance to Huntington's disease. Arch Biochem Biophys. 2003;410(1):1–6.

205. Dionísio PEA, Oliveira SR, Amaral JSJD, Rodrigues CMP. Loss of microglial parkin inhibits necroptosis and contributes to neuroinflammation. Mol Neurobiol. 2019;56(4):2990–3004.
206. Wang A, Zhong G, Ying M, Fang Z, Chen Y, Wang H, et al. Inhibition of NLRP3 inflammasome ameliorates LPS-induced neuroinflammatory injury in mice via PINK1/Parkin pathway. Neuropharmacology. 2024;257:110063.
207. Noureldeen ME, Shahin NN, Amin HAA, El-Sawalhi MM, Ghaiad HR. Parthenolide ameliorates 3-nitropropionic acid-induced Huntington's disease-like aberrations via modulating NLRP3 inflammasome, reducing microglial activation and inducing astrocyte shifting. Mol Med. 2024;30(1):158.
208. Staropoli JF, McDermott C, Martinat C, Schulman B, Demireva E, Abeliovich A. Parkin is a component of an SCF-like ubiquitin ligase complex and protects postmitotic neurons from kainate excitotoxicity. Neuron. 2003;37(5):735–49.
209. Chhangani D, Upadhyay A, Amanullah A, Joshi V, Mishra A. Ubiquitin ligase ITCH recruitment suppresses the aggregation and cellular toxicity of cytoplasmic misfolded proteins. Sci Rep. 2014;4(1):5077.
210. Latoszek E, Wiweger M, Ludwiczak J, Dunin-Horkawicz S, Kuznicki J, Czeredys M. Siah-1-interacting protein regulates mutated huntingtin protein aggregation in Huntington's disease models. Cell Biosci. 2022;12(1):34.
211. Qin Y, Chen L, Zhu W, Song J, Lin J, Li Y, et al. TRIM37 is a primate-specific E3 ligase for huntingtin and accounts for the striatal degeneration in Huntington's disease. Sci Adv. 2024;10(20):eadl2036.
212. Luo H, Cao L, Liang X, Du A, Peng T, Li H. Herp promotes degradation of mutant huntingtin: involvement of the proteasome and molecular chaperones. Mol Neurobiol. 2018;55(10):7652–68.
213. Cho K, Kim S, Choi SH. Suppressor of cytokine signaling 2 is induced in Huntington's disease and involved in autophagy. Biochem Biophys Res Commun. 2021;559:21–7.
214. Chhangani D, Nukina N, Kurosawa M, Amanullah A, Joshi V, Upadhyay A, Mishra A. Mahogunin ring finger 1 suppresses misfolded polyglutamine aggregation and cytotoxicity. Biochim Biophys Acta. 2014;1842(9):1472–84.
215. Pao PC, Huang NK, Liu YW, Yeh SH, Lin ST, Hsieh CP, et al. A novel RING finger protein, Znf179, modulates cell cycle exit and neuronal differentiation of P19 embryonal carcinoma cells. Cell Death Differ. 2011;18(11):1791–804.
216. Su TC, Lin SH, Lee PT, Yeh SH, Hsieh TH, Chou SY, et al. The sigma-1 receptor-zinc finger protein 179 pathway protects against hydrogen peroxide-induced cell injury. Neuropharmacology. 2016;105:1–9.
217. Wu CC, Lee PT, Kao TJ, Chou SY, Su RY, Lee YC, et al. Upregulation of Znf179 acetylation by SAHA protects cells against oxidative stress. Redox Biol. 2018;19:74–80.
218. Dong L, Xu M, Li Y, Xu W, Wu C, Zheng H, et al. SMURF1 attenuates endoplasmic reticulum stress by promoting the degradation of KEAP1 to activate NRF2 antioxidant pathway. Cell Death Dis. 2023;14(6):361.
219. Ndoja A, Reja R, Lee SH, Webster JD, Ngu H, Rose CM, et al. Ubiquitin ligase COP1 suppresses neuroinflammation by degrading c/EBPβ in microglia. Cell. 2020;182(5):1156–69.
220. Zheng SZ, Zhang XW, Song XM, Yang Z, Yao L, Tu PF, Zeng KW. Epoxymicheliolide directly targets histone H2B to inhibit neuroinflammation via recruiting E3 ligase RNF20. Pharmacol Res. 2022;177:106093.
221. Kurtishi A, Rosen B, Patil KS, Alves GW, Møller SG. Cellular proteostasis in neurodegeneration. Mol Neurobiol. 2019;56(5):3676–89.
222. Monsalvo-Maraver LA, Ovalle-Noguez EA, Nava-Osorio J, Maya-López M, Rangel-López E, Túnez I, et al. Interactions between the ubiquitin–proteasome system, Nrf2, and the Cannabinoidome as protective strategies to combat neurodegeneration: review on experimental evidence. Neurotox Res. 2024;42(2):18.

E3 Ubiquitin Ligases in Parkinson's Pathology

12

Sehar Usman, Surendar Ellappan, and Amal Chandra Mondal

12.1 Introduction

12.1.1 Prevalence and Characteristics of Parkinson's Disease (PD)

Parkinson's is a slowly progressive neurodegenerative disorder that affects over 8.5 million people worldwide, approximately 0.3% of the total population, with the prevalence of the disease 315 per 1,00,000 individuals globally [1, 2]. The prevalence of the disease increases with age. It is the second most common neurodegenerative disorder whose pervasiveness is expected to double in the coming three decades.

The disease is mainly characterized by the loss of the dopaminergic neurons, prominently in substantia niagra pars compacta (SNpc), a part of basal ganglia [3]. Basal ganglia are a collection of cell bodies (masses of grey matter) within the white matter of the cerebral hemisphere. It comprises four structures: striatum (made up of caudate nucleus and putamen), globus pallidus, substantia niagra, and subthalamic nuclei. Substantia niagra pars compacta (SNpc) contains neurons with a conspicuous amount of neuromelanin and thus appears particularly dark. SNpc is densely packed with the cell bodies of dopaminergic neurons. The axons or projections from the SNpc project to the striatum, primarily to the putamen. In PD, the dopaminergic neurons' axons in the striatum degenerate before their cell bodies. At the onset of PD symptoms, almost 60% of SNpc dopaminergic neurons and nearly 80% of dopamine neurotransmitters from putamen have already been lost [4].

The PD patient brain's autopsy reveals Lewy bodies in the affected neurons. Lewy bodies consist of the aggregation of various abnormal proteins, in which the

S. Usman · S. Ellappan · A. C. Mondal (✉)
Laboratory of Cellular and Molecular Neurobiology, School of Life Sciences, Jawaharlal Nehru University, New Delhi, India
e-mail: acmondal@mail.jnu.ac.in

S. Singh (ed.), *Ubiquitin Proteasome System in CNS Neurodegenerative Diseases*, https://doi.org/10.1007/978-981-95-6099-8_12

major contributor is α-synuclein, a cytosolic protein. These Lewy bodies are toxic to the cells and are considered the disease's pathological hallmark. Their presence in neurons accounts for the neurodegeneration and corresponds to the symptoms of the disease.

12.1.2 Symptoms Associated with PD

The substantia niagra plays a critical role in controlling motor movement of the body [5]. The death of dopaminergic neurons leads to a movement disorder. The classic motor symptoms of the disease are resting tremor, bradykinesia, rigidity, slurred speech, and postural deformities & instability [6]. Dysphagia, sialorrhea, micrographia, shuffling gait, respiratory disturbances, and ophthalmological abnormalities are secondary motor symptoms associated with PD [6]. In PD, the degeneration of neurons also occurs in the brain area other than SNpc, which accounts for the non-motor symptoms associated with the disease. These non-motor symptoms appear several years before the cardinal motor symptoms of the disease [7]. Depression, anxiety, visual hallucination, sleep disorder, sphincter dysfunction (urinary continence, nocturia), constipation, erectile dysfunction, cognitive decline, and memory loss [6, 7] are examples of non-motor symptoms associated with the disease.

12.1.3 Diagnosis of the PD

Currently, there is no standard test for the diagnosis of PD. It can be done mainly based on the clinical observation of the disease's cardinal, exclusionary, and associated symptoms. The diagnosis of PD is divided into two periods, i.e., the prodromal and the diagnostic period [8]. The Braak Hypothesis easily explains these periods. According to this hypothesis, the neuropathological progression of PD follows a specific pattern. Lewy bodies appear first in the lower brain region, including the medulla and olfactory bulb (stages 1 and 2), and over time PD progresses toward the higher brain region, such as substantia nigra pars compacta, midbrain, and basal forebrain (stages 3 and 4). The prodromal period is associated with stages 1 and 2 of the PD, which accounts for the onset of non-movement disorder, including rapid eye movement disorder, hyposmia, constipation, decreased sense of smell, urinary dysfunction, and depression. Stage 3 is associated with the degeneration of neurons in SNpc, which leads to the onset of classical Parkinson's motor disorder; PD is typically diagnosed at this stage. In advanced Parkinson's disease (stage 4 and above), the pathology progresses to the cerebral cortices and leads to cognitive impairment and hallucinations.

Once the physical examination of a person is done, the response to levodopa or other dopamine agonists is checked to confirm the PD. A person affected with PD must show at least 70–100% response to levodopa for 5 years or more. A person's family, medical history, and a DAT scan (to check dopamine transporter activity) can be a promising tool for the confirmation of the disease, too.

12.1.4 Genetic Changes Contribute to PD

Most PD cases are idiopathic (the exact cause of the PD is unknown). Still, factors like age, gender, genetic changes, environmental factors, compromised ubiquitin-proteasome system, increased oxidative stress, mitochondrial dysfunction, microglial activation, dopamine shortage, and aging increase the risk of PD [6]. Almost 15% of the cases belong to familial PD [9]. The mutation in genes, including SNCA, LRRK2, PRKN, PARK7, PINK1, and VPS35, is involved in the heritable form of the disease. Out of 15%, 5–10% of the disease cases are caused by an alteration in the sequence of a single gene (monogenic PD) [10].

12.1.4.1 Role of *SNCA* Gene in Etiology of PD

The *SNCA (*synuclein alpha) gene consists of 5 exons and is located on the long arm of chromosome 4 (4q22.1). It codes for the α-synuclein, a cytosolic protein expressed mainly in the brain, including substantia niagra, hippocampus, thalamus, and cerebellum [11]. α-synuclein is highly expressed in the pre-synaptic terminal near the synaptic vesicle and is involved in the docking and fusion of the vesicle (vesicle containing dopamine) to the active zone. It is also the main component of the Lewy bodies. α-synuclein is 140 amino acids long and consists of three regions: N-terminal amphipathic helical region (1–60 amino acids), a central hydrophobic non-amyloid beta component (NAC) region (61–95 amino acids), and C-terminus acidic region (97–140 amino acids).

SNCA duplication, triplication, and eight missense mutations (A30G, A30P, E46K, H50Q, G51D, A53E, A53T, and A53V) are reported to be associated with the autosomal dominant PD. [10] These missense variants are located in the N-terminal region of the protein [12]. Missense mutation alters the stability of the protein, and thus, the non-native conformation of the protein is prone to aggregation. Lewy bodies consist of deposits of many cellular proteins, majorly oligomers of misfolded α-synuclein protein. The multiplication and mutation of the α-synuclein are responsible for the reduced dopamine synthesis and release. α-synuclein acts as a negative regulator of the tyrosine hydroxylase (TH), an enzyme involved in dopamine synthesis [11]. α-synuclein inhibits the TH by activating the PP2A (protein phosphatase 2), an enzyme responsible for the dephosphorylation of the TH at Ser 40.

The oligomer of the α-synuclein binds to the N-terminal domain of the synaptobrevin-2, a SNARE complex protein, and prevents SNARE complex assembly [11]. Thus, it inhibits the fusion of vesicles containing dopamine with the membrane. As a result, dopamine degrades in the cytosol of the neuron into dihydroxyphenylacetaldehyde (DOPAL). DOPAL can form radicals and quinone species upon oxidation, which impairs mitochondrial structure and function. Reactive oxygen species (ROS) are responsible for DNA, protein, and phospholipid oxidation. The main target of ROS is mitochondria. ROS's compromised phospholipid of the mitochondria releases cytochrome c from the mitochondrial matrix. The cytochrome c is accountable for the caspase activation and, thus, apoptosis of the neuron. Also, the N-terminal domain of α-synuclein associated with respiratory chain complex I of mitochondria interferes with its function and leads to mitophagy [13]. The oligomers of α-synuclein also reduce the polymerization of tubulin, a cytoskeletal protein, into microtubules [11]. The Lewy bodies also lead to nuclear degradation. Thus, the ultimate result of Lewy bodies is low dopamine levels, mitochondrial dysfunction, depleted ATP, damaged DNA, and reduced axonal trafficking in the neurons.

12.1.4.2 Role of *LRRK2* Gene in PD

The leucine-rich repeat kinase 2 (*LRRK2*) gene is on the long arm of chromosome 12 (12q12) and consists of 51 exons. It codes for the multidomain, 2527 amino acids long LRRK2 protein. LRRK2 consists of seven domains, namely N-terminal armadillo (ARM), ankyrin (ANK), LRR domain, central Roc GTPase domain, COR domain, serine/threonine kinase domain, and C-terminal WD40 domain [14]. The center region of the LRRK2 constitutes the catalytic core of the protein, which includes two enzymatic domains (a GTPase and a kinase). LRRK2 protein is highly expressed in kidney, lung, and immune cells. It is moderately expressed in the brain, where the highest expression is observed in the dopamine-innervated areas, including the striatum, hippocampus, and cerebral cortex. LRRK2 protein is mainly present in the cytoplasm and localized to various membranous cell organelles. It interacts and modifies the microtubule organization and actin cytoskeleton to inhibit neurite growth [15]. It is also responsible for regulating autophagy and apoptosis. The downstream activator of the LRRK2 includes MAPKK (mitogen-activated protein kinase), such as MKK 3,6,4, and 7 [16]. MKK3 and MKK6 regulate the activity of p38 MAPKs, and MKK4 and MKK7 act as upstream regulators of JNK. P38 MAPKs and JNKs regulate various physiological processes, including apoptosis.

Seven-point mutations are reported in the LRRK2 protein, i.e., G2019S, I2020T, R1441C/G/H, Y1699C, and N1437H. LRRK2 mutations account for the autosomal dominant PD. G2019S mutation, where glycine 2019 is replaced by serine, is the prevalent one, which results in the constitutive LRRK2 kinase activity. This leads to the over-phosphorylation of the downstream MKKs. The G2019S mutation is reported to activate the MKK4-JNK-c-Jun mediated apoptotic pathway [17]. These seven mutations in the LRRK2 protein are located in different functional domains and are responsible for up to 5% of familial PD cases worldwide, with an average onset age of 55.2 years [18].

The *LRRK2* gene is expressed in the innate immune cells. The mutation in the *LRRK2* gene contributes to the microglia-mediated neuroinflammation in PD. Microglia are the resident immune cells of the CNS, which inspect the brain by extending and retracting their mobile processes. The primary function of the microglia is to recognize the antigen, redundant synapses, dead cells, and protein aggregates and help in their clearance [19]. The microglia in its inactive state consist of a rod-shaped soma and ramified processes. Microglia express various cell surface receptors and can sense the changing environment of the brain. One of the cell surface receptors of the microglia is the Toll-like receptor (TLR). α-synuclein interacts with the TLRs, mainly with TLR2 and TLR4, and induces microglia activation. The TLR2- and TLR4-mediated signaling activates the transcription factor nuclear factor kappa B (NF-κB) in microglia. The NF-κB translocates to the nucleus and acts as the transcription factor for producing proinflammatory cytokines such as interleukin (IL)-6, IL-1β, IL-18, and TNF-α. The burst of these cytokines is deleterious for the dopaminergic neurons, including increasing oxidative stress and enhancing apoptosis.

The activity of LRRK2 in microglia is upregulated after being induced by TLR2 or TLR4 stimulation (receptors of α-synuclein) [15]. It is reported that the expression of LRRK2 in microglia reduces the uptake of α-synuclein from the extracellular space [20]. Furthermore, LRRK2 also increases *SNCA* transcription by activating the MAPK/ERK cascade [21]. Thus, the increased level of α-synuclein and its oligomer interact more with the TLRs of microglia and contribute to the neuroinflammation that leads to neurodegeneration, a significant contributor to PD.

12.1.4.3 Role of *PRKN* Gene in PD

PRKN (Parkin RBR E3 ubiquitin ligase) gene, also known as *PARK2* gene, is located on the q arm of chromosome 6. The cytogenetic location of *PRKN* is 6q26. It codes for a protein called parkin. Parkin is a 465 amino acids-long protein that resides in the cytoplasm and adopts a compact structure like a coiled snake. It belongs to the family of RBR (Ring between ring) of E3 ubiquitin ligase and is responsible for tagging the substrate with ubiquitin, a 76 amino acid protein. Ubiquitination can signal autophagy, alter substrate location and activity, and tag the protein (damaged, misfolded, and excess protein) for degradation. The information regarding the ubiquitination process, types of ubiquitin ligases, families, and functions will be discussed in the next section.

Parkin is mainly expressed in the brain, kidney, and testis. It consists of 5 domains, including an N-terminus ubiquitin-like domain (Ubl) and four zinc-finger domains, i.e., RING0, RING1, IBR, and RING2. A REP linker is present between the IBR and RING2 domains. The Ubl domain is involved in substrate recognition, proteasome association, and regulation of cellular parkin level and activity [22]. The domains RING0 to RING2 contain two zinc ions, so eight zinc ions coordinate with each parkin protein. Zinc ions stabilize domain folding and properly regulate enzymatic activity [23]. The RING1 domain is the E2 ubiquitin ligase binding site. The IBR domain is conserved among the RBR family, but its function is still being researched; the REP linker is termed a parkin repressor element. It regulates the activity of parkin protein by occluding the E2 ligase binding site. The RING2 domain acts as a catalytic module consisting of catalytic cysteine (cys431), which forms a covalent thioester bond with the C-terminus of ubiquitin. The ubiquitin-conjugated E2 first binds to the RING1; then ubiquitin is transferred from E2 to the catalytic cysteine of RING2 and finally to a substrate lysine.

The Parkin protein's main target is mitochondria, where it regulates the mitophagy of dysfunctional mitochondria. The damaged or dysfunctional mitochondria lead to decreased ATP production, increased reactive oxygen species (ROS) burden, and reduced calcium buffering [24]. This all causes neuronal death; thus, mitochondrial dysfunction is one of the contributors to PD. Parkin works with *PINK1* (PTEN-induced putative kinase 1), a mitochondrial serine/threonine kinase with 581 amino acid residues. *PINK1* is found on the short arm of chromosome 1, with cytogenic location 1p36. Mitophagy is a multi-step process which proceed in an ordered manner. The sequence of events involved in mitophagy includes depolarization of the dysfucntional mitochondria, followed by stabilization or accumulation of *PINK1* on damaged mitochondria. *PINK1* then phosphorylates nearby ubiquitin molecules, which promotes the recruitment of Parkin from the cytosol. Subsequently, *PINK1* phosphorylates the Ubl domain of Parkin, leading to its full activation. As a result, parkin then ubiquitinates multiple outer mitochondrial membrane targets. This polyubiquitination acts as a tag for lysosomal-mediated degradation.

A mutation in *the PRKN* gene accounts for 15% of familial PD cases [25]. It results in autosomal recessive early onset Parkinson's disease (EOPD) with an average onset age of 31.2 years [18]. More than 120 mutations have been identified throughout the domains of parkin proteins [22], including deletions and duplications of exons, missense, nonsense, or splice site mutations. Missense mutations are mainly reported as they hinder Parkin protein function. In a deficient environment, dysfunctional mitochondria trigger inflammation. Neuroinflammation also acts as a significant contributor to the pathogenesis of PD.

12.1.4.4 Role of *PRK7* Gene in PD

PARK7 (Parkinsonism-associated deglaze) gene is located on the short arm of chromosome 1 with cytogenetic location 1p36.23. It encodes for a highly conserved 20KDa protein called Parkinson's protein disease seven or protein delicate DJ-1 [26]. However, it does not possess a delicate activity [27]. It consists of 189 amino acid residues. It is found in the high metabolic rate tissue, including the brain, where the highest expression is observed in the neurons, microglia, and astrocytes [27]. It is expressed ubiquitously in cells found predominantly in the cytosol, nucleus, and mitochondria. It presents as a dimer in its physiological state. It acts as an oxidative stress sensor, a molecular chaperone, a protease, a glyoxalase, an oncogene, an RNA-binding protein, a mitochondrial function regulator, and an autophagy regulator, and neutralizes reactive oxygen species (ROS) [28]. It contains cysteine residues at three sites: C45, C56, and C106; among them, C106 is highly susceptible to oxidative stress. Upon oxidation, the SH group of C106 is successively converted into sulfenic (SOH) and then into sulfinic (SO_2H) state [27]. SO_2H state of C106 represents the active form of the DJ-1. Further oxidation converts the C106 SO2H state into a sulfonic (SO3H) state and accounts for its loss of function.

PD is an oxidative stress-driven disease. Aggregating α-synuclein in neurons leads to ROS production, as discussed in Sect. 12.1.4.1. The binding of α-synuclein with the receptor of microglia leads to NADPH oxidase-mediated ROS production in them. The excess ROS production in PD pathogenesis leads to neurotoxicity in substantia nigra. The primary role of the DJ-1 is to protect the cells from oxidative stress-induced cell death. The activated or oxidized DJ-1 directly scavenges the ROS or stabilizes the transcription factor nrf2 (nuclear factor erythroid factor-E2 related factor 2). NRf2 acts as a master regulator of antioxidant response genes, which provide the defense against oxidative stress. Furthermore, the DJ-1 binds to the p47phox, one of the cytosolic components of the functional NADPH oxidase. The binding of DJ-1 with p47phox leads to its ubiquitination and, ultimately, its degradation. This results in reduced NADPH activity; thus, ROS production is reduced in microglia [27].

The neuroprotector DJ-1 physically binds to the monomeric and oligomeric α-synuclein both in vitro and in cells (Zondler et al.,2014). It has been studied that the increased expression of DJ-1 is associated with reduced α-synuclein dimerization and α-synuclein-dependent toxicity in cells (Zondler et al.,2014). Overexpression of DJ-1 also mitigates intracellular α-synuclein aggregates (N [29].). α-synuclein is degraded via chaperone-mediated autophagy (CMA). CMA includes lysosomal-mediated degradation of cytosolic proteins [30]. This process comprises recognizing the protein destined to degrade with chaperone HSC70, which delivers the protein to the lysosome-associated membrane protein type 2A (LAMP2A). LAMP2A, present on the surface of the lysosome, forms the translocation complex upon binding to the substrate and translocates the protein to the lysosomal lumen, where it is degraded by the lysosomal enzymes. DJ-1 deficiency downregulated the level of HSC70 and accelerated the degradation of LAMP2A in lysosomes, leading to the aggregation of α-synuclein [30]. DJ-1 also acts as a downstream mediator of PINK1/Parkin-mediated mitophagy [31].

Mutations in the *PARK7* gene are responsible for up to 2% of the autosomal recessive EOPD cases. Various amino acid substitutions E64D, A104T, D149A, E163K, L166P, and M26I [26], and oxidized DJ-1 are found in the brain of PD patients [32]. The lessened stability, folding, and dimerization ability characterise the mutated DJ-1. This all accounts for the inactive DJ-1 protein.

12.1.4.5 Role of *VPS35* Gene in PD

Vacuolar protein sorting 35 (VPS35) is located on the long arm of chromosome 16 with the cytogenetic location 16q11.2. It encodes for a 796 amino acid residue protein, which is almost ubiquitously expressed in human cells. It is expressed in various tissues, including the brain, heart, testis, spleen, and thymus. It localizes on the endosome and assembles with two peripheral membrane proteins, VPS26 and VPS29, to form a 150-kDa 1:1:1 heterotrimer [33]. This complex is mainly involved in retrograde transport (from endosome to Golgi) and is named the retromer complex. One of the roles of the VPS35 protein is to retrieve lysosome-associated membrane glycoprotein 2a (Lamp2a) from the endosome to the TGN (trans-Golgi network). Lamp2a is responsible for transporting cytosolic protein with a CMA (chaperon-mediated autophagy) motif to the lysosome for degradation [34]. Lamp2a is responsible for CMA-mediated degradation of α-synuclein [35].

Also, the VPS35 protein is responsible for the retrieval of CI-MPR (cation-independent mannose 6-phosphate receptor), whose function is to carry enzymes having mannose 6-phosphate moieties from TGN to the endosomes and eventually to lysosome. The VPS35 protein retrieves proteins like Lamp2a and CI-MPR to be delivered to the lysosome, thus protecting them from degradation. A mutation in the

VPS35 gene is responsible for the rare, late-onset, autosomal dominant familial PD. 14 missense *VPS35* mutations are found in PD patients in which p.Asp620Asn mutation is reported as pathogenic [33]. p.Asp620Asn mutation leads to impaired endosome-to-Golgi retrieval of Lamp2a. Thus, the mutation in the VPS35 protein impairs lysosomal function and autophagy, contributing to neuronal death.

12.1.5 Environmental Factors Contribute to PD

Almost 85% of PD cases are sporadic and may appear due to gene-environment interaction. Environmental exposure, like heavy metals and pesticides, can increase the risk of PD. This portion discusses the various environmental factors implicated in sporadic PD onset.

12.1.5.1 Association of Heavy Metals with PD

Exposure to excessive heavy metals like iron, lead, and copper was found to have a close association with PD [36]. Iron and copper are responsible for increasing the generation of reactive oxygen species (ROS) in the brain. The increased ROS reacts to deoxyribonucleic acid (DNA), proteins, and cell lipids, leading to their eventual dysfunction. Thus, the increased oxidative stress can lead to neuronal death. Lead exposure increases the risk of PD by two to threefold [37].

The other heavy metal is manganese (Mn), whose toxicity is either by occupational exposure or chronic liver failure. The brain possesses mechanisms that allow entry and accumulation of Manganese [38]. Chronic exposure to Mn results in manganism, a disorder that resembles some symptoms of PD. Globus pallidus and substantia nigra pars reticulata, caudate nucleus, and putamen are more prone to damage associated with Mn toxicity [38]. The known neurotoxin mercury is also reported to cause PD-like symptoms [37].

12.1.5.2 Pesticides as a Driver of PD

Significant available data suggest that pesticide exposure can lead to neurodegenerative disorders like PD. Various pesticides are found to be associated with PD. Still, among them, eight pesticides have been found to significantly increase the risk of disease, namely, rotenone, maneb, MPTP (1-methyl-4-phenyl-1,2,3,6-tetrahydropyridine), 2,4-Dichlorophenoxyacetic acid, paraquat, permethrin, dieldrin, and mancozeb [37].

MPTP is a lipophilic substance that can easily cross the blood-brain barrier (BBB). It is not toxic but converts into 1-methyl-4-phenylpyridinium (MPP^+) in the brain mainly via monoamine oxidase-B (MAO-B) of glial cells [39]. MPP^+ is released from the glial cells into the extracellular space and enters the dopaminergic neurons via the high-affinity dopamine transporter (DAT). MPP^+ is a polar compound and can't cross the BBB back. In neurons, the main target of the MPP^+ is the mitochondria, where it binds to complex I and inhibits NADH dehydrogenase [40]. This results in mitochondrial impairment, decreased ATP production, increased free radical production, mainly ROS, and reduced calcium buffering. The ROS generation leads to oxidative stress in neurons, which ultimately initiates the process of apoptosis.

The herbicides Paraquat, fungicides Maneb, and herbicides 2,4-Dichlorophenoxyacetic acid increase the formation of free radicals in dopaminergic neurons. Thus, it leads to oxidative stress-mediated neurodegeneration. Paraquat also causes the aggregation of parkin protein, an E3 ubiquitin ligase [41]. Rotenone (insecticides) also inhibits complex-I of mitochondria [42]. Mancozeb, a fungicide exposure, accounts for the dopaminergic neuronal structural damage. Permethrin exhibits neurotoxicity at high doses. High-dose exposure to permethrin leads to oxidative stress, mitochondrial complex I defect, and loss of dopamine in the striatum of rats [43].

12.2 Role of Ubiquitin-Mediated Degradation System in PD

The cellular proteins have to be maintained in their native structure at specific concentrations to perform their activity optimally. Any mutation at the gene level or the post-transcriptional level alters the stereochemistry, thus leading to the aberrant folding of the protein. Misfolded proteins like α-synuclein tend to aggregate in the cells. The hallmark feature of PD is aggregation of α-synuclein, which is neurotoxic and can lead to neurodegeneration. The intracellular clearance of misfolded α-synuclein can act as one of the methods of reducing PD pathogenesis. A cell possesses mechanisms to reduce the burden of misfolded proteins, including ubiquitin-mediated proteasomal degradation and lysosomal degradation.

Ubiquitin (Ub) is a 76-amino acid-long, highly conserved, and ubiquitously expressed protein in mammalian cells. The C-terminal glycine of Ub is covalently attached to the lysine residue of the targeted protein [44]. The ubiquitination process is catalyzed or regulated by E1 Ub-activating enzyme, E2 Ub-conjugating enzyme, and E3 Ub ligase [45]. Firstly, the E1 enzyme activates the ubiquitin molecule by adenylation of its C-terminal carboxylic group. Thus, this step is ATP-dependent. After that, E2 interacts with E3 ligase and identifies the targeted substrate, and lastly, the E3 ubiquitin ligase attaches the ubiquitin molecules to the lysine, serine, threonine, and cysteine residues to the specific substrate [46]. E3 enzyme can repeatedly add the ubiquitin molecules, and after the first ubiquitin is attached, the next round of ligation links the two Ub molecules in the form of a chain [47]. A new ubiquitin molecule is ligated to any of the attached ubiquitin molecule's seven lysine residues (K6, K11, K27, K29, K33, K48, or K63). The seven different types of ubiquitin linkages (K6, K11, K27, K29, K33, K48, or K63) accomplish distinct physiological functions. K6 linkage plays a role in DNA damage response, K11 linkage is associated with cell cycle, proteasomal degradation, or membrane trafficking, and K27 linkage is associated with protein secretion, DNA damage repair, and mitochondrial damage response. K29 linkage is associated with proteasomal degradation, innate immune response, and regulation of AMPK-related protein kinases; K33-linkages are related to the regulation of innate immune response; K48-linkages mainly aim to target substrates for 26S

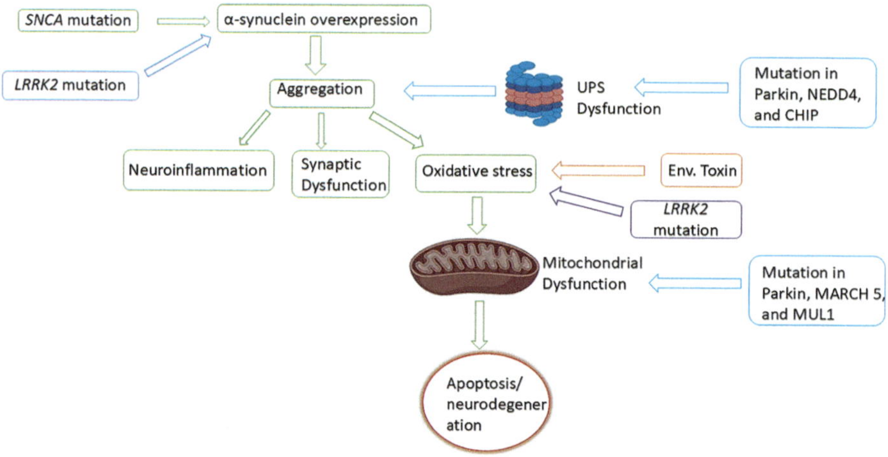

Fig. 12.1 Schematic representation of the role of the different E3 ubiquitin ligases in the etiology of Parkinson's disease. Parkin, NEDD4, and CHIP are the E3 ubiquitin ligases that ubiquitinate the α-synuclein to mark it for degradation via proteasomal or lysosomal system. Parkin, MARCH 5, and MUL1 are mainly involved in clearing dysfunctional mitochondria. The mutation in these E3 ligases results in protein aggregation and, ultimately, apoptosis of the neuron

proteasome-mediated degradation; and finally, K63 linkages are involved in DNA damage repair, cytokine signaling or autophagic degradation signaling [46]. Figure 12.1 summarizes the mechanisms involving E3 ubiquitin ligases in the pathogenesis of PD.

The overexpression or aggregation of misfolded α-synuclein leads to free radical stress, mitochondrial dysfunction, and synaptic impairment. The non-ubiquitinated α-synuclein can be degraded via autophagy, the mono-ubiquitinated α-synuclein can undergo proteasomal degradation, and the polyubiquitinated α-synuclein can be degraded via proteasomal or lysosomal system [48]. The ubiquitin molecule can attach to either of the nine lysine residues 6, 10, 12, 21, 23, 32, 34, 43, and 96 of α-synuclein. The lysine residue is preferred in different forms of α-synuclein (monomeric, oligomeric, and aggregated). In monomeric α-synuclein, the preferred sites are K21, K23, K32, and K34; on the other hand, the preformed fibrils are ubiquitinated at K6, K10, and K12 [49]. The reduced clearance of the misfolded α-synuclein in PD is either due to its increased expression or its reduced degradation via malfunctioning ubiquitin-proteasome or lysosome system. α-synuclein can undergo mono or polyubiquitination.

12.3 E3 Ubiquitin Ligases Involved in PD

12.3.1 Parkin

Parkin is a type of RBR E3 ubiquitin ligase. It ubiquitinates α-synuclein and promotes its degradation via the proteasomal or lysosomal system. It polyubiquitinates the protein via K27 linkage. Parkin also promotes the clearance of dysfunctional mitochondria. The details of the Parkin in the etiology of PD are already described above in Sect. 12.1.4.3.

12.3.2 NEDD4 Protein

The NEDD4 (neural precursor cell expressed developmentally downregulated protein 4) family is an E3 ubiquitin ligase belonging to the HECT family. It consists of an N-terminus C2 calcium/phospholipid-binding domain, 2–4 WW domains, and a conserved C-terminus HECT domain, which involves substrate recognition, binding with the substrate, and interacting with the E2 conjugation enzyme, respectively. NEDD4 family has 9 members, i.e., NEDD4, NEDD4L (NEDD4-like), AIP4 (atrophin-1 interacting protein 4), SMURF1 (SMAD-specific E3 ubiquitin–protein ligase 1), SMURF2, WWP1 (WW domain-containing E3 ubiquitin–protein ligase 1), WWP2, NEDL1 (NEDD4-like ubiquitin–protein ligase 1), and NEDL2. Among them, NEDD4, NEDD4L, SMURF1, and SMURF2 are able to ubiquitinate α-synuclein and promote its clearance via proteasomal or lysosomal degradation [48].

The *NEDD4* gene is located on chromosome 15, with the cytogenic location 15q21.3 in humans. It encodes for the NEDD4 or NEDD4–1 protein with a molecular weight of 120KDa. The expression of NEDD4 is found to increase in substantia niagra and locus coeruleus in the brains of PD patients, which suggests its role in the disease [48]. It has a neuroprotective role as it reduces the accumulation of α-synuclein. NEDD4 E3 ubiquitin ligases recognize the proline-rich area at the C-terminus of the α-synuclein via their tryptophan-rich (WW) domains. NEDD4 ubiquitinates α-synuclein on K21 and K96. This enzyme catalyzes the formation of K63-linked polyubiquitin chain on α-synuclein. The ubiquitinated α-synuclein is recognized by the ESCRT (endosomal sorting complex required for transport) complex and transferred to the late endosome, which promotes lysosomal degradation of α-synuclein [48].

The *NEDD4-2* gene is located on chromosome 18 with the cytogenic location 18q12.31. It codes for the protein NEDD4-2 or NEDD4L with a molecular weight of 120KDa. Both the NEDD4 and NEDD4L proteins express a 63% similarity. Also, the NEDD4L can equally ubiquitinate α-synuclein [50]. NEDD4 and NEDD4L perform their activity with the same E2 ligases Ubc4, UbcH5A, UbcH5B, UbcH5C, UbcH6, and UbcH7. Furthermore, the activity of NEDD4 and NEDD4L is regulated via autoinhibition. The autoinhibitory conformation is formed by the interaction of the C2 domain or WW domain with the HECT domain in NEDD4 and NEDD4L, respectively.

The mutation in the *NEDD4* gene leads to familial PD cases, but so far, the exact percentage of mutation has not been established.

12.3.3 CHIP

CHIP is the C-terminus of an Hsp70-interacting protein that plays a critical role in determining whether a protein is appropriately folded or targeted for degradation in the cell [51]. It is a crucial regulator of the Hsp70/Hsp90 chaperone system, participating in molecular triage. This involves the evaluation of misfolded or unstable proteins so that chaperones can attempt to refold them into their original structure [52]. In case of success, the hydrophobic regions are buried inside, and the protein escapes degradation. However, in terms of failure or the accumulation of misfolded proteins exceeding the cell's capacity, they are marked for degradation via proteasome by CHIP [51].

CHIP inhibits traditional chaperone activity and converts the chaperone complex into a ubiquitin ligase, tagging proteins for destruction. It interacts with both Hsp70 and Hsp90 chaperones, but in some instances, such as with the ErbB2 protein, it works in concert with Hsp70, rather than Hsp90, to ubiquitylate the target protein [53]. CHIP is incredibly efficient in targeting proteins that are either partially unfolded or fully misfolded due to exposed hydrophobic regions.

CHIP regulates the protein's stability in the context of the G2385R LRRK2 (leucine-rich repeat kinase 2) mutation [54]. The LRRK2 gene is one of the most common gene mutations directly impacting PD [55]. The G2385R variant binds more strongly to Hsp90 than wild-type LRRK2, indicating partial misfolding. While Hsp90 attempts to fold the mutant protein into an active conformation, the misfolded LRRK2 is frequently directed toward degradation. CHIP mediates this process in coordination with Hsp70, making the triage decision between folding and degradation. This suggests CHIP being a critical player in determining the fate of the G2385R LRRK2 variant, as it functionally opposes Hsp90's folding activity, driving the degradation of misfolded LRRK2.

CHIP has also emerged as a critical player in regulating mitochondrial dynamics, where mitochondrial dysfunction is a crucial factor in developing PD. Mutations or deficiencies in CHIP expression have been linked to several neurological disorders, highlighting its vital role in maintaining neuronal health. A study was done investigating the neuroprotective function of CHIP in both in vitro and in vivo models of PD [56]. In SH-SY5Y cells, overexpression of CHIP significantly enhanced cell viability and increased ATP production following exposure to the neurotoxin 1-methyl-4-phenylpyridinium (MPP+), a compound mimicking PD-like symptoms. Furthermore, in animal models, CHIP overexpression effectively mitigated the neurodegenerative effects of 1-methyl-4-phenyl-1,2,3,6-tetrahydropyridine (MPTP) exposure in mice, including improvements in movement impairments, motor coordination, and spontaneous locomotor activity, as well as preservation of dopaminergic neurons. The neuroprotective effect of CHIP appears to be mediated by its regulation of Drp1, a protein involved in mitochondrial fission, which was shown to be downregulated by CHIP, thus maintaining mitochondrial integrity in MPP+/MPTP-induced PD models.

Another protein, PTEN-induced putative kinase 1 (PINK1), regulates mitochondrial quality control and neuronal survival through its kinase activity, but its stability is tightly controlled, as shown by several reports [57, 58]. It is the second most common cause of autosomal recessive PD. Despite this, the molecular mechanisms that modulate PINK1 stability, including critical regulatory proteins, remain largely unknown. Recent studies have revealed that the ubiquitin E3 ligase CHIP promotes PINK1 ubiquitination, decreasing PINK1 steady-state levels [59]. Exposure of HEK293 and SH-SY5Y cells to the apoptosis inducer staurosporine significantly reduces PINK1 levels, an effect attributed to CHIP-mediated ubiquitination. Knocking down CHIP through siRNA showed reduced susceptibility to staurosporine-induced cell death, suggesting that CHIP negatively regulates PINK1's cytoprotective effects under apoptotic conditions, potentially contributing to PD pathogenesis.

In relation to these studies, mitochondrial quality control pathways involving proteins PINK1 and Parkin were upregulated in CHIP-deficient mice [60]. Notably, mitochondrial stress markers such as dynamin-related protein 1 (Drp1) are oxidised in a CHIP-dependent manner. Upon oxygen-glucose deprivation (OGD), CHIP localises from the cytosol to the mitochondria, enhancing mitochondrial integrity and bioenergetics.

CHIP's neuroprotective role has been well characterised in various models. It has been demonstrated that CHIP knockout (KO) mice show heightened sensitivity to oxidative stress and lipid peroxidation, accompanied by diminished antioxidant defences and energy deficits. In Drosophila models, overexpression of CHIP rescued mitochondrial dysfunction and dopaminergic neuronal degeneration in Pink1 and Parkin mutants. It also ameliorated several phenotypes, including muscle degeneration, reduced ATP content, and abnormal mitochondrial morphology [61]. These findings underscore CHIP's role in maintaining mitochondrial integrity downstream of Pink1 and Parkin pathways, reinforcing its importance in PD pathogenesis.

Interestingly, CHIPs also play a role in proteostasis, particularly in terms of regulating protein aggregation. CHIP degrades toxic oligomers of α-synuclein, a protein that is directly linked to PD [62]. Although CHIP promotes α-synuclein ubiquitination and clearance, its interaction with co-chaperones like Bcl-2-associated athanogene 5 (BAG5) can inhibit E3 ligase activity, suggesting the involvement of complex regulatory mechanisms [63]. Despite CHIP's capacity to clear α-synuclein aggregates, caution must be taken, as it may also lead to the degradation of essential proteins like tyrosine hydroxylase [62].

In endoplasmic reticulum (ER) stress, CHIP and UBE2D1 modulate the ubiquitination of ER-associated substrates under nitrosative stress conditions. Nitrosative stress interferes with ER-associated degradation (ERAD), prolonging ER stress and potentially exacerbating neurodegenerative processes. Disruption of this pathway has been highly implicated in PD [64].

In short, CHIP has many roles in the maintenance of neuronal function, contributing to, among others, mitochondrial balance, aggregation of proteins, and adaptation to stress. Its dysregulation, especially with regard to the pathology of this disorder, suggests its potential utility as a therapeutic target. However, the interactions with other chaperones, as well as the harmful or beneficial effects of CHIP intervention, remain essential points that should be addressed before such treatment.

12.3.4 MARCH5/MITOL

The outer mitochondrial membrane-localized E3 ubiquitin ligase MITOL/MARCHF5 is a mitochondrion-associated protein that is encoded by the *Mitol* gene [65]. Ubiquitylation of outer mitochondrial membrane (OMM) proteins is closely linked to familial PD, and a decrease in mitochondrial membrane potential triggers Parkin-mediated ubiquitylation of OMM proteins, targeting them for degradation. A study involving MARCH 5 (MITOL) reveals that it translocates from depolarised mitochondria to peroxisomes upon stimulation of mitophagy, which is mediated by peroxisomal biogenesis factors Pex3 and Pex16 and requires Parkin's E3 ligase activity. The lysine 268 (K268) on MARCH 5's C-terminus is ubiquitinated, facilitating its extraction via the p97/VCP complex. This allows the escape of specific proteins from mitochondria via the process of ubiquitination. It can be inferred that ubiquitination is not only meant for degradation but also acts as a sorting signal in some instances [66].

The loss of dopaminergic neurons in the substantia nigra is a hallmark of PD. Dimethylarginine dimethylaminohydrolase-1 (DDAH-1) is a compound that degrades asymmetric dimethylarginine (ADMA), an inhibitor of nitric oxide (NO) synthase, and has been linked to neurodegeneration. In a study involving DDAH-1, findings show its co-localisation with dopaminergic neurons and further knockout, leading to increased neuron loss and locomotor deficits following rotenone (ROT) exposure. Downregulation of DDAH-1's impaired MAM formation, which is linked to decreased MITOL phosphorylation, was reversed by DDAH-1 overexpression. In vitro studies indicated that DDAH-1 downregulation caused cell death and mitochondrial dysfunction, characterised by elevated ADMA and reduced NO levels, which were restored by L-arginine supplementation but failed to protect mitochondrial functions. These results suggest that DDAH-1 played a role in maintaining MAM integrity and mitochondrial function, highlighting its potential as a therapeutic target in PD [67].

PINK1 and PARKIN are critical genes in recessive familial PD. PINK1 phosphorylates Parkin and ubiquitin, activating Parkin to tag damaged mitochondrial proteins for degradation. Recent research has indicated that MARCH 5 is essential for Parkin recruitment and activation, with Parkin's substrate specificity not solely defined by amino acid sequences. The depletion of MARCH 5 delayed Parkin activation, which suggests that initial ubiquitylation by MARCH 5, followed by PINK1-mediated phosphorylation, amplified the response to mitochondrial damage [68].

Mutations in the LRRK2 gene, the most common genetic cause of familial PD, impair mitochondrial function. LRRK2 is crucial for ER-mitochondrial tethering, regulating E3 ligases like MARCH 5 through kinase-dependent interactions. The kinase-active LRRK2(G2019S) variant promotes the phosphorylation and activation of these ligases, enhancing the degradation of tethering proteins. In contrast, the kinase-dead LRRK2(D1994A) variant inhibits this process, increasing tethering protein levels. It generally binds to MARCH 5, MULAN, and Parkin through its N-terminal region, and further phosphorylation leads to them undergoing conformational changes from the E2 binding-incompetent state to the E2 binding-competent state. This highlights LRRK2's role in cell fate and PD pathogenesis, underscoring the interplay between MARCH 5, PINK1, and LRRK2 in mitochondrial health [69].

12.3.5 MUL1

Recent studies have pointed to mitochondrial dysfunction as a critical factor in PD pathogenesis, and MUL1 (Mitochondrial Ubiquitin Ligase 1) is a crucial player. MUL1 is vital in maintaining mitochondrial integrity, working in parallel with the PINK1/PARKIN pathway, which is crucial for mitochondrial quality control. Research in Drosophila models has shown that MUL1 can suppress phenotypes associated with PINK1 or PARKIN mutations. Specifically, MUL1 mediates the ubiquitin-dependent degradation of Mitofusin, a protein essential for mitochondrial fusion. In the absence of MUL1, the phenotypic effects of PINK1 and PARKIN mutations are exacerbated, leading to increased mortality and neuronal degeneration [70].

Beyond mitochondrial integrity, MUL1 also influences neuroprotection and circadian rhythms. In studies involving Drosophila models lacking functional MUL1, increased levels of reactive oxygen species (ROS) were observed alongside decreased levels of the antioxidant enzyme SOD1. These changes contributed to disrupted circadian rhythms and heightened neuronal vulnerability [71]. Furthermore, overexpression of MUL1 has shown neuroprotective effects against mitochondrial toxins, reinforcing its potential as a therapeutic target to mitigate neuronal loss in PD [72].

Research has indicated a significant association between genetic variants in the MUL1 gene and the risk of developing PD. A case-control study conducted on a Chinese cohort found that the single-nucleotide polymorphism (SNP) rs529974 was linked to increased susceptibility to PD. The T allele of this SNP was identified as a risk factor independent of gender, clinical features, and disease severity [73]. These findings underscore the potential of MUL1 genetic variants as biomarkers for early diagnosis and personalized treatment strategies in PD.

In addition, genetic studies of MUL1 variants and dementia in PD patients were performed, emphasizing the role of mitochondrial dysfunction in cognitive decline. In a microarray genomic study, the SNP rs3738128 in the MUL1 gene was significantly associated with dementia, suggesting that mitochondrial impairment could contribute to developing cognitive symptoms in PD [74]. These findings highlight the multifaceted role of MUL1 in the progression of PD and its potential influence on cognitive health.

A recent pilot study investigated the expression of MUL1 and other critical genes as potential biomarkers for monitoring PD progression. In MPTP-induced PD models, variations in the expression of MUL1 alongside PARKIN and PINK1 were observed throughout the disease course. These changes may serve as valuable biomarkers for early diagnosis and progression monitoring, ultimately aiding in the development of therapeutic interventions [75].

In conclusion, MUL1 E3 ligase plays a significant role in the pathogenesis of PD. It is involved in mitochondrial integrity, neuroprotection, and genetic associations with dementia and has potential as a therapeutic target. Hence, this area of research focus needs further attention. Understanding why alterations in the MUL1 gene and protein may be associated improves the chances of PD diagnosis and treatment, which facilitates health care for patients suffering from this disease.

12.4 E3 Ligases and PD Neurobiology

12.4.1 Oxidative Stress

Oxidative stress plays a crucial role in the neurodegeneration observed in PD, and E3 ubiquitin ligases have emerged as critical regulators in the response to cellular oxidative damage. Various E3 ubiquitin ligases, such as Parkin (PRKN), TRIM32, and FBXW7, have emerged as significant regulators of mitochondrial function, proteostasis, and neuronal survival under oxidative stress, influencing the progression of PD.

The E3 ubiquitin ligase Parkin, encoded by the *PRKN* gene, plays a pivotal role in mitochondrial quality control through the ubiquitination of outer mitochondrial membrane proteins and the initiation of mitophagy [76]. Loss-of-function mutations in *PRKN* are frequently associated with early-onset PD, and these mutations impair the ability of cells to remove damaged mitochondria, leading to increased oxidative stress and neuronal death [77]. Recent studies demonstrate that Parkin interacts with critical lipid regulators such as Sterol Regulatory Element-Binding Protein 2 (SREBP2) and lipoprotein lipase (LPL), regulating lipid homeostasis and mitigating oxidative stress under PD-related conditions [78].

TRIM32, another E3 ubiquitin ligase, has also been shown to sensitise dopaminergic neurons to oxidative stress-induced apoptosis by disrupting mitochondrial function [79]. Under PD stress conditions, TRIM32 translocates to the mitochondria, reducing mitochondrial complex-I activity, increasing reactive oxygen species (ROS) and decreasing cellular ATP levels. These effects contribute to neuronal apoptosis under oxidative stress, with the deletion of TRIM32's RING domain showing neuroprotective potential by rescuing complex-I activity and reducing ROS levels.

In the list of Ubiquitin ligases, FBXW7, yet another E3 ligase, regulates mitochondrial biogenesis and oxidative defences by targeting peroxisome proliferator-activated receptor gamma coactivator 1-alpha (PGC-1α) for degradation via the ubiquitin-proteasome system [80]. Decreased levels of FBXW7 have been associated with increased PGC-1α activity, promoting mitochondrial function and reducing oxidative stress in dopaminergic neurons. FBXW7 can be degraded by enhancing chaperone-mediated autophagy (CMA), thereby protecting neurons from MPTP-induced oxidative damage by boosting PGC-1α levels.

In addition to E3 ligases directly regulating mitochondrial quality control, other mechanisms of neuroprotection against oxidative stress have been identified. For instance, inhibiting the ubiquitin-specific protease 30 (USP30), an antagonist of Parkin, has been shown to promote mitophagy and reduce oxidative stress in neurons deficient in Parkin [81]. Similarly, therapeutic interventions like calcium channel blockers and iron chelators, such as felodipine and deferasirox, have demonstrated neuroprotective effects by modulating mitochondrial function, enhancing mitophagy, and reducing oxidative stress.

Altogether, the aberrant functioning of E3 ubiquitin ligases in PD gives rise to oxidative stress, mitochondrial impairment, and cellular death. These data suggest that therapeutic approaches aimed at E3 ligases and mitochondrial plasticity may be worthwhile to alleviate the neurodegenerative processes related to oxidative stress inherent to PD.

12.4.2 Mitochondrial Dysfunction

Mitochondrial dysfunction, alongside toxic protein aggregates, is central to the pathogenesis of neurodegenerative disorders such as PD. Research has identified the mitochondrial kinase PINK1 and the E3 ubiquitin ligase Parkin as critical players in mitochondrial quality control linked to PD. PINK1 accumulates on damaged mitochondria and subsequently recruits and activates Parkin, which tags mitochondrial proteins with ubiquitin to prime them for degradation. This PINK1/Parkin mitophagy pathway facilitates the removal of damaged mitochondria, thereby supporting cellular homeostasis and survival. However, disruptions in this pathway may lead to PD. The studies are still exploring whether the buildup of toxic Parkin substrates or impaired mitochondrial clearance drives pathology [82].

In response to the targeting of Parlin, Recent studies have identified the CISD family proteins CISD1 and CISD2, which contain iron-sulfur clusters [82]. In Drosophila models, Pink1 and parkin mutations lead to CISD accumulation, which impedes mitophagy and triggers neurodegenerative changes. Reducing CISD levels, whether genetically or through pharmacological inhibitors like rosiglitazone and NL-1, has been shown to enhance mitophagy, improve neuronal function, and alleviate PD phenotypes, suggesting a promising therapeutic approach targeting these proteins.

Further insights into E3 ligases highlight TRIM31, a mitochondrial E3 ubiquitin ligase that regulates the degradation of the mitochondrial channel VDAC1. Investigations on TRIM31-deficient mice demonstrated that its loss promotes dopaminergic neurodegeneration, with TRIM31 acting to ubiquitinate and degrade VDAC1, thereby reducing its associated stress on mitochondria. Enhancing TRIM31 activity in cellular and animal models mitigated PD-like symptoms, suggesting it is another potential therapeutic target [83].

Additional E3 ligase involvement has been observed in FBXO7, part of the SCF E3 ligase complex, where a novel FBXO7 mutation (L250P) impairs its function, reducing proteasomal activity and exacerbating mitochondrial dysfunction [84]. Likewise, FBW7β, another E3 ligase component, regulates PINK1 stability and maintains mitochondrial quality control [85].

These findings reveal that E3 ubiquitin ligases, such as Parkin, TRIM31, FBXO7, and FBW7β, are integral to mitochondrial regulation. Their dysfunction leads to a failure in mitochondrial quality control and contributes to neurodegeneration in PD, underscoring the importance of the therapeutic potential of targeting these pathways to restore mitochondrial health.

12.4.3 Neuroinflammation

Neuroinflammation is a hallmark of PD, with excessive inflammatory responses contributing to neuronal degeneration. E3 ubiquitin ligases, including Parkin, play pivotal roles in regulating neuroinflammatory pathways by mediating ubiquitination and degradation of critical inflammatory molecules. However, it plays a significant role in controlling inflammatory pathways independent of its mitophagy functions. For example, studies have shown that Parkin mediates the degradation of NLRP3 in microglia via K48-linked polyubiquitination, thereby limiting NLRP3 inflammasome activation and reducing neuroinflammation [86]. Loss of Parkin function exacerbates inflammatory responses in microglia, leading to enhanced assembly of NLRP3 inflammasomes, increased production of proinflammatory cytokines, and more significant dopaminergic neuron loss in the substantia nigra, ultimately resulting in motor impairment [87].

Further studies have identified that Parkin interacts with TRAF3, an adapter protein in innate immune signaling, promoting TRAF3 degradation via K48-linked polyubiquitination. This action downregulates proinflammatory signaling, suggesting Parkin's role as a negative regulator of antiviral immunity and highlighting its broader role in modulating immune responses in neuroinflammation [88].

E3 ligases such as Peli1 also regulate microglial activation in PD by promoting NF-κB and MAPK activation. Peli1's increased expression in the substantia nigra under inflammatory conditions exacerbates dopaminergic neuronal death, implicating Peli1 as a critical mediator of neuroinflammation [89]. Similarly, HACE1 has been shown to regulate Rac1 activity, thus modulating oxidative stress and inflammatory responses in microglia and SH-SY5Y neuronal cells in PD models [90].

In addition to this, Smurf1, an E3 ligase, is upregulated in neuroinflammatory conditions, promoting neuronal necroptosis through ubiquitination pathways, which are essential for inflammatory regulation post-LPS-induced neuroinflammation in animal models [91]. Together, these findings underline the potential for targeting E3 ubiquitin ligases to alleviate neuroinflammatory processes in PD.

12.4.4 Endoplasmic Reticulum Stress

Recent evidence points toward disruptions in the endoplasmic reticulum (ER) as a critical factor in PD pathogenesis [92]. Studies have shown that ER stress plays a significant role in PD, particularly in regulating protein misfolding, ubiquitination, and degradation [93].

Mutations in the FBXO7 gene have been found to cause early-onset autosomal recessive PD, with FBXO7 functioning as an adaptor in the SCF E3 ubiquitin ligase complex, promoting substrate ubiquitination. A recent study demonstrated that ubiquitin-specific protease 7 (USP7) stabilises FBXO7 by regulating Lys48-linked deubiquitination, preventing proteasomal degradation under ER stress, and thereby reducing apoptosis [94]. This indicates a possible therapeutic pathway for mitigating dopaminergic neuronal loss.

The interplay between mitophagy and ER stress also plays a critical role in PD. For instance, the transcription factor XBP1s is transactivated by PINK1 during ER stress, creating a feedback loop essential for mitophagy regulation [95]. This loop, when disrupted, contributes to neurodegeneration observed in PD.

Further evidence suggests that Parkin, a well-known PD gene, impacts dopaminergic neurons and regulates astrocyte ER stress responses. A study showed that Parkin knockdown in astrocytes results in elevated ER stress markers and increases neuronal vulnerability to apoptosis [96]. This highlights Parkin's broader role in managing cellular stress across various cell types.

Lastly, ubiquitin ligases such as RNF13 have been identified as critical modulators of ER stress, specifically by regulating the IRE1α-TRAF2-ASK1-JNK pathway. Silencing RNF13 alleviates motor dysfunction and dopamine neuronal damage in PD mouse models, emphasising the pathway's significance in PD pathology [93].

12.4.5 Apoptosis

The E3 ligases Parkin, TRIM32, and regulatory proteins such as Optineurin (OPTN) and USP30 are essential to selective autophagy (mitophagy) and apoptosis regulation, and disruptions in these proteins' functions can exacerbate apoptotic susceptibility, influencing PD progression.

In PD, selective autophagy of damaged mitochondria (mitophagy) helps prevent apoptosis by clearing dysfunctional mitochondria and mitigating oxidative stress.

OPTN is an autophagy receptor implicated in neurodegeneration and facilitates an unconventional pathway for mitophagy initiation [97]. Unlike typical autophagy receptors like NDP52, TAX1BP1, and p62, which recruit FIP200 and the ULK1/2 complex to initiate autophagosome formation, OPTN independently recruits the kinase TBK1, which directly binds the class III phosphatidylinositol 3-kinase complex I. This unique mitophagy pathway allows OPTN to bypass the conventional autophagy initiation mechanism and form autophagosomes around damaged mitochondria, potentially reducing apoptotic stress induced by mitochondrial dysfunction in PD.

Parkin, another E3 ubiquitin ligase, encoded by PARK2 gene, is vital for PD's apoptosis regulation. Mutations in PARK2 are associated with juvenile-onset PD, and the absence of Parkin increases susceptibility to oxidative stress-induced apoptosis. Studies have shown that in the absence of Parkin, the transcriptome of neural progenitor cells from PD patients exhibits widespread dysregulation in apoptosis-related pathways [98]. Remarkably, reintroducing Parkin normalises gene expression, particularly in mitochondrial function and apoptosis pathways, suggesting that Parkin supports cellular survival under PD conditions. Dopamine receptor D4, a gene linked to PD, was also among the most dysregulated, further implicating Parkin's regulatory role in dopaminergic neuron apoptosis.

Parkin also regulates apoptosis beyond PD contexts by interacting with Kindlin-2 (K2), a mitochondrial regulator associated with cell motility [99]. Parkin promotes the degradation of K2 through ubiquitination, reducing its stability and thereby modulating focal adhesion turnover and mitochondrial dynamics, which are essential for cellular motility and integrin activation. Though this interaction does not directly affect tumour cell apoptosis, it demonstrates Parkin's broader role in controlling cell viability via mitochondrial integrity.

The mitochondrial deubiquitinating enzyme USP30 is another critical factor in PD-associated apoptosis due to its role in opposing Parkin-mediated mitophagy. USP30 limits the clearance of damaged mitochondria, which can lead to apoptotic responses under oxidative stress [100]. Recent research highlights a novel peptide, Q14, that can inhibit USP30 activity through an autoinhibitory mechanism, enhancing autophagic flux and autophagosome formation by interacting with the LC3-interacting region (LIR) of Q14. This dual mechanism facilitates mitophagy, reducing the accumulation of damaged mitochondria and potentially lowering apoptotic signalling in PD.

TRIM32, another E3 ubiquitin ligase, has also been identified as a significant modulator of mitochondrial integrity under PD-associated oxidative stress. In PD models, TRIM32 expression sensitises dopaminergic SH-SY5Y cells to apoptosis through enhanced mitochondrial translocation and subsequent reduction of complex I activity, leading to elevated ROS levels and decreased ATP production [79]. TRIM32 further exacerbates the apoptotic response by downregulating the anti-apoptotic protein XIAP, amplifying neuronal apoptosis under oxidative stress. Consequently, modulating TRIM32 activity or restoring XIAP expression may be a therapeutic approach to mitigate apoptosis in PD.

These findings underscore the importance of E3 ligases, including Parkin and TRIM32, and regulatory factors like OPTN and USP30 in regulating apoptosis in PD. Dysregulation or loss of function in these proteins disrupts mitochondrial quality control, increases oxidative stress, and exacerbates apoptotic responses in dopaminergic neurons. Therapeutic strategies targeting these pathways, such as enhancing Parkin function, inhibiting USP30, or modulating TRIM32 offer promising avenues for reducing apoptosis and delaying PD progression.

12.5 Conclusion

The pathological hallmark of PD is the aggregation of α-synuclein. This cytosolic protein's oligomer or fibril forms have detrimental effects on the dopaminergic neurons, as they can lead to increased oxidative stress, reduced dopamine synthesis and secretion, synaptic dysfunction, neuroinflammation, and endoplasmic reticulum stress. The oxidative stress primarily affects mitochondrial health. It disrupts the mitochondrial membrane integrity, damages mitochondrial DNA, and impairs ATP production. The dysfunctional mitochondria lead to the apoptotic signal, hence neurodegeneration. The aggregates or abnormally folded α-synuclein can be removed from the cells via the ubiquitin-proteasome system (UPS). The UPS includes ubiquitin ligases and a 26S proteasome machinery. Ubiquitin ligases are recognized and mark the misfolded protein with a ubiquitin tag. The ubiquitinated protein undergoes degradation in the 26S machinery. Several ubiquitin ligases recognize misfolded α-synuclein, including Parkin, NEDD4, CHIP, MUL1, and MARCH 5, and maintain their homeostasis in the neurons. Parkin, MARCH5, and MUL1 can also recognize the dysfunctional mitochondria and remove them from the cells. Thus, the mutation in these specific E3 ligases contributes to the progression of PD. The enhancement or reactivation of these specific E3 ligase activities potentially offers a therapeutic approach to treating PD.

12.6 Future Perspective

The brief understanding of the ubiquitin-mediated degradation system has profoundly shaped the knowledge of Parkinson's disease (PD) pathogenesis. This understanding opens new therapeutic possibilities that may go beyond symptomatic relief and also address one of the underlying molecular mechanisms of neurodegeneration in PD. Targeting specific E3 ubiquitin ligases such as Parkin, NEDD4, MITOL, CHIP and MUL1 may enhance the clearance of toxic α-synuclein and will simultaneously improve mitochondrial function and neuronal survival.

Drug discovery efforts are underway to identify small-molecule modulators of these enzymes that regulate these ligases and explore their potential as personalized treatments. Additionally, alterations in these ligases' genetic and expression profiles may serve as biomarkers for early diagnosis and progression monitoring.

Gene editing technologies, including the CRISPR-Cas system, may be used to correct mutations in key PD-associated genes such as *PARK2*, *PINK1*, and *LRRK2*. These interventions can combine with mitochondrial-targeted therapy to halt or reverse the neurodegenerative process in PD. Furthermore, future research should emphasize the precise roles of different ubiquitin linkages, the interaction of E3 ligase and its substrate, and the downstream effects on neuronal health. A deeper mechanistic understanding will be essential for translating these molecular insights into safe, effective, and personalized therapeutic interventions for PD.

References

1. Kim DJ, Isidro-Pérez AL, Doering M, Llibre-Rodriguez JJ, Acosta I, Rodriguez Salgado AM, Pinilla-Monsalve GD, Tanner C, Llibre-Guerra JJ, Prina M. Prevalence and incidence of Parkinson's disease in Latin America: a meta-analysis. Mov Disord. 2024;39(1) https://doi.org/10.1002/mds.29682.
2. Surathi P, Jhunjhunwala K, Yadav R, Pal PK. Research in Parkinson's disease in India: a review. Ann Indian Acad Neurol. 2016;19(1) https://doi.org/10.4103/0972-2327.167713.
3. Wang J, Dai L, Chen S, Zhang Z, Fang X, Zhang Z. Protein–protein interactions regulating α-synuclein pathology. Trends Neurosci. 2024;47(3) https://doi.org/10.1016/j.tins.2024.01.002.
4. Schulz JB, Falkenburger BH. Neuronal pathology in Parkinson's disease. Cell Tissue Res. 2004;318(1) https://doi.org/10.1007/s00441-004-0954-y.
5. Sonne J, Reddy V, Beato MR. Neuroanatomy, substantia nigra. StatPearls 2020; (Dec 2018).
6. Jankovic J. Parkinson's disease: clinical features and diagnosis. J Neurol Neurosurg Psychiatry. 2008;79(4) https://doi.org/10.1136/jnnp.2007.131045.
7. Schapira AHV, Chaudhuri KR, Jenner P. Non-motor features of Parkinson disease. Nat Rev Neurosci. 2017;18(7) https://doi.org/10.1038/nrn.2017.62.
8. Armstrong MJ, Okun MS. Diagnosis and treatment of Parkinson disease: a review. JAMA. 2020;323(6) https://doi.org/10.1001/jama.2019.22360.
9. Tran J, Anastacio H, Bardy C. Genetic predispositions of Parkinson's disease revealed in patient-derived brain cells. NPJ Parkinsons Dis. 2020;6(1) https://doi.org/10.1038/s41531-020-0110-8.
10. Jia F, Fellner A, Kumar KR. Monogenic Parkinson's disease: genotype, phenotype, pathophysiology, and genetic testing. Genes. 2022;13(3) https://doi.org/10.3390/genes13030471.
11. Emamzadeh FN. Alpha-synuclein structure, functions, and interactions. J Res Med Sci. 2016;21(2) https://doi.org/10.4103/1735-1995.181989.
12. Brás J, Gibbons E, Guerreiro R. Genetics of synucleins in neurodegenerative diseases. Acta Neuropathol. 2021;141(4) https://doi.org/10.1007/s00401-020-02202-1.
13. Borsche M, Pereira SL, Klein C, Grünewald A. Mitochondria and Parkinson's disease: clinical, molecular, and translational aspects. J Parkinsons Dis. 2021;11(1) https://doi.org/10.3233/JPD-201981.
14. Usmani A, Shavarebi F, Hiniker A. The cell biology of LRRK2 in Parkinson's disease. Mol Cell Biol. 2021;41(5) https://doi.org/10.1128/mcb.00660-20.
15. Moehle MS, Webber PJ, Tse T, Sukar N, Standaert DG, Desilva TM, Cowell RM, West AB. LRRK2 inhibition attenuates microglial inflammatory responses. J Neurosci. 2012;32(5) https://doi.org/10.1523/JNEUROSCI.5601-11.2012.
16. Rui Q, Ni H, Li D, Gao R, Chen G. The role of LRRK2 in neurodegeneration of Parkinson disease. Curr Neuropharmacol. 2018;16(9) https://doi.org/10.2174/1570159x16666180222165418.
17. Chen CY, Weng YH, Chien KY, Lin KJ, Yeh TH, Cheng YP, Lu CS, Wang HL. (G2019S) LRRK2 activates MKK4-JNK pathway and causes degeneration of SN dopaminergic neurons

in a transgenic mouse model of PD. Cell Death Differ. 2012;19(10) https://doi.org/10.1038/cdd.2012.42.
18. Cruts M, Theuns J, Van Broeckhoven C. Locus-specific mutation databases for neurodegenerative brain diseases. Hum Mutat. 2012;33(9) https://doi.org/10.1002/humu.22117.
19. Gao C, Jiang J, Tan Y, Chen S. Microglia in neurodegenerative diseases: mechanism and potential therapeutic targets. Signal Transduct Target Ther. 2023;8(1) https://doi.org/10.1038/s41392-023-01588-0.
20. Maekawa T, Sasaoka T, Azuma S, Ichikawa T, Melrose HL, Farrer MJ, Obata F. Leucine-rich repeat kinase 2 (LRRK2) regulates α-synuclein clearance in microglia. BMC Neurosci. 2016;17(1) https://doi.org/10.1186/s12868-016-0315-2.
21. Carballo-Carbajal I, Weber-Endress S, Rovelli G, Chan D, Wolozin B, Klein CL, Patenge N, Gasser T, Kahle PJ. Leucine-rich repeat kinase 2 induces α-synuclein expression via the extracellular signal-regulated kinase pathway. Cell Signal. 2010;22(5) https://doi.org/10.1016/j.cellsig.2010.01.006.
22. Seirafi M, Kozlov G, Gehring K. Parkin structure and function. FEBS J. 2015;282(11) https://doi.org/10.1111/febs.13249.
23. Buetow L, Huang DT. Structural insights into the catalysis and regulation of E3 ubiquitin ligases. Nat Rev Mol Cell Biol. 2016;17(10) https://doi.org/10.1038/nrm.2016.91.
24. Norat P, Soldozy S, Sokolowski JD, Gorick CM, Kumar JS, Chae Y, Yağmurlu K, Prada F, Walker M, Levitt MR, Price RJ, Tvrdik P, Kalani MYS. Mitochondrial dysfunction in neurological disorders: exploring mitochondrial transplantation. npj Regener Med. 2020;5(1) https://doi.org/10.1038/s41536-020-00107-x.
25. Konovalova EV, Lopacheva OM, Grivennikov IA, Lebedeva OS, Dashinimaev EB, Khaspekov LG, Fedotova EY, Illarioshkin SN. Mutations in Parkinson's disease-associated PARK2 gene are accompanied by imbalance in programmed cell death systems. Acta Nat. 2015;7(4) https://doi.org/10.32607/20758251-2015-7-4-146-149.
26. Buneeva OA, Medvedev AE. DJ-1 protein and its role in the development of Parkinson's disease: studies on experimental models. Biochemistry (Mosc). 2021;86(6) https://doi.org/10.1134/S000629792106002X.
27. Lind-Holm Mogensen F, Scafidi A, Poli A, Michelucci A. PARK7/DJ-1 in microglia: implications in Parkinson's disease and relevance as a therapeutic target. J Neuroinflammation. 2023;20(1) https://doi.org/10.1186/s12974-023-02776-z.
28. Dolgacheva LP, Berezhnov AV, Fedotova EI, Zinchenko VP, Abramov AY. Role of DJ-1 in the mechanism of pathogenesis of Parkinson's disease. J Bioenerg Biomembr. 2019;51(3) https://doi.org/10.1007/s10863-019-09798-4.
29. Sharma N, Rao SP, Kalivendi SV. The deglycase activity of DJ-1 mitigates α-synuclein glycation and aggregation in dopaminergic cells: role of oxidative stress mediated downregulation of DJ-1 in Parkinson's disease. Free Radic Biol Med. 2019;135 https://doi.org/10.1016/j.freeradbiomed.2019.02.014.
30. Xu CY, Kang WY, Chen YM, Jiang TF, Zhang J, Zhang LN, Ding JQ, Liu J, Di Chen S. DJ-1 inhibits α-synuclein aggregation by regulating chaperone-mediated autophagy. Front Aging Neurosci. 2017;9(SEP) https://doi.org/10.3389/fnagi.2017.00308.
31. Imberechts D, Kinnart I, Wauters F, Terbeek J, Manders L, Wierda K, Eggermont K, Madeiro RF, Sue C, Verfaillie C, Vandenberghe W. DJ-1 is an essential downstream mediator in PINK1/parkin-dependent mitophagy. Brain. 2022;145(12) https://doi.org/10.1093/brain/awac313.
32. Repici M, Giorgini F. DJ-1 in Parkinson's disease: clinical insights and therapeutic perspectives. J Clin Med. 2019;8(9) https://doi.org/10.3390/jcm8091377.
33. Sassone J, Reale C, Dati G, Regoni M, Pellecchia MT, Garavaglia B. The role of VPS35 in the pathobiology of Parkinson's disease. Cell Mol Neurobiol. 2021;41(2) https://doi.org/10.1007/s10571-020-00849-8.
34. Qiao L, Hu J, Qiu X, Wang C, Peng J, Zhang C, Zhang M, Lu H, Chen W. LAMP2A, LAMP2B and LAMP2C: similar structures, divergent roles. Autophagy. 2023;19(11) https://doi.org/10.1080/15548627.2023.2235196.

35. Mak SK, McCormack AL, Manning-Bog AB, Cuervo AM, Di Monte DA. Lysosomal degradation of α-synuclein in vivo. J Biol Chem. 2010;285(18) https://doi.org/10.1074/jbc.M109.074617.
36. Gorell JM, Johnson CC, Rybicki BA, Peterson EL, Kortsha GX, Brown GG, Richardson RJ. Occupational exposures to metals as risk factors for Parkinson's disease. Neurology. 1997;48(3) https://doi.org/10.1212/WNL.48.3.650.
37. Ball N, Teo WP, Chandra S, Chapman J. Parkinson's disease and the environment. Front Neurol. 2019;10 https://doi.org/10.3389/fneur.2019.00218.
38. Rivera-Mancía S, Ríos C, Montes S. Manganese accumulation in the CNS and associated pathologies. Biometals. 2011;24(5) https://doi.org/10.1007/s10534-011-9454-1.
39. Pasquali L, Caldarazzo-Ienco E, Fornai F. MPTP neurotoxicity: actions, mechanisms, and animal modeling of parkinson's disease. In: Handbook of neurotoxicity, vol. 1. New York: Springer; 2014. https://doi.org/10.1007/978-1-4614-5836-4_170.
40. Rani L, Ghosh B, Ahmad MH, Mondal AC. Evaluation of potential neuroprotective effects of vanillin against MPP+/MPTP-induced dysregulation of dopaminergic regulatory mechanisms in SH-SY5Y cells and a mouse model of Parkinson's disease. Mol Neurobiol. 2023;60(8) https://doi.org/10.1007/s12035-023-03358-z.
41. Islam MS, Azim F, Saju H, Zargaran A, Shirzad M, Kamal M, Fatema K, Rehman S, Azad MAM, Ebrahimi-Barough S. Pesticides and Parkinson's disease: current and future perspective. J Chem Neuroanat. 2021;115 https://doi.org/10.1016/j.jchemneu.2021.101966.
42. Tanner CM, Kame F, Ross GW, Hoppin JA, Goldman SM, Korell M, Marras C, Bhudhikanok GS, Kasten M, Chade AR, Comyns K, Richards MB, Meng C, Priestley B, Fernandez HH, Cambi F, Umbach DM, Blair A, Sandler DP, Langston JW. Rotenone, paraquat, and Parkinson's disease. Environ Health Perspect. 2011;119(6) https://doi.org/10.1289/ehp.1002839.
43. Nasuti C, Brunori G, Eusepi P, Marinelli L, Ciccocioppo R, Gabbianelli R. Early life exposure to permethrin: a progressive animal model of Parkinson's disease. J Pharmacol Toxicol Methods. 2017;83 https://doi.org/10.1016/j.vascn.2016.10.003.
44. Sun M, Zhang X. Current methodologies in protein ubiquitination characterization: from ubiquitinated protein to ubiquitin chain architecture. Cell Biosci. 2022;12(1) https://doi.org/10.1186/s13578-022-00870-y.
45. Sharma B, Joshi D, Yadav PK, Gupta AK, Bhatt TK. Role of ubiquitin-mediated degradation system in plant biology. Front Plant Sci. 2016;7(June 2016) https://doi.org/10.3389/fpls.2016.00806.
46. Yang Q, Zhao J, Chen D, Wang Y. E3 ubiquitin ligases: styles, structures and functions. Mol Biomed. 2021;2(1) https://doi.org/10.1186/s43556-021-00043-2.
47. Snyder NA, Silva GM. Deubiquitinating enzymes (DUBs): regulation, homeostasis, and oxidative stress response. J Biol Chem. 2021;297(3) https://doi.org/10.1016/j.jbc.2021.101077.
48. Conway JA, Kinsman G, Kramer ER. The role of NEDD4 E3 ubiquitin–protein ligases in Parkinson's disease. Genes. 2022;13(3) https://doi.org/10.3390/genes13030513.
49. Abeywardana T, Lin YH, Rott R, Engelender S, Pratt MR. Site-specific differences in proteasome-dependent degradation of monoubiquitinated α-synuclein. Chem Biol. 2013;20(10) https://doi.org/10.1016/j.chembiol.2013.09.009.
50. Sugeno N, Hasegawa T, Tanaka N, Fukuda M, Wakabayashi K, Oshima R, Konno M, Miura E, Kikuchi A, Baba T, Anan T, Nakao M, Geisler S, Aoki M, Takeda A. Lys-63-linked ubiquitination by E3 ubiquitin ligase Nedd4-1 facilitates endosomal sequestration of internalized α-Synuclein. J Biol Chem. 2014;289(26) https://doi.org/10.1074/jbc.M113.529461.
51. McDonough H, Patterson C. CHIP: a link between the chaperone and proteasome systems. Cell Stress Chaperones. 2003;8(4) https://doi.org/10.1379/1466-1268(2003)008<0303:CALBTC>2.0.CO;2.
52. Wickner S, Maurizi MR, Gottesman S. Posttranslational quality control: folding, refolding, and degrading proteins. Science. 1999;286(5446) https://doi.org/10.1126/science.286.5446.1888.
53. Xu W, Marcu M, Yuan X, Mimnaugh E, Patterson C, Neckers L. Chaperone-dependent E3 ubiquitin ligase CHIP mediates a degradative pathway for c-ErbB2/Neu. Proc Natl Acad Sci U S A. 2002;99(20) https://doi.org/10.1073/pnas.202365899.

54. Rudenko IN, Kaganovich A, Langston RG, Beilina A, Ndukwe K, Kumaran R, Dillman AA, Chia R, Cookson MR. The G2385R risk factor for Parkinson's disease enhances CHIP-dependent intracellular degradation of LRRK2. Biochem J. 2017;474(9) https://doi.org/10.1042/BCJ20160909.
55. Singleton AB, Farrer MJ, Bonifati V. The genetics of Parkinson's disease: progress and therapeutic implications. Mov Disord. 2013;28(1) https://doi.org/10.1002/mds.25249.
56. Hu Z, Mao C, Wang H, Zhang Z, Zhang S, Luo H, Tang M, Yang J, Yuan Y, Wang Y, Liu Y, Fan L, Zhang Q, Yao D, Liu F, Schisler JC, Shi C, Xu Y. CHIP protects against MPP+/MPTP-induced damage by regulating Drp1 in two models of Parkinson's disease. Aging. 2021;13(1) https://doi.org/10.18632/aging.202389.
57. Deng H, Dodson MW, Huang H, Guo M. The Parkinson's disease genes pink1 and parkin promote mitochondrial fission and/or inhibit fusion in Drosophila. Proc Natl Acad Sci U S A. 2008;105(38) https://doi.org/10.1073/pnas.0803998105.
58. Yang Y, Gehrke S, Imai Y, Huang Z, Ouyang Y, Wang JW, Yang L, Beal MF, Vogel H, Lu B. Mitochondrial pathology and muscle and dopaminergic neuron degeneration caused by inactivation of Drosophila Pink1 is rescued by Parkin. Proc Natl Acad Sci U S A. 2006;103(28) https://doi.org/10.1073/pnas.0602493103.
59. Yoo L, Chung KC. The ubiquitin E3 ligase CHIP promotes proteasomal degradation of the serine/threonine protein kinase PINK1 during staurosporine-induced cell death. J Biol Chem. 2018;293(4) https://doi.org/10.1074/jbc.M117.803890.
60. Palubinsky AM, Stankowski JN, Kale AC, et al. CHIP Is an Essential Determinant of Neuronal Mitochondrial Stress Signaling. Antioxid Redox Signal. 2015;23(6):535–549. https://doi.org/10.1089/ars.2014.6102.
61. Chen J, Xue J, Ruan J, Zhao J, Tang B, Duan R. Drosophila CHIP protects against mitochondrial dysfunction by acting downstream of Pink1 in parallel with Parkin. FASEB J. 2017;31(12) https://doi.org/10.1096/fj.201700011R.
62. Dimant H, Zhu L, Kibuuka LN, Fan Z, Hyman BT, McLean PJ. Direct visualization of CHIP-mediated degradation of alpha-synuclein in vivo: implications for PD therapeutics. PLoS One. 2014;9(3) https://doi.org/10.1371/journal.pone.0092098.
63. Kalia LV, Kalia SK, Chau H, Lozano AM, Hyman BT, McLean PJ. Ubiquitinylation of α-synuclein by carboxyl terminus hsp70-interacting protein (chip) is regulated by bcl-2-associated athanogene 5 (bag5). PLoS One. 2011;6(2) https://doi.org/10.1371/journal.pone.0014695.
64. Fujikawa K, Nakahara K, Takasugi N, Nishiya T, Ito A, Uchida K, Uehara T. S-Nitrosylation at the active site decreases the ubiquitin-conjugating activity of ubiquitin-conjugating enzyme E2 D1 (UBE2D1), an ERAD-associated protein. Biochem Biophys Res Commun. 2020;524(4) https://doi.org/10.1016/j.bbrc.2020.02.011.
65. Yonashiro R, Ishido S, Kyo S, Fukuda T, Goto E, Matsuki Y, Ohmura-Hoshino M, Sada K, Hotta H, Yamamura H, Inatome R, Yanagi S. A novel mitochondrial ubiquitin ligase plays a critical role in mitochondrial dynamics. EMBO J. 2006;25(15) https://doi.org/10.1038/sj.emboj.7601249.
66. Koyano F, Yamano K, Kosako H, Kimura Y, Kimura M, Fujiki Y, Tanaka K, Matsuda N. Parkin-mediated ubiquitylation redistributes MITOL/March5 from mitochondria to peroxisomes. EMBO Rep. 2019a;20(12) https://doi.org/10.15252/embr.201947728.
67. Zhao Y, Shen W, Zhang M, Guo M, Dou Y, Han S, Yu J, Cui M, Zhao Y. DDAH-1 maintains endoplasmic reticulum-mitochondria contacts and protects dopaminergic neurons in Parkinson's disease. Cell Death Dis. 2024a;15(6):399. https://doi.org/10.1038/s41419-024-06772-w.
68. Koyano F, Yamano K, Kosako H, Tanaka K, Matsuda N. Parkin recruitment to impaired mitochondria for nonselective ubiquitylation is facilitated by MITOL. J Biol Chem. 2019b;294(26) https://doi.org/10.1074/jbc.RA118.006302.
69. Toyofuku T, Okamoto Y, Ishikawa T, Sasawatari S, Kumanogoh A. LRRK 2 regulates endoplasmic reticulum–mitochondrial tethering through the PERK-mediated ubiquitination pathway. EMBO J. 2020;39(2) https://doi.org/10.15252/embj.2018100875.

70. Yun J, Puri R, Yang H, Lizzio MA, Wu C, Sheng Z-H, Guo M. MUL1 acts in parallel to the PINK1/parkin pathway in regulating mitofusin and compensates for loss of PINK1/parkin. Elife. 2014;3 https://doi.org/10.7554/elife.01958.
71. Doktór B, Damulewicz M, Pyza E. Effects of MUL1 and PARKIN on the circadian clock, brain and behaviour in Drosophila Parkinson's disease models. BMC Neurosci. 2019a;20(1) https://doi.org/10.1186/s12868-019-0506-8.
72. Doktór B, Damulewicz M, Pyza E. Overexpression of mitochondrial ligases reverses rotenone-induced effects in a Drosophila model of Parkinson's disease. Front Neurosci. 2019b;13(FEB) https://doi.org/10.3389/fnins.2019.00094.
73. Taximaimaiti R, Li H. MUL1 gene polymorphisms and Parkinson's disease risk. Acta Neurol Scand. 2019;139(5) https://doi.org/10.1111/ane.13081.
74. Jo S, Park KW, Hwang YS, Lee SH, Ryu HS, Chung SJ. Microarray genotyping identifies new loci associated with dementia in parkinson's disease. Genes. 2021;12(12) https://doi.org/10.3390/genes12121975.
75. Shi L, Huang C, Luo Q, Xia Y, Liu H, Li L, Liu W, Ma W, Fang J, Tang L, Zeng W, Chen Z. Pilot study: molecular risk factors for diagnosing sporadic Parkinson's disease based on gene expression in blood in MPTPinduced rhesus monkeys. Oncotarget. 2017;8(62) https://doi.org/10.18632/oncotarget.22348.
76. Uoselis L, Lindblom R, Lam WK, Küng CJ, Skulsuppaisarn M, Khuu G, Nguyen TN, Rudler DL, Filipovska A, Schittenhelm RB, Lazarou M. Temporal landscape of mitochondrial proteostasis governed by the UPRmt. Sci Adv. 2023;9(38) https://doi.org/10.1126/sciadv.adh8228.
77. El Kodsi DN, Tokarew JM, Sengupta R, Lengacher NA, Chatterji A, Nguyen AP, Boston H, Jiang Q, Palmberg C, Pileggi C, Holterman CE, Shutinoski B, Li J, Fehr TK, LaVoie MJ, Ratan RR, Shaw GS, Takanashi M, Hattori N, et al. Parkin coregulates glutathione metabolism in adult mammalian brain. Acta Neuropathol Commun. 2023;11(1) https://doi.org/10.1186/s40478-022-01488-4.
78. Tang W, Thundyil J, Yin Lim GG, Tng TJW, Zhang Yeow SQ, Nair A, Chai C, Yao TP, Lim KL. Parkin regulates neuronal lipid homeostasis through SREBP2-lipoprotein lipase pathway—implications for Parkinson's disease. Hum Mol Genet. 2023;32(9) https://doi.org/10.1093/hmg/ddac297.
79. Goyani S, Shinde A, Shukla S, Saranga MV, Currim F, Mane M, Singh J, Roy M, Gohel D, Chandak N, Vasiyani H, Singh R. Enhanced translocation of TRIM32 to mitochondria sensitizes dopaminergic neuronal cells to apoptosis during stress conditions in Parkinson's disease. FEBS J. 2024;291(12) https://doi.org/10.1111/febs.17065.
80. Eleuteri S, Wang B, Cutillo G, Zhang Fang TS, Tao K, Qu Y, Yang Q, Wei W, Simon DK. PGC-1α regulation by FBXW7 through a novel mechanism linking chaperone-mediated autophagy and the ubiquitin-proteasome system. FEBS J. 2024; https://doi.org/10.1111/febs.17276.
81. Okarmus J, Agergaard JB, Stummann TC, Haukedal H, Ambjørn M, Freude KK, Fog K, Meyer M. USP30 inhibition induces mitophagy and reduces oxidative stress in parkin-deficient human neurons. Cell Death Dis. 2024;15(1) https://doi.org/10.1038/s41419-024-06439-6.
82. Martinez A, Sanchez-Martinez A, Pickering JT, Twyning MJ, Terriente-Felix A, Chen PL, Chen CH, Whitworth AJ. Mitochondrial CISD1/CISD accumulation blocks mitophagy and genetic or pharmacological inhibition rescues neurodegenerative phenotypes in Pink1/parkin models. Mol Neurodegener. 2024;19(1) https://doi.org/10.1186/s13024-024-00701-3.
83. Zhao Z, Song X, Wang Y, Yu L, Huang G, Li Y, Zong R, Liu T, Ji Q, Zheng Y, Liu B, Zhu Q, Chen L, Gao C, Liu H. E3 ubiquitin ligase TRIM31 alleviates dopaminergic neurodegeneration by promoting proteasomal degradation of VDAC1 in Parkinson's disease model. Cell Death Differ. 2024b;31(11):1410–21. https://doi.org/10.1038/s41418-024-01334-1.
84. Navarro E, Esteras N. A new mutation in the Parkinson's-related FBXO7 gene impairs mitochondrial and proteasomal function. FEBS J. 2024;291(12):2562–4. https://doi.org/10.1111/febs.17155.

85. Jeon SJ, Chung KC. The SCF-FBW7β E3 ligase mediates ubiquitination and degradation of the serine/threonine protein kinase PINK1. J Biol Chem. 2024;300(4):107198. https://doi.org/10.1016/j.jbc.2024.107198.
86. Yan YQ, Zheng R, Liu Y, Ruan Y, Lin ZH, Xue NJ, Chen Y, Zhang BR, Pu JL. Parkin regulates microglial NLRP3 and represses neurodegeneration in Parkinson's disease. Aging Cell. 2023;22(6) https://doi.org/10.1111/acel.13834.
87. Wasner K, Smajic S, Ghelfi J, Delcambre S, Prada-Medina CA, Knappe E, Arena G, Mulica P, Agyeah G, Rakovic A, Boussaad I, Badanjak K, Ohnmacht J, Gérardy JJ, Takanashi M, Trinh J, Mittelbronn M, Hattori N, Klein C, et al. Parkin deficiency impairs mitochondrial DNA dynamics and propagates inflammation. Mov Disord. 2022;37(7) https://doi.org/10.1002/mds.29025.
88. Xin D, Gu H, Liu E, Sun Q. Parkin negatively regulates the antiviral signaling pathway by targeting TRAF3 for degradation. J Biol Chem. 2018;293(31) https://doi.org/10.1074/jbc.RA117.001201.
89. Dai D, Yuan J, Wang Y, Xu J, Mao C, Xiao Y. Peli1 controls the survival of dopaminergic neurons through modulating microglia-mediated neuroinflammation. Sci Rep. 2019;9(1) https://doi.org/10.1038/s41598-019-44573-w.
90. Zang, C. xia, Wang, L., Yang, H. yu, Shang, J. mei, Liu, H., Zhang, Z. hong, Ju, C., Yuan, F. yu, Li, F. yuan, Bao, X. qi, & Zhang, D. (2022). HACE1 negatively regulates neuroinflammation through ubiquitylating and degrading Rac1 in Parkinson's disease models. Acta Pharmacol Sin, 43(2). doi: https://doi.org/10.1038/s41401-021-00778-2.
91. Shao L, Liu X, Zhu S, Liu C, Gao Y, Xu X. The role of Smurf1 in neuronal necroptosis after lipopolysaccharide-induced neuroinflammation. Cell Mol Neurobiol. 2018;38(4) https://doi.org/10.1007/s10571-017-0553-6.
92. Omura T, Matsuda H, Nomura L, Imai S, Denda M, Nakagawa S, Yonezawa A, Nakagawa T, Yano I, Matsubara K. Ubiquitin ligase HMG-CoA reductase degradation 1 (HRD1) prevents cell death in a cellular model of Parkinson's disease. Biochem Biophys Res Commun. 2018;506(3) https://doi.org/10.1016/j.bbrc.2018.10.094.
93. Ji M, Niu S, Guo J, Mi H, Jiang P. Silencing RNF13 alleviates Parkinson's disease – like problems in mouse models by regulating the endoplasmic reticulum stress–mediated IRE1α-TRAF2-ASK1-JNK pathway. J Mol Neurosci. 2020;70(12) https://doi.org/10.1007/s12031-020-01599-4.
94. Lee SH, Chung KC. USP7 attenuates endoplasmic reticulum stress-induced apoptotic cell death through deubiquitination and stabilization of FBXO7. PLoS One. 2023;18(10) https://doi.org/10.1371/journal.pone.0290371.
95. El Manaa W, Duplan E, Goiran T, Lauritzen I, Vaillant Beuchot L, Lacas-Gervais S, Morais VA, You H, Qi L, Salazar M, Ozcan U, Chami M, Checler F, Alves da Costa C. Transcription- and phosphorylation-dependent control of a functional interplay between XBP1s and PINK1 governs mitophagy and potentially impacts Parkinson disease pathophysiology. Autophagy. 2021;17(12) https://doi.org/10.1080/15548627.2021.1917129.
96. Singh K, Han K, Tilve S, Wu K, Geller HM, Sack MN. Parkin targets NOD2 to regulate astrocyte endoplasmic reticulum stress and inflammation. Glia. 2018;66(11) https://doi.org/10.1002/glia.23482.
97. Nguyen TN, Sawa-Makarska J, Khuu G, Lam WK, Adriaenssens E, Fracchiolla D, Shoebridge S, Bernklau D, Padman BS, Skulsuppaisarn M, Lindblom RSJ, Martens S, Lazarou M. Unconventional initiation of PINK1/Parkin mitophagy by Optineurin. Mol Cell. 2023;83(10) https://doi.org/10.1016/j.molcel.2023.04.021.
98. Lebedeva O, Poberezhniy D, Novosadova E, Gerasimova T, Novosadova L, Arsenyeva E, Stepanenko E, Shimchenko D, Volovikov E, Anufrieva K, Illarioshkin S, Lagarkova M, Grivennikov I, Tarantul V, Nenasheva V. Overexpression of Parkin in the neuronal progenitor cells from a patient with Parkinson's disease shifts the transcriptome towards the normal state. Mol Neurobiol. 2023;60(6) https://doi.org/10.1007/s12035-023-03293-z.
99. Yeon M, Bertolini I, Agarwal E, Ghosh JC, Tang HY, Speicher DW, Keeney F, Sossey-Alaoui K, Pluskota E, Bialkowska K, Plow EF, Languino LR, Skordalakes E, Caino MC, Altieri

DC. Parkin ubiquitination of Kindlin-2 enables mitochondria-associated metastasis suppression. J Biol Chem. 2023;299(6) https://doi.org/10.1016/j.jbc.2023.104774.
100. Qin X, Wang R, Xu H, Tu L, Chen H, Li H, Liu N, Wang J, Li S, Yin F, Xu N, Li Z. Identification of an autoinhibitory, mitophagy-inducing peptide derived from the transmembrane domain of USP30. Autophagy. 2022;18(9) https://doi.org/10.1080/15548627.2021.2022360.

Part III

Therapeutic Options

Natural Products in the Regulation of Ubiquitin-Proteasome System and Neurodegenerative Disorders

13

Priyanka Barman and Sukesh R. Bhaumik

13.1 Introduction

The ubiquitin-proteasome system (UPS) is evolutionarily conserved among eukaryotes [1–4]. It plays an important role in regulating cellular protein homeostasis via targeted degradation. In the UPS pathway, the target protein gets ubiquitylated through its direct interaction with an E3 ubiquitin ligase following the sequential actions of the E1 activating and E2 conjugates enzymes and is subsequently targeted for degradation by the 26S proteasome [1–4]. The 26S proteasome consists of a barrel-shaped 20S core particle (CP) and a 19S regulatory particle (RP). The 20S CP has the proteolytic activity, while the 19S RP recognizes K48 (lysine 48)-linked polyubiquitylated-protein, and then feeds to the 20S CP for degradation in an ATPase-dependent fashion [1–4]. The 20S CP is composed of the four layers of heptamer rings [two α (α1-α7) and two β (β1–β7) rings] that are organized as α-β-β-α [1–5]. The catalytic activity of the 20S CP resides within three components in the β ring. These components possess the chymotrypsin-like (β5), trypsin-like (β2), and post-acidic or caspase-like (β1) proteolytic activities that cleave proteins after hydrophobic, basic and acidic residues, respectively [1, 2, 6, 7].

Through targeted degradation, UPS is involved in protein quality control. Thus, the dysfunction of the UPS is implicated in the pathogenesis of tumors and cardiovascular, neurodegenerative, and other diseases [8–15]. Pathologies in neurodegenerative diseases are characterized by the accumulation of aggregation-prone proteins such as α-synuclein in Parkinson's disease (PD), amyloid β and tau in Alzheimer's disease (AD), and mutant huntingtin (HTT) containing polyglutamine (polyQ) repeat in Huntingtin disease (HD) [16]. These aggregates can interact with the

P. Barman · S. R. Bhaumik (✉)
Division of Biochemistry and Molecular Biology, Department of Biomedical Sciences, Southern Illinois University School of Medicine, Carbondale, IL, USA
e-mail: sbhaumik@siumed.edu

S. Singh (ed.), *Ubiquitin Proteasome System in CNS Neurodegenerative Diseases*, https://doi.org/10.1007/978-981-95-6099-8_13

proteasome and alter its activity negatively, thereby impairing proteasome activity [17]. Such impaired activity would lead to progressive accumulation of these proteins in neuronal cells, hence contributing to the progression and severity of the diseases. Further, neuroinflammation, oxidative damage, apoptosis and other factors also contribute to the progression of the neurodegenerative disorders. Importantly, inflammation, oxidative stress and apoptosis are regulated by the UPS/proteasome [18–22]. Thus, UPS/proteasome plays crucial roles in regulating in inflammation, oxidative stress, apoptosis, protein aggregation and neuronal cell survival, and hence neuronal/brain health. Therefore, the compounds/molecules that can modulate the UPS/proteasomal activity would have neuroprotective functions and can be useful in treating neurodegenerative diseases. Several natural products have been shown to regulate the proteasomal/UPS activity with impact on inflammation, oxidative stress, apoptosis and cell survival for neuroprotection, and thus have therapeutic potentials for neurodegenerative disorders. Here, we discuss several natural products such as curcumin, epigallocatechin gallate (EGCG), genistein, quercetin and resveratrol in the regulation of UPS/proteasome and neurodegenerative diseases.

13.2 Curcumin in the Regulation of Cellular Processes

Curcumin, a polyphenolic compound, is extracted from the rhizomes of Curcuma longa (Turmeric). It exerts diverse biological functions such as anti-inflammatory, anti-oxidant, anti-bacterial, anti-viral, neuroprotective, chemo-preventive, and anti-cancer activities, and has attained potential therapeutic values in treating various diseases such as cancer, cardiovascular and neurodegenerative disorders. Curcumin controls the activity of the 26S proteasome and oxidative stress by regulating the transcription factor, NRF2 (Nuclear factor erythroid 2-related factor 2), as described below.

NRF2 undergoes polyubiquitylation and 26S proteasomal degradation in the cytoplasm under normal physiological conditions. Such targeted degradation of NRF2 occurs via its association with KEAP1 (Kelch-like ECH associated protein 1), which is recognized by CUL3-RBX1 to form the KEAP1-CUL3-RBX1 E3 ubiquitin ligase complex for polyubiquitylation and 26S proteasomal degradation of NRF2 [23–27]. However, upon oxidative stress, KEAP1 gets oxidized and phosphorylated [23, 26, 27]. Such modification changes KEAP1's conformation, which loses interaction with NRF2, thereby releasing NRF2 to be translocated to the nucleus [23, 26–28]. In the nucleus, NRF2 forms a heterodimer with sMAF (Small musculoaponeurotic fibrosarcoma) and binds to AREs (Anti-oxidant response elements) of various genes involved in anti-oxidant response [e.g., HO-1 (Heme oxygenase-1), NQO1 (NADPH:quinine oxidoreductase-1), GST (Glutathione S-transferases), subunits of gamma-glutamylcysteine synthetase, SOD (Superoxide dismutase), and GPx (Glutathione peroxidase)] [23, 29, 30]. Such binding of NRF2 promotes transcription of the anti-oxidant genes that maintain cellular homeostasis by neutralizing reactive oxygen species.

Shin et al. [31] investigated the role of curcumin on the activation/stabilization of NRF2 and expression of the NRF2 target genes in mice. Treatment of mouse cells with curcumin stabilized NRF2 by inhibiting its ubiquitination and 26S proteasomal degradation, and hence increased expression of the NRF2 target gene. However, cells transfected with KEAP1 point-mutated at C151 (Cysteine 151) with S (Serine) showed reduced transcription of the NRF2-target genes in response to curcumin treatment [31]. This is due to the fact that curcumin binds to KEAP1 at C151 to impair NRF2-KEAP1 interaction and subsequent release of NRF2 to be translocated to the nucleus for transcription of the target genes. Thus, when KEAP1 is point-mutated at C151, curcumin cannot stabilize NRF2, and hence the expression of the NRF2 target genes is impaired. Therefore, curcumin plays a critical role in the stabilization of NRF2 (and subsequent expression of the target genes) by preventing its ubiquitylation and 26S proteasomal degradation. Further, another study showed that curcumin activates NRF2 through PKCδ (Protein kinase C-delta)-mediated p62 phosphorylation at S351 [32–34]. Such phosphorylation of p62 impairs NRF2-KEAP1 interaction, thus releasing NRF2 for transcription of the target genes, including p62 (that subsequently stabilizes NRF2 in a positive feedback loop) [32–34]. Thus, curcumin increases the level of NRF2 by disrupting the NRF2-KEAP1 interaction (directly by interacting with KEAP1 and indirectly via phosphorylation of p62), which, in turn, promotes the expression of the anti-oxidant genes, thereby exhibiting anti-oxidant activity. Further, an increased level of NRF2 in the presence of curcumin enhances the expression of the 26S proteasome components, and hence its assembly and activity [35, 36]. Increased proteasomal activity would promote degradation through heat shock proteins (HSPs), thus preventing protein aggregation to improve neuronal/brain health. Therefore, curcumin has neuroprotective activity via enhanced expression of the proteasome subunits and anti-oxidant genes.

Further, curcumin regulates inflammation by controlling the activity of NF-kB [Nuclear factor kappa B; a heterodimer of p50 and p65 (RelA)]. It inhibits pro-inflammatory cytokine IL-1β (Interleukin-1β)-induced NF-κB activation, and hence inflammation. Such downregulation of NF-kB activity occurs via the inhibition of phosphorylation of I-kB that binds to NF-kB and blocks NF-kB's NLS (Nuclear localization signal), thus impairing NF-kB's translocation from cytoplasm to the nucleus for transcription of the target genes. Curcumin inhibits the activity of I-kB kinase that phosphorylates I-kB for ubiquitylation and 26S proteasomal degradation [37–40]. Impaired phosphorylation of I-kB by curcumin inhibits I-kB's ubiquitylation and subsequent proteasomal degradation. Thus, in response to curcumin treatment, I-kB remains bound to NF-kB in the cytoplasm, resulting in impaired translocation of NF-kB to the nucleus. Such a defect in nuclear translocation would impair transcription of the NF-kB target/regulated genes, including the ones associated with inflammation (e.g., cytokines, chemokines, adhesion molecules, and acute phase proteins) [41]. Further, decreased NF-kB activity is anti-apoptotic via downregulation of apoptosis gene expression and stimulation of cell survival/anti-apoptotic genes. Therefore, curcumin has anti-inflammatory and anti-apoptotic functions via downregulation of the NF-kB activity. Further, curcumin exhibits these effects by interacting with and triggering the PPAR-γ [Peroxisome proliferator-activated receptor-γ, a transcription activator that binds to PPAR-responsive regulatory elements as a heterodimer with RXR (Retinoid X receptor) to regulate transcription of the target genes] pathway that promotes ubiquitylation and 26S proteasomal degradation of the NF-kB component, p65, via PPAR-γ's E3 ubiquitin ligase activity (independently of its transactivation function), thereby suppressing NF-kB pathway and target gene expression [42].

Curcumin also regulates inflammatory response through the JAK/STAT (Janus kinase/ Signal transducer and activator of transcription) pathway via the regulation of STAT and SOCS (Suppressor of cytokine signaling; antagonist of JAK/STAT pathway) [43–45]. It binds to STAT and JAK, and inhibits phosphorylation of JAK/STAT, thus attenuating the inflammatory response. Further, curcumin also protects neurons via its ability to upregulate sirtuin 1 (SIRT1). SIRT1 is a NAD^+-dependent deacetylase and is involved in the deacetylation of histone and non-histone proteins (including transcription factors), and regulates immune functions [46]. In addition to the above mechanisms, curcumin has also been shown to regulate the activities of other transcription factors (e.g., ATF3, AP-1, STAT3 and EGR-1), and hence gene expression and associated cellular processes, as described elsewhere [47–49].

13.3 Neuroprotective Functions of Curcumin

A number of studies supported the neuroprotective functions of curcumin against neurodegenerative diseases, as described below. In AD, two major pathological hallmarks are extracellular amyloid β plaques and intraneuronal neurofibrillary tangles [50]. Curcumin prevents amyloid β aggregation. It binds to amyloid β to block aggregation and fibril formation in vitro as well as in vivo [51]. Curcumin at a low dose effectively disaggregates amyloid β, and impedes fibril and oligomer formation. Further, curcumin treatment helps to improve learning and memory, prevents degradation of synapses, and inhibits overproduction and accumulation of amyloid β in AD brains by downregulating the expression of BACE1 (Beta-site amyloid precursor protein cleaving enzyme 1 that processes amyloid precursor protein to amyloid β in inducing AD) [52–54]. Curcumin can also reduce amyloid β-induced neurotoxicity by inhibiting oxidative damage and regulating ERK (Extracellular signal-regulated kinase that is overactive in AD). Further, as described above, curcumin shows decreased inflammatory response and oxidative stress. It also stimulates AMPK (AMP-activated protein kinase that is decreased in the brains of AD patients with reduced mitochondrial biosynthesis and functions) [55, 56]. These observations support the therapeutic potential of curcumin for AD patients.

In addition to interacting with amyloid β, curcumin can also bind to other amyloid proteins, p-tau (Phosphorylated tau) and α-synuclein [57]. Mutations in α-synuclein (i.e., A30P, A53T and E46K; A, Alanine; P, Proline; T, Threonine; and E, Glutamic acid) cause PD [58]. Using the PC12 cells expressing A53T α-synuclein under induction, Liu et al. [59] showed that curcumin provides protection against A53T α-synuclein-induced cell death. Further, curcumin reduces mutant α-synuclein-induced intracellular reactive oxygen species level, mitochondrial depolarization, cytochrome c release (Mitochondrial membrane protein that is released to cytosol under apoptotic stimulation), and activation of caspase-3 (an effector of apoptosis) and caspase-9 (an initiator of intrinsic apoptosis) [59]. These implicate curcumin as a neuroprotective agent for A53T α-synuclein-linked PD. In addition, using human neuroblastoma SH-SY5Y cells following treatment with MPTP (1-methyl-4-phenyl-tetrahydropyridinium) ion and curcumin, Wu et al. [60] revealed that curcumin provides protection against dopaminergic neuron damage caused by MPTP. Further, curcumin treatment improves the behavior of the rotenone-induced mouse model of PD [61].

Aggregation of amyloidogenic proteins such as amyloid β, tau, HTT and α-synuclein is inhibited by HSPs [62]. HSPs promote ubiquitination and 26S proteasomal degradation of the aggregated or misfolded proteins [63]. Curcumin enhances the expression of HSPs [64]. Elevated levels of HSPs prevent aggregation, hence attenuating disease progression [65–69]. Curcumin has beneficial roles in HD with improved neuropathology and transcription in CAG 140 (CAG, codon for Q or glutamine) knock-in mice [70]. Further, curcumin and its combination with piperidine improve behavioral and neurological disorders in quinolinic acid-induced HD rats [71]. Furthermore, curcumin has been shown to suppress the degeneration of photoreceptor neurons in the Drosophila HD model. In addition, curcumin antagonizes polyQ-induced motor neuronal dysfunction [72].

Curcumin also has beneficial effects in ALS. For example, curcumin removes TDP-43 (TAR DNA binding protein 43; a pathological protein in ALS)-induced excitability in ALS [73]. Further, dimethoxy curcumin decreases mitochondrial dysfunction in ALS [74]. In addition, curcumin binds to SOD1 to reduce fibrillation, leading to smaller and unstructured aggregates with less toxic effects [75]. Taken together, the above studies support the therapeutic potential of curcumin for neurodegenerative diseases.

There are a number of clinical studies on curcumin for neurodegenerative diseases. In a six-month randomized, placebo-controlled, double-blind, pilot clinical trial in China, 34 AD patients were investigated in three subgroups where curcumin was administered at doses including 4, 1 or 0 g of curcumin once daily [76]. When comparing between placebo and intervention groups, no considerable difference in cognitive function or plasma isoprostane iPF2α-III and Amyloid β40 levels was observed. Curcumin administration at a high dose did not have adverse side effects in AD patients [76]. However, there was a marked increase in vitamin E level. Likewise, in another randomized, double blind, placebo-controlled clinical study using curcumin C3 complex for 36 people with mild-to-moderate AD, Ringman et al. [77] did not find a significant difference in outcomes between curcumin-treated and placebo groups. However, in a 12-month, double-blind, randomized, placebo-controlled clinical trial of 54 probable ALS patients in Iran with nanocurcumin treatment (80 mg/day), a significant improvement in outcomes in terms of survival and mechanical ventilation was observed [78]. Another pilot trial performed in Italy with 42 ALS patients showed less oxidative stress in response to curcumin treatment, suggesting a delay in ALS progression [79]. Although there have been a number of clinical studies on curcumin, various reasons, such as poor bioavailability of curcumin, selection of cohorts, and differences between animal models and patients, might have contributed to the outcomes.

13.4 Neuroprotective Functions of EGCG

EGCG is the main bioactive polyphenol in green tea [80]. It has been demonstrated to inhibit the proteolytic function of the 26S proteasome and associated pathways (e.g., NRF2 and NF-kB) to combat neuroinflammation, oxidation and apoptosis, as described above for curcumin [39]. Moreover, it also modulates ubiquitylation by regulating E2 conjugates and E3 ligase enzymes [39]. Further, EGCG impairs protein aggregates via direct interaction to provide neuroprotection [81]. In addition, EGCG prevents tau phosphorylation [82]. Phosphorylation of tau is associated with mitochondrial dysfunction. Thus, EGCG has therapeutic implications in neurological diseases. There have been clinical trials on EGCG for neurodegenerative disorders. EGCG has been shown to reduce the measurable molecular impairment in PD [83]. A phase 2 clinical trial initiated in Spain assessed the effects of 12 months of treatment of EGCG on various cognitive outcomes in 84 people with Down's syndrome between the ages of 14 and 29 years. Participants with ECGC treatment showed significant improvement on the tests of memory, executive function and attention [84]. In another clinical trial using randomized, double-blind, parallel group, placebo-controlled patients with possible or probable multiple system atrophy in Germany with 48 weeks of EGCG treatment (1.2 g per day), there was no improvement, but rather side effects in some patients [85]. Likewise, a phase 2 clinical trial in patients with multiple sclerosis following 36-month treatment of EGCG (up to 1.2 g per day) did not show improvement as well as side effects [86]. Poor bioavailability of EGCG, selection of cohorts, differences between animal models and patients, and other factors might have contributed to the outcomes in the above clinical trials.

13.5 Neuroprotective Functions of Genistein

Genistein, an isoflavone, is predominantly found in soybean [87, 88]. This natural product can cross the blood–brain barrier, and exerts neuroprotective activity against neurodegenerative diseases such as AD, PD and HD [87, 88]. Like curcumin and EGCG, genistein possesses a number of activities, including anti-oxidant and anti-inflammatory properties and improves spatial memory and learning [87, 88]. It prevents the accumulation of amyloid β protein and suppresses amyloid β deposition by upregulating Apolipoprotein E (as Apolipoprotein E has a strong affinity to amyloid β for ingestion via endocytosis with its receptor, thereby reducing its extracellular level) [89, 90]. Further, it prevents neuronal fiber entanglement by decreasing the hyperphosphorylation of microtubule-associated protein, tau [89]. Genistein also suppresses mitochondrial dysfunction and reduces caspase-3 overexpression, thereby impairing apoptosis [88, 89]. It also has an anti-estrogenic property, as it resembles estrogen [88]. Further, genistein is found to inhibit the chymotrypsin-like activity of the proteasome in vitro and in vivo, thus 26S proteasome activity [91].

As described above, genistein has beneficial roles for neuronal/brain health. Indeed, Yi et al. [92] showed that genistein alleviates amyloid β-induced neurotoxicity in the neuroblastoma cell line (SH-SY5Y), thus providing a new strategy for the treatment of neurodegenerative diseases. Further, Li et al. [93] showed neuroprotective effects of genistein on hypoxic-ischemic brain damage by attenuating oxidative stress and neuroinflammation through modulating NRF2 and NF-κB signaling pathways. Further, Gao et al. [94] have shown that genistein improves learning and memory in AD rats through reduction of the neuronal loss in the hippocampus. Such decrease in neuronal loss is mediated via inhibition of mitochondrial apoptotic proteins [e.g., caspase-3, Bax (Bcl-2-associated X protein; a pro-apoptotic factor) and cytochrome c]. Further, in a double-blind, placebo-controlled, bicentric clinical trial using 120 mg/day of genistein for 12 months on 24 prodromal AD patients, Vina et al. [95] found a delay in the onset of the disease with significant improvement in genistein-treated participants. Moreover, Zhao et al. [96] have shown the neuroprotective function of genistein in the transgenic mouse model (SOD1-G93A; G, Glycine) of ALS, implicating it as a promising agent for the treatment of the ALS patients. Further, their study revealed that genistein downregulates the production of inflammatory mediators and improves gliosis in the spinal cord of the transgenic mouse model of ALS with prolonged lifespan [96]. Thus, a number of studies implicated the beneficial roles of genistein in neurodegenerative diseases.

13.6 Neuroprotective Functions of Quercetin

Quercetin is a flavonoid (a polyphenol containing 3-hydroxyl flavone structure) that can be found in fruits (apples, berries and grapes), tea, onion and green vegetables [97]. It possesses the anti-cancer, anti-viral, cardioprotective and anti-inflammatory activities [98, 99]. Further, quercetin can suppress α-synuclein fibrillization to potentially prevent the accumulation of protein aggregates in the brain [100]. It can also activate above-described NRF2-ARE pathway to provide anti-oxidant activity and neuroprotective functions [101]. It reduces NF-kB p65 ubiquitylation, and hence increases its stability. However, quercetin also inhibits the proteolytic activities of all three catalytic subunits of the proteasome, and thus impairs transcription of the NF-kB target genes, leading to its anti-inflammatory function [39, 102]. Intriguingly, Martin-Aragon et al. [103] reported the role of quercetin in enhancing proteasomal activity. Likewise, Chakraborty et al. [104] revealed that quercetin improves UPS. It is not clear how quercetin has both stimulatory and inhibitory effects on UPS. Further, quercetin promotes frataxin-mediated PINK1/Parkin-dependent mitochondrial autophagy (mitophagy) [39]. It inhibits ubiquitin-like protein MNSFβ (Monoclonal non-specific suppressor factor beta) and HSC70 siRNA, thus regulating inflammation [39]. Moreover, quercetin regulates sirtuins for neuroprotection [39]. Quercetin promotes the activity of SIRT1 that reduces the level of reactive oxygen species in neural and glial cells [105], hence promoting neuronal/brain health. Several in vitro and in vivo studies suggested the efficacy of this natural product in treating neurodegenerative diseases. For example, Zu et al. [106] found the neuroprotective effects of quercetin in AD. Bayazid and Lim [101] revealed that quercetin modulates the NRF2/HO1 signaling pathway for neuroprotection against oxidative stress-mediated damage. Ho et al. [107] implicated that quercetin increases mitochondrial biogenesis and reduces free radicals in neuronal cells, thus providing a protective effect against neuronal damage. Additionally, several studies in the animal models of PD showed neuroprotective effects of quercetin [108]. Thus, a number of studies support the neuroprotective functions of quercet.

13.7 Neuroprotective Functions of Resveratrol

Resveratrol is a natural polyphenol. It is found in grapes, berries, peanuts and wine [109]. It has shown strong anti-inflammatory and anti-oxidative activities by regulating proteasome/UPS [39]. Thus, resveratrol has neuroprotective functions. Resveratrol downregulates proteasomal subunit expression [39]. Resveratrol an also inhibit USP22 (Ubiquitin protease that reverses ubiquitylation) expression [39, 110]. By reducing proteasome/UPS activity, resveratrol impairs the NF-kB signaling pathway, which contributes to its anti-inflammatory functions. Further, resveratrol activates SIRT1 and regulates immune functions [39]. It also inhibits the insulin-like growth factor-1 (IGF-1)/serine/threonine kinase (Akt)/Wnt pathways, and hence downstream events [39, 111–113].

Studies on neuroprotection reveal the roles of resveratrol in maintaining cognitive functions with improvement of neurological disorders due to its anti-oxidative and anti-inflammatory properties [39]. Further, resveratrol impairs expression of lipopolysaccharide (LPS; a pathogen-derived inflammatory stimulus)-derived pro-inflammatory molecules in microglial cells, lowers neuroinflammation and increases adaptive immunity in AD patients [39, 114]. Further, resveratrol reduces amyloid β plaque and reactive oxygen species, and improves mitochondrial function and cell survival (by activating the SIRT1 pathway) in AD patients [114, 115]. A phase 2 randomized and placebo-controlled clinical trial with mid-to-moderate cases of AD showed that resveratrol was detectable in CSF (Cerebrospinal fluid) with a decline of amyloid β40 in CSF and plasma of AD patients [114, 115]. Further, resveratrol has been shown to reduce metalloprotease 9 (an inflammation marker) in CSF and modulate neuroinflammation in AD subjects [114, 115]. These observations implicate the therapeutic potential of resveratrol in AD patients. Further, resveratrol modulates mRNA and protein levels of apoptotic regulators, Bax and Bcl-2 (B-cell lymphoma 2; an anti-apoptotic protein), to increase cell survival [114].

Further, resveratrol shows neuroprotective roles in MPTP-generated PD mice [116, 117]. This occurs via scavenging free radicals. Moreover, resveratrol treatment decreases MPTP-induced OH (Hydroxyl radical) production, motor coordination destruction, and neuronal death, and thus protects PD mice [116, 117]. Similarly, resveratrol lowers oxidative damage in the PD animal model [117]. Further, resveratrol also reduces apoptosis in neuroblastoma cells [114]. Moreover, resveratrol may reduce mitochondrial dysfunction via functional impairment of parkin (encoded by the *PARK2* gene), which has an important role in ubiquitylation and increases cell survival [118]. Mutations in *PARK2* contribute to mitochondrial dysfunction and neuronal death in PD [118].

Resveratrol has also been shown to improve motor and cognitive impairments through inhibition of cyclooxygenase-1 in the HD model [119]. It also protects against the toxicity of mutant polyQ HTT via activation of SIRT1 [119]. Mutant polyQ HTT has been implicated to trigger neurodegeneration via mitochondrial dysfunction, apoptosis and oxidative stress [119]. However, resveratrol decreases such toxic effects of mutant polyQ HTT by increasing SIRT1, which interacts with and deacetylates p53 [119, 120]. Deacetylated-p53 has less activity, and the p53-mediated apoptosis is reduced following resveratrol treatment. Otherwise, activated p53 would increase mitochondrial oxidation/dysfunction in HD [119, 121]. Further, resveratrol would activate the SIRT1-PGC1α (Peroxisome proliferator-activated receptor γ co-activator-1 alpha) pathway that controls the expression of the anti-oxidant enzymes to lower reactive oxygen species (and hence oxidative stress), leading to improved mitochondrial functions [118, 119, 122]. Likewise, resveratrol has beneficial roles in slowing down mitochondrial dysfunction and associated motor neuron damage in ALS [123–127]. In agreement, resveratrol has been shown to decrease the level of reactive oxygen species in ALS mice. Further, activation of PGC1α in response to resveratrol treatment decreases oxidative stress, increases neuronal survival, and improves mitochondrial functions and motor performance in the ALS cell model [126, 127]. Thus, various studies support potential therapeutic roles of resveratrol for neurodegenerative disorders.

13.8 Concluding Remarks

Neurodegenerative diseases are caused by various factors such as neuroinflammation, oxidative stress and protein aggregates. Natural products have shown promising therapeutic potential for the treatment of neurodegenerative disorders. We have described above a set of natural products with functions in neuroprotection through various mechanisms/pathways. We have also discussed the roles of these natural products in the regulation of UPS/proteasome, which plays important roles in the regulation of the processes associated with the maintenance of neuronal/brain health. However, to provide the simplistic view of the field in this book chapter, the details of various regulatory pathways and the mechanisms-of-action of the natural products in neurodegenerative disorders are not thoroughly described. Nonetheless, we have concisely described a set of natural products in the regulation of UPS/proteasome and associated/other cellular processes involved in neuroprotection with therapeutic potentials for neurodegenerative diseases, which would be an important source for researchers working (or planning to work) in this field.

Acknowledgments The work in the Bhaumik laboratory was supported by the grants from American Heart Association (25AIREA1416730), National Cancer Institute (1R15CA290595-01A1), Simmons Cancer Institute (Team science grant) and Southern Illinois University School of Medicine (Research seed and near-miss grants).

References

1. Bhaumik SR, Malik S. Diverse regulatory mechanisms of eukaryotic transcriptional activation by the proteasome complex. Crit Rev Biochem Mol Biol. 2008;43:419–33.
2. Frankland-Searby S, Bhaumik SR. The 26S proteasome complex: an attractive target for cancer therapy. Biochim Biophys Acta. 2012;1825:64–76.
3. Bhaumik SR. Distinct regulatory mechanisms of eukaryotic transcriptional activation mediated by SAGA and TFIID. Biochim Biophys Acta. 2011;1809:97–108.
4. Durairaj G, Garg P, Bhaumik SR. Nuclear export of mRNA and its regulation by ubiquitylation. RNA Biol. 2009;6:531–5.
5. Groll M, Dtizel L, Lowe J, Stock D, Bochtler M, Wolf DH, Huber R. Structure of the 20S proteasome from yeast at 2.4 a resolution. Nature. 1997;386:463–71.
6. Heinemeyer W, Fischer M, Krimmer T, Stachon U, Wolf DH. The active sites of the eukaryotic 20S proteasome and their involvement in subunit precursor processing. J Biol Chem. 1997;272:25200–9.
7. Groll M, Heinemeyer W, Jager S, Ulrich T, Bochtler M, Wolf DH, Huber R. The catalytic sites of 20S proteasomes and their role in subunit maturation: a mutational and crystallographic study. Proc Natl Acad Sci. 1999;96:10975–83.
8. Popovic D, Vucic D, Dikic I. Ubiquitination in disease pathogenesis and treatment. Nat Med. 2014;20:1242–53.
9. Zeng L, Zhang X, Huang Z, Song S, Li M, Wang T, Sun A, Ge J. Ubiquitin proteasome system in cardiac fibrosis. J Adv Res. 2025;76:501–10.
10. Hwang JT, Lee A, Kho C. Ubiquitin and ubiquitin-like proteins in cancer, neurodegenerative disorders, and heart diseases. Int J Mol Sci. 2022;23:5053.
11. Barac YD, Emrich F, Krutzwakd-Josefson E, Schrepfer S, Sampaio LC, Willerson JT, Robbins RC, Ciechanover A, Mohr FW, Aravot D, Taylor DA. The ubiquitin-proteasome system: a potential therapeutic target for heart failure. J Heart Lung Transplant. 2017;36:708–14.
12. Bagde PH, Kandpal M, Rani A, Kumar S, Mishra A, Jha HC. Proteasomal dysfunction in cancer: mechanistic pathways and targeted therapies. J Cell Biochem. 2025;126:e70000.
13. Awan AB, Osman MJA, Khan OM. Ubiquitination enzymes in cancer, cancer immune evasion, and potential therapeutic opportunities. Cells. 2025;14:69.
14. Huang Y, Gao Y, Lin Z, Miao H. Involvement of the ubiquitin-proteasome system in the regulation of the tumor microenvironment and progression. Genes Dis. 2024;12:101240.
15. Atta H, Kassem DH, Kamal MM, Hamdy NM. Targeting the ubiquitin proteasome system in cancer stem cells. Trends Cell Biol. 2025;35:97–101.
16. Rodolfo C, Campello S, Cecconi F. Mitophagy in neurodegenerative diseases. Neurochem Int. 2018;117:156–66.
17. Thibaudeau TA, Anderson RT, Smith DM. A common mechanism of proteasome impairment by neurodegenerative disease-associated oligomers. Nat Commun. 2018;9:1097.
18. Wójcik C. Regulation of apoptosis by the ubiquitin and proteasome pathway. J Cell Mol Med. 2002;6:25–48.
19. Qureshi N, Morrison DC, Reis J. Proteasome protease mediated regulation of cytokine induction and inflammation. Biochim Biophys Acta. 2012;1823:2087–93.
20. Goetzke CC, Ebstein F, Kallinich T. Role of proteasomes in inflammation. J Clin Med. 2021;10:1783.
21. Qiu M, Chen J, Li X, Zhuang J. Intersection of the ubiquitin–proteasome system with oxidative stress in cardiovascular disease. Int J Mol Sci. 2022;23:12197.

22. Ding Q, Dimayuga E, Keller JN. Proteasome regulation of oxidative stress in aging and age-related diseases of the CNS. Antioxid Redox Signal. 2006;8:163–72.
23. Yu C, Xiao JH. The Keap1-Nrf2 system: a mediator between oxidative stress and aging. Oxid Med Cell Longev. 2021;2021:6635460.
24. Canning P, Sorrell FJ, Bullock AN. Structural basis of Keap1 interactions with Nrf2. Free Radic Biol Med. 2015;88(Pt B):101–7.
25. Adamson RJ, Payne NC, Bartual SG, Mazitschek R, Bullock AN. Structural and biochemical characterization establishes a detailed understanding of KEAP1-CUL3 complex assembly. Free Radic Biol Med. 2023;204:215–25.
26. Zipper LM, Mulcahy RT. The Keap1 BTB/POZ dimerization function is required to sequester Nrf2 in cytoplasm. J Biol Chem. 2002;277:36544–52.
27. Wei S, Pei Y, Wang Y, et al. Role of human Keap1 S53 and S293 residues in modulating the binding of Keap1 to Nrf2. Biochimie. 2019;158:73–81.
28. Kobayashi A, Kang MI, Okawa H, Ohtsuji M, Zenke Y, Chiba T, Igarashi K, Yamamoto M. Oxidative stress sensor Keap1 functions as an adaptor for Cul3-based E3 ligase to regulate proteasomal degradation of Nrf2. Mol Cell Biol. 2004;24:7130–9.
29. Sajadimajd S, Khazaei M. Oxidative stress and cancer: the role of Nrf2. Curr Cancer Drug Targets. 2018;18:538–57.
30. Yuan H, Xu Y, Luo Y, Wang NX, Xiao JH. Role of Nrf2 in cell senescence regulation. Mol Cell Biochem. 2012;476:247–59.
31. Shin JW, Chun KS, Kim DH, Kim SJ, Kim SH, Cho NC, Na HK, Surh YJ. Curcumin induces stabilization of Nrf2 protein through Keap1 cysteine modification. Biochem Pharmacol. 2020;173:113820.
32. Park JY, Sohn HY, Koh YH, Jo C. Curcumin activates Nrf2 through PKCδ-mediated p62 phosphorylation at Ser351. Sci Rep. 2021;11:8430.
33. Shah SZA, Zhao D, Hussain T, Sabir N, Mangi MH, Yang L. p62-Keap1-NRF2-ARE pathway: a contentious player for selective targeting of autophagy, oxidative stress and mitochondrial dysfunction in prion diseases. Front Mol Neurosci. 2018;11:310.
34. Ichimura Y, Waguri S, Sou YS, Kageyama S, Hasegawa J, Ishimura R, Saito T, Yang Y, Kouno T, Fukutomi T, Hoshii T, Hirao A, Takagi K, Mizushima T, Motohashi H, Lee MS, Yoshimori T, Tanaka K, Yamamoto M, Komatsu M. Phosphorylation of p62 activates the Keap1-Nrf2 pathway during selective autophagy. Mol Cell. 2013;51:618–31.
35. Kwak MK, Wakabayashi N, Greenlaw JL, Yamamoto M, Kensler TW. Antioxidants enhance mammalian proteasome expression through the Keap1-Nrf2 signaling pathway. Mol Cell Biol. 2003;23:8786–94.
36. Jana NR, Dikshit P, Goswami A, Nukina N. Inhibition of proteasomal function by curcumin induces apoptosis through mitochondrial pathway. J Biol Chem. 2004;279:11680–5.
37. Ambegaokar SS, Wu L, Alamshahi K, et al. Curcumin inhibits dose-dependently and time-dependently neuroglial cell proliferation and growth. Neuroendocrinol Lett. 2003;24:469–73.
38. Seo EJ, Fischer N, Efferth T. Phytochemicals as inhibitors of NF-κB for treatment of Alzheimer's disease. Pharmacol Res. 2018;129:262–73.
39. Gu W, Wu G, Chen G, Meng X, Xie Z, Cai S. Polyphenols alleviate metabolic disorders: the role of ubiquitin-proteasome system. Front Nutr. 2024;11:1445080.
40. Jobin C, Bradham CA, Russo MP, Juma B, Narula AS, Brenner DA, Balfour Sartor R. Curcumin blocks cytokine-mediated NF-κB activation and proinflammatory gene expression by inhibiting inhibitory Factor I-κB kinase activity. J Immunol. 1999;163:3474–83.
41. Liu T, Zhang L, Joo D, et al. NF-κB signaling in inflammation. Sig Transduct Target Ther. 2017;2:17023.
42. Hou Y, Moreau F, Chadee K. PPARγ is an E3 ligase that induces the degradation of NFκB/p65. Nat Commun. 2012;3:1300.
43. Porro C, Cianciulli A, Trotta T, Lofrumento DD, Panaro MA. Curcumin regulates anti-inflammatory responses by JAK/STAT/SOCS signaling pathway in BV-2 microglial cells. Biology (Basel). 2019;8:51.

44. Kim HY, Park EJ, Joe E-h, Jou I. Curcumin suppresses janus kinase-STAT inflammatory signaling through activation of Src homology 2 domain-containing tyrosine phosphatase 2 in brain microglia. J Immunol. 2003;171:6072–9.
45. Ashrafizadeh M, Rafiei H, Mohammadinejad R, Afshar EG, Farkhondeh T, Samarghandian S. Potential therapeutic effects of curcumin mediated by JAK/STAT signaling pathway: a review. Phytother Res. 2020;34:1745–60.
46. Zendedel E, Butler AE, Atkin SL, Sahebkar A. Impact of curcumin on sirtuins: a review. J Cell Biochem. 2018;119:10291–300.
47. Kong D, Zhang Z, Chen L, Huang W, Zhang F, Wang L, et al. Curcumin blunts epithelial-mesenchymal transition of hepatocytes to alleviate hepatic fibrosis through regulating oxidative stress and autophagy. Redox Biol. 2020;36:101600.
48. Liczbiński P, Michałowicz J, Bukowska B. Molecular mechanism of curcumin action in signaling pathways: review of the latest re-search. Phytother Res. 2020;34:1992–2005.
49. Aggarwal BB, Kumar A, Bharti AC. Anticancer potential of curcumin: preclinical and clinical studies. Anticancer Res. 2003;23:363–98.
50. Zhang YW, Thompson R, Zhang H, Xu H. APP processing in Alzheimer's disease. Mol Brain. 2011;4:3.
51. Yang F, Lim GP, Begum AN, Ubeda OJ, Simmons MR, Ambegaokar SS, Chen PP, Kayed R, Glabe CG, Frautschy SA, Cole GM. Curcumin inhibits formation of amyloid beta oligomers and fibrils, binds plaques, and reduces amyloid in vivo. J Biol Chem. 2005;280:5892–901.
52. Zheng K, Dai X, Xiao N, Wu X, Wei Z, Fang W, Zhu Y, Zhang J, Chen X. Curcumin ameliorates memory decline via inhibiting BACE1 expression and β-amyloid pathology in 5×FAD transgenic mice. Mol Neurobiol. 2017;54:1967–77.
53. Huang P, Zheng N, Zhou HB, Huang J. Curcumin inhibits BACE1 expression through the interaction between ERβ and NFκB signaling pathway in SH-SY5Y cells. Mol Cell Biochem. 2020;463:161–73.
54. Hamaguchi T, Ono K, Yamada M. Curcumin and Alzheimer's disease. CNS Neurosci Ther. 2010;16:285–97.
55. Qiao P, Ma J, Wang Y, Huang Z, Zou Q, Cai Z, Tang Y. Curcumin prevents neuroinflammation by inducing microglia to transform into the M2-phenotype via CaMKKβ-dependent activation of the AMP-activated protein kinase signal pathway. Curr Alzheimer Res. 2020;17:735–52.
56. Cheng J, Zhou Y, Qiao H, Jiang H, Fan Y. Curcumin protects from LPS-induced activation of astrocytes via AMPK pathway. Neuroreport. 2023;34:748–58.
57. Jerom JP, Madhukumar S, Nair RH, Narayanan SP. Anti-amyloid potential of some phytochemicals against Aβ-peptide and α-synuclein, tau, prion, and Huntingtin protein. Drug Discov Today. 2023;28:103802.
58. Bisaglia M, Mammi S, Bubacco L. Structural insights on physiological functions and pathological effects of α -synuclein. FASEB J. 2009;23:329–40.
59. Liu Z, Yu Y, Li X, Ross CA, Smith WW. Curcumin protects against A53T alpha-synuclein-induced toxicity in a PC12 inducible cell model for Parkinsonism. Pharmacol Res. 2011;63:439–44.
60. Wu Y, Liang S, Xu B, Zhang R, Xu L. Protective effect of curcumin on dopamine neurons in Parkinson's disease and its mechanism. J Zhejiang Univ (Med Sci). 2018;47:480–6.
61. Motawi TK, Sadik NAH, Hamed MA, et al. Potential therapeutic effects of antagonizing adenosine A_{2A} receptor, curcumin and niacin in rotenone-induced Parkinson's disease mice model. Mol Cell Biochem. 2020;465:89–102.
62. Törner R, Kupreichyk T, Hoyer W, Boisbouvier J. The role of heat shock proteins in preventing amyloid toxicity. Front Mol Biosci. 2022;9:1045616.
63. Wyttenbach A, Arrigo AP. The role of heat shock proteins during neurodegeneration in Alzheimer's, Parkinson's and Huntington's disease. In: Madame Curie Bioscience Database [Internet]. Austin: Landes Bioscience; 2000-2013.
64. Ma QL, Zuo X, Yang F, Ubeda OJ, Gant DJ, Alaverdyan M, Teng E, Hu S, Chen PP, Maiti P, Teter B, Cole GM, Frautschy SA. Curcumin suppresses soluble tau dimers and corrects

molecular chaperone, synaptic, and behavioral deficits in aged human tau transgenic mice. J Biol Chem. 2013;288:4056–65.
65. Söti C, Csermely P. Chaperones and aging: role in neurodegeneration and in other civilizational diseases. Neurochem Int. 2002;41:383–9.
66. Carmichael J, Chatellier J, Woolfson A, Milstein C, Fersht AR, Rubinsztein DC. Bacterial and yeast chaperones reduce both aggregate formation and cell death in mammalian cell models of Huntington's disease. Proc Natl Acad Sci U S A. 2000;97:9701–5.
67. Cummings CJ, Mancini MA, Antalffy B, DeFranco DB, Orr HT, Zoghbi HY. Chaperone suppression of aggregation and altered subcellular proteasome localization imply protein misfolding in SCA1. Nat Genet. 1998;19:148–54.
68. Hughes RE, Olson JM. Therapeutic opportunities in polyglutamine disease. Nat Med. 2001;7:419–23.
69. Krobitsch S, Lindquist S. Aggregation of huntingtin in yeast varies with the length of the polyglutamine expansion and the expression of chaperone proteins. Proc Natl Acad Sci U S A. 2000;97:1589–94.
70. Hickey MA, Zhu C, Medvedeva V, Lerner RP, Patassini S, Franich NR, Maiti P, Frautschy SA, Zeitlin S, Levine MS, Chesselet MF. Improvement of neuropathology and transcriptional deficits in CAG 140 knock-in mice supports a beneficial effect of dietary curcumin in Huntington's disease. Mol Neurodegener. 2012;7:12.
71. Singh S, Kumar P. Neuroprotective activity of curcumin in combination with piperine against quinolinic acid induced neurodegeneration in rats. Pharmacology. 2016;97:151–60.
72. Chongtham A, Agrawal N. Curcumin modulates cell death and is protective in Huntington's disease model. Sci Rep. 2016;6:18736.
73. Dong H, Xu L, Wu L, Wang X, Duan W, Li H, Li C. Curcumin abolishes mutant TDP-43 induced excitability in a motoneuron-like cellular model of ALS. Neuroscience. 2014;272:141–53.
74. Lu J, Duan W, Guo Y, Jiang H, Li Z, et al. Mitochondrial dysfunction in human TDP-43 transfected NSC34 cell lines and the protective effect of dimethoxy curcumin. Brain Res Bull. 2012;89:185–90.
75. Bhatia NK, Srivastava A, Katyal N, Jain N, Khan MA, Kundu B, Deep S. Curcumin binds to the pre-fibrillar aggregates of Cu/Zn superoxide dismutase (SOD1) and alters its amyloidogenic pathway resulting in reduced cytotoxicity. Biochim Biophys Acta. 2015;1854:426–36.
76. Baum L, Lam CW, Cheung SK, Kwok T, Lui V, Tsoh J, Lam L, Leung V, Hui E, Ng C, Woo J, Chiu HF, Goggins WB, Zee BC, Cheng KF, Fong CY, Wong A, Mok H, Chow MS, Ho PC, Ip SP, Ho CS, Yu XW, Lai CY, Chan MH, Szeto S, Chan IH, Mok V. Six-month randomized, placebo-controlled, double-blind, pilot clinical trial of curcumin in patients with Alzheimer disease. J Clin Psychopharmacol. 2008;28:110–3.
77. Ringman JM, Frautschy SA, Teng E, Begum AN, Bardens J, Beigi M, Gylys KH, Badmaev V, Heath DD, Apostolova LG, Porter V, Vanek Z, Marshall GA, Hellemann G, Sugar C, Masterman DL, Montine TJ, Cummings JL, Cole GM. Oral curcumin for Alzheimer's disease: tolerability and efficacy in a 24-week randomized, double blind, placebo-controlled study. Alzheimer's Res Ther. 2012;4:43.
78. Ahmadi M, Agah E, Nafissi S, Jaafari MR, Harirchian MH, Sarraf P, et al. Safety and efficacy of nanocurcumin as add-on therapy to riluzole in patients with amyotrophic lateral sclerosis: a pilot randomized clinical trial. Neurotherapeutics. 2018;15:430–8.
79. Chico L, Ienco EC, Bisordi C, Lo Gerfo A, Petrozzi L, Petrucci A, Mancuso M, Siciliano G. Amyotrophic lateral sclerosis and oxidative stress: a double-blind therapeutic trial after curcumin supplementation. CNS Neurol Disord Drug Targets. 2018;17:767–79.
80. Sae-tan S, Grove KA, Lambert JD. Weight control and prevention of metabolic syndrome by green tea. Pharmacol Res. 2011;64:146–54.
81. Payne A, Nahashon S, Taka E, Adinew GM, Soliman KFA. Epigallocatechin-3-gallate (EGCG): new therapeutic perspectives for neuroprotection, aging, and neuroinflammation for the modern age. Biomolecules. 2022;12:371.

82. Sonawane SK, Chidambaram H, Boral D, et al. EGCG impedes human Tau aggregation and interacts with Tau. Sci Rep. 2020;10:12579.
83. Wang Y, Wu S, Li Q, Lang W, Li W, Jiang X, Wan Z, Chen J, Wang H. Epigallocatechin-3-gallate: a phytochemical as a promising drug candidate for the treatment of Parkinson's disease. Front Pharmacol. 2022;13:977521.
84. de la Torre R, de Sola S, Hernandez G, Farré M, Pujol J, Rodriguez J, Espadaler JM, Langohr K, Cuenca-Royo A, Principe A, Xicota L, Janel N, Catuara-Solarz S, Sanchez-Benavides G, Bléhaut H, Dueñas-Espín I, Del Hoyo L, Benejam B, Blanco-Hinojo L, Videla S, Fitó M, Delabar JM, Dierssen M. TESDAD study group. Safety and efficacy of cognitive training plus epigallocatechin-3-gallate in young adults with Down's syndrome (TESDAD): a double-blind, randomised, placebo-controlled, phase 2 trial. Lancet Neurol. 2016;15:801–10.
85. Jellinger KA. Epigallocatechin gallate in multiple system atrophy (PROMESA). Ann Transl Med. 2019;7(Suppl 8):S278.
86. Rust R, Chien C, Scheel M, Brandt AU, Dörr J, Wuerfel J, Klumbies K, Zimmermann H, Lorenz M, Wernecke KD, Bellmann-Strobl J, Paul F. Epigallocatechin gallate in progressive MS: a randomized, placebo-controlled trial. Neurol Neuroimmunol Neuroinflamm. 2021;8:e964.
87. Li R, Robinson M, Ding X, Geetha T, Al-Nakkash L, Broderick TL, Babu JR. Genistein: a focus on several neurodegenerative diseases. J Food Biochem. 2022;46:e14155.
88. Sharma D, Singh V, Kumar A, Singh TG. Genistein: a promising ally in combating neurodegenerative disorders. Eur J Pharmacol. 2025;991:177273.
89. Sahab Uddin M, Tanvir Kabir M. Emerging signal regulating potential of genistein against Alzheimer's disease: a promising molecule of interest. Front Cell Dev Biol. 2019;7:197.
90. Duan X, Li Y, Xu F, Ding H. Study on the neuroprotective effects of Genistein on Alzheimer's disease. Brain Behav. 2021;11:e02100.
91. Kazi A, Daniel KG, Smith DM, Kumar NB, Dou QP. Inhibition of the proteasome activity, a novel mechanism associated with the tumor cell apoptosis-inducing ability of genistein. Biochem Pharmacol. 2003;66:965–76.
92. Yi S, Chen S, Xiang J, Tan J, Huang K, Zhang H, Wang Y, Heng W. Genistein exerts a cell-protective effect via Nrf2/HO-1//PI3K signaling in Ab25-35-induced Alzheimer's disease models in vitro. Folia Histochem Cytobiol. 2021;59:49–56.
93. Li Y, Zhang JJ, Chen RJ, Chen L, Chen S, Yang XF, Min JW. Genistein mitigates oxidative stress and inflammation by regulating Nrf2/HO-1 and NF-kB signaling pathways in hypoxin-ischemic brain damage in neonatal mice. Ann Transl Med. 2022;10:32.
94. Gao H, Lei X, Ye S, Ye T, Hua R, Wang G, Song H, Zhou P, Wang Y, Cai B. Genistein attenuates memory impairment in Alzheimer's disease via ERS-mediated apoptotic pathway in vivo and in vitro. J Nutr Biochem. 2022;109:109118.
95. Viña J, Escudero J, Baquero M, Cebrián M, Carbonell-Asíns JA, Muñoz JE, Satorres E, Meléndez JC, Ferrer-Rebolleda J, Cózar-Santiago MDP, Santabárbara-Gómez JM, Jové M, Pamplona R, Tarazona-Santabalbina FJ, Borrás C. Genistein effect on cognition in prodromal Alzheimer's disease patients. The GENIAL clinical trial. Alzheimer's Res Ther. 2022;14:164.
96. Zhao Z, Fu J, Li S, Li Z. Neuroprotective effects of genistein in a SOD1-G93A transgenic mouse model of amyotrophic lateral sclerosis. J Neuroimmune Pharmacol. 2019;14:688–96.
97. Andres S, Pevny S, Ziegenhagen R, Bakhiya N, Schäfer B, Hirsch-Ernst KI, et al. Safety aspects of the use of quercetin as a dietary supplement. Mol Nutr Food Res. 2018;62:447.
98. Das D, Paul A, Lahiri A, Adak M, Maity SK, Sarkar A, et al. Proteasome dysfunction under compromised redox metabolism dictates liver injury in NASH through ASK1/PPARγ binodal complementary modules. Redox Biol. 2021;45:102043.
99. Gnoni GV, Paglialonga G, Siculella L. Quercetin inhibits fatty acid and triacylglycerol synthesis in rat-liver cells. Eur J Clin Invest. 2009;39:761–8.
100. Zhu M, Han S, Fink AL. Oxidized quercetin inhibits alpha-synuclein fibrillization. Biochim Biophys Acta. 2013;1830:2872–81.
101. Bayazid AB, Lim BO. Quercetin is an active agent in berries against neurodegenerative diseases progression through modulation of Nrf2/HO1. Nutrients. 2022;14:5132.

102. Dosenko VE, Nagibin VS, Tumanovskaia LV, Viu Z, Moĭbenko AA. The influence of quercetin on the activity of purified 20S, 26S proteasome and proteasomal activity in isolated cardiomyocytes. Biomed Khim. 2006;52:138–45.
103. Martín-Aragón S, Jiménez-Aliaga KL, Benedí J, Bermejo-Bescós P. Neurohormetic responses of quercetin and rutin in a cell line over-expressing the amyloid precursor protein (APPswe cells). Phytomed Int J Phytother Phytopharm. 2016;23:1285–94.
104. Chakraborty J, Rajamma U, Jana N, Mohanakumar KP. Quercetin improves the activity of the ubiquitin-proteasomal system in 150Q mutated huntingtin-expressing cells but exerts detrimental effects on neuronal survivability. J Neurosci Res. 2015;93:1581–91.
105. Cui Z, Zhao X, Amevor FK, Du X, Wang Y, Li D, Shu G, Tian Y, Zhao X. Therapeutic application of quercetin in aging-related diseases: SIRT1 as a potential mechanism. Front Immunol. 2022;13:943321.
106. Zu G, Sun K, Li L, et al. Mechanism of quercetin therapeutic targets for Alzheimer disease and type 2 diabetes mellitus. Sci Rep. 2021;11:22959.
107. Ho CL, Kao NJ, Lin CI, Cross TL, Lin SH. Quercetin increases mitochondrial biogenesis and reduces free radicals in neuronal SH-SY5Y cells. Nutrients. 2022;14:3310.
108. Chiang MC, Tsai TY, Wang CJ. The potential benefits of quercetin for brain health: a review of anti-inflammatory and neuroprotective mechanisms. Int J Mol Sci. 2023;24:6328.
109. Peñalver P, Belmonte-Reche E, Adán N, Caro M, Mateos-Martín ML, Delgado M, et al. Alkylated resveratrol prodrugs and metabolites as potential therapeutics for neurodegenerative diseases. Eur J Med Chem. 2018;146:123–38.
110. Shi JX, Wang QJ, Li H, Huang Q. Silencing of USP22 suppresses high glucose-induced apoptosis, ROS production and inflammation in podocytes. Mol Biosyst. 2016;12:1445–56.
111. Genovese T, Impellizzeri D, D'Amico R, Cordaro M, Peritore AF, Crupi R, Gugliandolo E, Cuzzocrea S, Fusco R, Siracusa R, Di Paola R. Resveratrol inhibition of the WNT/β-catenin pathway following discogenic low back pain. Int J Mol Sci. 2022;23:4092.
112. Hioki T, Kawabata T, Sakai G, Fujita K, Kuroyanagi G, Matsushima-Nishiwaki R, et al. Resveratrol suppresses insulin-like growth factor I-induced osteoblast migration: attenuation of the p44/p42 MAP kinase pathway. Biosci Biotechnol Biochem. 2020;84:2428–39.
113. Li P, Liang ML, Zhu Y, Gong YY, Wang Y, Heng D, Lin L. Resveratrol inhibits collagen I synthesis by suppressing IGF-1R activation in intestinal fibroblasts. World J Gastroenterol. 2014;20:4648–61.
114. Bastianetto S, Ménard C, Quirion R. Neuroprotective action of resveratrol. Biochim Biophys Acta (BBA) - Mol Basis Dis. 2015;1852:1195–201.
115. Moussa C, Hebron M, Huang X, et al. Resveratrol regulates neuro-inflammation and induces adaptive immunity in Alzheimer's disease. J Neuroinflammation. 2017;14:1.
116. Liu Q, Zhu D, Jiang P, Tang X, Lang Q, Qinyi Y, Zhang S, Che Y, Feng X. Resveratrol synergizes with low doses of L-DOPA to improve MPTP-induced Parkinson disease in mice. Behav Brain Res. 2019;367:10–8.
117. Su C-F, Jiang L, Zhang X-W, Iyaswamy A, Li M. Resveratrol in rodent models of Parkinson's disease: a systematic review of experimental studies. Front Pharmacol. 2021;12:644219.
118. Ferretta A, Gaballo A, Tanzarella P, Piccoli C, Capitanio N, Nico B, Annese T, Di Paola M, Dell'Aquila C, De Mari M, Ferranini E, Bonifati V, Pacelli C, Cocco T. Effect of resveratrol on mitochondrial function: implications in parkin-associated familiar Parkinson's disease. Biochim Biophys Acta (BBA) - Mol Basis Dis. 2014;1842:902–15.
119. Danışman B, Ercan Kelek S, Aslan M. Resveratrol in neurodegeneration, in neurodegenerative diseases, and in the redox biology of the mitochondria. Psychiatry Clin Psychopharmacol. 2023;33:147–55.
120. Parker JA, Arango M, Abderrahmane SE, et al. Resveratrol rescues mutant polyglutamine cytotoxicity in nematode and mammalian neurons. Nat Genet. 2005;37:349–50.
121. Zhou S, Kachhap S, Singh KK. Mitochondrial impairment in p53-deficient human cancer cells. Mutagenesis. 2003;18:287–92.
122. Wang N, Luo Z, Jin M, et al. Exploration of age-related mitochondrial dysfunction and the anti-aging effects of resveratrol in zebrafish retina. Aging (Albany NY). 2019;11:3117–37.

123. Pinilla-González V, Montecinos-Barrientos B, Martin-Kommer C, Chichiarelli S, Saso L, Rodrigo R. Exploring antioxidant strategies in the pathogenesis of ALS. Open Life Sci. 2024;19:20220842.
124. Novak V, Rogelj B, Župunski V. Therapeutic potential of polyphenols in amyotrophic lateral sclerosis and frontotemporal dementia. Antioxidants. 2021;10:1328.
125. Mancuso R, del Valle J, Modol L, Martinez A, Granado-Serrano AB, Ramirez-Núñez O, Pallás M, Portero-Otin M, Osta R, Navarro X. Resveratrol improves motoneuron function and extends survival in SOD1G93A ALS mice. Neurotherapeutics. 2014;11:419–32.
126. Ho DJ, Calingasan NY, Wille E, Dumont M, Beal MF. Resveratrol protects against peripheral deficits in a mouse model of Huntington's disease. Exp Neurol. 2010;225:74–84.
127. Kim D, Nguyen MD, Dobbin MM, Fischer A, Sananbenesi F, Rodgers JT, Delalle I, Baur JA, Sui G, Armour SM, Puigserver P, Sinclair DA, Tsai LH. SIRT1 deacetylase protects against neurodegeneration in models for Alzheimer's disease and amyotrophic lateral sclerosis. EMBO J. 2007;26:3169–79.

Therapeutic Potential of HDAC Inhibitors in Neurodegenerative Diseases

14

Abhay Kumar Singh, Yemgadda Goutham Sudhan, Sudharshanam Lakshmi Chandana, Ashok Iyaswamy, and Siva Sundara Kumar Durairajan

14.1 Introduction

Neurodegenerative diseases such as Alzheimer's disease (AD), Parkinson's disease (PD), and Huntington's disease (HD) are the result of the gradual loss of neurons, which leads to extensive cognitive and motor impairments. The present approaches are more geared toward alleviating the symptoms than addressing the cause of the disease; hence, the people who have the diseases need therapies that are more focused on the mechanisms of the disease [96, 170]. Histone deacetylases (HDACs) have shown potential in therapy development. These enzymes remove acetyl groups from histones, causing chromatin condensation and gene silencing [153]. Chronic hypoacetylation due to aberrant HDAC activity is known to occur in neurodegenerative diseases, leading to neuronal loss and dysfunction [9, 28, 136]. While in neurodegenerative diseases, HDAC dysregulation leads to hypoacetylation that triggers neuronal death and dysfunction [9, 28, 136]. HDAC inhibitors (HDACIs) might help to achieve normal acetylation levels and thus promote the reactivation of

A. K. Singh · Y. G. Sudhan · S. L. Chandana
Molecular Mycology and Neurodegenerative Disease Research Laboratory, Department of Microbiology, Central University of Tamil Nadu, Neelakudi, Tiruvarur, India

A. Iyaswamy
Mr. & Mrs Ko Chi-Ming Centre for Parkinson's Disease Research, School of Chinese Medicine, Hong Kong Baptist University, Hong Kong, SAR, China

Department of Biochemistry, Karpagam Academy of Higher Education, Coimbatore, India

S. S. K. Durairajan (✉)
Molecular Mycology and Neurodegenerative Disease Research Laboratory, Department of Microbiology, Central University of Tamil Nadu, Neelakudi, Tiruvarur, India

School of Biomedical Sciences, Li Ka Shing Faculty of Medicine, The University of Hong Kong, Hong Kong, SAR, China
e-mail: d.sivasundarakumar@cutn.ac.in

S. Singh (ed.), *Ubiquitin Proteasome System in CNS Neurodegenerative Diseases*, https://doi.org/10.1007/978-981-95-6099-8_14

protective genes. They increase histone acetylation and improve genes' expression, which is critical to neuronal survival, synapses, and memory [170]. HDACIs also influence neuroinflammation, reduce oxidative stress and modify synaptic plasticity as well [96, 170]. Recent studies aim at synthesizing isoform-selective compounds in an effort to avoid the general side effects that come with broad-spectrum inhibition. Unlike nonselective inhibitors, such an approach minimizes interference with other cellular activities while selectively modulating certain dysregulated HDACs in neurodegenerative diseases [90, 96]. HDACIs can easily penetrate the blood-brain barrier and are hence valuable for nervous system disorders [44]. It makes them a relatively rational therapeutic approach since they target the causes of neurodegeneration, not only symptoms. Further investigation into the mode of action and the therapeutic application is essential for treating neurodegenerative illnesses.

14.2 Classification and Types of HDACs

There are 18 HDACs in the mammalian HDAC family that are divided into four major classes based on yeast HDAC and cofactor dependency homology [104]. These classes are Class I, Class II, Class IIa and IIb), Class III (sirtuins), and Class IV. Structurally, each class has distinct physiological specificities, localization, and biological functions that contribute to diverse cellular processes [133]. HDAC1, HDAC2, HDAC3 and HDAC8 are class I HDACs homologous to yeast Rpd3. Shukla & Tekwani [133] suggest that these isoforms are primarily found in the nucleus and ubiquitously expressed in multiple tissue types and are pivotal in regulating the cell cycle, apoptosis, and transcription repression. HDAC1 and HDAC2 are frequently members of multiprotein co-repressor complexes, including Sin3, NuRD and CoREST, and factorially influence chromatin structure and gene expression [86]. Inflammatory responses and metabolic regulation are mediated by HDAC3, a part of the nuclear receptor co-repressor (NCoR) complex [56]. HDAC8 shows unique substrate specificity and participates in smooth muscle differentiation and cohesion of sister chromatids during mitosis [104] (Fig. 14.1, Table 14.1).

Class II HDACs are divided into Class IIa (HDAC4, HDAC5, HDAC7, HDAC9) and Class IIb (HDAC6, HDAC10) based on structural and functional homologies. Although they are highly conserved, Class IIa HDACs possess a highly conserved N terminus that mediates interactions with transcription factors such as MEF2 and can shuttle between the nucleus and cytoplasm in response to cellular signals [22, 105]. These isoforms are also very involved in tissue-specific functions, including muscle differentiation and cardiac development [85, 142]. HDAC6 and HDAC10 are class IIb HDACs cytoplasmically localized and have specialized roles. Its two catalytic domains and the ubiquitin-binding zinc finger allow HDAC6 to function in protein degradation via the aggresome pathway [83]. In addition, in cytoskeletal dynamics and cell motility, HDAC6 is also involved in tubulin deacetylation [23]. Although less explored, HDAC10 involves DNA damage response and autophagy regulation [165].

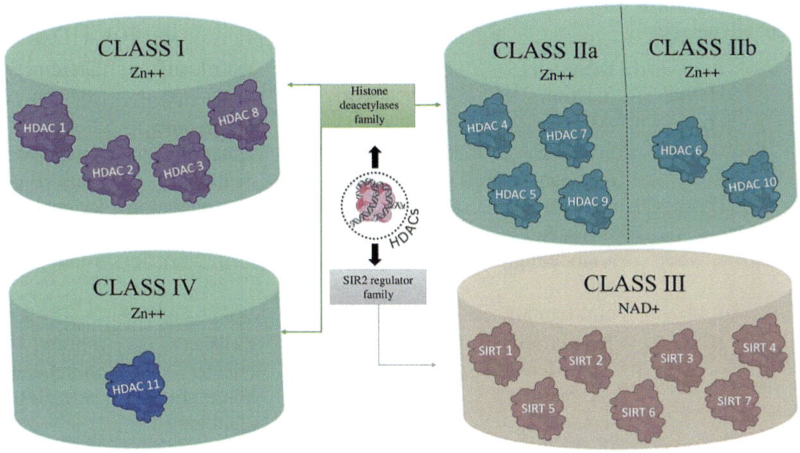

Fig. 14.1 Four classes of histone deacetylases (HDACs). Class I contains HDAC1-3 and 8 (shown in purple), Class IIa includes HDAC4, 5, 7, and 9, and Class IIb contains HDAC6 and 10 (both in blue). All these classes require Zn++ as a cofactor. Class III comprises seven sirtuins (SIRT1-7, shown in pink) uniquely using NAD+ as a cofactor. Class IV consists of a single member, HDAC11 (in dark blue). The central diagram shows the regulatory relationship between these enzymes through the histone deacetylase family and SIR2 regulator family pathways

NAD+-dependent deacetylases class III HDACs, known as sirtuins (SIRT1-7), have been characterized with NAD+-dependent activities and diverse subcellular localization: nucleus (SIRT1, SIRT6, SIRT7), cytoplasm (SIRT2), and mitochondria (SIRT3, SIRT4, SIRT5) [64]. Sirtuins perform critical functions of metabolic regulation, genomic stability, and stress responses. One example is SIRT1, which regulates cellular ageing and longevity through deacetylating transcription factors p53, FOXO and NFκB [135]. A mitochondrial sirtuin, SIRT3, regulates energy metabolism and oxidative stress by targeting enzymes, including MnSOD and IDH2 [75]. Sirtuins are attractive targets for therapeutic intervention in ageing and metabolic disorders as these functions. Class IV is a single-member HDAC11, structurally similar to Class I and II HDACs [21]. HDAC11 is expressed predominantly in the brain, heart and immune cells, suggesting a specific role in neural and immune function. Its role in regulating interleukin-10 (IL-10) expression indicates that HDAC11 is involved in immune modulation and inflammatory responses [157]. Although HDAC11 has been proposed as necessary, the least characterized member of the HDAC family holds potential for further exploration. HDACs participate in and modulate many aspects of cellular activity that involve chromatin remodeling, gene expression, and regulation of protein stability. HDAC dysregulation is implicated in several diseases, including cancer, neurodegenerative disorders and metabolic syndromes [67, 127]. Developing selective HDACIs with minimal off-target effects while maximizing therapeutic efficacy depends on understanding the classification and specific function of HDAC isoforms [1, 82].

Table 14.1 Classical HDAC enzymes: biological functions, physiological implications, and knockout phenotypes

Class	HDAC	Biological function	Physiological implications	Knockout/key phenotype	Structural insights
I	HDAC1	Deacetylates histones; regulates gene expression, chromatin remodeling, cell proliferation	Neurodegenerative disorders, cancer development	Embryonic lethal; impairs neuronal differentiation	Forms complexes with HDAC2; critical in transcriptional repression
	HDAC2	Transcriptional regulation, chromatin remodeling	Liver diseases, leukemia, neurodegenerative disorders	Reduces nuclear huntingtin pathology	Forms complexes with HDAC1
	HDAC3	Maintains chromatin structure, regulates transcription	Metabolism regulation, circadian rhythms	Causes infertility in mice	Targets oxidative stress and cellular metabolism
	HDAC8	Deacetylates histones and non-histone proteins	Cancer progression, cardiac fibrosis	Delays neuroblastoma tumor growth	Distinct catalytic activity
IIa	HDAC4	Transcriptional repression, chromatin remodeling	Tumorigenesis, ischemic stroke	Protects against muscle atrophy	Nuclear-cytoplasmic shuttling
	HDAC5	Epigenetic gene expression regulation	Cancer, inflammatory lung disease	Develops enlarged hearts under pressure	Critical in cardiac development
	HDAC7	Chromatin remodeling, gene expression	Cancer, inflammation, angiogenesis	Embryonic lethal at E11.0	Immune cell development
	HDAC9	Developmental and immune function regulation	Cardiovascular diseases, cancer	Reduces fat mass	Regulates neuron activity
IIb	HDAC6	Protein acetylation regulation	Cancer, cellular stress response	Increases tubulin acetylation	Two catalytic domains
	HDAC10	Angiogenesis, gene expression	Cancer progression	Exacerbates lung cancer tumorigenesis	Influences ERK signaling

(continued)

Table 14.1 (continued)

Class	HDAC	Biological function	Physiological implications	Knockout/key phenotype	Structural insights
III (Sirtuins)	SIRT1	Cellular metabolism, stress response	Longevity, diabetes	Regulates metabolic proteins	Insulin sensitivity modulation
	SIRT3	Mitochondrial metabolism	Metabolic health, aging	Mitochondrial protein regulation	Oxidative stress resistance
	SIRT6	Genome stability, metabolism	Ageing, cancer	DNA repair regulation	Telomere maintenance
IV	HDAC11	Immune system regulation	Inflammatory responses	Unique structure	Most recently discovered HDAC

14.3 Functions of HDACs in the Nervous System

HDACs contribute to neurodevelopment, synaptic learning, and balanced neuronal function. Class I HDACs, especially HDAC 1 and HDAC 2, are essential for cortical, hippocampal or cerebellar development [57, 163]. These enzymes modulate the differentiation and survival of neuronal precursors; homozygous HDAC1/2 null mice display profound phenotypes and die within hours after birth [84]. In mature neurons, HDAC2 inhibits the activity of the synaptic plasticity and memory formation genes, while HDAC1 regulates only a small number of these genes [45].

Clinically relevant Class IIa HDACs, which include HDAC4 and HDAC5, regulate activity-dependent synaptic plasticity and neuroprotection [73, 128]. These HDACs can also translocate between the nucleus and cytoplasm based on neuronal activation and are thought to play a role in mediating gene expression profiles necessary for aspects of synaptic plasticity and neuroprotection [159]. HDAC6, a class IIb HDAC, is associated explicitly with neuronal structure and function since it modifies cytoskeletal proteins by removing acetyl groups from α-tubulin and other cytoplasmic proteins [13, 23].

Sirtuins (Class III HDACs) that include Sirt 1 and 3 have been reported to be involved in neuronal metabolism and viability. SIRT1 directly interacts with the DNA and modifies FOXO3 and PGC-1α transcription factors to upregulate antioxidant gene expressions for neuronal protection under oxidative stress. Recent studies suggest that SIRT1 phosphorylation reduces the factors contributing to several neurodegenerative diseases, including Huntington's disease and Amyotrophic lateral sclerosis (ALS) in rodent models. SIRT3 is a mitochondrial sirtuin that modulates oxidative phosphorylation and ROS removal to elongate neurons' energy balance [10].

The HDAC11, the only member of Class IV, appeared to modulate IL-10 expression associated with immune regulation and neuroinflammation. Experimental research in mice shows that downregulation of HDAC11 decreases neuroinflammation and, therefore, it may be viewed as a potential therapeutic target in neurodegenerative diseases [157].

Reduced or increased levels of HDACs are associated with neurological disorders such as schizophrenia, autism, and Alzheimer's disease. Animal studies support the antidepressant application of HDACIs. For example, the function of cognition is determined to be ameliorated in mouse models for AD and Rett syndrome through regulating synaptic plasticity and neuroinflammation after HDACIs treatment [55, 125]. These findings highlight HDACs as promising targets for neurological and psychiatric interventions

14.4 Role of HDACs in Neurodegenerative Processes

HDAC products are thus an essential component of neurodegenerative processes, and the abnormalities in the functioning of these proteins are shown to be a necessary factor for various neurodegenerative diseases. Several molecular mechanisms of the regulation of histone proteins and other proteins necessary for neuronal function have been identified through epigenetic studies connected with HDACs [125]. Further understanding of HDAC's role in AD has revealed that the overexpression of HDAC2 specifically inhibits neuroplasticity genes and memory formation by changing chromatin accessibility [40]. However, recent studies have established that HDAC3 can directly react with protein phosphatase 1 to regulate tau hyperphosphorylation, and an HDAC6 inhibitor shows excellent potential in targeting both tau and amyloid-β pathology due to the stabilization of microtubules and activation of autophagy [15, 154]. Novel studies have identified dynamic nuclear-cytoplasmic shuttling of HDAC4, which regulates synaptic plasticity by modulating MEF2 and CREB-dependent gene transcription according to its phosphorylation state, thus controlling memory formation [128]. In Parkinson's disease, advances have gone beyond the initial discovery that HDAC6 is involved in α-synuclein aggregation and HDAC4 has been found to play a significant role in dopaminergic neuron survival through BDNF signaling [159]. Recent studies have shown that SIRT3 activation possesses a potent neuroprotective effect against MPTP-induced dopaminergic neuron decline, enhancing mitochondrial biogenesis and alleviating oxidative stress caused by direct SOD2 deacetylation by SIRT3 [131]. This field has radically shifted with the identification that HDAC4 modulates mutant huntingtin aggregation through unknown protein-protein interactions involving the SUMO pathway [97].

Selective HDAC3 inhibition has recently been identified as an unprecedented and highly effective approach for promoting nuclear and cytoplasmic huntingtin clearance through enhanced proteostasis or protein degradation pathways [61]. Novel developments in the field of proteomics and single-cell sequencing have identified new functions for HDAC and its relationship with protein homeostasis, particularly a complex interplay between HDAC6 and the heat shock proteins in aggregate removal [6], as well the identification of SIRT1 as a master regulator of autophagy through deacetylation of components of the ULK1/Atg XIII complex

[78]. HDAC4 has been employed to unveil the fundamental roles of stress granules, revealing important mechanisms of how cells respond to stress in neurodegenerative diseases ([133, 156]. Recent advances in mitochondrial physiology have unveiled a more regulatory role of SIRT3 involving dynamic aspects of mitochondria and their biogenesis through the modulation of PGC-1α [167]. Recent studies that have highlighted the role of HDAC1 in mediating neuronal metabolic flexibility disturbed to glucose transporter expression and insulin signaling have paved the way for new therapeutic strategies [41], as well as the identification of new interconnections between class IIa HDACs and mitochondrial proteins on respiratory chain complex assembly [92]. Furthermore, targeted drugs that selectively inhibit specific HDAC isoforms are more likely to achieve greater penetration through the blood-brain barrier and fewer side effects [129]; there have also been significant studies on multiple therapies that target various pathways in which HDAC is involved and have even been synergistic in preclinical models of neurodegenerative disorders [41, 129]

14.5 HDAC Inhibitors (HDACIs)

HDACIs have gained importance as potential drugs for neurodegenerative diseases as they are available in numerous structural categories and have varying selectivity to different aspects of the multifaceted disorders (Fig. 14.2). Four main structural classes of HDACIs have been identified, including hydroxamates, cyclic peptides, aliphatic acids and benzamides; each class possesses different chemical properties

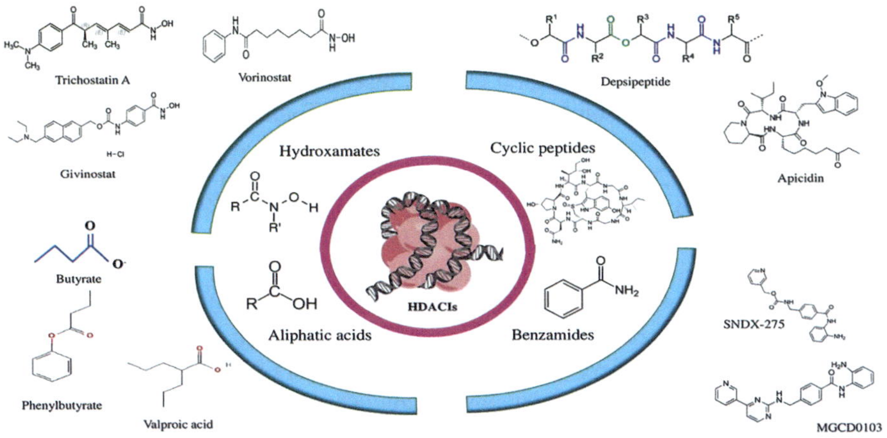

Fig. 14.2 A comprehensive diagram of histone deacetylase (HDAC) inhibitor classes arranged in a circular layout. At the center, HDACs are depicted in pink with associated DNA helices. Four major structural classes are displayed around the perimeter: hydroxamates (including Trichostatin A, Vorinostat, and Givinostat), cyclic peptides (featuring Depsipeptide and Apicidin), benzamides (showing MGCD0103 and SNDX-275), and aliphatic acids (represented by Butyrate, Phenylbutyrate, and Valproic acid). Each class shows detailed chemical structures of representative compounds with their characteristic functional groups

and zinc-binding groups specific to modulating HDAC activity. Trichostatin A (TSA) and Vorinostat (SAHA) are examples of hydroxamates that act as potent and highly selective HDACIs that have been shown to provide promising results in experimental models of neurodegenerative diseases [89]. These compounds generally include a zinc-binding fragment, a spacer, and a cap structure capable of engaging with the rim of the HDAC catalytic domain [164]. Hydroxamates have recently emerged as promising drug candidates because of their high nanomolar profile and ability to inhibit multiple HDAC isoforms selectively; treatments that affect global histone acetylation may benefit neurodegenerative diseases.

Small cyclic peptides like Romidepsin (FK228) create a unique structural regimen for HDAC inhibition. Their macrocyclic structure provides selectivity for specific HDAC isoforms, which may be advantageous for designing more effective therapies [104]. Accordingly, cyclic peptide derivatives based on HDACIs have shown their neuroprotective effects in the cellular models of AD and HD, which proposed the potential drugs as inhibitors of neurodegeneration [170]. Aliphatic acids are among the oldest HDACIs on the market; short-chain fatty acids include Butyrate, Phenylbutyrate and Valproic acid. Hydroxamate derivatives, in general, are less effective than hydroxamates, but numerous positive effects have been observed in animal models of neurodegenerative diseases, including modulation of cognitive abilities and neuroinflammation [104]. The safety profile of some aliphatic acids, like Valproic acid, may ease their clinical repositioning for neurodegenerative diseases. Select members like Entinostat (MS-275) and Mocetinostat (MGCD0103) are more selective for class I HDACs, making these compounds more selective for HDAC inhibition [103]. It may be especially the case when some of the HDAC isoforms have been found to play a role in certain neurodegenerative diseases. Some benzamide-based HDACIs have been put forward as possible therapeutic drugs for AD as well as PD, as evidenced by experimental animal studies [171]. In addition to these main structural classes, research on new HDACIs with other structural characteristics constantly enriches the new treatment spectrum. For example, Tubastatin A, the selective HDAC6 inhibitor, showed neuroprotection in the AD model and ALS [132]. This type of isoform-selective inhibitor shows that targeting certain HDACs can be effective in neurodegenerative diseases.

The pharmacological characterization of HDACIs by describing their selectivity profiles can contribute to understanding possible therapeutic uses (Table 14.2). Classwide HDACIs like vorinostat and panobinostat have wide-ranging effects of histone acetylation but can be linked to side effects [76]. On the other hand, class-selective inhibitors selectively act on particular classes of HDACs, which may be effective in selectively abrogating pathways implicated with neurodegeneration. For instance, class I-selective inhibitors such as MS-275 may be more applicable in diseases where class I HDACs are the most dominant [169]. Several current generations of HDACIs serve as isoform-selective inhibitors, which remain the most selective agents in modulating neurodegenerative diseases. In particular, the emergence of selective HDAC6 inhibitors, including Tubastatin A, has stemmed from the identification of the contribution of HDAC6 to tau pathogenesis and axonal transport impairments in AD and other tauopathies [164]. As with other isoforms such as

Table 14.2 Histone deacetylase inhibitors as potential protective compounds in neurodegenerative disorders

Classification	Examples	Specificity to HDAC	Treated disorders	Citations
Aliphatic acids	Butyrate/sodium butyrate (SB, NaB), phenyl butyrate (PBA), butyric acid prodrugs (BA3G, VX563), valproic acid (VPA)	Class I/IIa	PD, HD, AD, ALS, SMA, FRD	[29, 66]
Hydroxamates	Suberoylanilide hydroxamic acid (SAHA; vorinostat), Trichostatin A (TSA), LBH-589 (panobinostat), Tubacin	Class I/II/IV, HDAC6	PD, HD, AD, ALS, SMA, FRDA	[8, 14]
Benzamides	MS-275, SNX-275 (entinostat), K560, Tubastatin A, HDACi 4b, HDACi 106, RGFP966	HDAC1/9/11, HDAC1/2, HDAC6, HDAC1/2/3, HDAC3	PD, HD, AD, FRDA	[8, 70, 149]
Cyclic peptides	Romidepsin (FK228), Apicidin	HDAC1/2, HDAC2/3	PD	[19, 74, 111]
Mercaptoacetamides	W2	Class II	AD	[143]
Sirtuin ligands	AGK-2, selisistat, nicotinamide	SIRT2, SIRT1, Class III	PD, HD, AD, FRDA	[102, 137]

PD Parkinson's disease, *HD* Huntington's disease, *AD* Alzheimer's disease, *ALS* amyotrophic lateral sclerosis, *SMA* spinal muscular atrophy, *FRDA* Friedreich's ataxia

HDAC8, selective chemicals like PCI-34051 might be beneficial in particular neurodegenerative situations. However, recent approaches to HDACIs have also touched upon using bispecific molecules that target both enzymes and other activities. Targeting multiple aspects of the disease simultaneously is one of the strategies that will be discussed in detail; this includes multitarget-directed ligands (MTDLs), which include inhibition of HDAC but also intrinsically can act as antioxidants or neuroprotectors [103]. These agents were developed to simultaneously act on multiple pathologic processes, which might provide additional therapeutic advantages. The structural classes and selectivity profiles of HDACIs make it possible to create a range of drugs for different aspects of neurodegenerative diseases. With a continued understanding of the function of different HDAC isoforms in neurodegeneration, the therapeutic function of more selective and effective HDAC inhibitors may be forthcoming. Further improvement of these compounds and better insight into disease processes and mechanisms suggest progress towards improved treatments for fatal neurodegenerative conditions.

14.5.1 Mechanisms of Action

HDACIs affect in various ways, mainly by blocking the deacetylase action of HDACs that cause a rise in histone and nonhistone protein acetylation [24]. These inhibitors chelate with a zinc molecule in the active site of zinc-dependent HDACs (Classes I, II, and IV), making them inactive [91]. The action of HDACIs is based on shifting the rate of histone acetylation and deacetylation to favor chromatin relaxation and activation of neuroprotective gene expression [14]. Due to the structural flexibility of the HDACIs, different levels of selectivity are possible. Classified inhibitors, like vorinostat, act pan-HDAC. On the other hand, class-selective or isoform-selective, there are more selective therapies [69]. For instance, MGCD0103 targets only HDAC classes I and IV and may have minimal side effects [33].

14.6 Pharmacological Properties

HDACIs have multifaceted pharmacological profiles relevant to their therapeutic effects for many diseases, including neurodegenerative diseases, cancer, and other diseases with an underlying epigenetic contribution. These compounds bind to the zinc ion at the active site of zinc-dependent HDACs from Class I, II, and IV, reducing their deacetylase activity and enhancing histone and non-histone protein acetylation [162]. Such a change in acetylation status leads to chromatin decondensation. It thereby enhances gene expression, especially genes that play protective functions to neurons or genes involved in processes that oppose the disease advancement [9, 28]. Among the pharmacokinetic characteristics of HDACIs, the capability to cross the blood-brain barrier is considered one of the most essential features. This property is necessary for their efficiency in treating neurodegenerative disorders with therapeutic targets in the CNS. Lipophilic inhibitors of class I and II HDACs, including vorinostat and valproic acid, successfully penetrate the BBB, producing changes in gene expression and other cellular activities within the brain [49]. This penetration capability is highly desirable in situations such as PD and AD since one has to deliver a neuroprotector to the CNS. There are differences in the pharmacokinetics of these drugs due to the differences in their chemical properties. Specifically, only some HDACIs like the SAHA (vorinostat) have a relatively short half-life, which means that the substance is quickly cleared from the body and, therefore, requires a repeat dosage to reach therapeutic levels [68]. Other inhibitors may have a longer half-life, meaning they are taken less often, which translates to better patient compliance and convenience in treating chronic diseases.

14.6.1 Neuroprotective Effects of HDAC Inhibitors

14.6.1.1 Epigenetic Modulation

Neuroprotection through HDACIs is mainly due to epigenetic mechanisms. By suppressing histone deacetylase enzymes, these compounds enhance histone acetylation, which weakens chromatin compactness, making genes important in neuronal survival, synaptic plasticity, and cognition be expressed [41]. Such modulation of epigenetic marks has been demonstrated to increase the production of brain-derived neurotrophic factor (BDNF), which is known to play an important role in neuronal survival, plasticity and neurogenesis [72]. Besides the BDNF, HDAC also brings a change in transcription factors such as p53, NF-κB, and STAT3 for deacetylation, along with its role in homeostasis, cell cycle regulation, inflammation and apoptosis [3, 46]. These diverse impacts on transcription factors imply that HDACIs have potential therapeutic value in treating neurodegenerative diseases by modulating various pathways affected by these diseases [14]. Recent studies have pointed out the use of HDACIs in treating directional depression, the epigenetic modification of which may affect the transcriptional profile of the brain [35].

14.6.1.2 Neurotrophic Effects

HDACIs have proven to have profound neurotrophic effects, including neuron growth, differentiation, and survival. These compounds also upregulate the neurotrophic factors like BDNF or the nerve growth factor that plays a significant role in the health and functioning of neurons [14, 66]. Stimulation of neurogenesis and enhancement of synaptic plasticity by HDACIs are quite valuable; these may be partial reasons for cognition augmentation in neurodegenerative diseases [31, 41]. In addition, HDACIs have recently been proven to modulate other neurotrophic factors, such as GDNF and CNTF, that promote neuronal survival and growth [14, 66, 158]. These effects are not selective for one class of compounds; multiple classes of HDACIs have been shown to upregulate neurotrophic factors, suggesting that the class may have broad therapeutic utility for neurodegenerative diseases.

14.6.1.3 Anti-inflammatory Properties

Neuroinflammation is a hallmark of many neurodegenerative diseases, and HDACIs have shown potent anti-inflammatory effects. These compounds suppress the activation of microglia and astrocytes, reducing the production of pro-inflammatory cytokines and chemokines, thereby attenuating the inflammatory response in the central nervous system (CNS) [125, 141]. HDACIs also modulate the expression of genes involved in the inflammatory response, such as NF-κB, which further contributes to their anti-inflammatory properties [80, 141]. In addition, HDACIs can inhibit the production of pro-inflammatory mediators like prostaglandins and leukotrienes through their interaction with enzymes such as COX-2 and 5-LOX [50, 79]. They also influence the expression of anti-inflammatory genes such as IL-10 and TGF-β, which play protective roles in inflammation resolution [39, 130]. The ability of HDACIs to counteract neuroinflammation suggests their potential to treat conditions where inflammation exacerbates neuronal damage, such as multiple sclerosis, AD, and PD [30, 141]. Neuroinflammation is a pathophysiological characteristic of most neurodegenerative diseases, and HDACIs have been demonstrated to possess remarkable anti-inflammatory properties. These compounds decrease the microglial and astrogliosis activation, reduce the output of pro-inflammatory cytokines and chemokines and alleviate microglia-astrocyte-induced neuroinflammatory response in the CNS [125, 141]. They also regulate the expression of other genes that play a role in inflammation, including NF-κB and exhibit a modulatory effect on inflammatory cytokines [80, 141]. Furthermore, HDACIs could suppress the production of prostaglandins and leukotrienes by modulating the enzymes, including COX-2 and 5-LOX [50, 79]. They also modulate genes involved in resolving inflammation, including the IL-10 and TGF-β genes considered to have anti-inflammatory effects [39, 130]. The benefits of HDACIs to overcome neuroinflammation give temptation for treating diseases where inflammation causes neuronal death, like multiple sclerosis, AD, and PD [30, 141].

14.6.1.4 Antioxidant Effects

HDACIs have been described to have antioxidant potential and, therefore, are thought to aid in neuroprotection. They also act to increase antioxidant genes, which help to prevent and alleviate oxidative stress in neurons and thus may counteract the oxidative process linked to neurodegenerative mechanisms [116, 133]. HDACIs might also modulate the expression of other endogenous antioxidant factors, including the Nrf2 pathway critical to antioxidant protection in cells [87, 170]. The signaling cascade after being invoked through HDACIs is the Nrf2 pathway, which induces antioxidant enzymes in the cell, such as HO-1, SOD and GST, which enhance the antioxidant potential of the cell [62, 121]. These antioxidant effects have been elaborated in later studies where pathways regulated by HDACIs are not only involved in suppressing oxidation but also modulating cell resistance to numerous forms of damage. For instance, the phenylhydroxamate HDAC6 inhibitors are also reported to work as activators of both Nrf2 and HIF-1α, including neuroprotection by a three-pronged approach involving HDAC inhibition, HIF-1α stabilization and Nrf2 activation [63]. Therefore, HDACIs could serve as an approach to neuroprotection with diverse targets in the pathogenesis of neurodegenerative diseases. Due to their capacity to regulate epigenetic events, enhance neurotrophic impact, decrease inflammation, and possibly mitigate oxidative stress, they have been proposed for use in the treatment of numerous neurological diseases [14, 41, 125]. They also may find uses in other pathological states that involve changes in epigenomic gene profiles, such as cancer, cardiovascular diseases, and traumatic brain injury (TBI) [31, 40, 46]. The precise mechanisms and specific HDAC classes targeted by these inhibitors continue to be elucidated, and ongoing research aims to improve their efficacy, selectivity, and safety for clinical use.

14.6.2 HDAC Inhibitors in Specific Neurodegenerative Diseases

14.6.2.1 Alzheimer's Disease

HDACIs have been tested in different animal models of AD and have shown different extents of efficacy. Sodium butyrate treatment increased H3 and H4 acetylation levels, restoration of cognitive deficits and synaptic plasticity in p25-expressing mice [17, 32, 107]. In Tg2576 AD mice, daily 4-phenylbutyrate injections reversed spatial memory deficits and normalized tau hyperphosphorylation in the hippocampus [115]. TSA, VPA, SAHA, and other HDACIs have been shown to increase contextual memory and promote synapse remodeling in fear-conditioned AD mouse models [32, 34, 45, 70, 114]; the novel mercaptoacetamide-based compound W2 reduced amyloid plaque load and decreased tau phosphorylation in APP/PSI-21 and 3xTG-AD mice [143]. In AD mice, Tubastatin A and ACY-1215 HDAC6 inhibition improved behavioral deficits, suppressed tau hyperphosphorylation and reduced amyloid-β load [168]. A Class III HDAC, nicotinamide, blocked all memory deficits and reduced tau pathologies in 3xTg-AD mice [42]. APP23 transgenic AD mice substantiate that early treatment with valproic acid reduced the Aβ plaque number and memory loss [99]. Particular HDACs involved in AD development have also been identified. In AD mice, HDAC2 knockdown improved hippocampus-dependent memory performance and synaptic plasticity (Guan et al. 2009). HDAC2 overexpression decreased dendritic spine density and impaired memory formation (Guan et al. 2009). CA1 deletion of HDAC3 improved cognition and elevated expression of genes associated with cognition [95]. In contrast, in APPswe/PS1dE9 mice, HDAC3 overexpression elevated Aβ levels and decreased dendritic spine density [4].

HDAC6 has been discovered to cause robust neurotoxicity and depletion returned learning- and memory-impaired AD models to normalcy by repairing impaired axonal transport [38, 168]. Downregulation of SIRT6 resulted in tau hyperphosphorylation as a protective factor in AD [144]. Other recent strategies in treating AD have involved the dual antagonists of phosphodiesterase five and HDAC [112, 113]. An acridine-based HDACIs were rationally designed as a multi-targeting agent for HDAC and acetylcholinesterase; the compound possesses selectivity to HDAC6 and potent activity against Aβ-aggregation [150]. A novel derivative of protopine, bromo-protopine, regulates tau aggregation by modulating HDAC6-HSP90 chaperone activity and improves memory by activating the Ras-GRF1/ERK pathway [27, 138, 139], while protopine promotes the proteasomal degradation of pathological tau in AD models through HDAC6 inhibition [26, 138, 139]. A newly synthesized tacrine-hydroxamate pan HDACi showed inhibitory activity of Aβ1–42 self-aggregation and Aβ fibrils disaggregation [160]. Valproate treatment at different stages of AD pathology effectively reduced the level of NF-kB and IL-1β in the plasma of Tg6799 AD mice and increased hippocampal NGF levels. VPA also reduced escape latencies of Tg6799 mice at both early and late symptomatic phases of AD [99]. Kilgore et al. demonstrated that sodium butyrate, valproate or vorinostat for chronic treatment restored contextual memory severely in APPswe/PS1dE9 transgenic mice model for AD [70]. Although HDACIs have therapeutic potential in AD, evaluating tau acetylation is essential. Post-translational modification of soluble tau through acetylation has been demonstrated to inhibit its degradation and consequent accumulation in cells. Resetting histone acetylation in AD treatment needs to be appropriate [59].

14.6.2.2 Parkinson's Disease

Parkinson's disease is caused by dopaminergic neuron loss in the substantia nigra. HDACs have been proposed as a potential new therapeutic target for PD. In the substantia nigra microglia of PD subjects, HDAC2 was upregulated [146]. HDACI, such as phenylbutyrate, has been found to prevent or reduce dopamine depletion and neuronal loss in MPTP-lesioned PD mice model [37]. In familial PD, the α-synuclein mutations have been linked to HDACIs such as vorinostat and sodium butyrate, showing protective effects of α-synuclein toxicity [71]. The functions of HDACIs are multifaceted since the compounds can be involved in developing a disease in one instance and can serve as a treatment in another. HDAC 1 and 2 inhibitor K560 attenuated the MPTP-induced cell death both in vitro and in vivo [12]. Pinho et al. [110] showed that M275 (HDAC 1 inhibitor) and Tubastatin A (HDAC 6 inhibitor) improved motor impairments. Valproic acid (VPA) has been shown to attenuate the effects of 6-OHDA-induced toxicity in dopaminergic cells by inhibiting apoptotic caspase proteins [54]. Valproate, sodium butyrate, and phenylbutyrate (PBA), the HDACIs, have shown neuroprotective effects in PD models. They enhance histone acetylation, stimulate neurotrophic factors (GDNF, BDNF), decrease caspase-3 activity, decrease oxidative stress, and inhibit neuroinflammation [51, 129]. PBA also decreased α-synuclein plaque formation and increased DJ-1 protein levels in MPTP-induced mice [37]. These showed that TSA improved histone acetylation and had a beneficial effect on MPTP-induced models; however, other studies have claimed that TSA promoted dopaminergic neuronal death [152]. Vorinostat (SAHA) alleviates MPP+-induced cell death in dopaminergic cell lines [71]. Apicidin and other more recent and potent inhibitors include romidepsin and K560, which are prospective candidates for PD treatment [12]. HDAC6 contributes to inclusion formation in cytoplasm and autophagy and may have a neuroprotective effect on α-synuclein [81, 82]. However, inhibition of tubacin induced defects in aggresome formation and cell death [102]. SIRT2 suppression enhanced α-tubulin content, and SIRT2-shRNA protected dopaminergic neurons in vitro and Drosophila PD models [102]. Although initial studies on the use of HDACIs in PD treatment are encouraging, their potential actions differ with regard to cell type and tissue targeting. Nevertheless, more functional studies are required to elucidate their actions and possible application as drugs in PD.

14.6.2.3 Huntington's Disease

HD is a rare, inherited, neurodegenerative disorder due to the expansion of CAG repeats in the huntingtin gene. Histone hypoacetylation is a key pathogenic mechanism manifested by transcriptional dysregulation. HDACIs have been reported to have therapeutic efficacy in numerous HD models. McFarland et al. [94] showed reduced H3 hypoacetylation throughout the whole genome in the HD mice; Dompierre et al. [25] showed that HDA6 inhibitor enhanced microtubular transport abnormality. Pimelic diphenylamine reversed H3 hypoacetylation and improved motor deficits in transgenic HD mice [149]. Neuroprotective effects of SAHA and sodium butyrate have been shown in a Drosophila and R6/2 mouse model [29, 140]. Vorinostat and TSA enhanced BDNF vesicular trafficking by attenuating HDAC6 activity and α-tubulin deacetylation [25]. HDAC3 inhibitor, RGFP966, ameliorated motor dysfunction in the N171-82Q CAG transgenic mice [61], while LBH589, a broad-spectrum HDACI, also mitigated striatal neuronal atrophy in R6/2 mice (Chopra et al. 2016; [134]).

In their study, Gardian and colleagues noted that phenylbutyrate had neuroprotective roles in HD-N171-82Q therapeutic mice [36], enhancing their lifespan and decreasing striatal atrophy. Jia et al. [60] demonstrated that class I inhibitors exert neuroprotective effects in regulating HDAC1 and HDAC3. Pan-HDAC inhibitors like SAHA, sodium butyrate and PBA have also shown effectiveness in other HD models, ameliorating motor dysfunction, and increasing lifespan [29, 65, 93, 100, 140]. PBA has also been demonstrated in the first clinical trials with HD patients [52]. HDAC1 and HDAC3 have been implicated in HD pathophysiology out of all the HDACs. In multiple HD models, benzamide-derived compounds targeting these isoforms, HDACi4b and HDACi136, have been beneficial [148]. Body weight, motor function, and cognitive decline are improved, and these inhibitors reduce huntingtin aggregates. However, there are conflicting results concerning the relative importance of HDAC1 versus HDAC3 in HD pathology [60]. HDACIs considered non-histone targets, like HDAC4 and HDAC6, are also possibly involved in the pathogenesis and treatment of HD [14]. HDACIs considered non-histone targets, like HDAC4 and HDAC6, are also possibly involved in the pathogenesis and treatment of HD [14]. The NAD + -dependent Class III HDACs, sirtuins, are potential targets, though some studies reported neuroprotective roles of both SIRT1 inhibition and activation in HD model organisms [58, 135, 147]. Ongoing clinical trials of HDACIs in HD patients are underway, and PBA has displayed good safety and tolerability outcomes in the phase I and II trials [52, 155]. Alternative strategies, for example, ketogenic diets to drive endogenous β-hydroxybutyrate production, are being explored [151].

14.6.2.4 Amyotrophic Lateral Sclerosis (ALS)

ALS is a devastating neurodegenerative disorder characterized by motor neuron degeneration, resulting in progressive muscular atrophy, paralysis and demise within 2 to 5 years postdiagnosis [120]. It manifests in two primary forms: intermittent (90–95%) and hereditary (5–10%), the latter associated with the mutation of the superoxide dismutase 1 (SOD1) gene [117]. Multiple cellular dysfunctions, including glutamate-mediated excitotoxicity, apoptosis, oxidative stress and disrupted axonal transport, have been associated with the pathogenesis of ALS [5, 16]. Recent studies have suggested that aberrant transcription and altered transcriptomic features may be major contributors to ALS development [47, 48, 88, 101]. As histone deacetylase inhibitors (HDACIs) may regulate transcriptional dysregulation, research has been directed towards studying these compounds to treat ALS. Experimental studies using rodent models have supported the potentially promising therapeutic effects of HDACis in preventing the occurrence and improving signs of the disease [119]. For example, VPA and PBA had neuroprotective properties due to the regulation of histone acetylation and increased motor neuron viability [122]. PBA combined with riluzole, the only approved ALS drug, improved outcomes, although the mechanisms underlying this improvement have not been elucidated [20]. However, preclinical outcomes were promising with VPA, but effects in clinical trials were less so, with no significant prolongation of survival in ALS patients [108]. On the other hand, PBA was found to be effective when initiated early in transgenic ALS mice and well-tolerated in phase II trials of ALS therapy [18, 124]. Moreover, trichostatin (TSA) rescued muscle atrophy and neuromuscular degeneration in ALS mouse models by promoting histone acetylation and neuroprotective effects [166]. The role of HDAC4 in ALS is very significant, as the knockdown accelerated disease onset in the mouse models, raising issues with broad-spectrum HDACis [109]. At the same time, therapeutic strategies combining SAHA with HDAC1/3 inhibitors and arimoclomol have demonstrated the capacity to prevent nuclear degradation and promote DNA repair in motor neurons harboring FUS mutations [76]. However, the precise mechanisms of HDACis' neuroprotection remain to be elucidated, and further studies are needed to optimize further clinical applications [145]. While HDACis may provide therapeutic avenues in ALS, their complex roles in transcriptional regulation and neuroprotection underscore the requirement for targeted approaches in ALS to assess the diverse aspects of pathophysiology. Since HDAC inhibition targets various cellular systems, it is imperative to establish these connections to further the development of ALS therapeutic strategies.

14.7 Challenges and Future Perspectives

There is a general focus on isoform-selective HDAC inhibitors for the development of therapeutic strategies for neurodegenerative diseases, including AD, PD, and HD, which are diseases with progressive neuronal degeneration and complex pathological processes [76, 161]. Pan-HDAC inhibitors have demonstrated neuroprotection, but as nonselective, they cause off-target effects, toxicity, and poor therapeutic windows [14]. As an alternative, selective HDAC inhibitors have emerged, aiming at specific isoforms involved in neurodegeneration. In AD models, HDAC2 inhibitors, including W2, decreased Aβ levels and attenuated the cognitive deficit [161]. Likewise, the HDAC3-specific inhibitor RGFP966 showed efficacy in protecting synaptic plasticity from amyloid β oligomers [70]. These results demonstrate the promise of selectively targeting tumor-specific isoforms to reduce toxicity and increase therapeutic gain.

HDAC inhibitors are challenging to balance efficacy and safety owing to the high degree of homology between HDAC isoforms, resulting in off-target effects and systemic toxicity [97, 126]. The enhancement of isoform specificity proposed is currently accomplished through structure-guided design and computational modelling [43]. For example, the HDAC6-selective inhibitors, such as compound 9, exhibit a 14-fold selectivity over HDAC1, minimizing off-target toxicities while maintaining efficacy [97]. Other novel approaches include the use of PROTACs and allosteric modulators in the selective targeting of enzymes. PROTACs provoke selective degradation of the desired HDAC isoforms and significantly decrease the number of side effects compared to traditional inhibitors [7]. Furthermore, the covalent inhibitors that target the isoform-specific residues have also been analyzed to improve specificity [77].

The multifactorial etiology of neurodegenerative diseases makes combination therapies a comprehensive treatment approach. HDAC inhibitors, combined with neuroprotective compounds or drugs against oxidative damage, inflammation, or proteostasis, have additive effects in animal models [76, 161]. For example, the neuroprotective and cognitive benefits of AD models combining HDAC inhibitors and anti-inflammatory agents were enhanced [14]. Combination therapies can be rationally designed from systems biology and network pharmacology approaches with identification of synergistic targets and avoidance of adverse interactions [53]. For instance, the simultaneous treatment with HDAC inhibitors and autophagy-promoting agents in HD models decreases the formation of mutant huntingtin aggregates and ameliorates phenotypes [97].

Improving BBB permeability is essential for the use of HDAC inhibitors for CNS diseases. Low BBB passage hampers the effectiveness of most FDA-approved HDACIs, including SAHA [97, 123]. To overcome this challenge, some possible approaches include chemical alterations, pro-drug approaches, and delivery system improvements up to that of lipid nanoparticles and exosomes [118]. Preclinical studies showed that benzoheterocycle-based HDAC inhibitors, such as 9b, penetrated the BBB better than SAHA [97]. Based on BBB penetrating transporters like transferrin or glutathione, prodrug strategies increase brain delivery with reduced systemic exposure [2, 161]. HDAC inhibitors can be delivered to the brain in a targeted manner by emerging nanotechnologies, polymeric micelles and liposomes, improving efficacy and decreasing toxicity [118]. These advances are a significant step toward overcoming BBB-associated challenges.

Designing agents selective for particular HDAC isoforms implies greater insight into the involvement of these enzymes in neurodegenerative mechanisms. Combining structural biology with bioinformatics and high-throughput screening propels the discovery process towards specific and safe inhibitors [76]. Emerging technologies such as artificial intelligence and machine learning opened up new avenues for drug design and isoform-specific interactions [11]. To settle the questions of the effectiveness and non-toxicity of HDAC inhibitors, strong preclinical and clinical projects are needed. Multidisciplinary teamwork will enhance the translation processes of potential drugs focused on HDACs; challenges exist in this therapeutic area [76]. Thus, isoform-specific HDAC inhibitors represent a promising therapeutic approach for treating neurodegenerative diseases with currently unmet medical needs. Through multiplatform structures, different ways of delivery, and modern computational analysis, one can depict the true potential of these compounds.

14.8 Conclusion

HDAC inhibitors have been identified "as potential therapeutics for neurodegenerative diseases, showing neuroprotection, cognitive enhancement, and the modulation of major pathological processes in multiple animal models[106, 161]. Specifically, isoform-specific inhibitors, including HDAC2, HDAC3, and HDAC6 inhibitors, have shown promise in addressing specific aspects of neurodegenerative pathology while minimizing off-target effects [98, 161]. The potential of using HDAC inhibitors as a therapeutic approach for guiding neurodegenerative diseases is established by the fact that these inhibitors affect several pathological processes, such as alteration of protein aggregation, neuroinflammation, and synaptopathy [106, 161]. However, translating preclinical successes into clinical trials presents a critical step to achieving the advantage of HDAC inhibitors as a neurodegenerative disease treatment [98]. Improvements in BBB penetration and isoform selectivity, exemplified by T-518 and compound 9b, suggest that novel compounds are a promising avenue for clinical development[98]. Regarding future direction, more effort is needed to precisely define the functions of individual members of the HDAC family in the neurodegenerative process, new selective drugs with better BBB permeability should be developed, and a synergistic approach should be explored to increase the therapeutic effect and decrease the side effects[106, 161]. However, translating preclinical successes into clinical trials presents a critical step to achieving the advantage of HDAC inhibitors as a neurodegenerative disease treatment [98].

References

1. Bae J, Hideshima T, Tai YT, Song Y, Richardson P, Raje N, Munshi NC, Anderson KC. Histone deacetylase (HDAC) inhibitor ACY241 enhances anti-tumour activities of antigen-specific central memory cytotoxic T lymphocytes against multiple myeloma and solid tumours. Leukemia. 2018;32(9):1932–47. https://doi.org/10.1038/s41375-018-0062-8.
2. Banks WA. From blood-brain barrier to blood-brain interface: new opportunities for CNS drug delivery. Nat Rev Drug Discov. 2016;15(4):275–92. https://doi.org/10.1038/nrd.2015.21.
3. Bantscheff M, Hopf C, Savitski MM, Dittmann A, Grandi P, Michon AM, Schlegl J, Abraham Y, Becher I, Bergamini G, Boesche M, Delling M, Dümpelfeld B, Eberhard D, Huthmacher C, Mathieson T, Poeckel D, Reader V, Strunk K, Sweetman G, Kruse U, Neubauer G, Ramsden NG, Drewes G. Chemoproteomics profiling of HDAC inhibitors reveals selective targeting of HDAC complexes. Nat Biotechnol. 2011;29(3):255–65. https://doi.org/10.1038/nbt.1759.
4. Bardai FH, D'Mello SR. Selective toxicity by HDAC3 in neurons: regulation by Akt and GSK3β. J Neurosci. 2011;31(5):1746–51.
5. Bilsland LG, Sahai E, Kelly G, Golding M, Greensmith L, Schiavo G. Deficits in axonal transport precede ALS symptoms in vivo. Proc Natl Acad Sci. 2010;107(47):20523–8. https://doi.org/10.1073/pnas.1006869107.
6. Boyault C, Zhang Y, Fritah S, Caron C, Gilquin B, Kwon SH, Garrido C, Yao TP, Vourc'h C, Matthias P, Khochbin S. HDAC6 controls major cell response pathways to cytotoxic accumulation of protein aggregates. Genes Dev. 2007;21(17):2172–81. https://doi.org/10.1101/gad.436407.
7. Burslem GM, Crews CM. Proteolysis-targeting chimeras as therapeutics and tools for biological discovery. Cell. 2020;181(1):102–14. https://doi.org/10.1016/j.cell.2019.11.031.
8. Butler KV, Kalin J, Brochier C, Vistoli G, Langley B, Kozikowski AP. Rational design and simple chemistry yield a superior, neuroprotective HDAC6 inhibitor, tubastatin A. J Am Chem Soc. 2010;132(31):10842–6.
9. Butler R, Bates GP. Histone deacetylase inhibitors as therapeutics for polyglutamine disorders. Nat Rev Neurosci. 2006;7(10):784–96.

10. Cheng A, Yang Y, Zhou Y, Maharana C, Lu D, Peng W, Liu Y, Wan R, Marosi K, Misiak M, Bohr VA, Mattson MP. Mitochondrial SIRT3 mediates adaptive responses of neurons to exercise and metabolic and excitatory challenges. Cell Metab. 2016;23(1):128–42. https://doi.org/10.1016/j.cmet.2015.10.013.
11. Chen H, Engkvist O, Wang Y, Olivecrona M, Blaschke T. The rise of deep learning in drug discovery. Drug Discov Today. 2018;23(6):1241–50. https://doi.org/10.1016/j.drudis.2018.01.039.
12. Choong CJ, Sasaki T, Hayakawa H, Yasuda T, Baba K, Hirata Y, Uesato S, Mochizuki H. A novel histone deacetylase 1 and 2 isoform-specific inhibitor alleviates experimental Parkinson's disease. Neurobiol Aging. 2016;37:103–16. https://doi.org/10.1016/j.neurobiolaging.2015.10.001.
13. Chopra V, Quinti L, Kim J, Vollor L, Narayanan KL, Edgerly C, Cipicchio PM, Lauver MA, Choi SH, Silverman RB, Ferrante RJ, Hersch S, Kazantsev AG. The sirtuin two inhibitor AK-7 is neuroprotective in Huntington's disease mouse models. Cell Rep. 2012;2(6):1492–7. https://doi.org/10.1016/j.celrep.2012.11.001.
14. Chuang DM, Leng Y, Marinova Z, Kim HJ, Chiu CT. Multiple roles of HDAC inhibition in neurodegenerative conditions. Trends Neurosci. 2009;32(11):591–601. https://doi.org/10.1016/j.tins.2009.06.002.
15. Cook C, Carlomagno Y, Gendron TF, Dunmore J, Scheffel K, Stetler C, Davis M, Dickson D, Jarpe M, DeTure M, Petrucelli L. Acetylation of the KXGS motifs in tau is a critical determinant in modulation of tau aggregation and clearance. Hum Mol Genet. 2014;23(1):104–16. https://doi.org/10.1093/hmg/ddt402.
16. Corona JC, Tovar-y-Romo LB, Tapia R. Glutamate excitotoxicity and therapeutic targets for amyotrophic lateral sclerosis. Expert Opin Ther Targets. 2007;11(11):1415–28. https://doi.org/10.1517/14728222.11.11.1415.
17. Cruz JC, Kim D, Moy LY, Dobbin MM, Sun X, Bronson RT, Tsai LH. p25/cyclin-dependent kinase 5 induces production and intraneuronal accumulation of amyloid beta in vivo. J Neurosci. 2006;26(41):10536–41.
18. Cudkowicz ME, Andres PL, Macdonald SA, Bedlack RS, Choudry R, Brown RH, Zhang H, Schoenfeld DA, Shefner J, Matson S, Matson WR, Ferrante RJ. Phase 2 study of sodium phenylbutyrate in ALS. Amyotroph Lateral Scler. 2009;10(2):99–106.
19. Darkin-Rattray SJ, Gurnett AM, Myers RW, Dulski PM, Crumley TM, Allocco JJ, Cannova C, Meinke PT, Colletti SL, Bednarek MA, Singh SB, Goetz MA, Dombrowski AW, Polishook JD, Schmatz DM. Apicidin: a novel antiprotozoal agent that inhibits parasite histone deacetylase. Proc Natl Acad Sci. 1996;93(23):13143–7. https://doi.org/10.1073/pnas.93.23.13143.
20. Del Signore SJ, Amante DJ, Kim J, Stack EC, Goodrich S, Cormier K, Smith K, Cudkowicz ME, Ferrante RJ. Combined riluzole and sodium phenylbutyrate therapy in transgenic amyotrophic lateral sclerosis mice. Amyotroph Lateral Scler. 2009;10(2):85–94. https://doi.org/10.1080/17482960802226148.
21. Deubzer HE, Schier MC, Oehme I, Lodrini M, Haendler B, Sommer A, Witt O. HDAC11 is a novel drug target in carcinomas. Int J Cancer. 2013;132(9):2200–8. https://doi.org/10.1002/ijc.27876.
22. Di Giorgio E, Brancolini C. Regulation of class IIa HDAC activities: it is not only matter of subcellular localization. Epigenomics. 2016;8(2):251–69. https://doi.org/10.2217/epi.15.106.
23. Ding G, Liu HD, Huang Q, Liang HX, Ding ZH, Liao ZJ, Huang G. HDAC6 promotes hepatocellular carcinoma progression by inhibiting P53 transcriptional activity. FEBS Lett. 2013;587(7):880–6. https://doi.org/10.1016/j.febslet.2013.02.001.
24. Dokmanovic M, Clarke C, Marks PA. Histone deacetylase inhibitors: overview and perspectives. Mol Cancer Res. 2007;5(10):981–9. https://doi.org/10.1158/1541-7786.MCR-07-0324.
25. Dompierre JP, Godin JD, Charrin BC, Cordelières FP, King SJ, Humbert S, Saudou F. Histone deacetylase 6 inhibition compensates for the transport deficit in Huntington's disease by increasing tubulin acetylation. J Neurosci. 2007;27(13):3571–83. https://doi.org/10.1523/JNEUROSCI.0037-07.2007.

26. Durairajan SSK, Li M, Malampatti S, Zhang Y, Liu LF, Song J, Chen LL, Zeng Y, Senapati S. Protopine, a promising novel histone deacetylase 6 inhibitor reduces tauopathy in in vitro and in vivo. Neurodegener Dis. 2015;15(Suppl. 1):579. https://doi.org/10.1159/000381736.
27. Durairajan SS, Reddy R, Shetty SG, Iyaswamy A, Kammella AK, Krishnamoorthi SK, Zhang H, Li M. Bromo-protopine, a novel derivative of protopine with improved bioavailability and bioactivity, degrades tau aggregation through modulation of hdac6-hsp90 chaperonic activity and improves memory via stimulation of the ras-Grf1/Erk pathway. Alzheimers Dement. 2017;13(7):P1575–6.
28. Eckschlager T, Plch J, Stiborova M, Hrabeta J. Histone Deacetylase Inhibitors as Anticancer Drugs. Int J Mol Sci. 2017;18(7):1414. https://doi.org/10.3390/ijms18071414. PMID: 28671573; PMCID: PMC5535906.
29. Ferrante RJ, Kubilus JK, Lee J, Ryu H, Beesen A, Zucker B, Smith K, Kowall NW, Ratan RR, Luthi-Carter R, Hersch SM. Histone deacetylase inhibition by sodium butyrate chemotherapy ameliorates the neurodegenerative phenotype in Huntington's disease mice. J Neurosci. 2003b;23(28):9418–27. https://doi.org/10.1523/JNEUROSCI.23-28-09418.2003.
30. Fessler EB, Chibane FL, Wang Z, Chuang DM. Potential roles of HDAC inhibitors in mitigating ischemia-induced brain damage and facilitating endogenous regeneration and recovery. Curr Pharm Des. 2013;19(28):5105–20. https://doi.org/10.2174/1381612811319280009.
31. Fischer A, Sananbenesi F, Mungenast A, Tsai LH. Targeting the correct HDAC(s) to treat cognitive disorders. Trends Pharmacol Sci. 2010;31(12):605–17. https://doi.org/10.1016/j.tips.2010.09.003.
32. Fischer A, Sananbenesi F, Wang X, Dobbin M, Tsai LH. Recovery of learning and memory is associated with chromatin remodelling. Nature. 2007;447(7141):178–82.
33. Fournel M, Bonfils C, Hou Y, Yan PT, Trachy-Bourget MC, Kalita A, Liu J, Lu AH, Zhou NZ, Robert MF, Gillespie J, Wang JJ, Ste-Croix H, Rahil J, Lefebvre S, Moradei O, Delorme D, Macleod AR, Besterman JM, Li Z. MGCD0103, a novel isotype-selective histone deacetylase inhibitor, has broad spectrum antitumor activity in vitro and in vivo. Mol Cancer Ther. 2008;7(4):759–68. https://doi.org/10.1158/1535-7163.MCT-07-2026.
34. Francis YI, Fa M, Ashraf H, Zhang H, Staniszewski A, Latchman DS, Arancio O. Dysregulation of histone acetylation in the APP/PS1 mouse model of Alzheimer's disease. J Alzheimer's Dis. 2009;18(1):131–9.
35. Fuchikami M, Yamamoto S, Morinobu S, Okada S, Yamawaki Y, Yamawaki S. The potential use of histone deacetylase inhibitors in the treatment of depression. Prog Neuropsychopharmacol Biol Psychiatry. 2016;64:320–4. https://doi.org/10.1016/j.pnpbp.2015.03.010.
36. Gardian G, Browne SE, Choi DK, Klivenyi P, Gregorio J, Kubilus JK, Ryu H, Langley B, Ratan RR, Ferrante RJ, Beal MF. Neuroprotective effects of phenylbutyrate in the N171-82Q transgenic mouse model of Huntington's disease. J Biol Chem. 2005;280(1):556–63. https://doi.org/10.1074/jbc.M410210200.
37. Gardian G, Yang L, Cleren C, Calingasan NY, Klivenyi P, Beal MF. Neuroprotective effects of phenylbutyrate against MPTP neurotoxicity. Neuromolecular Med. 2004;5(3):235–41. https://doi.org/10.1385/NMM:5:3:235.
38. Govindarajan N, Rao P, Burkhardt S, Sananbenesi F, Schlüter OM, Bradke F, Lu J, Fischer A. Reducing HDAC6 ameliorates cognitive deficits in a mouse model for Alzheimer's disease. EMBO Mol Med. 2013;5(1):52–63.
39. Grabiec AM, Krausz S, de Jager W, Burakowski T, Groot D, Sanders ME, Prakken BJ, Maslinski W, Eldering E, Tak PP, Reedquist KA. Histone deacetylase inhibitors suppress inflammatory activation of rheumatoid arthritis patient synovial macrophages and tissue. J Immunol. 2011;184(5):2718–28. https://doi.org/10.4049/jimmunol.0901467.
40. Gräff J, Rei D, Guan JS, Wang WY, Seo J, Hennig KM, Nieland TJ, Fass DM, Kao PF, Kahn M, Su SC, Samiei A, Joseph N, Haggarty SJ, Delalle I, Tsai LH. An epigenetic blockade of cognitive functions in the neurodegenerating brain. Nature. 2012;483(7388):222–6. https://doi.org/10.1038/nature10849.
41. Gräff J, Tsai LH. Histone acetylation: molecular mnemonics on the chromatin. Nat Rev Neurosci. 2013;14(2):97–111. https://doi.org/10.1038/nrn3427.

42. Green KN, Steffan JS, Martinez-Coria H, Sun X, Schreiber SS, Thompson LM, LaFerla FM. Nicotinamide restores cognition in Alzheimer's disease transgenic mice via a mechanism involving sirtuin inhibition and selective reduction of Thr231-phosphotau. J Neurosci. 2008;28(45):11500–10.
43. Grozinger CM, Schreiber SL. Deacetylase enzymes: biological functions and the use of small-molecule inhibitors. Chem Biol. 2002;9(1):3–16. https://doi.org/10.1016/s1074-5521(02)00092-3.
44. Gu X, Zhang H, Jiao M, Han B, Zhang Z, Li J, Zhang Q. Histone deacetylase 6 inhibitors with blood-brain barrier penetration as a potential strategy for CNS-disorders therapy. Eur J Med Chem. 2022;227:113924.
45. Guan JS, Haggarty SJ, Giacometti E, Dannenberg JH, Joseph N, Gao J, Nieland TJ, Zhou Y, Wang X, Mazitschek R, Bradner JE, DePinho RA, Jaenisch R, Tsai LH. HDAC2 negatively regulates memory formation and synaptic plasticity. Nature. 2009;459(7243):55–60. https://doi.org/10.1038/nature07925.
46. Haberland M, Montgomery RL, Olson EN. The many roles of histone deacetylases in development and physiology: implications for disease and therapy. Nat Rev Genet. 2009;10(1):32–42. https://doi.org/10.1038/nrg2485.
47. Hadano S, Hand CK, Osuga H, Yanagisawa Y, Otomo A, Devon RS, Miyamoto N, Showguchi-Miyata J, Okada Y, Singaraja R, Figlewicz DA, Kwiatkowski T, Hosler BA, Sagie T, Skaug J, Nasir J, Brown RH, Scherer SW, Rouleau GA, Ikeda JE. A gene encoding a putative GTPase regulator is mutated in familial amyotrophic lateral sclerosis 2. Nat Genet. 2001a;29(2):166–73. https://doi.org/10.1038/ng1001-166.
48. Hadano S, Yanagisawa Y, Skaug J, Fichter K, Nasir J, Martindale D, Koop BF, Scherer SW, Nicholson DW, Rouleau GA, Ikeda J, Hayden MR. Cloning and characterization of three novel genes, ALS2CR1, ALS2CR2, and ALS2CR3, in the juvenile amyotrophic lateral sclerosis (ALS2) critical region at chromosome 2q33-q34: candidate genes for ALS2. Genomics. 2001b;71(2):200–13. https://doi.org/10.1006/geno.2000.6392.
49. Hahnen E, Hauke J, Tränkle C, Eyüpoglu IY, Wirth B, Blümcke I. Histone deacetylase inhibitors: possible implications for neurodegenerative disorders. Expert Opin Investig Drugs. 2008;17(2):169–84. https://doi.org/10.1517/13543784.17.2.169.
50. Halili MA, Andrews MR, Labzin LI, Schroder K, Matthias G, Cao C, Lovelace E, Reid RC, Le GT, Hume DA, Irvine KM, Matthias P, Fairlie DP, Sweet MJ. Differential effects of selective HDAC inhibitors on macrophage inflammatory responses to the Toll-like receptor 4 agonist LPS. J Leukoc Biol. 2010;87(6):1103–14. https://doi.org/10.1189/jlb.0509363.
51. Harrison IF, Dexter DT. Epigenetic targeting of histone deacetylase: therapeutic potential in Parkinson's disease? Pharmacol Ther. 2013;140(1):34–52. https://doi.org/10.1016/j.pharmthera.2013.05.010.
52. Hogarth P, Lovrecic L, Krainc D. Sodium phenylbutyrate in Huntington's disease: a dose-finding study. Mov Disord. 2007;22(13):1962–4. https://doi.org/10.1002/mds.21632.
53. Hopkins AL. Network pharmacology: the next paradigm in drug discovery. Nat Chem Biol. 2008;4(11):682–90. https://doi.org/10.1038/nchembio.118.
54. Hsu SW, Hsu PC, Chang WS, Yu CC, Wang YC, Yang JS, Tsai FJ, Chen KY, Tsai CW, Bau DT. Protective effects of valproic acid on 6-hydroxydopamine-induced neuroinjury. Environ Toxicol. 2020;35(8):840–8. https://doi.org/10.1002/tox.22920. Epub 2020 Mar 13
55. Janczura KJ, Volmar CH, Sartor GC, Rao SJ, Ricciardi NR, Lambert G, Brothers SP, Wahlestedt C. Inhibition of HDAC3 reverses Alzheimer's disease-related pathologies in vitro and in the 3xTg-AD mouse model. Proc Natl Acad Sci. 2018;115(47):E11148–57. https://doi.org/10.1073/pnas.1805436115.
56. Jasim SA, Altalbawy FMA, Abohassan M, Oghenemaro EF, Bishoyi AK, Singh RP, Kaur P, Sivaprasad GV, Mohammed JS, Hulail HM. Histone deacetylases (HDACs) roles in inflammation-mediated diseases; current knowledge. Cell Biochem Biophys. Advance online publication. 2024; https://doi.org/10.1007/s12013-024-01587-0.
57. Jaworska J, Ziemka-Nalecz M, Zalewska T. Histone deacetylases 1 and 2 are required for brain development. Int J Dev Biol. 2015;59(4–6):171–7. https://doi.org/10.1387/ijdb.150071tz.

58. Jeong H, Cohen DE, Cui L, Supinski A, Savas JN, Mazzulli JR, Yates JR, Bordone L, Guarente L, Krainc D. Sirt1 mediates neuroprotection from mutant huntingtin by activation of the TORC1 and CREB transcriptional pathway. Nat Med. 2011;18(1):159–65. https://doi.org/10.1038/nm.2559.
59. Jeong H, Shin S, Lee JS, Lee SH, Baik JH, Lim S, Kim YK. Pan-HDAC inhibitors promote tau aggregation by increasing the level of acetylated tau. Int J Mol Sci. 2019;20(17):4283.
60. Jia H, Kast RJ, Steffan JS, Thomas EA. Selective histone deacetylase (HDAC) inhibition imparts beneficial effects in Huntington's disease mice: implications for the ubiquitin–proteasomal and autophagy systems. Hum Mol Genet. 2012;21(24):5280–93. https://doi.org/10.1093/hmg/dds379.
61. Jia H, Morris CD, Williams RM, Loring JF, Thomas EA. HDAC inhibition imparts beneficial transgenerational effects in Huntington's disease mice via altered DNA and histone methylation. Proc Natl Acad Sci. 2015;112(1):E56–64. https://doi.org/10.1073/pnas.1415195112.
62. Jiang S, Deng C, Lv J, Fan C, Hu W, Di S, Yan X, Ma Z, Liang Y, Yang Y. Nrf2 weaves an elaborate network of neuroprotection against stroke. Mol Neurobiol. 2017;54(2):1440–55. https://doi.org/10.1007/s12035-016-9707-7.
63. Jimenez-Sanchez M, Licitra F, Underwood BR, Rubinsztein DC. Huntington's disease: mechanisms of pathogenesis and therapeutic strategies. Cold Spring Harb Perspect Med. 2015;7(7):a024240. https://doi.org/10.1101/cshperspect.a024240.
64. Kane AE, Sinclair DA. Sirtuins and NAD+ in the development and treatment of metabolic and cardiovascular diseases. Circ Res. 2018;123(7):868–85. https://doi.org/10.1161/CIRCRESAHA.118.312498.
65. Kazantsev A, Walker HA, Slepko N, Bear JE, Preisinger E, Steffan JS, Zhu YZ, Gertler FB, Housman DE, Marsh JL, Thompson LM. A bivalent Huntingtin binding peptide suppresses polyglutamine aggregation and pathogenesis in Drosophila. Nat Genet. 2002;30(4):367–76.
66. Kazantsev AG, Thompson LM. Therapeutic application of histone deacetylase inhibitors for central nervous system disorders. Nat Rev Drug Discov. 2008;7(10):854–68. https://doi.org/10.1038/nrd2681.
67. Kelly RD, Cowley SM. The physiological roles of histone deacetylase (HDAC) 1 and 2: complex co-stars with multiple leading parts. Biochem Soc Trans. 2013;41(3):741–9. https://doi.org/10.1042/BST20130010.
68. Kelly WK, Marks PA. Drug insight: Histone deacetylase inhibitors--development of the new targeted anticancer agent suberoylanilide hydroxamic acid. Nat Clin Pract Oncol. 2005;2(3):150–7. https://doi.org/10.1038/ncponc0106.
69. Khan N, Jeffers M, Kumar S, Hackett C, Boldog F, Khramtsov N, Qian X, Mills E, Berghs SC, Carey N, Finn PW, Collins LS, Tumber A, Ritchie JW, Jensen PB, Lichenstein HS, Sehested M. Determination of the class and isoform selectivity of small-molecule histone deacetylase inhibitors. Biochem J. 2008;409(2):581–9. https://doi.org/10.1042/BJ20070779.
70. Kilgore M, Miller CA, Fass DM, Hennig KM, Haggarty SJ, Sweatt JD, Rumbaugh G. Inhibitors of class 1 histone deacetylases reverse contextual memory deficits in a mouse model of Alzheimer's disease. Neuropsychopharmacology. 2010;35(4):870–80. https://doi.org/10.1038/npp.2009.197.
71. Kontopoulos E, Parvin JD, Feany MB. α-Synuclein acts in the nucleus to inhibit histone acetylation and promote neurotoxicity. Hum Mol Genet. 2006;15(20):3012–23. https://doi.org/10.1093/hmg/ddl243.
72. Koppel I, Timmusk T. Differential regulation of Bdnf expression in cortical neurons by class-selective histone deacetylase inhibitors. Neuropharmacology. 2013;75:106–15. https://doi.org/10.1016/j.neuropharm.2013.07.015.
73. Krishna K, Behnisch T, Sajikumar S. Inhibition of histone deacetylase 3 restores amyloid-β oligomer-induced plasticity deficit in hippocampal CA1 pyramidal neurons. J Alzheimer's Dis. 2016;51(3):783–91. https://doi.org/10.3233/JAD-150838.
74. Kritzer JA, Hamamichi S, McCaffery JM, Santagata S, Naumann TA, Caldwell KA, Caldwell GA, Lindquist S. Rapid selection of cyclic peptides that reduce alpha-synuclein toxicity

in yeast and animal models. Nat Chem Biol. 2009;5(9):655–63. https://doi.org/10.1038/nchembio.193.
75. Kumar S, Lombard DB. Functions of the sirtuin deacylase SIRT5 in normal physiology and pathobiology. Crit Rev Biochem Mol Biol. 2018;53(3):311–34. https://doi.org/10.1080/10409238.2018.1458071.
76. Kuta R, Larochelle N, Fernandez M, Pal A, Minotti S, Tibshirani M, St Louis K, Gentil BJ, Nalbantoglu JN, Hermann A, Durham HD. Depending on the stress, histone deacetylase inhibitors act as heat shock protein co-inducers in motor neurons and potentiate arimoclomol, exerting neuroprotection through multiple mechanisms in ALS models. Cell Stress Chaperones. 2020;25(1):173–91. https://doi.org/10.1007/s12192-019-01064-1.
77. Lauffer BE, Mintzer R, Fong R, Mukund S, Tam C, Zilberleyb I, Flicke B, Ritscher A, Fedorowicz G, Vallero R, Ortwine DF, Gunzner J, Modrusan Z, Neumann L, Koth CM, Lupardus PJ, Kaminker JS, Heise CE, Steiner P. Histone deacetylase (HDAC) inhibitor kinetic rate constants correlate with cellular histone acetylation but not transcription and cell viability. J Biol Chem. 2013;288(37):26926–43. https://doi.org/10.1074/jbc.M113.490706.
78. Lee IH, Cao L, Mostoslavsky R, Lombard DB, Liu J, Bruns NE, Tsokos M, Alt FW, Finkel T. A role for the NAD-dependent deacetylase Sirt1 in the regulation of autophagy. Proc Natl Acad Sci. 2008;105(9):3374–9. https://doi.org/10.1073/pnas.0712145105.
79. Leoni F, Fossati G, Lewis EC, Lee JK, Porro G, Pagani P, Modena D, Moras ML, Pozzi P, Reznikov LL, Siegmund B, Fantuzzi G, Dinarello CA, Mascagni P. The histone deacetylase inhibitor ITF2357 reduces production of pro-inflammatory cytokines in vitro and systemic inflammation in vivo. Mol Med. 2005;11(1–12):1–15.
80. Leus NG, Zwinderman MR, Dekker FJ. Histone deacetylase 3 (HDAC 3) as emerging drug target in NF-κB-mediated inflammation. Curr Opin Chem Biol. 2016;33:160–8. https://doi.org/10.1016/j.cbpa.2016.06.019.
81. Li G, Jiang H, Chang M, Xie H, Hu L. HDAC6 α-tubulin deacetylase: a potential therapeutic target in neurodegenerative diseases. J Neurol Sci. 2011;304(1–2):1–8. https://doi.org/10.1016/j.jns.2011.02.017.
82. Li G, Tian Y, Zhu WG. The roles of histone deacetylases and their inhibitors in cancer therapy. Front Cell Dev Biol. 2020;8:576946. https://doi.org/10.3389/fcell.2020.576946.
83. Li T, Zhang C, Hassan S, Liu X, Song F, Chen K, Zhang W, Yang J. Histone deacetylase 6 in cancer. J Hematol Oncol. 2018;11(1):111. https://doi.org/10.1186/s13045-018-0654-9.
84. Li Y, Seto E. HDACs and HDAC inhibitors in cancer development and therapy. Cold Spring Harb Perspect Med. 2016;6(10):a026831. https://doi.org/10.1101/cshperspect.a026831.
85. Liu F, Hon GC, Villa GR, Turner KM, Ikegami S, Yang H, Ye Z, Li B, Kuan S, Lee AY, Zanca C, Wei B, Lucey G, Jenkins D, Zhang W, Barr CL, Furnari FB, Cloughesy TF, Yong WH, Gahman TC, Shiau AK, Cavenee WK, Ren B, Mischel PS. EGFR mutation promotes glioblastoma through epigenome and transcription factor network remodeling. Mol Cell. 2015;60(2):307–18. https://doi.org/10.1016/j.molcel.2015.09.002.
86. Liu H. The roles of histone deacetylases in kidney development and disease. Clin Exp Nephrol. 2021;25(3):215–23.
87. Mai A, Valente S, Meade S, Carafa V, Tardugno M, Nebbioso A, Galmozzi A, Mitro N, De Fabiani E, Altucci L, Kazantsev A. Study of 1,4-dihydropyridine structural scaffold: discovery of novel sirtuin activators and inhibitors. J Med Chem. 2017;52(15):5496–504. https://doi.org/10.1021/jm9008289.
88. Malaspina A, de Belleroche J. Spinal cord molecular profiling provides a better understanding of amyotrophic lateral sclerosis pathogenesis. Brain Res Rev. 2004;45(3):213–29. https://doi.org/10.1016/j.brainresrev.2004.04.002.
89. Marks PA. Discovery and development of SAHA as an anticancer agent. Oncogene. 2007;26(9):1351–6. https://doi.org/10.1038/sj.onc.1210204.
90. Marks PA. The clinical development of histone deacetylase inhibitors as targeted anticancer drugs. Expert Opin Investig Drugs. 2010;19(9):1049–66.
91. Marks PA, Xu WS. Histone deacetylase inhibitors: potential in cancer therapy. J Cell Biochem. 2009;107(4):600–8. https://doi.org/10.1002/jcb.22185.

92. Martin, S.D., Connor, T., Sanigorski, A. et al. Class IIa HDACs inhibit cell death pathways and protect muscle integrity in response to lipotoxicity. Cell Death Dis 14, 787 (2023). https://doi.org/10.1038/s41419-023-06319-5
93. McCampbell A, Taye AA, Whitty L, Penney E, Steffan JS, Fischbeck KH. Histone deacetylase inhibitors reduce polyglutamine toxicity. Proc Natl Acad Sci. 2001;98(26):15179–84.
94. McFarland KN, Das S, Sun TT, Leyfer D, Xia E, Sangrey GR, Kuhn A, Luthi-Carter R, Clark TW, Sadri-Vakili G, Cha JHJ. Genome-wide histone acetylation is altered in a transgenic mouse model of Huntington's disease. PLoS One. 2012;7(7):e41423. https://doi.org/10.1371/journal.pone.0041423.
95. McQuown SC, Barrett RM, Matheos DP, Post RJ, Rogge GA, Alenghat T, Mullican SE, Jones S, Rusche JR, Lazar MA, Wood MA. HDAC3 is a critical negative regulator of long term memory formation. J Neurosci. 2011;31(2):764–74.
96. Meka ST, Bojja SL, Kumar G, Birangal SR, Rao CM. Novel HDAC inhibitors provide neuroprotection in MPTP-induced Parkinson's disease model of rats. Eur J Pharmacol. 2023;959:176067. https://doi.org/10.1016/j.ejphar.2023.176067.
97. Mielcarek M, Landles C, Weiss A, Bradaia A, Seredenina T, Inuabasi L, Osborne GF, Wadel K, Touller C, Butler R, Robertson J, Franklin SA, Smith DL, Park L, Marks PA, Wanker EE, Olson EN, Luthi-Carter R, van der Putten H, Beaumont V, Bates GP. HDAC4 reduction: a novel therapeutic strategy to target cytoplasmic huntingtin and ameliorate neurodegeneration. PLoS Biol. 2013;11(11):e1001717. https://doi.org/10.1371/journal.pbio.1001717.
98. Mielcarek M, Zielonka D, Carnemolla A, Marcinkowski JT, Guidez F. HDAC4 as a potential therapeutic target in neurodegenerative diseases: a summary of recent achievements. Front Cell Neurosci. 2015;9:42.
99. Noh H, Seo H. Age-dependent effects of valproic acid in Alzheimer's disease (AD) mice are associated with nerve growth factor (NGF) regulation. Neuroscience. 2014;266:255–65.
100. Nucifora FC Jr, Sasaki M, Peters MF, Huang H, Cooper JK, Yamada M, Takahashi H, Tsuji S, Troncoso J, Dawson VL, Dawson TM, Ross CA. Interference by huntingtin and atrophin-1 with cbp-mediated transcription leading to cellular toxicity. Science. 2001;291(5512):2423–8. https://doi.org/10.1126/science.1056784.
101. Olsen MK, Roberds SL, Ellerbrock BR, Fleck TJ, McKinley DK, Gurney ME. Disease mechanisms revealed by transcription profiling in SOD1-G93A transgenic mouse spinal cord. Ann Neurol. 2001;50(6):730–40.
102. Outeiro TF, Kontopoulos E, Altmann SM, Kufareva I, Strathearn KE, Amore AM, Volk CB, Maxwell MM, Rochet JC, McLean PJ, Young AB, Abagyan R, Feany MB, Hyman BT, Kazantsev AG. Sirtuin 2 inhibitors rescue alpha-synuclein-mediated toxicity in models of Parkinson’s disease. Science. 2007;317(5837):516–9. https://doi.org/10.1126/science.1143780.
103. Papa A, Cursaro I, Pozzetti L, Contri C, Cappello M, Pasquini S, Carullo G, Ramunno A, Gemma S, Varani K, Butini S, Campiani G, Vincenzi F. Pioneering first-in-class FAAH-HDAC inhibitors as potential multitarget neuroprotective agents. Arch Pharm (Weinheim). 2023;356(12):e2300410. https://doi.org/10.1002/ardp.202300410.
104. Park Y, Yu S, Hwang SY, Seo H. Molecular targets of histone deacetylase inhibitors in neurodegeneration and neuroprotection. Mol Cell Toxicol. 2024;21:1.
105. Parra M, Verdin E. Regulatory signal transduction pathways for class IIa histone deacetylases. Curr Opin Pharmacol. 2010;10(4):454–60. https://doi.org/10.1016/j.coph.2010.04.004.
106. Pascoal TA, Chamoun M, Lax E, Wey HY, Shin M, Ng KP, Kang MS, Mathotaarachchi S, Benedet AL, Therriault J, Lussier FZ, Schroeder FA, DuBois JM, Hightower BG, Gilbert TM, Zürcher NR, Wang C, Hopewell R, Chakravarty M, Savard M, Thomas E, Mohaddes S, Farzin S, Salaciak A, Tullo S, Cuello AC, Soucy JP, Massarweh G, Hwang H, Kobayashi E, Hyman BT, Dickerson BC, Guiot MC, Szyf M, Gauthier S, Hooker JM, Rosa-Neto P. (11C) Martinostat PET analysis reveals reduced HDAC I availability in Alzheimer’s disease. Nat Commun. 2022;13(1):4171. https://doi.org/10.1038/s41467-022-30653-5.

107. Patrick GN, Zukerberg L, Nikolic M, de la Monte S, Dikkes P, Tsai LH. Conversion of p35 to p25 deregulates Cdk5 activity and promotes neurodegeneration. Nature. 1999;402(6762):615–22.
108. Piepers S, Veldink JH, de Jong SW, van der Tweel I, van der Pol WL, Uijtendaal EV, Schelhaas HJ, Scheffer H, de Visser M, de Jong JM, Wokke JH, Groeneveld GJ, van den Berg LH. Randomized sequential trial of valproic acid in amyotrophic lateral sclerosis. Ann Neurol. 2009;66(2):227–34. https://doi.org/10.1002/ana.21620.
109. Pigna E, Simonazzi E, Sanna K, Bernadzki KM, Proszynski T, Heil C, Palacios D, Adamo S, Moresi V. Histone deacetylase 4 protects from denervation and skeletal muscle atrophy in a murine model of amyotrophic lateral sclerosis. EBioMedicine. 2019;40:717–32. https://doi.org/10.1016/j.ebiom.2019.01.038.
110. Pinho BR, Reis SD, Guedes-Dias P, Leitão-Rocha A, Quintas C, Valentão P, Andrade PB, Santos MM, Oliveira JM. Pharmacological modulation of HDAC1 and HDAC6 in vivo in a zebrafish model: therapeutic implications for Parkinson's disease. Pharmacol Res. 2016;103:328–39. https://doi.org/10.1016/j.phrs.2015.11.024.
111. Pojani E, Barlocco D. Romidepsin (FK228), a histone deacetylase inhibitor and its analogues in cancer chemotherapy. Curr Med Chem. 2021;28(7):1290–303. https://doi.org/10.2174/0929867327666200203113926.
112. Rabal O, Sánchez-Arias JA, Cuadrado-Tejedor M, de Miguel I, Pérez-González M, García-Barroso C, Ugarte A, Estella-Hermoso de Mendoza A, Sáez E, Espelosin M, Ursua S, Haizhong T, Wei W, Musheng X, Garcia-Osta A, Oyarzabal J. Design, synthesis, and biological evaluation of first-in-class dual acting histone deacetylases (HDACs) and phosphodiesterase 5 (PDE5) inhibitors for the treatment of Alzheimer's disease. J Med Chem. 2016;59(19):8967–9004.
113. Rabal O, Sánchez-Arias JA, Cuadrado-Tejedor M, de Miguel I, Pérez-González M, García-Barroso C, Ugarte A, Estella-Hermoso de Mendoza A, Sáez E, Espelosin M, Ursua S, Haizhong T, Wei W, Musheng X, Garcia-Osta A, Oyarzabal J. Design, synthesis, biological evaluation and in vivo testing of dual phosphodiesterase 5 (PDE5) and histone deacetylase 6 (HDAC6)-selective inhibitors for the treatment of Alzheimer's disease. Eur J Med Chem. 2018;150:506–24.
114. Ricobaraza A, Cuadrado-Tejedor M, Marco S, Pérez-Otaño I, García-Osta A. Phenylbutyrate rescues dendritic spine loss associated with memory deficits in a mouse model of Alzheimer disease. Hippocampus. 2012;22(5):1040–50.
115. Ricobaraza A, Cuadrado-Tejedor M, Pérez-Mediavilla A, Frechilla D, Del Río J, García-Osta A. Phenylbutyrate ameliorates cognitive deficit and reduces tau pathology in an Alzheimer's disease mouse model. Neuropsychopharmacology. 2009;34(7):1721–32.
116. Rodrigues R, Lourenço JC, Almeida SM, Nunes C. Mitochondrial-targeted antioxidants as a therapeutic strategy for protecting against oxidative stress and mitochondrial dysfunction in neurodegenerative diseases. Antioxidants. 2020;9(10):1020.
117. Rosen DR, Siddique T, Patterson D, Figlewicz DA, Sapp P, Hentati A, Donaldson D, Goto J, O'Regan JP, Deng HX. Mutations in Cu/Zn superoxide dismutase gene are associated with familial amyotrophic lateral sclerosis. Nature. 1993;362(6415):59–62. https://doi.org/10.1038/362059a0.
118. Rosenberg GA. Neurological diseases in relation to the blood–brain barrier. J Cereb Blood Flow Metab. 2012;32(7):1139–51. https://doi.org/10.1038/jcbfm.2011.197.
119. Rouaux C, Panteleeva I, Rene F, Gonzalez de Aguilar JL, Echaniz-Laguna A, Dupuis L, Menger Y, Boutillier AL, Loeffler JP. Sodium valproate exerts neuroprotective effects in vivo through CREB-binding protein-dependent mechanisms but does not improve survival in an amyotrophic lateral sclerosis mouse model. J Neurosci. 2007;27(21):5535–45. https://doi.org/10.1523/JNEUROSCI.1139-07.2007.
120. Rowland LP, Shneider NA. Amyotrophic lateral sclerosis. N Engl J Med. 2001;344(22):1688–700.

121. Rushworth SA, Megson IL. Existing and potential therapeutic uses for N-acetylcysteine: the need for conversion to intracellular glutathione for antioxidant benefits. Pharmacol Ther. 2014;141(2):150–9. https://doi.org/10.1016/j.pharmthera.2013.09.006.
122. Ryu H, Ferrante RJ. Translational therapeutic strategies in amyotrophic lateral sclerosis. Mini Rev Med Chem. 2007;7(2):141–50. https://doi.org/10.2174/138955707779802570.
123. Ryu H, Lee J, Olofsson BA, Mwidau A, Dedeoglu A, Escudero M, Flemington E, Azizkhan-Clifford J, Ferrante RJ, Ratan RR. Histone deacetylase inhibitors prevent oxidative neuronal death independent of expanded polyglutamine repeats via an Sp1-dependent pathway. Proc Natl Acad Sci. 2003;100(7):4281–6. https://doi.org/10.1073/pnas.0737363100.
124. Ryu H, Smith K, Camelo SI, Carreras I, Lee J, Iglesias AH, Dangond F, Cormier KA, Cudkowicz ME, Brown RH, Ferrante RJ. Sodium phenylbutyrate prolongs survival and regulates expression of anti-apoptotic genes in transgenic amyotrophic lateral sclerosis mice. J Neurochem. 2005;93(5):1087–98. https://doi.org/10.1111/j.1471-4159.2005.03077.x.
125. Saha RN, Pahan K. HATs and HDACs in neurodegeneration: a tale of disconcerted acetylation homeostasis. Cell Death Differ. 2006;13(4):539–50. https://doi.org/10.1038/sj.cdd.4401769.
126. Saha S, Panigrahi DP, Patil S, Bhutia SK. Autophagy in health and disease: a comprehensive review. Biomed Pharmacother. 2018;104:485–95. https://doi.org/10.1016/j.biopha.2018.05.007.
127. Sanaei M, Kavoosi F. Histone deacetylases and histone deacetylase inhibitors: molecular mechanisms of action in various cancers. Adv Biomed Res. 2019;8:63. https://doi.org/10.4103/abr.abr_142_19.
128. Sando R, Gounko N, Pieraut S, Liao L, Yates J, Maximov A. HDAC4 governs a transcriptional program essential for synaptic plasticity and memory. Cell. 2012;151(4):821–34. https://doi.org/10.1016/j.cell.2012.09.037.
129. Sharma S, Sarathlal KC, Taliyan R. Epigenetics in neurodegenerative diseases: the role of histone deacetylases. CNS Neurol Disord Drug Targets. 2019;18(1):11–8. https://doi.org/10.2174/1871527317666181004155136.
130. Shakespear MR, Halili MA, Irvine KM, Fairlie DP, Sweet MJ. Histone deacetylases as regulators of inflammation and immunity. Trends Immunol. 2011;32(7):335–43. https://doi.org/10.1016/j.it.2011.04.001.
131. Shi H, Deng HX, Gius D, Schumacker PT, Surmeier DJ, Ma YC. Sirt3 protects dopaminergic neurons from mitochondrial oxidative stress. Hum Mol Genet. 2017;26(10):1915–26. https://doi.org/10.1093/hmg/ddx100.
132. Shi MQ, Xu Y, Fu X, Pan DS, Lu XP, Xiao Y, Jiang YZ. Advances in targeting histone deacetylase for treatment of solid tumors. J Hematol Oncol. 2024;17(1):37. https://doi.org/10.1186/s13045-024-01551-8.
133. Shukla S, Tekwani BL. Histone deacetylases inhibitors in neurodegenerative diseases, neuroprotection and neuronal differentiation. Front Pharmacol. 2020;11:537. https://doi.org/10.3389/fphar.2020.00537.
134. Siebzehnrübl FA, Raber KA, Urbach YK, Schulze-Krebs A, Canneva F, Moceri S, Habermeyer J, Achoui D, Gupta B, Steindler DA, Stephan M, Nguyen HP, Bonin M, Riess O, Bauer A, Aigner L, Couillard-Despres S, Paucar MA, Svenningsson P, Osmand A, Andreew A, Zabel C, Weiss A, Kuhn R, Moussaoui S, Blockx I, Van der Linden A, Cheong RY, Roybon L, Petersén Å, von Hörsten S. Early postnatal behavioral, cellular, and molecular changes in models of Huntington disease are reversible by HDAC inhibition. Proc Natl Acad Sci U S A. 2018;115(37):E8765–74.
135. Singh CK, Chhabra G, Ndiaye MA, Garcia-Peterson LM, Mack NJ, Ahmad N. The role of Sirtuins in antioxidant and redox signaling. Antioxid Redox Signal. 2018;28(8):643–61. https://doi.org/10.1089/ars.2017.7290.
136. Singh R, Rathore AS, Dilnashin H, et al. HAT and HDAC: enzyme with contradictory action in neurodegenerative diseases. Mol Neurobiol. 2024;61:9110–24.
137. Smith MR, Syed A, Lukacsovich T, Purcell J, Barbaro BA, Worthge SA, Wei SR, Pollio G, Magnoni L, Scali C, Massai L, Franceschini D, Camarri M, Gianfriddo M, Diodato E, Thomas R, Gokce O, Tabrizi SJ, Caricasole A, Landwehrmeyer B, Menalled L, Murphy C,

Ramboz S, Luthi-Carter R, Westerberg G, Marsh JL. A potent and selective Sirtuin 1 inhibitor alleviates pathology in multiple animal and cell models of Huntington's disease. Hum Mol Genet. 2014;23(11):2995–3007.
138. Sreenivasmurthy SG, Iyaswamy A, Krishnamoorthi S, Senapati S, Malampati S, Zhu Z, Su CF, Liu J, Guan XJ, Tong BC, Cheung KH, Tan JQ, Lu JH, Durairajan SSK, Song JX, Li M. Protopine promotes the proteasomal degradation of pathological tau in Alzheimer's disease models via HDAC6 inhibition. Phytomedicine. 2022a;96:153887.
139. Sreenivasmurthy SG, Iyaswamy A, Krishnamoorthi S, Reddi RN, Kammala AK, Vasudevan K, Senapati S, Zhu Z, Su CF, Liu J, Guan XJ, Chua KK, Cheung KH, Chen H, Zhang HJ, Zhang Y, Song JX, Kumar Durairajan SS, Li M. Bromo-protopine, a novel protopine derivative, alleviates tau pathology by activating chaperone-mediated autophagy for Alzheimer's disease therapy. Front Mol Biosci. 2022b;9:1030534. https://doi.org/10.3389/fmolb.2022.1030534.
140. Steffan JS, Bodai L, Pallos J, Poelman M, McCampbell A, Apostol BL, Kazantsev A, Schmidt E, Zhu YZ, Greenwald M, Kurokawa R, Housman DE, Jackson GR, Marsh JL, Thompson LM. Histone deacetylase inhibitors arrest polyglutamine-dependent neurodegeneration in Drosophila. Nature. 2001;413(6857):739–43. https://doi.org/10.1038/35099568.
141. Suh HS, Choi S, Khattar P, Choi N, Lee SC. Histone deacetylase inhibitors suppress the expression of inflammatory and innate immune response genes in human microglia and astrocytes. J Neuroimmune Pharmacol. 2010;5(4):521–32. https://doi.org/10.1007/s11481-010-9192-0.
142. Sun XY, Qu Y, Ni AR, Wang GX, Huang WB, Chen ZP, Lv ZF, Zhang S, Lindsay H, Zhao S, Li XN, Feng BH. Novel histone deacetylase inhibitor N25 exerts anti-tumor effects and induces autophagy in human glioma cells by inhibiting HDAC3. Oncotarget. 2017;8(43):75232–42. https://doi.org/10.18632/oncotarget.20744.
143. Sung YM, Lee T, Yoon H, DiBattista AM, Song JM, Sohn Y, Moffat EI, Turner RS, Jung M, Kim J, Hoe HS. Mercaptoacetamide-based class II HDAC inhibitor lowers Aβ levels and improves learning and memory in a mouse model of Alzheimer's disease. Exp Neurol. 2013;239:192–201. https://doi.org/10.1016/j.expneurol.2012.10.005.
144. Szegő ÉM, Outeiro TF, Kazantsev AG. Sirtuins in brain and neurodegenerative disease. In: Guarente L, Mostoslavsky R, Kazantsev A, editors. Introductory review on sirtuins in biology, aging, and disease. Academic; 2018. p. 175–95. https://doi.org/10.1016/B978-0-12-813499-3.00013-7.
145. Taes I, Timmers M, Hersmus N, Bento-Abreu A, Van Den Bosch L, Van Damme P, Auwerx J, Robberecht W. Hdac6 deletion delays disease progression in the SOD1G93A mouse model of ALS. Hum Mol Genet. 2013;22(9):1783–90. https://doi.org/10.1093/hmg/ddt028.
146. Tan Y, Delvaux E, Nolz J, Coleman PD, Chen S, Mastroeni D. Upregulation of histone deacetylase 2 in laser capture nigral microglia in Parkinson's disease. Neurobiol Aging. 2018;68:134–41.
147. Tang BL. Sirt1 and the mitochondria. Mol Cells. 2016;39(2):87–95. https://doi.org/10.14348/molcells.2016.2318.
148. Thomas EA. Involvement of HDAC1 and HDAC3 in the pathology of polyglutamine disorders: therapeutic implications for selective HDAC1/HDAC3 inhibitors. Pharmaceuticals. 2014;7(6):634–61.
149. Thomas EA, Coppola G, Desplats PA, Tang B, Soragni E, Burnett R, Gao F, Fitzgerald KM, Borok JF, Herman D, Geschwind DH, Gottesfeld JM. The HDAC inhibitor 4b ameliorates the disease phenotype and transcriptional abnormalities in Huntington's disease transgenic mice. Proc Natl Acad Sci U S A. 2008;105(40):15564–9. https://doi.org/10.1073/pnas.0804249105.
150. Tseng HJ, Lin MH, Shiao YJ, Yang YC, Chu JC, Chen CY, Chen YY, Lin TE, Su CJ, Pan SL, Chen LC, Wang CY, Hsu KC, Huang WJ. Synthesis and biological evaluation of acridine-based histone deacetylase inhibitors as multitarget agents against Alzheimer's disease. Eur J Med Chem. 2020;192:112193.
151. Valor LM, Guiretti D. What's wrong with epigenetics in Huntington's disease? Neuropharmacology. 2014;80:103–14.

152. Wang Y, Wang X, Liu L, Wang X. HDAC inhibitor trichostatin A-inhibited survival of dopaminergic neuronal cells. Neurosci Lett. 2009;467(3):212–6. https://doi.org/10.1016/j.neulet.2009.10.037.
153. Watson CT, Kuppuswamy A, Ledford JG, Sukumari-Ramesh J. The role of HDAC3 in inflammation: mechanisms and therapeutic implications. Front Immunol. 2024;15 https://doi.org/10.3389/fimmu.2024.1419685.
154. Wen Y, Planel E, Herman M, Figueroa HY, Wang L, Liu L, Lau LF, Yu WH, Duff KE. Interplay between cyclin-dependent kinase 5 and glycogen synthase kinase 3 beta mediated by neuregulin signaling leads to differential effects on tau phosphorylation and amyloid precursor protein processing. J Neurosci. 2008;28(10):2624–32. https://doi.org/10.1523/JNEUROSCI.5245-07.2008.
155. Westerberg G, Chiesa JA, Andersen CA, Diamanti D, Horne L, Karila P, Kjellberg M, Magnusson O. Safety, pharmacokinetics, pharmacogenomics and QT concentration-effect modelling of the SirT1 inhibitor selisistat in healthy volunteers. Br J Clin Pharmacol. 2015;79(3):477–91.
156. Wolozin B, Ivanov P. Stress granules and neurodegeneration. Nat Rev Neurosci. 2019;20(11):649–66. https://doi.org/10.1038/s41583-019-0222-5.
157. Woods DM, Woan KV, Cheng F, Sodré AL, Wang D, Wu Y, Wang Z, Chen J, Powers J, Pinilla-Ibarz J, Yu Y, Zhang Y, Wu X, Zheng X, Weber J, Hancock WW, Seto E, Villagra A, Yu XZ, Sotomayor EM. T cells lacking HDAC11 have increased effector functions and mediate enhanced alloreactivity in a murine model. Blood. 2017;130(2):146–55. https://doi.org/10.1182/blood-2016-08-731505.
158. Wu X, Chen PS, Dallas S, Wilson B, Block ML, Wang CC, Kinyamu H, Lu N, Gao X, Leng Y, Chuang DM, Zhang W, Lu RB, Hong JS. Histone deacetylase inhibitors up-regulate astrocyte GDNF and BDNF gene transcription and protect dopaminergic neurons. Int J Neuropsychopharmacol. 2008;11(8):1123–34. https://doi.org/10.1017/S1461145708009024.
159. Wu Y, Hou F, Wang X, Kong Q, Han X, Bai B. Aberrant expression of histone deacetylases 4 in cognitive disorders: molecular mechanisms and a potential target. Front Mol Neurosci. 2016;9:114. https://doi.org/10.3389/fnmol.2016.00114.
160. Xu A, He F, Zhang X, Li X, Ran Y, Wei C, James Chou C, Zhang R, Wu J. Tacrine-hydroxamate derivatives as multitarget-directed ligands for the treatment of Alzheimer's disease: design, synthesis, and biological evaluation. Bioorg Chem. 2020;98:103721.
161. Xu K, Dai XL, Huang HC, Jiang ZF. Targeting HDACs: a promising therapy for Alzheimer's disease. Oxid Med Cell Longev. 2011;2011:143269. https://doi.org/10.1155/2011/143269.
162. Xu WS, Parmigiani RB, Marks PA. Histone deacetylase inhibitors: molecular mechanisms of action. Oncogene. 2007;26(37):5541–52. https://doi.org/10.1038/sj.onc.1210620.
163. Yamakawa H, Cheng J, Penney J, Gao F, Rueda R, Wang J, Yamakawa S, Kritskiy O, Gjoneska E, Tsai LH. The transcription factor Sp3 cooperates with HDAC2 to regulate synaptic function and plasticity in neurons. Cell Rep. 2017;20(6):1319–34. https://doi.org/10.1016/j.celrep.2017.07.044.
164. Yang SS, Zhang R, Wang G, Zhang YF. The development prospection of HDAC inhibitors as a potential therapeutic direction in Alzheimer's disease. Transl Neurodegener. 2017;6:19. https://doi.org/10.1186/s40035-017-0089-1.
165. Yang Y, Huang Y, Wang Z, Wang HT, Duan B, Ye D, Wang C, Jing R, Leng Y, Xi J, Chen W, Wang G, Jia W, Zhu S, Kang J. HDAC10 promotes lung cancer proliferation via AKT phosphorylation. Oncotarget. 2016;7(37):59388–401. https://doi.org/10.18632/oncotarget.10673.
166. Yoo YE, Ko CP. Treatment with trichostatin A initiated after disease onset delays disease progression and increases survival in a mouse model of amyotrophic lateral sclerosis. Exp Neurol. 2011;231(1):147–59. https://doi.org/10.1016/j.expneurol.2011.06.003. Epub 2011 Jun 25. PMID: 21712032
167. Zhang J, Xiang H, Liu J, Chen Y, He RR, Liu B. Mitochondrial Sirtuin 3: new emerging biological function and therapeutic target. Theranostics. 2020;10(18):8315–42. https://doi.org/10.7150/thno.45922.

168. Zhang L, Liu C, Wu J, Tao JJ, Sui XL, Yao ZG, Xu YF, Huang L, Zhu H, Sheng SL, Qin C. Tubastatin A/ACY-1215 improves cognition in Alzheimer's disease transgenic mice. J Alzheimer's Dis. 2014;41(4):1193–205.
169. Zhang ZY, Schluesener HJ. Oral administration of histone deacetylase inhibitor MS-275 ameliorates neuroinflammation and cerebral amyloidosis and improves behavior in a mouse model. J Neuropathol Exp Neurol. 2013;72(3):178–85.
170. Ziemka-Nalecz M, Jaworska J, Sypecka J, Zalewska T. Histone deacetylase inhibitors: a therapeutic key in neurological disorders? J Neuropathol Exp Neurol. 2018;77(10):855–70.
171. Ziemka-Nałęcz M, Zalewska T. Neuroprotective effects of histone deacetylase inhibitors in brain ischemia. Acta Neurobiol Exp. 2014;74(4):383–95. https://doi.org/10.55782/ane-2014-2001.

Part IV

Experimental Tools

Animal Models of CNS Neurodegenerative Disease/Research Implications of Animal Models of CNS Neurodegenerative Diseases

15

Angel P. Godad, Tejas Bhatia, Muqtada Shaikh, and Gaurav Doshi

Abbreviations

11C-CHDI-180R PET	Carbon-11 C-Huntington Disease Imaging-180R Positron Emission Tomography
3-NPA	3-Nitropropionic Acid
5-HT	Serotonin
AIS	Acute Immobilization Stress
ALS	Amyotrophic Lateral Sclerosis
AM	Animal Model
AMPA	α-amino-3-hydroxy-5-methyl-4-isoxazole propionic Acid
ASC	Antidepressant Sensitive Catalepsy
AUD	Alcohol Use Disorder
AWA	Animal Welfare Act
BBB	Blood-Brain Barrier
BDNF	Brain-Derived Neurotrophic Factor
BDNF	brain-derived neurotrophic factor
C+A	C16-angiopoietin 1
CAG	Cytosine-Adenine-Guanine
CBA	Control Balb/c Anemia
CDM	Chronic Behavioral Despair Model
CNS	Central Nervous System

A. P. Godad · T. Bhatia · M. Shaikh
SVKM's Dr. Bhanuben Nanavati College of Pharmacy, Mumbai, India

G. Doshi (✉)
Department of Pharmacology, SVKM's Dr. Bhanuben Nanavati College of Pharmacy, Mumbai, India
e-mail: gaurav.doshi@bncp.ac.in

S. Singh (ed.), *Ubiquitin Proteasome System in CNS Neurodegenerative Diseases*, https://doi.org/10.1007/978-981-95-6099-8_15

CPCSEA	Committee for the Purpose and Control and Supervision of Experiments on Animals
CRISPR-Cas9	Clustered Regularly Interspaced Short Palindromic Repeats-associated Protein 9
CRS	Chronic Restraint Stress
CSF	Cerebrospinal Fluid
CUMS	Chronic Unpredictable Mild Stress
CWG-cPIP	Cyclic Pyrrole-Imidazole Peptide
DA	Dopamine
DEX	Dexamethasone
DMN	Default Mode Network
EA	Electroacupuncture
EAE	Experimental Autoimmune Encephalomyelitis
EP	Europinidin
FELASA	Federation of European Laboratory Animal Science Associations
fMIP	N-formyl methionyl-leucyl-phenylalanine
FUS	Fused in Sarcoma
GABA	Gamma-Aminobutyric Acid
GWI	Gulf War Illness
HD	Huntington's Disease
HDBOI	Huntington's Disease Burden of Illness
Homer1a	Homer Protein 1a
HTT	Huntingtin
IACUC	Institutional Animal Care and Use Committee.
IKKβ	IκB Kinase Beta
JAK2	Janus Kinase 2
KLK6	Kallikrein 6
KLK6-KO	Kallikrein 6 Knockout
LDLR	low-density lipoprotein receptors
LPC	Lysophosphatidylcholine
LRRK2	Leucine-Rich Repeat Kinase 2
mHTT	Mutant Huntington (Huntington's Disease)
mPFC	Medial Prefrontal Cortex
MS	Multiple Sclerosis
NAc-DBS	Nucleus Accumbens Deep Brain Stimulation
NDDs	Neurodegenerative Diseases
NF-κB	Nuclear Factor-Kappa B
NSDUH	National Survey on Drug Use and Health
NT	Nutritional Ketosis
OBX	Olfactory Bulbectomy
OD	Oligodendrocytes
OECD	Organisation for Economic Co-operation and Development
OPCs	Oligodendrocyte Precursor Cells
OT	Oxytocin

PBMCs	Peripheral Blood Mononuclear Cells
PCSK9	Proprotein Convertase Subtilisin/Kexin Type 9
Per2	Period Circadian Protein 2
PET	Positron Emission Tomography
PGC-1α	Peroxisome Proliferator-Activated Receptor Gamma Coactivator 1 Alpha
PSIL	Psilocybin
REV-ERB	Nuclear receptor
REV-ERB	Nuclear Receptor Subfamily 1, Group D, Member 1
ROR	Retinoic Acid Receptor-Related Orphan Receptor
SNRI	Selective Norepinephrine Reuptake Inhibitors
SOD1	Superoxide Dismutase 1
SPT	Sinking Platform Test
SSRIs	Selective Serotonin Reuptake Inhibitors
ST	Static Training
taVNS	Transcutaneous Auricular Vagus Nerve Stimulation
TCA	Tricyclic Antidepressants
TDP-43	TAR DNA-Binding Protein 43
TYROBP/DAP12	TYRO Protein Tyrosine Kinase Binding Protein/DNAX-activating Protein of 12 kDa
VTA	Ventral Tegmental Area

15.1 Introduction

Neurodegeneration is marked by the gradual deterioration of functional neurons, often accompanied by neuroinflammation, which can lead to serious clinical deficits [1, 2]. Neurodegenerative diseases (NDDs) are disorders characterized by the gradual loss of structure and function in the nervous system [3]. Some of these diseases are depression disorder, multiple sclerosis (MS), Huntington's disease (HD), and alcohol use disorder (AUD) [4–8]. The progressive loss of neurons associated with these conditions can severely affect cognitive abilities, motor skills, and overall quality of life [9]. It is really important to study these diseases because they are degenerative and also because the treatment and management of these diseases are so complex. One of the most prevalent types of mental illness is depression, which afflicts millions of individuals worldwide. According to the Global Burden of Diseases, Injuries, and Risk Factors Study (GBD) 2019, depressive disorders accounted for the largest share of the mental health burden, contributing up to 37.3% of the total burden [10, 11]. Depression affects not only the individual, but also their social networks, the general public health care system, and society in general [11]. MS is another NDD that attacks the central nervous system (CNS) and can lead to such ailments as difficulty in movement, fatigue, and cognitive deterioration [5]. Currently, over 2.8 million people worldwide are affected by MS [12]. Prevalence studies of MS in China indicate an exponential increase and predict that the numbers will peak sometime between the years 2025 and 2029 and then rise dramatically over the next 25 years [13]. Additionally, HD is an autosomal dominant neurodegenerative disorder characterized by progressive motor dysfunction and chorea, cognitive decline, and neuropsychiatric symptoms [14]. According to the Huntington's Disease Burden of Illness (HDBOI) study, the prevalence of HD in Europe and North America is a significant public health issue and more research and support are warranted [15]. AUDs are intricate, long-lasting conditions characterized by repeated episodes that result in enduring changes to neurotransmitter systems. Notably, alterations in the glutamatergic system play a crucial role in both the onset and persistence of ethanol dependence [16, 17]. The National Institute on Alcohol Abuse and Alcoholism states that alcohol use disorder (AUD) affects almost 11% of Americans [18]. One of the major areas of research in CNS is the application of the animal model (AM) to shed some light on some of the enigmas of neurological diseases. These models can be employed to understand the biology of disease and to identify potential targets of intervention since they can mimic human disease. They offer a wealth of knowledge regarding the course of a disease and how well treatment works, not to mention the entire genetic and environmental factors that contribute to CNS disorders. These models allow us to better understand how the brain works and thus lead to the creation of new drugs that can help patients. We are going to look into the AM used in the study of some of the major NDDs that affect a large number of people in the world, including depression, MS, AUD, and HD. The main point of this chapter is to examine the progression of these diseases and how the AM helps us better understand these processes, with the overall goal of finding treatments for NDDs.

15.1.1 Importance of Animal Model

Diseases of the CNS are a major burden of disease worldwide, including such neuropsychiatric illnesses as schizophrenia, mood disorders, anxiety disorders, Alzheimer's disease, Parkinson's disease, and MS. These diseases form a bewildering assortment of maladies that strike at the very core of the organ that makes us human, the brain, altering its structure and function and crippling in their wake cognitive, developmental and often motor skills. Despite advancements in neuroscience, our understanding of the molecular and cellular mechanisms underlying CNS diseases remains limited, which significantly hampers the development of effective treatments. The AM is integral to CNS disease research because it permits the investigation of pathogenic pathways that would never be found through human studies alone. These model systems offer the ability to investigate certain aspects of disease pathophysiology in a controlled setting, especially those that result from genetic manipulation, toxic exposure, or surgical procedures. For instance, researchers investigating the formation of amyloid-beta plaques and tau protein tangles (the two most notable features of Alzheimer's disease) have employed transgenic mice carrying human genetic mutations known to give rise to the disease, such as mutations in the presenilin or amyloid precursor protein (APP) genes. Other mechanisms that have been explained by these models include synaptic dysfunction and neuroinflammation, which are both possible treatment targets [19]. AMs are so important in studying the etiology of CNS diseases. In their regulated environments, researchers can delve into the molecular, cellular, and behavioral components of these intricate diseases, such as MS, Parkinson's, and Alzheimer's. Because the information gained from these models usually far surpasses that which can be learned from human studies, research on disease progression and treatment success can be done. The cardinal features of CNS diseases are reproduced in animal models, and the pathophysiological mechanisms of these diseases are then studied [20]. These models are so important because they help to explain the complex mechanisms of CNS disorders, provide a way to test treatments and monitor the course of disease in a controlled setting. The AM of neurodegeneration, inflammation, and remyelination, especially in light of conditions such as MS [21]. Preclinical studies using these models can be used to identify new therapeutic targets and to test new therapies [22]. Even better, animal models offer increased translational validity for preclinical research, help explain the mechanics of disease, and permit the experimentation of treatments [23]. To sum it all up, AMs are the key to knowing how CNS diseases work and how to treat them. These vary from more primitive creatures such as worms to more complex organisms such as rats and companion animals [24, 25]. The main AM for MS is experimental autoimmune encephalomyelitis (EAE), and this has paved the way for the development of numerous FDA-approved drugs [26].

There are notable differences in the prevalence of anxiety, substance use disorders, Parkinson's disease, Alzheimer's disease, and schizophrenia among various groups and environments. Planning and intervention efforts for public health require an understanding of these rates. Anxiety affects up to 31% of people with Parkinson's disease, although it is common in many other populations as well [27]. Anxiety disorders are particularly common among opioid injectors; reports of social anxiety disorder and panic disorder are 40.5% and 41.2%, respectively [28]. With notable increases in age-standardized incidence rates seen in places like East Asia, the prevalence of Parkinson's disease has grown globally. An increasing public health risk is shown by the significantly high frequency of Parkinson's disease in high-SDI locations. Age-standardized disability-adjusted life years (DALYs) are rising in Alzheimer's disease and other dementias, indicating a substantial strain on healthcare systems [29]. Substance use disorders are common; research shows that 4.0% of people in a biobank had tobacco use disorder (TUD) and 19.5% of people had alcohol use disorder (AUD) [30]. Psychiatric comorbidities were present in 88.1% of opioid injectors, with a notable overlap with drug use disorders [28]. A large percentage of people worldwide suffer from migraines, a common neurological condition. According to estimates, migraine affects 14.4% of people globally, with prevalence rates varying between 9% and 18% [31]. About 47 million individuals, or almost 15% of the population, suffer from migraines in the United States alone [32].

The AM is crucial when investigating CNS ailments because it offers valuable information about disease processes and possible treatments. They make it easier for researchers to investigate complicated neurological diseases, including schizophrenia, Parkinson's disease, and Alzheimer's, by enabling the replication of human-like pathological states. Optogenetics and genetic engineering are cool tools that allow for the very select manipulation of disease processes in AMs [33]. These methods also offer a superb proving ground for novel therapies, thereby bridging the gap between basic research and medical practice [22]. These animal models are far more reliable subjects of study when the molecular and behavioral features of the human CNS pathologies are manifested in them [33]. Before advancing to human clinical trials, animal models provide a crucial environment for evaluating the safety and efficacy of new medications [22]. One example of these models is transgenic mice, and through their use, many of the genetic bases of neuropsychiatric disorders have been worked out, and therefore, the mechanisms of disease are better understood [34]. But this research needs these AMs to understand the complicated relationship between the neurons, the glial cells, and the blood-brain barrier (BBB). This will enable researchers to investigate the roles these factors play in CNS pathophysiology, particularly neurodegenerative disease. These days, transgenic models have helped us understand the composition of the BBB and its vital role in maintaining brain homeostasis, especially when compromised during disease states like Parkinson's and Alzheimer's disease [35]. They also serve as a useful tool for researchers to examine how endothelial cells and glial cells (astrocytes and microglia) communicate to regulate blood flow and maintain barrier integrity [36].

15.2 Animal Models for CNS Neurodegenerative Diseases

15.2.1 Alzheimer's Disease (AD)

Memory loss and cognitive decline are hallmarks of Alzheimer's Disease (AD), a progressive neurodegenerative illness that is mainly caused by certain pathological characteristics. The illness is characterized by the buildup of neurofibrillary tangles (NFTs) made of hyperphosphorylated tau protein and amyloid-β (Aβ) plaques, which impair neuronal function and cause cell death. The main pathogenic characteristics of AD are as follows: Aβ peptide deposits outside of cells that interfere with cell communication and cause neuroinflammation, Hyperphosphorylated tau protein clumps inside cells that disrupt neuronal transport mechanisms, and neuronal damage results from chronic inflammation in the brain, which is frequently made worse by the development of plaques and tangles. Cognitive decline is correlated with increased oxidative damage to proteins, lipids, and DNA, which is common in AD [37].

Research on Alzheimer's disease (AD) relies heavily on animal models, which offer valuable insights into the illness's processes and possible treatment approaches. Transgenic mice, which express mutations linked to familial AD, and more recent models that make use of gene editing technologies are the most often used models. The pathogenesis of AD, such as the development of amyloid plaque and neurofibrillary tangles, is better understood by researchers through these models. The main features of the popular and promising AM for AD research are listed below.

Transgenic Mice: Models such as PDAPP, Tg2576, and 5XFAD exhibit human mutations in presenilin 1 (PS1) and amyloid precursor protein (APP), resulting in amyloid buildup and cognitive impairments [38]. Human AD-related genes are substituted for mouse genes in knock-in/knock-out models, which offer more physiologically relevant insights without overexpression problems [39]. Heavy metals like aluminum and chemicals like streptozotocin cause neurodegeneration and resemble the pathophysiology of AD. Novel models, such as oAβ25–35 and APP knock-in rats, are being created to more accurately mimic the pathophysiology of AD in humans [40]. Another technique produces neurotoxic peptides and induces early AD-like disease without plaque formation by using adeno-associated viruses to express mutant APP and PS1 genes [41]. Even while these models have improved our understanding of AD, they frequently fall short of accurately capturing the disease's complexity, especially when it comes to neurodegeneration and behavioral symptoms. This demonstrates the continuous need for novel models that more accurately capture the complex character of AD [42].

Transgenic mice are essential models for researching Alzheimer's dementia (AD) because they can replicate important clinical traits and behavioral signs of the illness. Researchers may examine disease causes and possible treatments by using these models, which mostly use genetic changes to express human mutations associated with familial AD. Amyloid beta (Aβ) plaque development and cognitive impairment are caused by transgenic mice's frequent overexpression of mutations in genes such as APP and PS1 [39]. For example, the 5xFAD model shows neuroinflammation and early-onset amyloid deposition, with notable behavioral deficits by 4 months [43]. Other models, such as Thy1-ApoE4, demonstrate Aβ aggregation and neurofibrillary tangles while simulating sporadic AD without APP mutations [44]. Knock-in models, which substitute human sequences for mouse genes, are recent developments that offer more physiologically relevant insights into AD pathophysiology [38]. Transgenic models are useful, but they do not accurately represent all facets of AD, including neurodegeneration and tau pathology, which are essential for a thorough comprehension of the illness [45].

Another potential method to induce AD is stereotactic surgery. Using a precise procedure called stereotactic surgery, drugs are injected into certain brain areas to imitate AD in animals [46]. An intrahippocampal injection of amyloid beta peptide is a popular method that is especially helpful for short-term research concentrating on its effects in certain brain regions [47].

AD knock-in models offer important new information on the pathophysiology and course of the illness. Transgenic mice's overexpression artifacts are avoided in these models. APP knock-in models in mice exhibit a higher likelihood of glutamate release before plaque development, with microglial responses not occurring until a significant plaque burden is developed [48]. Similar to human AD, app knock-in mice have tau pathology, Aβ plaques, neuronal death, and cognitive impairments [49]. Diffuse Aβ aggregates and cognitive abnormalities are seen in the hAβ-KI mouse model, which expresses humanized Aβ without familial AD mutations and may be indicative of sporadic AD [50]. The development of AD depends heavily on microglia, and the App knock-in phenotype is exacerbated by partial loss [48]. These knock-in models offer physiologically applicable resources for comprehending AD processes and evaluating possible treatments [49]. The potential AM for AD has been listed in Table 15.1.

15.2.2 Parkinson's Disease (PD)

Degeneration of dopaminergic neurons in the substantia nigra is the hallmark of Parkinson's disease (PD), a common neurodegenerative illness that mainly affects older persons. Such degeneration causes bradykinesia, stiffness, resting tremor, and postural instability among the motor symptoms. A hallmark of PD pathogenesis is aggregates of α-synuclein, forming Lewy bodies that contribute to neuronal malfunction and death. The main pathogenic characteristic of dopamine insufficiency is caused by the loss of dopaminergic neurons in the striatum. Affected neurons

Table 15.1 Animal models for Alzheimer's disease

Sr. no	Type of animal model	Mechanism	Specific outcome	Reference
1.	Transgenic mice			
	PDAPP	Increased APP expression due to the Arctic mutation.	Amyloid plaque formation and cognitive impairments	[39]
	Tg2576	Human APP overexpression caused by the Swedish mutation	Memory impairment and the early development of amyloid pathology	[39]
	5XFAD	APP and PS1 mutation co-expression	Increased neurodegeneration and the production of amyloid plaque	[39]
2.	Knock-in/knock-out mice	Using CRISPR gene editing to target genes linked to AD	More precise sporadic AD modeling that avoids overexpression	[39]
3.	Chemical induction	Amyloid-beta oligomer injection into the brain	Induction of cognitive decline and neuronal loss	[51]
4.	Phenomimetic models			
	oAβ25-35	Oligomeric Aβ25-35 peptide administration	Induction of memory impairments and synaptic dysfunction	[52]
	APP	APP transgenic mice are used to simulate the pathophysiology of AD.	Alterations in behavior that mimic the symptoms of AD	[52]
5.	Stereotactic models	Stereotactic surgery for the injection of specific substances	Neuronal loss and behavioral abnormalities particular to a region	[46]

Abbreviations: *APP* Amyloid Precursor Protein, *CRISPR* Clustered Regularly Interspaced Short Palindromic Repeats, *PDAPP* PD Amyloid Precursor Protein, *Tg2576* Transgenic 2576, *5XFAD* Five Familial Alzheimer's Disease Mutations

contain characteristic aggregates of phosphorylated α-synuclein called Lewy bodies, which help in the course of the disease [53]. Bradykinesia, Rigidity, Resting Tremor & Postural Instability are the motor symptoms associated with PD. Bradykinesia is a hallmark symptom characterized by slowness of movement. Resting Tremor is involuntary shaking that typically occurs when the muscles are at rest [54]. Rigidity is muscle stiffness that can lead to discomfort and reduced mobility. Postural instability is the difficulty in maintaining balance, increasing the risk of falls [55].

To comprehend Parkinson's disease (PD) and create possible treatments, the AM is essential. The mechanism and symptoms of the disease are replicated using a variety of models, such as invertebrate, genetic, and neurotoxin-induced models. Neurotoxin-induced models: Often used in rats and mice, 6-hydroxydopamine (6-OHDA) mimics the symptoms of late-stage Parkinson's disease by inducing dopaminergic cell degeneration [56]. Mostly utilized in primate models, MPTP exposure causes a fast loss of dopaminergic neurons that closely resembles Parkinson's disease in humans. Additionally, substances like paraquat and rotenone are used to investigate PD processes [57]. Genetic Models: These models are essential for researching the early stages of the disease because they concentrate on the aggregation of α-synuclein, a characteristic of PD [58]. The LRRK2 and PINK1 models offer insights into hereditary types of Parkinson's disease by linking genetic abnormalities in these genes to familial PD [57]. Drosophila and Caenorhabditis elegans are useful models for preliminary research because they enable quick genetic modification and screening of medicinal substances [57]. An essential theoretical framework for researching PD is the 6-hydroxydopamine (6-OHDA) model, which is especially useful for comprehending dopaminergic neuron loss and its consequences. By creating specific lesions in the nigrostriatal pathway, this model enables researchers to mimic early-stage PD and explore the disease's motor and non-motor symptoms. 6-OHDA mimics the pathogenesis of Parkinson's disease by specifically targeting dopaminergic neurons and causing their degeneration. Research indicates that different 6-OHDA dosages cause varying degrees of behavioral abnormalities and neuronal death, underscoring the drug's usefulness in simulating early-stage PD [56]. The degree of dopaminergic neuron loss is correlated with the model's notable motor deficits, which include decreased locomotion and changed exploratory behavior. The model's applicability for examining the whole range of PD symptoms is highlighted by research showing that 6-OHDA lesions also result in emotional and cognitive impairments [59].

One more key theoretical framework for researching PD is the MPTP model, which simulates the neurodegenerative processes of PD by administering the neurotoxin 1-methyl-4-phenyl-1,2,3,6-tetrahydropyridine (MPTP). Researchers can investigate the course of dopaminergic neuron degeneration and related behavioral alterations using this model, which offers insights into possible treatment approaches. When MPTP is converted to MPP+, it specifically targets and kills dopaminergic neurons in the substantia nigra, resulting in symptoms that resemble PD [60]. Different regimens (acute, sub-acute, and chronic) can be used to administer the model, and each one produces unique behavioral and neurochemical results.

Research has demonstrated that the administration of MPTP causes notable changes in metabolic connections, motor abnormalities, and cognitive impairments, all of which are indicative of the complexity of Parkinson's disease. The usefulness of the model in comprehending PD symptomatology is demonstrated by behavioral tests, which show that the intensity of motor and non-motor symptoms changes with the administration schedule [61]. Key molecular pathways implicated in neurodegeneration, such as oxidative stress and apoptosis, have been identified by transcriptomic analysis in MPTP-treated mice. These findings are crucial for comprehending the course of PD [62]. Table 15.2 below discusses the important aspects of the AM used for PD.

Table 15.2 Animal models for Parkinson's disease

Sr. no	Type of animal model	Mechanism	Specific outcomes	Reference
1.	6-Hydroxydopamine (6-OHDA)	Neurotoxin that selectively targets dopaminergic neurons	Motor impairments are caused by the death of nigrostriatal dopaminergic neurons	[63]
		Dopamine/noradrenaline transporters mediate intracellular accumulation; oxidative stress causes cytotoxicity and impairs mitochondrial function	Neuronal loss that occurs quickly, followed by fiber degeneration	[64]
		Severe motor impairments as well as non-motor comorbidities including cognitive impairments, depression, and anxiety-like behavior are caused by bilateral injections in the substantia nigra or dorsal striatum	Recurrence of non-motor comorbidities, including circadian rhythm disruption, cognitive impairment, depression-anxiety-like behavior, and poor olfactory discrimination	[64]
2.	MPTP	Damages dopaminergic neurons in the substantia nigra pars compacta and striatum by causing oxidative stress, mitochondrial apoptosis, inflammation, excitotoxicity, and inclusion body formation	Dopaminergic neuronal damage is caused by selective neurotoxicity, which successfully simulates Parkinson's disease symptoms even if it does not exactly replicate PD symptomatology	[65]

(continued)

Table 15.2 (continued)

Sr. no	Type of animal model	Mechanism	Specific outcomes	Reference
		A key factor in MPTP toxicity is the opening of the mitochondrial permeability transition pore; DAT and VMAT-2 condition susceptibility to MPTP neurotoxicity	PD research is being revitalized, which creates fresh avenues for the creation of innovative treatment molecules	[66]
3.	Agricultural Pesticides (Rotenone)	Reactive oxygen species (ROS) are produced in greater quantities when rotenone inhibits Complex I of the electron transport chain; paraquat produces ROS through redox cycling, which causes DNA damage and lipid peroxidation	Both pesticides promote dopaminergic neuronal death; paraquat triggers the JNK-caspase pathway; rotenone results in nigral lesions that can be reversed by apomorphine	[67]
4	Alpha-Synuclein	Protein aggregation and aberrant folding are caused by mutations. Lewy bodies	Heightened vulnerability to neurodegeneration	[68]
5.	LRRK2	Mutations that cause gain-of-function increase kinase activity, which affects cellular signaling pathways	Causes neurodegeneration and is connected to the autosomal dominant pattern of inheritance	[69]
6.	PINK1	Controls mitophagy to ensure that damaged mitochondria are removed; flaws interfere with autophagic flow	Defects interfere with the autophagic flow, while mitophagy ensures the removal of damaged mitochondria. Early-onset Parkinson's disease is linked to an increased mitochondrial load that causes cell death	[70]
7.	Parkin	Mutations affect the activity of E3 ubiquitin ligase, which marks damaged proteins for destruction	Autosomal recessive Parkinson's disease is linked to the accumulation of defective proteins that cause neurodegeneration	[71]

Abbreviations: *MPTP* 1-Methyl-4-phenyl-1,2,3,6-tetrahydropyridine, *LRRK2* Leucine-Rich Repeat Kinase 2, *PINK1* PTEN-Induced Putative Kinase 1

15.2.3 Huntington's Disease (HD)

HD is a hereditary NDD that requires the use of the AM to understand its underlying mechanism and potential therapeutic intervention. This study presents a novel focus on the retina as a research area in HD by using the AM. The researchers obtained important information regarding retinal impairment in HD using the AM. They examined the effects of HD on the retina using two transgenic mouse models, R6/1 and zQ175. This study identified anomalies in the structure and function of the retina, indicating that retinal changes may occur earlier in HD progression than previously recognized. The research also discovered neuroinflammation and glial activation in the retinas of HD mouse models, suggesting that the retina does not simply serve as a passive component to the neurodegenerative process associated with HD, but instead plays an active role. Prolonged stimulation of glial cells can lead to blindness because these cells damage the retinal neurons [72]. The research offers a different look at HD and implies that the retina could be a very meaningful model for neuroinflammatory processes. There is no cure for HD, and the only treatment for it right now is to try to relieve the symptoms and slow down the progress of the disease. Alharbi used male Wistar rats as a model for HD (induced by 3-Nitropropionic acid (3-NPA) to produce HD-like symptoms) to examine the effects of Europinidin (EP). The results showed that EP did indeed improve motor and cognitive functions and also greatly decreased the neurotoxic effects of 3-NPA. It increases the activity of important enzyme complexes, improves cellular energy production, restores mitochondrial function, and lowers oxidative stress—all of which are critical for neurodegenerative illnesses. Also, EP balances neurotransmitter levels and reduces neuroinflammation, so it may serve as a good treatment [73]. In another study, researchers used R6/2 transgenic mice to create neurospheres to study HD. They isolated neural progenitor cells from the brains of these mice, creating neurospheres that mimic the disease characteristics. This innovative approach allowed for a detailed examination of the pathological features associated with HD, such as the presence of expanded CAG (Cytosine-Adenine-Guanine) repeats in the Huntington gene, resulting in protein aggregation. Neurospheres derived from R6/2 mice exhibited many CAG repeats, a hallmark of HD. This genetic mutation causes the production of mutant huntingtin protein, which, for some reason, tends to misfold and aggregate, thereby causing cellular malfunction and death. They employed CRISPR-Cas9 (Clustered Regularly Interspaced Short Palindromic Repeats-associated protein 9) technology to upregulate the expression of neuroprotective factors such as brain-derived neurotrophic factor (BDNF) and PGC-1α (Peroxisome Proliferator-Activated Receptor Gamma Coactivator 1 Alpha), and simultaneously downregulate polyglutamine aggregation. So, fixing this genetic defect would increase cellular robustness and, in turn, alleviate the neurodegenerative aspects of HD [74]. Also, the research examined the TYROBP/DAP12 (TYRO protein tyrosine kinase binding protein/DNAX-activating protein of 12 kDa) gene in HD mice (Q175 genetically modified mice that model HD-like phenotypes). They showed that the knockout of TYROBP/DAP12 caused drastic alterations in microglial function, the brain's immune cells that are implicated in neuroinflammation. These mice, null for TYROBP/DAP12, showed less activation than control mice, which suggests a decreased proinflammatory gene expression. Erasure of TYROBP/DAP12 leads to the downregulation of many inflammatory markers, among them components of the complement cascade. In all, TYROBP/DAP12-deficient microglia showed decreased activation of inflammatory pathways; therefore, TYROBP/DAP12 is a key regulator of microglial activation and inflammation in HD [75]. The R6/2 mouse model, which is genetically engineered to express the human HTT (Huntingtin) gene, has been used to model HD. The mice displayed motor impairments and behavioral abnormalities characteristic of HD, consisting of motor coordination problems and hind limb clasping. CWG-cPIP (Cyclic Pyrrole-Imidazole Peptide) has been shown to have ameliorative effects on motor dysfunction and other HD-related symptoms and therefore may be a viable treatment option. The treatment reversed the changed gene expression, enhanced synaptic function, and affected the cognitive process often seen in HD mice for the better. It also triggered an immune response, which points to inflammatory processes playing a part in the progression of HD [76]. Research using the zQ175DN mouse model has shown that as HD progresses brain perfusion and vascular dysfunction occur. The researchers identified early changes in cerebral blood flow and vascular function in the zQ175DN mice, with an increase in blood flow in specific brain regions at around 3 months of age. But as the mice aged, this blood flow started to diminish, suggesting a change in vascular reactivity. This indicates that the initial increase in blood flow may act as a compensatory mechanism during the early stages of neurodegeneration, ultimately leading to a decline in blood flow in the later stages of the disease [77]. 11C-CHDI-180R PET (Carbon-11 C-Huntington Disease Imaging-180R Positron Emission Tomography) imaging is a new technique that provides important clues about the pathology of HD, the use of 11C-CHDI-180R PET imaging in a rhesus macaque model of HD. They determined that this method very well images and measures mutant Huntington (mHTT) aggregates in the brain, a hallmark of HD. The macaque model displayed both behavioral and neurological deficits as do HD patients. 11C-CHDI-180R PET imaging makes it possible to trace mHTT aggregates in real-time in different areas of the brain and helps us understand mHTT accumulation dynamics and how that corresponds to behavioral deficits [78]. In light of recent advancements in imaging technology, a study using 3D electron tomography and zQ175 knock-in mice found that the structure of polysomes in the striatal neurons of these mice became increasingly compacted over time, indicating ribosome stalling. This change coincides with the emergence of HD-related symptoms, suggesting a direct link between protein synthesis machinery changes and the disease progression. The study also found an increase in the expression of ribosome stalling relief factors, such as EIF5A2, in the striatal samples from the HD model, suggesting that neurons are coping with disruption in protein synthesis [79]. Table 15.3 gives a detailed summary of AMs used in HD.

Table 15.3 Animal models for Huntington's disease

Sr no	Disease	Name of animal model	Specific outcomes	Reference
1	Huntington's disease (HD)	HD Transgenic Mice Models (R6/1 and zQ175)	Gilila activation and neuroinflammation are involved in retinal changes linked to HD, and the retina can serve as a key model for studying neuroinflammation in HD	[72]
2	HD	3-Nitropropionic Acid (3-NPA)-Induced HD Rat Model	Europinidin effectively reduces 3-NPA-induced cognitive deficits in HD rats	[73]
3	HD	R6/2 Transgenic HD Mice Model	The study confirmed the presence of a genetic mutation linked to HD in the R6/2 Mice Model. Additionally, the findings suggest that gene editing could mitigate the detrimental effects of HD in this model	[74]
4	HD	Q175 (Genetically Modified) Transgenic HD Mice Model	The TYROBP/DAP12 (TYRO protein tyrosine kinase binding protein/ DNAX-activating protein of 12 kDa) gene is a crucial regulator of microglial activation and inflammation in HD	[75]
5	HD	R6/2 Transgenic HD Mice Model	The model exhibited motor deficits and behavioral alteration; however, treatment with CWG-cPIP (Cyclic Pyrrole-Imidazole Peptide) led to improvements in motor function	[76]
6	HD	zQ175DN HD Mice Model	This model contributes to our understanding of neurovascular coupling in the context of neurodegeneration related to HD dynamics	[77]
7	HD	Rhesus Macaque Model of HD	The model exhibited behavioral and neurological impairments similar to those in HD patients, with 11C-CHDI-180R PET (Carbon-11 C-Huntington Disease Imaging-180R Positron Emission Tomography) providing insight into mHTT (mutant Huntington) accumulation dynamics and its correlation with behavioral deficits	[78]
8	HD	zQ175 Knock-in HD Mice Model	Impaired polysomal structural integrity in striatal neurons potentially hinders protein translation and contributes to neuronal dysfunction in HD	[79]

Abbreviations: *HD* Huntington's Disease, *3-NPA* 3-Nitropropionic Acid, *mHTT* Mutant Huntingtin, *CWG-cPIP* Cyclic Pyrrole-Imidazole Peptide, *TYROBP/DAP12* TYRO Protein Tyrosine Kinase Binding Protein/DNAX-Activating Protein of 12 kDa, *11C-CHDI-180R PET* Carbon-11 C-Huntington Disease Imaging-180R Positron Emission Tomography

15.2.4 Amyotrophic Lateral Sclerosis (ALS)

ALS is a life-threatening neurodegenerative disease with a pathophysiological characteristic of ongoing loss of brain and spinal cord motor neurons, with a progression towards extreme weakness, atrophy, and eventual paralysis. ALS takes a deep toll in terms of daily life and general well-being, with no satisfactory therapy yet developed. It is critical to comprehend pathophysiology in creating effective therapies for ALS. ALS primarily involves upper and lower motor neurons, with atrophy and paralysis of the involved muscles subsequently following [80]. Abnormal aggregation of TDP-43 protein is a characteristic of ALS, in both motor neurons and skeletal muscle tissue, and its contribution to disease pathology is proposed [81, 82]. Genetic mutations in SOD1 cause familial ALS and contribute to toxicity in motor neurons [83]. Inflammation in the central nervous system hastens the loss of motor neurons and implicates glia in its role [80]. Mitochondrial function defects cause oxidative stress and a lack of energy in motor neurons and contribute to ongoing degeneration [83].

Animal models have a key role in studying ALS, for in them, one can: Investigate disease progression and mechanism, test potential therapeutic interventions, such as gene therapy and therapies with stem cells [84]. Evaluate the efficacy of new drugs in a laboratory setting, and provide information regarding disease processes in humans. While significant advances have been made in terms of knowing about ALS, its pathophysiological complexity and individual-to-patient variation complicate the development of universally effective therapies. There is an ongoing necessity for investigation to sort out such a complex disease and maximize therapeutic benefit. Animal models play a significant role in investigating amyotrophic lateral sclerosis (ALS) and in creating future therapies. There have been numerous types of models, ranging from invertebrate animals to mammals, developed for studying pathogenesis and therapeutic intervention for the disease. In the following sections, common animal models being utilized in ALS studies have been discussed. Invertebrate Models like C. elegans offer ease and rapid genetic manipulation, with access to investigating ALS-related gene studies. Other models like Drosophila, useful for studying the loss of motor neurons and for testing drugs for effectiveness with its short life and genetic manipulability, and Zebrafish, which provides a transparent embryo model, in which in vivo visualization of motor neuron development and decay can occur [85, 86].

Rodent Models that include Mice—The most prevalent model, namely transgenic mice with familial mutations in ALS" as "The most prevalent rodent model is transgenic mice with familial mutations in ALS. It reproduces the most significant pathologic features but can miss specific aspects of disease in humans [87, 88], and Rats that offer increased brain volumes for surgical interventions and behavior testing, in addition to in-mice studies [89]. Larger Animal Models like Pigs and nonhuman primates have neurodegenerative phenotypes most similar to human ALS and disclose disease mechanism information not obtainable in smaller model species [87]. While these models have aided in advancement in studies of ALS, the challenge of recreating the sporadic form of disease, comprising most cases, continues to dominate. In future studies, work can draw upon investigations in these disparate models to develop a fuller picture and therapies.

Animal models for Amyotrophic Lateral Sclerosis (ALS) have been particularly important for testing many therapeutic interventions, most prominently interventions for mutations in SOD1 and TDP-43 and neuroprotective therapies. Transgenic animals have shed information about the efficacy of interventions for SOD1 and TDP-43 mutations, antioxidants, stem cell therapy, and anti-inflammatory therapies. Antisense oligonucleotide (ASO) therapy, in the form of tofersen, targets SOD1 mRNA for suppression of toxic protein expression. In animals, such therapy is effective, and for its application in treating ALS, it gained FDA approval [90]. Mouse models with TDP-43 mutations have aided pathophysiological investigation and therapeutic testing for decreasing TDP-43, with individual therapies under development [85]. Antioxidants have been studied in experimental animals for effectiveness in countering oxidative damage, a driving mechanism in ALS pathology. Studies have determined compounds with the potential to minimize oxidative species. Stem cell therapies are being explored in ALS models to assess their ability to replace lost motor neurons and promote regeneration. Preclinical studies indicate potential benefits, although clinical translation remains challenging [80]. Animal models have confirmed that neuroinflammatory processes can be suppressed, and disease progression can be decelerated. For instance, drugs that inhibit the MAPK pathway have therapeutic effectiveness in models of ALS, most prominently in female subjects [91]. While these therapies have potential, barriers to extrapolating information derived from animals necessitate follow-up studies in a quest to maximize therapy regimens and success.

The challenges in ALS model animals arise mainly through species variation in disease progression, insufficiency in disease model reproduction for sporadic ALS, and a lack of translationally relevant preclinical model requirements. Overcoming these limitations is important for developing therapeutic approaches. Various animal models, such as rats, mice, and nonhuman primates, have considerable variation in the progression of ALS, and translation of observations to humans is therefore challenging [89]. For instance, transgenic model animals including SOD1 and C9orf72 have disease processes with specific pathology that can not necessarily model pathology in humans [85]. Most existing studies have focused on familial forms of ALS, with little information regarding sporadic ALS, which is responsible for most cases. Recent efforts have supplemented cerebrospinal fluid-delivered models with added fidelity to model sporadic ALS, but confirmation continues [92]. The integration of iPSCs with model animals is proposed to make them even more relevant for translation. Multiomic approaches have identified specific molecular networks in human ALS that can inform new and improved animal model development. Conversely, while such impediments are significant, ongoing refinement of new approaches and models, such as that of CSF, brings hope for increasingly accurate representations of ALS, and with them, therapeutic breakthroughs [85, 91]. Table 15.4 gives a detailed list of AMs used in ALS.

Table 15.4 Animal models for amyotrophic lateral sclerosis

Sr No	Type of animal models	Mechanism	Specific outcome	Reference
1.	SOD1 transgenic mice	Overexpression of SOD1 gene with mutations responsible for degenerating motor neurons	Progressive motor deficits, muscle atrophy, and shortened lifespan, mimic ALS pathology	[87]
2.	TDP-43 model	Expression of mutated TDP-43 responsible for its misplacement and aggregation in neuronal cells	Motor neuron loss, myopathic weakness, and cytoplasmic inclusions similar to ALS	[93]
3.	FUS model	Expression of mutated FUS with its cytoplasmic aggregation and depletion in the nucleus	Motor dysfunction, neuronal loss, and inclusions positive for FUS in motor neurons	[93]
4.	BMAA model	Administration of β-N-methylamino-L-alanine (BMAA), a neurotoxin, inducing excitotoxicity	Motor neuron loss, impairments in behavior, and excitotoxicity through glutamate	[89]
5.	Kainic acid and excitotoxicity model	Injection of kainic acid to stimulate glutamate receptors, causing excitotoxic neuronal injury	Acute motor neuron loss, seizures, and neurodegeneration via overexcitation of neural circuits	[94]
6.	iPSC-derived motor neurons	Differentiation of induced pluripotent stem cells into motor neurons for studying mechanisms of ALS	Recapitulation of phenotypes for ALS, including protein aggregation and neuronal loss	[94]

Abbreviations: *SOD1* Superoxide Dismutase 1, *TDP-43* TAR DNA-Binding Protein 43, *FUS* Fused in Sarcoma, *BMAA* β-N-methylamino-L-alanine, *MAPK* Mitogen-Activated Protein Kinase, *ASO* Antisense Oligonucleotide, *iPSC* Induced Pluripotent Stem Cells, *C9orf72* Chromosome 9 Open Reading Frame 72

15.3 Animal Models for Other Brain Disorders

15.3.1 Depression

Depression is a serious mental illness that affects millions of people worldwide it is characterized by sadness, loss of interest in activities, and difficulty performing daily tasks [95]. The variability in patient responses to existing antidepressants like Tricyclic antidepressants (TCA), Selective serotonin reuptake inhibitors (SSRIs), and selective norepinephrine reuptake inhibitors (SNRIs) [95–98] highlights the urgent need for new treatments, making the AM essential for understanding depression and discovering more effective therapies. The most widely used AMs in depression research are integral to the progress that is being made in elucidating the disorder's mechanisms and possible treatments. Of these, the Chronic Unpredictable Mild Stress (CUMS) paradigm is particularly interesting, for it can simulate many of the behavioral and physiological aspects of depression seen in man [99]. Recent studies have demonstrated that electroacupuncture (EA) at the ST36 acupoint can effectively prevent astrocyte atrophy in the prefrontal cortex, which in turn reduces depressive-like behaviors in mice exposed to CUMS. This indicates that EA could be an effective nonpharmacological treatment for depression by restoring ezrin expression, highlighting the role of astrocytes as key targets in depression research [100].

In a related study, Song et al. examined the effect of NAc-DBS (nucleus accumbens deep brain stimulation) on depression-like behaviors and the role of dopaminergic neurons located in the VTA (ventral tegmental area), which is an important part of the brain's reward circuitry. Long-term stress can eventually lead to a reduction in dopaminergic function, anhedonia, and lack of motivation. NAc-DBS bypasses this problem by simply elevating GABA (Gamma-Aminobutyric Acid) in the VTA, disinhibiting, and making the dopaminergic neurons more excitable. The restoration of this normal function in the reward pathway can help to relieve depressive symptoms [101]. Similarly, scientists have discovered that when the olfactory bulbectomy (OBX) is performed on mice, it throws off the balance between GABA and glutamate neurotransmitter systems. The result is an overstimulation of the excitatory system, which then kills the cell. Also, the eradication of the OBX leads to decreased production of neurotrophic factors and increased levels of calcium, which leads to tau protein hyperphosphorylation, which is linked to neurodegeneration. The same study also found deficiencies in the acetylcholinergic system and hormonal imbalances in OBX mice [102]. There is a growing amount of evidence that depression is linked to inflammation and that targeting neuroinflammation could be a good way to treat depression. Transcutaneous auricular vagus nerve stimulation (taVNS) is a non-invasive method of stimulating the auricular branch of the vagus nerve, which plays a vital role in controlling inflammation and mood. taVNS has been shown to significantly decrease the expression of pro-inflammatory cytokines, and hence the depressive-like behaviors found by wang et al. It has been shown that the anti-inflammatory effect of taVNS is mediated through the α7 nicotinic acetylcholine receptors, which aid in the regulation of neuroinflammation. Also, taVNS upregulates the Janus kinase (JAK2) signaling pathway, which has a neuroprotective effect and increases the production of neurotrophic factors. The fact that taVNS can alter the rats' behavior suggests that taVNS directly affects their mood and behavior [103]. Similarly, a study using rats exposed to chronic restraint stress (CRS) found that chronic stress activates the TLR4 signaling pathway, triggering an inflammatory response that worsens depression symptoms. Electroacupuncture applied to the Baihui (GV20) and Yintang (EX-HN3) acupoints in these rats, however, led to better behavioral recovery and a significant suppression in the expression of TLR4 and another downstream signaling molecule, which in turn resulted in decreased pro-inflammatory cytokines and increased anti-inflammatory cytokines. This neuroinflammation reduction is associated with better behavioral results, meaning less depressive-like behaviors [104]. In a related study on Gulf War Illness (GWI), a chronic multi-syndrome disorder affecting approximately 25–32% of Gulf War veterans, researchers revealed that exposure to permethrin and environmental pollutants can alter behavior and brain function. Permethrin exposure stimulates immune cells and microglia, causing a neuroinflammatory reaction that impairs regular brain activity and intensifies mood disorders. Microglial morphology and function are altered by prolonged exposure to permethrin and stress, which reduces their capacity to maintain neuronal health and increases depressive-like behaviors in mice, providing valuable insights into biological mechanisms underlying GWI [105]. In another study, researchers investigated the impact of dexamethasone (DEX) on cognitive function and bioenergetic signaling pathways in a rat model of depression by administering it to pregnant rats during late pregnancy. The study specifically evaluated the effect of acute immobilization stress on the adult male offspring, focusing on brain chemistry indicators and cognitive function. The findings demonstrated a complex interaction between prenatal glucocorticoid exposure and stress responses, as evidenced by cognitive deficits, changes in orexin A and B levels, alterations in the AMPK-SIRT1-PGC1α signaling pathway, modification in the mineralocorticoid receptor/glucocorticoid receptor ratio, and elevated levels of HDAC5 in rats that received DEX during pregnancy [106]. Xie et al. studied the effect of Static Training (ST) on the AM and explored its mechanistic aspects through methods such as pole climbing, hind-limb suspension, isometric concentration weight bearing, and electrical stimulation. The study found that ST enhances muscle function, strength, and endurance. Additionally, it influences mood regulation by increasing serotonin and dopamine levels, potentially alleviating depressive symptoms. Overall, the findings indicate ST's therapeutic efficacy in promoting physical and mental health [107]. Depression is also associated with disrupted circadian rhythm. A study found that in the medial prefrontal cortex (mPFC) of a mouse model displaying depressive-like behaviors, induced Chronic Behavioral Despair Model (CDM), the expression of circadian negative loop regulators and positive clock regulators increased and decreased. The antidepressant ketamine reversed these clock disturbances. The effects of ketamine and depression-like symptoms depend on the functioning of the mPFC clock. While activating REV-ERB (Nuclear receptor) increased depressive-like behaviors and reduced the effectiveness of ketamine, silencing the Per2 (Period Circadian Protein 2) gene in the mPFC produced effects similar to those of antidepressants. By raising the expression of the plasticity protein Homer1a (Homer protein 1a), increasing the amounts of synaptic AMPA (α-amino-3-hydroxy-5-methyl-4-isoxazolepropionic acid) receptors, and encouraging slow wave activity, the positive clock modulator ROR (Retinoic Acid Receptor-Related Orphan Receptor) produced effects akin to those of antidepressants [108]. Additionally, the Sinking Platform Test (SPT) provides a valuable model for investigating persistence in the context of depression. In the SPT test, the mice were placed in a water-filled tank with a visible platform above the water level. They are trained to climb a platform and escape. Sometimes they have failed trials, in which they must find a new platform since the old one is flooded. The mice take part in a 5-minute failure trial phase following the training phase to gauge their persistence in trying to escape despite numerous failures.

Throughout the test, researchers counted the number of platforms they attempted to climb, which serves as an indicator of their persistence and motivation [109]. The response to treatment and stressors and their related outcomes are described in Table 15.5.

15.3.2 Anxiety

Approximately 34% of individuals in the US suffer from anxiety disorders at some point in their lives [110]. They include a range of conditions, such as obsessive-compulsive disorder, post-traumatic stress disorder, panic disorder, social anxiety disorder, and generalized anxiety disorder [111]. Excessive worry, dread, and

Table 15.5 Animal model responses to treatment and stressors in depression

Sr no	Treatment and stressors	Outcomes	Reference
1.	Electroacupuncture in chronic unpredictable mild stress (CUMS) mice	The treatment restored astrocytic morphology and improved behavioral outcomes in stressed mice	[100]
2.	Nucleus accumbens deep brain stimulation (NAc-DBS) in CUMS mice	NAc-DBS enhances dopaminergic neurons in the ventral tegmental area and increases GABA (Gamma-Aminobutyric Acid) levels	[101]
3.	Olfactory bulbectomy (OBX) mice	OBX mice can be used to study neurodegenerative diseases and depression	[102]
4.	Transcutaneous auricular vagus nerve stimulation (taVNS) in lipopolysaccharide (LPS) rats	taVNS enhances mood and behaviors in rats	[103]
5.	Acupuncture in chronic restrain stress (CRS) rats	Acupuncture improved depressive symptoms in rats exposed to CRS	[104]
6.	Chronic permethrin exposure in mice	Permethrin exposure increases depression-like behaviors in mice	[105]
7.	Acute immobilization stress (AIS) in rats	AIS significantly impaired cognitive function in rats	[106]
8.	Static training (ST) animal model	ST positively affects mood regulation and can potentially reduce depressive symptoms	[107]
9.	Chronic behavioral despair model (CDM) in mice	CDM disrupts circadian rhythm in mice, leading to depression-like behavioral changes	[108]
10.	Sinking platform test (SPT) in mice	The SPT measures persistence in mice and reveals behavioral changes associated with depression	[109]

Abbreviations: *CUMS* Chronic Unpredictable Mild Stress, *NAc-DBS* Nucleus accumbens deep brain stimulation, *OBX* Olfactory bulbectomy, *taVNS* Transcutaneous Auricular Vagus Nerve Stimulation, *LPS* Lipopolysaccharide, *CRS* chronic restrain stress, *AIS* Acute immobilization stress, *ST* Static Training, *CDM* Chronic Behavioral Despair Model, *SPT* Sinking Platform Test, *GABA* Gamma-Aminobutyric Acid

avoidance behaviors are the hallmarks of these diseases, and physical symptoms frequently accompany them [110]. Children and teenagers are also affected by anxiety disorders, which may result in long-term mental health issues [112]. Elevated Plus Maze, Vogel Conflict Test, Social Interaction Test, and Separation-Induced Vocalizations (SIV) are some models that can induce anxiety in rodents. To research anxiety disorders and create novel therapies, The AM is essential. Often used as a model for generalized anxiety disorder is the Carioca significant-freezing (CHF) rat line, which displays significant trait anxiety [113]. Another strategy is to socially isolate mice, which mimics anxiety-inducing circumstances in real life and increases aggressiveness and anxiety-like behavior [114]. Rats are also subjected to electric foot shocks to elicit anxiety-like symptoms; an ideal intensity of 1.0 mA has been discovered [115]. The pathogenic underpinnings of anxiety disorders, which entail intricate interactions between genetic, psychological, social, and environmental variables, may be studied by researchers owing to these models [116]. The Elevated Plus-Maze (EPM), Light/Dark Box (LDB), and Open Field (OF) tests are the most widely used models, and they are mostly conducted with rodents like rats and mice. These models measure exploration in anxiogenic vs non-anxiogenic situations to evaluate anxiety-related behaviors [117]. Furthermore, developmental models emphasize the significance of age in anxiety reactions, showing that young rats react differently to fear as adults [118]. Models of panic disorder concentrate on how people respond to dangers by using methods like electrical stimulation and exposure to real predators [119]. Numerous important insights into the underlying causes and possible therapies for anxiety disorders are highlighted by recent research on the AM of the condition. Research has demonstrated that increased excitability of GABA neurons in the ventral tegmental area (VTA) is associated with stress-induced anxiety in rats, indicating that targeting these neurons may be a useful treatment approach [120]. Moreover, research on the effects of long-term moderate noise exposure on anxiety-like behaviors has shown notable changes to the lateral amygdala's neuronal circuit [121].

Because of its unique behavioral and neurobiological traits, the Carioca substantial freezing rat line is a useful model for anxiety disorders. This model is useful for researching anxiety processes because it shows clear anxiety-like behaviors associated with malfunctioning fear circuitry and changes in important brain regions, especially in the Carioca High-conditioned Freezing (CHF) rats. Strong anxiety-like behaviors and a model for researching anxiety disorders are displayed by the Carioca High-Freezing (CHF) rat strain, which was carefully selected for high anxiety-like features [122]. Neurochemically, the glutamatergic system is altered in CHF rats, especially in the ventral hippocampus, where the expression of PSD-95, GluN1, and GluA1 subunits is up and that of GluN2A subunit is decreased [123]. These rats have abnormal fear circuitry, which includes changes to the serotonergic system and important brain regions. Carioca Low-Conditioned Freezing (CLF) rats exhibit hyperactivity and impulsivity, whereas CHF rats exhibit significant anxiety-like behaviors [122]. In auditory fear training, CLF rats show quick extinction while CHF rats show prolonged freezing responses, suggesting different anxiety characteristics [124]. Compared to CLF rats, CHF rats have lower levels of Fos expression in locations such as the prelimbic cortex and higher levels in parts of the brain linked to anxiety, such as the locus coeruleus. The CHF model selectively breeds rats for high-freezing behavior, especially in response to conditioned fear stimuli like foot shocks, resulting in a genetic predisposition towards heightened anxiety responses, enabling researchers to study neural and behavioral patterns associated with anxiety-like traits [125]. Elevated levels of corticosterone, the primary stress hormone in rodents, are observed in CHF rats, a model for studying stress-related disorders due to the hyperactivity of the hypothalamic-pituitary-adrenal axis [126].

A popular behavioral test for evaluating anxiety in rats is the elevated plus-maze (EPM), which focuses on the tension between the natural dread of open areas and the desire to explore. The EPM causes anxiety through several behavioral and neurobiological mechanisms. EPM is a behavioral test used to evaluate anxiety-like behaviors in mice and rats, utilizing their innate aversion to open spaces, which leads to more time spent in enclosed areas of the maze. The EPM is a plus-shaped apparatus with two open and closed arms, designed to test rodents' natural preference for dark, enclosed spaces due to their instinctual fear of predation and heights. The maze consists of two open and two closed arms arranged perpendicularly, with a central platform [127]. For five minutes, the rats' behavior is observed while positioned at the intersection of their arms, facing an open arm. Additional ethological parameters, such as rears and head dips, can also be observed [128]. Anxiety is measured by the number of entries and time spent in open versus closed arms, with increased open-arm activity indicating reduced anxiety. The raised plus maze is regarded as dependable, cost-effective, and able to yield legitimate effects in a single session [129]. This test takes advantage of rodents' innate dislike of open areas [130]. Rodents naturally avoid open arms due to the EPM's architecture, which reflects an underlying fear reaction rather than a taught behavior [131]. Rodents exhibit elevated anxiety levels during EPM testing by having higher heart rates in open arms and lower rates in closed arms. These behavioral reactions are correlated with these physiological changes, indicating a complex interaction between mental and physical states [132]. On the other hand, even if the EPM is a useful tool for measuring anxiety, other researchers contend that factors like past experience with the maze may affect how behaviors are interpreted, perhaps resulting in taught avoidance rather than natural anxiety reactions. This emphasizes how crucial it is to evaluate experimental circumstances when assessing rode anxiety carefully.

One well-known animal model for studying rodent anxiety is the Vogel conflict test, which focuses on the approach-avoidance conflict paradigm. This test assesses how rats make decisions when presented with an unpleasant stimulus (like foot shock) and a rewarding stimulus (like food). Animals may exhibit more anxiety-like behaviors as a result of the fundamental process of anxiety, which is the tension between the desire for a reward and the dread of possible punishment. The conflict that rats must balance between avoiding danger and pursuing rewards makes them more anxious [133]. The test gauges some behavioral traits, including vigor, delay, and response choice, which represent the rats' anxiety levels [134]. Despite its effectiveness, the Vogel conflict test has limits in its construct validity since it only captures the surface-level symptoms of anxiety rather than the underlying neurological causes [135]. According to research, the ventral medial prefrontal cortex (vMPFC) is essential for controlling anxiety reactions in the VCT. Punished licks increase when the vMPFC is inhibited with lidocaine or cobalt chloride, suggesting anxiolytic action [136]. Furthermore, because blocking NMDA receptors or inhibiting nitric oxide signaling has anxiolytic-like effects, the local glutamatergic and nitrergic systems in the vMPFC are engaged in anxiety regulation [137]. Interestingly, alprazolam can counteract the proconflict impact that a short carbon dioxide inhalation causes in the VCT, indicating a possible role for GABA-mediated transmission in anxiety reactions [138].

A popular technique for creating anxiety-like behaviors in rat models is restraint stress, which sheds light on the neurological processes that underlie anxiety disorders. This model is a useful tool for researching anxiety since it has been demonstrated to impact a variety of brain circuits and metabolic processes. Certain brain areas, like the amygdala and the medial prefrontal cortex (mPFC), which are essential for controlling anxiety, are activated by restraint stress. Enhanced neural activity linked to anxiety reactions is shown by increased c-Fos expression in certain regions [139]. Long-term stress from constraint raises circulating lipocalin 2 (LCN2), which interferes with cortical processes and causes anxiety-like behaviors [140]. The social interaction test and separation-induced vocalization (SIV) are prominent AMs used to study anxiety. These models simulate social stressors and measure behavioral responses, providing insights into the underlying mechanisms of anxiety disorders. Social Interaction Test follows the mechanism that evaluates social behavior by placing animals in a controlled environment where they can interact with a conspecific. Reduced interaction time indicates anxiety-like behavior. It has been used to evaluate the effectiveness of anxiolytic medications, showing that compounds such as mGluR antagonists can lessen mice's anxiety-related behaviors [141]. SIV involves isolating animals from their social group, leading to distress vocalizations. This model captures both panic responses and behavioral despair over time [142]. Research indicates that responses to separation stress are mediated by distinct brain circuits, especially in the anterior cingulate cortex [143].

15.3.3 Schizophrenia

A complex mental illness, schizophrenia is typified by a variety of symptoms that seriously hinder social, emotional, and cognitive functioning. It usually appears in late adolescence or early adulthood and affects about 1% of the world's population. Schizophrenia has a complex etiology that includes genetic, neurodevelopmental, and neurochemical elements. As explained in more detail in the next section, the primary symptoms can be divided into positive, negative, and cognitive symptoms. The main mechanism includes changes in the glutamate and GABA systems as well as dopaminergic dysfunction, especially in subcortical areas. The genesis of the condition has been linked to several genes, including DTNBP1 and NRG1, suggesting a substantial genetic component. Gray and white matter abnormalities are common in patients, and the changes become more noticeable after psychosis starts [144]. Hallucinations, delusions, and disordered thinking are examples of positive symptoms that indicate an excess or distortion of typical functions. Decreased emotional expressiveness, a lack of drive, and social disengagement are examples of negative symptoms that indicate a decline in typical emotional and behavioral processes [145]. Attention, working memory, and executive functioning impairments are prevalent and impact day-to-day functioning [144].

The AM is important in understanding the complex etiology of schizophrenia and in developing effective therapies. There have been various models of the illness developed, each offering a unique insight into the neurological mechanisms underpinning the disorder. The apomorphine-susceptible rat model exhibits aspects of dopaminergic dysfunction evident in schizophrenia, the symptoms of which are hyperlocomotion and disrupted prepulse inhibition (PPI). Spontaneous Hypertension (SHR) rats exhibit behavioral abnormalities and impairments in PPI that resemble schizophrenia patients. Brattleboro (BRAT)-reared rats present neurobehavioral features of schizophrenia, including cognitive flexibility deficits [146]. NMDA antagonist-using models regularly exhibit lower levels of parvalbumin (PV) in the hippocampus and prefrontal cortex, which is congruent with observations in individuals with schizophrenia [147]. The neurodevelopmental model of maternal immune activation results in behavioral and neuroanatomical alterations that are pertinent to schizophrenia [148].

An essential tool for researching schizophrenia is the NMDA receptor (NMDAR) antagonist model, which reflects the pathophysiological underpinnings of the disorder. By causing NMDAR hypofunction, this model simulates some symptoms associated with schizophrenia, such as cognitive impairments and psychotic behaviors. NMDAR antagonists such as phencyclidine (PCP) and dizocilpine (MK-801) cause NMDARs to become hypofunctional, which changes neurotransmission, especially in GABAergic interneurons and dopaminergic pathways [149]. Research indicates that NMDAR antagonism impairs the medial prefrontal cortex-hippocampus connection, which is essential for memory processing [150]. In models of schizophrenia, NMDAR antagonists also affect cerebral microcirculation, suggesting a more extensive effect on brain function [151]. Table 15.6 gives a detailed summary of AMs in schizophrenia.

15.3.4 Multiple Sclerosis

The study of MS is critical and the AM is the key because it allows scientists to understand the disease's complex mechanism and immune-mediated process that leads to neuroinflammation and demyelination. Because researchers can induce these diseases in animals, they can study the pathophysiology of MS and test treatment approaches. According to the researchers, one of the best animal models to simulate the autoimmune nature of MS is Experimental Autoimmune Encephalomyelitis (EAE), and they discovered that NF-κB (nuclear factor-kappa B) has a pivotal role in immunoregulation in the CNS. The deletion of IKKβ, a critical

Table 15.6 Animal models for schizophrenia

Sr no	Type of animal model	Mechanism	Specific outcome	Reference
1.	NMDA receptor antagonist	Causes NMDA receptor hypofunction, simulating the symptoms of schizophrenia	Behavioral changes such as hyperlocomotion and cognitive deficits	[152]
2.	Dopaminergic dysfunction	Positive symptoms are modeled by increased dopamine production in the dorsal striatum	Hyperlocomotion, elevated dopamine release, and compromised pre-pulse inhibition (PPI)	[153]
3.	DISC1 mutant	Neurodevelopment and synaptic function are impacted by the genetic disruption of DISC1	Cognitive impairment, PPI changes, and schizophrenia-like behavioral abnormalities	[154]

Abbreviations: *NMDA* N-Methyl-D-Aspartate, *PPI* Pre-Pulse Inhibition, *DISC1* Disrupted in Schizophrenia 1, *DA* Dopamine

kinase for the NF-κB pathway, considerably impacted the development of EAE. Earlier onset and greater severity of the disease were observed in CNS macrophages lacking IKKβ, indicating that NF-κB signaling in these cells guards against neuroinflammation and demyelination. Deletion of IKKβ changed some intrinsic CNS processes (that is, microglial activation and neuronal excitability) but did not seem to affect any peripheral immunological responses [155]. In a related study, researchers found that certain LRRK2 (Leucine-Rich Repeat Kinase 2) inhibitors could be used to treat demyelinating disease because they encourage remyelination. This inhibitor greatly promoted the proliferation and differentiation of oligodendrocyte precursor cells (OPCs) in an EAE mouse model. The study demonstrated that treatment with an LRRK2 inhibitor reduced clinical severity and improved motor performance in the affected mice [156]. More and more researchers are utilizing the AM as a tool for understanding the workings of MS, and it shows great promise in the field of such a complicated disease. One study involved transplanting peripheral blood mononuclear cells (PBMCs) from patients with MS into B2m-NOG mice, which then displayed immune responses akin to those observed in human MS. The results showed that human T cells, specifically CD8 T cells, migrate into the CNS, which characterizes MS. The infiltration, which is vital to neuroinflammation, results in the formation of inflammatory plaques in both the brain and spinal cord [157]. In a study, mice were exposed to systemic inflammation through lipopolysaccharide injections, which led to significant changes in the localization of Kallikrein 6 (KLK6) in oligodendrocytes (OD). During inflammation, KLK6 shifted to paranodal regions, indicating its role in the inflammatory response and demyelination process. Additionally, KLK6 expression in spinal cord white matter increased, indicating its involvement in the structural changes occurring in the nervous system [158]. Also, lysophosphatidylcholine (LPC) was used to develop a new AM of MS by injecting it into the corpus callosum and cerebellar white matter. This unique, dual injection site permitted researchers to fully study the effect of demyelination while still modelling the clinical manifestations of MS, including motor and bladder deficits. The LPC model allowed us to study remyelination progress, where oligodendrocyte precursor cells divide, move to the areas of injury, differentiate into mature OD, and finally remyelinate axons [159]. Along with this model, the researchers also developed a computerized bioinformatic imaging method to examine the loss and repair of grey matter in old mice. This method offers an accurate and unbiased measure of cellular alterations. They found in old mice with cuprizone-induced demyelination, remyelination was impaired, which points to local reasons. The results also demonstrated how glial cell activation, specifically that of astrocytes and microglia, plays a role in the process of remyelination. Unlike white matter structures, significant astrogliosis was temporary, but microgliosis continued into the remyelination phase [160]. Moreover, the utilization of N-formyl methionyl-leucyl-phenylalanine (fMIP)-modified nano-carriers for achieving C16-angiopoietin 1 (C + A) transmission via blood-circulation into the CNS was mapped in the Lewis rat EAE model. The data indicated that compared with traditional methods, this new delivery system significantly upregulates the therapeutic outcomes. Specifically, the delivery capacity of fMIP-nano-carriers to C + A in the CNS was impressive, which contributed as an essential factor for MS treatment. Furthermore, in this direction, through targeting, inflammation is decreased, axonal damage occurs, and the blood-brain barrier (BBB) is strengthened. Unlike other ways of administration, the fMIP-nano-carriers expedited the solubility and bioactivity of C + A, which increased endothelial uptake in the BBB gap and decreased BBB inflammation in the MS. In addition, the apoptosis rate of neurons and demyelination demonstrated remarkable interjection selection in the fMIP-nano-carriers treatments as compared to that in the traditional system. Thus, fMIP-nano-carriers technology showed a positive result in MS treatment [161].

15.3.5 Migraine

Migraine is a neurovascular, debilitating disease usually manifested by paroxysmal headaches. It is most of the time accompanied by phonophobia, photophobia, and nausea. Current research underlines calcitonin gene-related peptide and neuroinflammation as part of its etiology. A grasp of these pathways is important for developing successful treatments. Neuroinflammation has been increasingly recognized as an important factor in migraine pathogenesis, and particularly, increased levels of cytokines such as TNF-α, IL-1β, and IL-6 during migraine attacks have been observed. According to some studies, cortical spreading depolarization (CSD) may extend the migraine cycle by setting up inflammatory reactions [162]. Glial cells, such as astrocytes and microglia, are important in this process, especially in the trigeminal nucleus caudalis, which processes pain [163]. A neuropeptide called CGRP plays a key role in migraine attacks by mediating neurogenic inflammation and vasodilation [164]. Monoclonal antibodies and gepants, two CGRP-targeted treatments, have demonstrated effectiveness in lowering the frequency and intensity of migraine attacks [165].

The AM presents various employments in migraine research toward the understanding of the biology of the disease and the creation of possible remedies. Each one of these hypotheses describes a different aspect of migraine mechanisms, ranging from chemical stimulation to genetic changes. Chemical Models Inducing Nitro-glycerine: Nitro-glycerine is considered one of the most common drugs for the induction of chronic migraine models in mice with obvious symptoms of photophobia and mechanical pain hypersensitivity [166]. Models of Calcitonin Gene-Related Peptides (CGRP): Its importance in migraine pathogenesis is highlighted because CGRP injection in mice has been demonstrated to produce migraine-like symptoms [167]. Models of Familial Hemiplegic Migraine (FHM): Researchers can examine the molecular mechanisms behind migraine vulnerability using genetic mouse models that possess mutations linked to FHM, with a specific emphasis on cortical spreading depression [168]. The Cortical Spreading Depression model is a common model used to investigate the neurophysiological alterations that occur during migraine attacks and replicate the aura phase of migraine [169]. Chemical Irritant Models that apply irritants to the dura mater, such as inflammatory soup and Complete Freund's Adjuvant, help mimic both acute and chronic migraine symptoms [170]. The role of the trigeminovascular system in migraine is the subject of trigeminal activation models, which shed light on neurogenic inflammation and pain mechanisms [171].

The nitroglycerin-induced animal model of migraine has gained wide acceptance as a generally acceptable experimental framework for simulating human migraine pathophysiology by producing features of migraine-like headaches. It is exceptionally useful in chronological studies aimed at the etiology of chronic migraine and testing potential therapies. Nitroglycerin will be used in this procedure to induce migraine-like symptoms in rats; behavioral abnormalities described include cephalic allodynia, enhanced pain sensitivity, and anxiety-like behavior. NTG induces activation specifically in those brain regions implicated in pain processing, such as the somatosensory cortex and trigeminal nucleus caudalis, thereby representing an important model system for migraine pathophysiology research. Glial cells may be implicated in migraine pathogenesis since astrocyte activation was increased following NTG [172]. The NTG model is a useful tool for drug testing and discovery since it accurately mimics the features of human migraines. When predicting the effectiveness of antimigraine treatments, it has demonstrated good sensitivity and specificity [173]. Table 15.7 gives detailed information about AMs used for MS and migraine.

15.3.6 Alcohol Use Disorder (AUD)

AUD is a significant public health challenge requiring innovative research approaches. Recent AM studies have provided clinical insights into the neurological mechanisms underlying AUD. A previous study found that prolonged exposure significantly affects the expression and activity of serotonin (5-HT) and dopamine (DA) receptors in two mice strains, CBA (Control Balb/c Anemia) and ASC

Table 15.7 Animal models for multiple sclerosis and migraine

Sr no	Disease	Name of animal model	Specific outcomes	Reference
1.	Multiple sclerosis (MS)	Experimental Autoimmune Encephalomyelitis (EAE) Mice Model	Understanding the specific functions of NF-κB in EAE could offer valuable insights into the cognitive decline observed in MS	[155]
2.	MS	EAE Mice Model	The EAE model revealed that treatment with Leucine-Rich Repeat Kinase 2 inhibitors in mice has shown potential for repairing myelin damage associated with MS	[156]
3.	MS	B2m-NOG Humanized Mice Model	The B2m-NOG mice engrafted with peripheral blood mononuclear cells from patients with MS exhibited an immune response similar to that of humans with MS	[157]
4.	MS	KLK6-KO (Kallikrein 6 Knockout) Mice Model of Systemic Inflammation	Lipopolysaccharide injections increased Kallikrein 6 (KLK6) expression in the spinal cord. KLK6-KO mice did not exhibit changes in nodal and paranodal structures, whereas wild-type mice did	[158]
5.	MS	LPC (lysophosphatidylcholine)-Induced Demyelination Model in Mice	The LPC-induced model effectively replicates the key cellular characteristics of MS, including remyelination	[159]
6.	MS	Cuprizone-Induced Model in Aged Mice	Automated image analysis improves understanding of demyelination and remyelination mechanisms in aged mice	[160]
7.	MS	EAE Model Using Lewis Rat	N-formyl methionyl-leucyl-phenylalanine (fMIP)-modified nano-carriers reduce inflammation and neuronal damage in rats with MS	[161]
8.	Migraine			
A.	Nitroglycerin induced	Releases nitric oxide, which causes an episode similar to a migraine	Neuronal and behavioral results in rodents	[174]

(continued)

Table 15.7 (continued)

Sr no	Disease	Name of animal model	Specific outcomes	Reference
B	Calcitonin gene-related peptide (CGRP)	Both cerebral and peripheral sensitization result from CGRP release	Symptoms were similar to migraines in test participants	[175]
C.	Cortical spreading depression	Neurovascular alterations result from spreading depolarization	Disrupted sensory processing and aura	[176]
D.	Chemical irritant model	Headache pathways are triggered by chemical irritants	Both sensory and cognitive inputs are altered	[177]
E.	Trigeminal activation	Trigeminal pathway activation results in migraine symptoms	Headache and pain responses induced	[178]

Abbreviations: *EAE* Experimental Autoimmune Encephalomyelitis, *NF-κB* Nuclear Factor-Kappa B, *KLK6* Kallikrein 6, *LPC* Lysophosphatidylcholine, *B2m-NOG* Beta-2 Microglobulin Knockout NOG, *CGRP* Calcitonin Gene-Related Peptide

(Antidepressant Sensitive Catalepsy). ASC mice (which are genetically prone to depression-like behavior) displayed unique variations in receptor activity as opposed to the CBA mice. Administration of alcohol resulted in a downregulation of the mRNA levels for the 5-HT2A receptor in the frontal cortex of CBA mice, while ASC mice displayed an altered expression of both 5-HT and DA receptor gene expression. A MASH diet combined with alcohol intake has been shown to worsen liver injury and inflammation, of which the latter is associated with AUD. Also, it has been proven that a high-fat diet with a lot of drinking seriously screws up the liver of a male C57BL/6J mouse. The MASH diet, which is a diet high in fats, cholesterol, and sugars, when combined with daily alcohol intake, resulted in a large amount of liver damage when compared to either treatment alone. These results corresponded with significantly raised levels of liver injury markers and liver enzymes, as well as inflammatory chemokines, plus augmented neutrophil infiltration, all pointing to a severe inflammatory response [179]. Additionally, the study examined the function of oxytocin (OT) in AUD in adult male baboons to determine its effect on alcohol intake. Well, this hasn't been very widely researched in this particular AM. The study found that administering OT reduced alcohol intake, suggesting that it can modulate drinking behavior by disrupting the learned association between alcohol and its rewarding effect. The findings also provide insights into how OT interacts with the brain reward system, particularly the mesolimbic dopamine system, which is essential for processing reward and reinforcing behaviors, including alcohol consumption. By affecting this system, OT could reduce the reinforcing effects of alcohol, leading to decreased consumption [180]. Another study looked at the impact of psilocybin (PSIL) on AUD in female Wistar rats. PSIL changed the connectivity in the default mode network (DMN), which is the self-referential thought network. The results revealed that PSIL induced a decrease in functional connectivity within the DMN, which correlated with the severity of alcohol relapse in rats. PSIL's effect may be caused by its affinity for serotonin receptors, namely 5-HT2A, which regulate mood and cognitive activity. This engagement could cause neuroplasticity and aid in rehabilitation from addiction [181]. Another study investigated the efficacy of nutritional ketosis (NT) as a therapeutic approach for AUD using female C57BL/6J mice as an AM. Research has found that ketogenic diets and ketone esters can help alleviate withdrawal symptoms from alcohol dependence. The diet reduces depressive-like behaviors in mice and improves neurochemical balance by restoring levels of neurotransmitters disrupted in individuals with alcohol dependence. Alcohol consumption leads to a metabolic shift in the brain, characterized by increased acetate metabolism and decreased glucose metabolism, resulting in energy depletion and negative emotional states. The ketogenic diet promotes the production of ketone bodies, which serve as an alternative energy source to the brain. The benefits of NT are not solely dependent on high ketone levels but could fuel various metabolic pathways, improving brain function and mood [182]. Additionally, Alirocumab, a PCSK9 (Proprotein Convertase Subtilisin/Kexin Type 9) inhibitor, presents a novel therapeutic strategy for managing AUD that reduces inflammation and oxidative stress in the brains of rats. The researchers found that Chronic alcohol exposure significantly increases PCSK9 levels in the brain, thereby elevating oxidative stress and neuroinflammation. Alirocumab treatment reduces oxidative stress markers and inflammation in the brain while simultaneously increasing the expression of low-density lipoprotein receptors (LDLR), which regulate cholesterol levels. This implies that PCSK9 has significant ramifications for the pathophysiology of AUD because PCSK9 inhibition allows LDLR recycling, which in turn promotes greater removal of low-density lipoprotein cholesterol from the bloodstream, which can, in turn, decrease its buildup in the brain [183]. Furthermore, another study using a C57BL/6J mouse model revealed notable disparities in alcohol consumption patterns between male and female mice due to acute postoperative pain. In an experiment, male mice drank much more alcohol, following laparotomy (a surgical procedure causing acute pain), than they did before, but female mice did not drink more following the procedure. This discovery shows that there is a sex difference in pain reactivity and how it relates to alcohol consumption. This suggests that the sensation of pain changes the mice's reaction to alcohol, possibly for the better in terms of the rewards that alcohol brings. Also, Pain can affect the development of functional tolerance to alcohol, that is, as time goes by, the mice would need increasingly larger doses of alcohol to achieve the same effect [184].

This study provides valuable insights into the effect of acute pain on alcohol sensitivity and consumption, indicating that individuals with a history of pain may be at a higher risk of developing AUD.

15.3.7 Substance Use Disorder

The chronic illness known as substance use disorder (SUD) is typified by obsessive drug seeking and use that causes severe impairment or suffering. The intricate neurobiological alterations and behavioral patterns that underlie addiction are mostly caused by reward processes. It is essential to comprehend these systems to create solutions that work. Reinforcement Learning: Goal-directed conduct frequently gives way to habitual behavior in SUDs, when impulsive choices put short-term gains ahead of long-term effects [185]. Drug usage reduces unpleasant emotional states, such as withdrawal sensations, and addiction may be thought of as a shift from impulsivity (positive reward) to compulsivity (negative reinforcement). Compulsive drug-seeking behavior is exacerbated by dysregulation in reward pathways (e.g., reduced dopamine) and stress systems (e.g., increased corticotropin-releasing factor) [186].

The AM is essential for comprehending substance use disorders (SUD), especially when examining the neurological and behavioral processes that underlie addiction. Treatment tactics have advanced as a result of the development of several models, each with its applications and methodology. Forced Exposure is the system that uses alcohol vapor inhalation to cause dependency. Voluntary Consumption is done by giving rats an option between water and alcohol; the two-bottle choice technique mimics how people drink [187]. The Self-Administration Model is a model where animals deliberately give themselves medications in an attempt to replicate the actions of human addicts. Environmental signals or stress might have an impact on them [188]. Addiction-like behaviors in rodents are used to investigate behaviors like drug-seeking, resilience to extinction, and a preference for drugs over natural rewards. Models have been created [189].

Researchers can examine the reinforcing qualities of substances by using the self-administration model, a crucial experimental technique for studying substance use disorders. In this paradigm, animals usually replicate human drug-seeking behavior by engaging in particular activities to get drug infusions. By surgically inserting a catheter into the jugular vein, animals may self-administer medicines like cocaine by engaging in operant behaviors like pulling levers or sticking their noses. This technique is known as intravenous self-administration or IVSA [190]. The Operant Social Self-Administration approach emphasizes the relevance of social aspects in addiction by including social rewards and giving animals the option to choose between drug administration ion and social engagement [191]. Table 15.8 provides information about AMs in SUD and AUD.

15.4 Future Perspectives for Animal Models (AM)

In this chapter, we explored the relationship between NDDs and different AMs implicated in these disorders. Exploration of depression in AMs serves as a critical avenue for understanding the disorder's complex pathophysiology. Animal model's depression shares similarities and differences in pathophysiological findings.

Table 15.8 Animal models for SUD and AUD

Sr no	Disease	Name of animal model	Specific outcomes	Reference
1.	Alcohol use disorder (AUD)	Antidepressant Sensitive Catalepsy (ASC) and CBA (Control Balb/c Anemia) Mice Model, in AUD	The study revealed that prolonged alcohol exposure in CBA and ASC mice affected serotonin and dopamine receptor activity and locomotor activity, indicating that genetic predisposition affects alcohol response	[192]
2.	AUD	MASH Diet and Alcohol-Induced Mice Model	The MASH diet and alcohol combination significantly increased liver injury, intensified inflammation, and contributed to liver damage in mice	[179]
3.	AUD	Oxytocin (OT) Effects on Alcohol Consumption in Adult Male Baboons	OT significantly reduces alcohol intake in baboons and positively influences the mesolimbic dopamine pathway	[180]
4.	AUD	Psilocybin (PSIL) Effect on Female Wistar Rats Model	PSIL has potential neuroplasticity and addiction recovery mechanisms.	[181]
5.	AUD	C57BL/6J Mice Model for Investigating Nutritional Ketosis (NT) in AUD	The NT diet restored neurotransmitters that were disrupted by chronic alcohol exposure in mice and enhanced energy metabolism in the brain	[182]
6.	AUD	PCSK9 (Proprotein Convertase Subtilisin/ Kexin Type 9) Inhibition in AUD Rat Model	The results indicate that PCSK9 inhibition can protect neuronal health in AUD	[183]
7.	AUD	C57BL/6J Mice Model of Laparotomy Surgery	This study indicates that acute pain in male mice can enhance the rewarding effects of alcohol, potentially leading to increased consumption and altered alcohol sensitivity	[184]
8.	Substance use disorder			
A.	Self-administration model	Drug-seeking behavior is evaluated via operant conditioning	Increased drug use and resumption following cessation	[193]
B.	Conditioned place preference (CPP)	Learning via association between environmental signals and drug exposure	Preference for contexts with drugs, suggesting a reward link	[194]

(continued)

Table 15.8 (continued)

Sr no	Disease	Name of animal model	Specific outcomes	Reference
C.	Forced exposure model	Pharmacological exposure that is not continuous to evaluate neurobiological alterations	Addiction-related behavioral sensitization and neuroadaptations	[195]
D.	Withdrawal models	Evaluation of withdrawal symptoms after stopping drugs	Increased drug-seeking behavior, hyperalgesia, and anxiety-like behaviors	[196]

Abbreviations: *AUD* Alcohol Use Disorder, *ASC* Antidepressant Sensitive Catalepsy, *CBA* Control Balb/c Anemia, *PSIL* Psilocybin, *NT* Nutritional Ketosis, *PCSK9* Proprotein Convertase Subtilisin/Kexin Type 9

Neurotransmitter imbalance is common with altered serotonin and dopamine levels observed across models. Chronic stress is a common factor in depression models, leading to neuroinflammation and changes in brain structure and function. Different models may highlight distinct mechanisms underlying depression, such as prefrontal cortex astrocyte atrophy and ventral tegmental dopaminergic dysfunction. Genetic variability is also present in some models, such as genetically predisposed mice, which show unique receptor activity changes in response to stressors or treatments. The effectiveness of treatments can vary significantly across models, with DBS having varying effects on depressive symptoms depending on the specific model used. Similarly, MS is studied using various animal models, revealing similarities and differences in pathophysiological findings. The EAE model mimics the autoimmune characteristic of MS, demonstrating significant neuroinflammatory responses and demyelination. Behavioral changes, such as motor deficits and cognitive impairments, are also observed in EAE. However, specific modes reveal differences in pathophysiological findings. The LPC-induced demyelination model replicates key cellular characteristics of MS by directly inducing demyelination in the CNS, highlighting the intrinsic repair mechanism of oligodendrocytes and the potential for remyelination. The cuprizone-induced model is useful for studying demyelination in aged mice, allowing researchers to investigate age-related factors in MS progression. The fMlP-modified nano-carrier EAE model demonstrates how specific interventions can alter the disease course, showing that targeted therapies can reduce inflammation and neuronal damage, indicating that the therapeutic response can vary significantly depending on the model used. Furthermore, various AMs have been used to study HD. The R6/1 and zQ175 transgenic mouse models exhibit motor deficits and cognitive impairments similar to HD. Moreover, they showed the accumulation of mHTT protein, leading to cellular dysfunction and apoptosis. Neuroinflammation and glial activation are common features, indicating that the neuroinflammatory process plays a significant role in HD progression. However, the R6/1 model exhibits more pronounced motor deficits and earlier symptom onset than the zQ175 model. The 3-NPA model provides a different perspective on HD pathophysiology by inducing HD-like symptoms through mitochondrial dysfunction, highlighting the role of energy metabolism and oxidative stress in HD progression. The EP treatment model revealed that therapeutic interventions yielded varying results across the model, highlighting the importance of selecting an appropriate model for testing potential therapies. Similarly, the AUD AM exhibits different pathological findings. Chronic alcohol exposure causes notable alterations in the expression and activity of neurotransmitter systems, especially serotonin and dopamine, which are constantly seen in AUD models. Increased levels of inflammatory markers and cytokines are indicative of inflammatory responses, which are linked to neuroinflammation and liver damage. This finding underscores the significance of neuroinflammatory processes in the pathophysiology of AUD. Genetic predisposition also plays a role in the varying responses to alcohol. ASC mice with a genetic predisposition to depressive-like behavior exhibit distinct changes in receptor activity compared to control strains like CBA mice. The mechanism of liver injury also contributes to the study of liver pathology related to AUD. However, therapeutic responses to potential interventions can vary significantly across AUD models, with studies involving OT in adult male baboons indicating a potential therapeutic avenue for AUD. Overall, these discrepancies highlight the complexity of NDDs and indicate that no single model can fully encapsulate the multifaceted nature of diseases. Understanding these similarities and differences is crucial for developing targeted therapies and advancing our knowledge of NDD pathophysiology.

Animal models discussed for CNS neurodegenerative and psychiatric disorders have indeed provided valuable insights into the multifaceted biological and behavioral processes underlying these diseases. Each one, in this regard, offers unique views on the mechanisms of the disease, symptomatology, and therapeutic potentials. For instance, the transgenic mice and neurotoxin-induced models of Alzheimer's and Parkinson's diseases point to critical pathways in amyloid and α-synuclein aggregation, respectively, while underlining the value of molecular precision in mimicking human pathology. Animal models of anxiety, including the Elevated Plus Maze and Vogel Conflict Test, have pointed out the role of neurochemical modulation in stress-related behaviors, thus illustrating how alterations in GABAergic and serotonergic systems contribute to anxiety phenotypes.

The use of genetic and pharmacological models in Substance Use Disorder and Schizophrenia further supports the involvement of specific neurotransmitter systems, including the dopamine system and NMDA receptor pathways, in the disease etiology. These models, if using illustration, underline powerful frameworks enabling investigation into the mechanisms of addiction and cognitive deficits that are very important to understand in the context of relapse and resistance to treatment. Finally, migraine models that study mechanisms such as CGRP and neuroinflammation further help to elucidate neurovascular contributions to pain pathways and provide new points of intervention in the management of chronic migraine.

15.5 Ethical Considerations in CNS Disease Modeling

Ethical considerations in modeling CNS disease should focus on welfare concerns and the 3Rs: replacement, reduction, and refinement. Researchers should aim to replace AMs with alternative methods, such as in vitro systems or computational and mathematical models, to minimize the use of animals while achieving statistically significant results. In this context, Li et al. have developed an advanced mathematical modeling technique to investigate AUD. This model integrates fractal and fractional calculus to better understand the complex dynamics of alcohol addiction. The model revealed that the behavior of heavy drinkers significantly affects potential drinkers, emphasizing the importance of understanding these interactions when developing effective treatment strategies [197]. This approach allows researchers to capture behaviors and interactions that traditional models may overlook. Similarly, the development of other computational and mathematical models will help elucidate NDDs, further minimizing the reliance on animal models. Refinement involves improving animal welfare through improved housing, handling, and experimental procedures. Regulatory and institutional guidelines include compliance with relevant regulations, institutional review by institutional animal care committees, and transparency in reporting methodologies. The ethical issue in modeling disease in the CNS, and in studies with animals in general, involves a high level of demand for ethical approval and following stringent guidelines. Foremost is a consideration for animals' welfare, supported by the 3Rs principle—replacement, reduction, and refinement—and its objective of minimizing misery and maximizing ethics in studies [198]. Regulatory frameworks for compliance have been constructed, and alternative approaches, such as the use of brain organoids, have been encouraged to minimize the use of animals. All work with animals must have ethical approval for compliance with welfare requirements. Guidelines are prepared and executed by institutes and governing agencies for controlling the humane use of animals in experiments [199]. The 3Rs Principle, which includes Replacement involving alternative strategies for not utilizing animals, Reduction to minimize the use of animals in experiments, and Refinement that enhances the processes for minimizing pain and distress, is widely recommended [200]. Human brain organoids and in vitro approaches are in development for application in a quest to replace animal models, overcoming ethical concerns and providing relevant information [199]. While the ethical basis for researching animals is strong, an alternative view is that concern for animals' welfare can hinder scientific progression, with scientists potentially less willing to undertake studies that involve animals [201]. In the USA, the Animal Welfare Act (AWA) regulates the use of animals in experiments, with a requirement for an Institutional Animal Care and Use Committee (IACUC) responsible for safeguarding animals' welfare in institutions. The National Institutes of Health (NIH) releases supplementary guidance with a strong focus on the 3Rs: Replacement, Reduction, and Refinement of the use of animals in experiments. For the UK, the Animals (Scientific Procedures) Act 1986 mandates that any work with animals must be licensed and that ethical review processes must be conducted. The UK adheres to European Union laws, such as high welfare requirements and justification for animals in experiments [202]. In India, the Committee for the Purpose and Control and Supervision of Experiments on Animals (CPCSEA) regulates experiments with animals under ethics and welfare standards. Indian guidelines prefer 3Rs, promoting alternatives to testing wherever possible [203]. The OECD and WHO provide key guidance that helps harmonize welfare standards for animals, and in supporting nations in collaboration towards improving welfare and care for animals in a laboratory setting, the OECD and WHO also prioritize strong consultation with stakeholders in developing welfare guidance that covers specific region-by-region requirements [204]. The Federation of European Laboratory Animal Science Associations (FELASA) offers in-depth training in ethical care and the use of laboratory animals. Early-career researchers benefit from structured courses in having a heightened awareness of ethics and conduct in research [205]. Advancements in organoids, computational simulations, and in vitro platforms have real alternatives for traditional animal models, potentially minimizing ethical concerns with animals [206].

15.6 Conclusion

Animal models have played a critical role in extending our present understanding of CNS neurodegenerative disease, offering significant information about disease pathology, disease progression, and therapeutic intervention potential. Through them, researchers have been in a position to identify molecular mechanisms, evaluate the efficacy of drugs, and explore new therapeutic modalities. Despite not being in a position to replicate full disease complexity in humans, continued model refinement, in combination with new tools including iPSCs and computational modeling, holds promise for closing gaps in translation. Looking ahead, integration with state-of-the-art genetic and imaging tools will make them even more predictive, and with it, ever more personalized and targeted therapeutic interventions will follow. Integration between studies in animals, humans, and artificial intelligence-guided analysis will drive the rapid discovery of new therapies and improved patient care. Ultimately, the continued development of animal models, in conjunction with computational technology and patient-derived cellular platforms, will be responsible for revolutionizing our care and understanding of CNS disease. Together, these will pave the way for new, focused, and personalized therapeutic interventions, bringing us ever nearer to lessening the burden of neurodegenerative disease in global healthcare.

References

1. Höglund K, Salter H. Molecular biomarkers of neurodegeneration. Expert Rev Mol Diagn. 2013;13:845–61. https://doi.org/10.1586/14737159.2013.850033.
2. Stephenson J, Nutma E, van der Valk P, Amor S. Inflammation in CNS neurodegenerative diseases. Immunology. 2018;154:204–19. https://doi.org/10.1111/imm.12922.

3. Archer T, Kostrzewa RM, Beninger RJ, Palomo T. Staging neurodegenerative disorders: structural, regional, biomarker, and functional progressions. Neurotox Res. 2011;19:211–34. https://doi.org/10.1007/s12640-010-9190-2.
4. Nguyen HD. Resveratrol, endocrine disrupting chemicals, neurodegenerative diseases and depression: genes, transcription factors, microRNAs, and sponges involved. Neurochem Res. 2023;48:604–24. https://doi.org/10.1007/s11064-022-03787-7.
5. Jakimovski D, Bittner S, Zivadinov R, Morrow SA, Benedict RH, Zipp F, Weinstock-Guttman B. Multiple sclerosis. Lancet. 2024;403:183–202. https://doi.org/10.1016/S0140-6736(23)01473-3.
6. Harroud A. The genetic basis of multiple sclerosis severity suggests central nervous system involvement (S5.005). Neurology. 2022;98 https://doi.org/10.1212/WNL.98.18_supplement.3650.
7. Xiao F, He Z, Wang S, Li J, Fan X, Yan T, Yang M, Yang D. Regulatory mechanism of circular RNAs in neurodegenerative diseases. CNS Neurosci Ther. 2024;30 https://doi.org/10.1111/cns.14499.
8. Crews FT, Sarkar DK, Qin L, Zou J, Boyadjieva N, Vetreno RP. Neuroimmune function and the consequences of alcohol exposure. Alcohol Res. 2015;37(331–41):344–51.
9. Wu Y, Ma B, Liu C, Li D, Sui G. Pathological involvement of protein phase separation and aggregation in neurodegenerative diseases. Int J Mol Sci. 2024;25:10187. https://doi.org/10.3390/ijms251810187.
10. Global, regional, and national burden of 12 mental disorders in 204 countries and territories, 1990–2019: a systematic analysis for the Global Burden of Disease Study 2019. Lancet Psychiatry. 2022;9:137–50. https://doi.org/10.1016/S2215-0366(21)00395-3.
11. Wilhelm M, Bauer S, Feldhege J, Wolf M, Moessner M. Alleviating the burden of depression: a simulation study on the impact of mental health services. Epidemiol Psychiatr Sci. 2024;33:e19. https://doi.org/10.1017/S204579602400012X.
12. Walton C, King R, Rechtman L, Kaye W, Leray E, Marrie RA, Robertson N, La Rocca N, Uitdehaag B, van der Mei I, Wallin M, Helme A, Angood Napier C, Rijke N, Baneke P. Rising prevalence of multiple sclerosis worldwide: insights from the atlas of MS, third edition. Mult Scler J. 2020;26:1816–21. https://doi.org/10.1177/1352458520970841.
13. Wang H, Zhang X, Li H, Sun Z, Zhong Y. Gender differences in the burden of multiple sclerosis in China from 1990 to 2019 and its 25-year projection: an analysis of the Global Burden of Diseases Study. Health Sci Rep. 2023;6 https://doi.org/10.1002/hsr2.1738.
14. Novak MJU, Tabrizi SJ. Huntington's disease. BMJ. 2010;340:c3109. https://doi.org/10.1136/bmj.c3109.
15. Rodríguez-Santana I, Mestre T, Squitieri F, Willock R, Arnesen A, Clarke A, D'Alessio B, Fisher A, Fuller R, Hamilton JL, Hubberstey H, Stanley C, Vetter L, Winkelmann M, Doherty M, Wu Y, Finnegan A, Frank S. Economic burden of Huntington disease in Europe and the USA: results from the Huntington's Disease Burden of Illness study. Eur J Neurol. 2023;30:1109–17. https://doi.org/10.1111/ene.15645.
16. Kalivas PW. The glutamate homeostasis hypothesis of addiction. Nat Rev Neurosci. 2009;10:561–72. https://doi.org/10.1038/nrn2515.
17. Alhaddad H, Wong W, Abou-Gharbia M, Childers W, Melenski E, Bell RL, Sari Y. Effects of a novel Beta lactam compound, MC-100093, on the expression of glutamate transporters/receptors and ethanol drinking behavior of alcohol-preferring rats. J Pharmacol Exp Ther. 2022;383:208–16. https://doi.org/10.1124/jpet.122.001147.
18. Kranzler HR. Overview of alcohol use disorder. Am J Psychiatry. 2023;180:565–72. https://doi.org/10.1176/appi.ajp.20230488.
19. Hall AM, Roberson ED. Mouse models of Alzheimer's disease. Brain Res Bull. 2012;88:3–12. https://doi.org/10.1016/j.brainresbull.2011.11.017.
20. Dhanasekaran S, Arunadevi R, Nandan M, Chandrababu K, Baby Chakrapani PS, Sivan U. Animal model for Alzheimer's disease, Parkinson's disease, And schizophrenia. In: animal models in research. Singapore: Springer Nature Singapore; 2024. p. 283–317. https://doi.org/10.1007/978-981-97-0048-6_11.

21. Gharagozloo M, Mace JW, Calabresi PA. Animal models to investigate the effects of inflammation on remyelination in multiple sclerosis. Front Mol Neurosci. 2022;15 https://doi.org/10.3389/fnmol.2022.995477.
22. Metwally E, Hussain T, Ahmed MF. Editorial: animal models for basic and applied research in neuroscience. Front Vet Sci. 2024;10 https://doi.org/10.3389/fvets.2023.1356032.
23. Moon C. New insights into and emerging roles of animal models for neurological disorders. Int J Mol Sci. 2022;23:4957. https://doi.org/10.3390/ijms23094957.
24. Gomes ED, Barata-Antunes S, Teixeira-Castro A, Assunção-Silva RC, Marques CR, Monteiro S, Teixeira FG, Fernandes AM, Silva NA. Animal models of central nervous system disorders. In: Handbook of innovations in central nervous system regenerative medicine. Elsevier; 2020. p. 621–50. https://doi.org/10.1016/B978-0-12-818084-6.00018-0.
25. Partridge B, Rossmeisl JH. Companion animal models of neurological disease. J Neurosci Methods. 2020;331:108484. https://doi.org/10.1016/j.jneumeth.2019.108484.
26. Smith P. Animal models of multiple sclerosis. Curr Protoc. 2021;1 https://doi.org/10.1002/cpz1.185.
27. Yassine S, Almarouk S, Gschwandtner U, Auffret M, Fuhr P, Verin M, Hassan M. Electrophysiological signatures of anxiety in Parkinson's disease. Transl Psychiatry. 2024;14:66. https://doi.org/10.1038/s41398-024-02745-x.
28. Kar H, Gania AM, Bandy A, Dar N u d, Rafiq F. Psychiatric comorbidities and concurrent substance use among people who inject drugs: a single-centre hospital-based study. Sci Rep. 2023;13:19053. https://doi.org/10.1038/s41598-023-45633-y.
29. Huang Y, Li Y, Pan H, Han L. Global, regional, and national burden of neurological disorders in 204 countries and territories worldwide. J Glob Health. 2023;13:04160. https://doi.org/10.7189/jogh.13.04160.
30. Coombes B, Oesterle T, Jennings M, Fennessy B, Pazdernik V, Batzler A, Frank J, Karpyak V, Ruiz JAS, Ryu E, Charney A, Sanchez-Roige S, Biernacka J. A multi-PRS approach to predicting substance use disorders and associated comorbidities. Eur Neuropsychopharmacol. 2023;75:S9–S10. https://doi.org/10.1016/j.euroneuro.2023.08.026.
31. Petrova M, Besedina O, Hare B, Gupta A, Lakhan S. Comprehensive analysis of migraine burden and treatment trends: a retrospective cohort study of recent large multiple commercial payer databases. (P9-12.005). Neurology. 2024;102 https://doi.org/10.1212/WNL.0000000000208156.
32. Narouze S. Why pain physicians should consider becoming headache specialists: bridging the gap in patient care. Reg Anesth Pain Med. 2024;49:77–8. https://doi.org/10.1136/rapm-2023-105060.
33. Krut' VG, Kalinichenko AL, Maltsev DI, Jappy D, Shevchenko EK, Podgorny OV, Belousov VV. Optogenetic and chemogenetic approaches for modeling neurological disorders in vivo. Prog Neurobiol. 2024;235:102600. https://doi.org/10.1016/j.pneurobio.2024.102600.
34. Hyman SE. Use of mouse models to investigate the contributions of CNVs associated with schizophrenia and autism to disease mechanisms. Curr Opin Genet Dev. 2021;68:99–105. https://doi.org/10.1016/j.gde.2021.03.004.
35. Goncalves A, Antonetti DA. Transgenic animal models to explore and modulate the blood brain and blood retinal barriers of the CNS. Fluids Barriers CNS. 2022;19:86. https://doi.org/10.1186/s12987-022-00386-0.
36. Gao L, Pan X, Zhang JH, Xia Y. Glial cells: an important switch for the vascular function of the central nervous system. Front Cell Neurosci. 2023;17 https://doi.org/10.3389/fncel.2023.1166770.
37. Plascencia-Villa G, Perry G. Neuropathologic changes provide insights into key mechanisms of Alzheimer disease and related dementia. Am J Pathol. 2022;192:1340–6. https://doi.org/10.1016/j.ajpath.2022.07.002.
38. Wilcock DM, Lamb BT. The importance of continuing development of novel animal models of Alzheimer's disease and Alzheimer's disease and related dementias. Alzheimers Dement. 2024;20:5078–9. https://doi.org/10.1002/alz.14105.

39. Zhong MZ, Peng T, Duarte ML, Wang M, Cai D. Updates on mouse models of Alzheimer's disease. Mol Neurodegener. 2024;19:23. https://doi.org/10.1186/s13024-024-00712-0.
40. Dhapola R, Kumari S, Sharma P, HariKrishnaReddy D. Insight into the emerging and common experimental in-vivo models of Alzheimer's disease. Lab Anim Res. 2023;39:33. https://doi.org/10.1186/s42826-023-00184-1.
41. Audrain MM, Fol RR, Bemelmans A, Deglon N, Hantraye P, Cartier N, Braudeau J. P1-031: amyloid cascade induction in an AAV-based mouse model of Alzheimer's disease. Alzheimers Dement. 2014;10 https://doi.org/10.1016/j.jalz.2014.05.266.
42. Sinha S, Wal P, Goudanavar P, Divya S, Kimothi V, Jyothi D, Sharma MC, WAL A. Research on Alzheimer's disease (AD) involving the use of in vivo and in vitro models and mechanisms. Cent Nerv Syst Agents Med Chem. 2024;24 https://doi.org/10.2174/0118715249293642240522054929.
43. Pádua MS, Guil-Guerrero JL, Lopes PA. Behaviour hallmarks in Alzheimer's disease 5xFAD mouse model. Int J Mol Sci. 2024;25:6766. https://doi.org/10.3390/ijms25126766.
44. Qian Z, Wang Z, Li B, Meng X, Kuang Z, Li Y, Yang Y, Ye K. Thy1-ApoE4/C/EBPβ double transgenic mice act as a sporadic model with Alzheimer's disease. Mol Psychiatry. 2024;29:3040–55. https://doi.org/10.1038/s41380-024-02565-x.
45. Gao Y, Wang Y, Lei H, Xu Z, Li S, Yu H, Xie J, Zhang Z, Liu G, Zhang Y, Zheng J, Wang J-Z. A novel transgenic mouse line with hippocampus-dominant and inducible expression of truncated human tau. Transl Neurodegener. 2023;12:51. https://doi.org/10.1186/s40035-023-00379-5.
46. Tekin E. Stereotactic surgery and its application in Alzheimer's disease rat models. J Surg Med. 2024;8:90–2. https://doi.org/10.28982/josam.7643.
47. Facchinetti R, Bronzuoli MR, Scuderi C. An animal model of Alzheimer disease based on the intrahippocampal injection of amyloid β-peptide (1–42). Methods Mol Biol. 2018;1727:343–52. https://doi.org/10.1007/978-1-4939-7571-6_25.
48. Benitez DP, Jiang S, Wood J, Wang R, Hall CM, Peerboom C, Wong N, Stringer KM, Vitanova KS, Smith VC, Joshi D, Saito T, Saido TC, Hardy J, Hanrieder J, De Strooper B, Salih DA, Tripathi T, Edwards FA, Cummings DM. Knock-in models related to Alzheimer's disease: synaptic transmission, plaques and the role of microglia. Mol Neurodegener. 2021;16:47. https://doi.org/10.1186/s13024-021-00457-0.
49. Pang K, Jiang R, Zhang W, Yang Z, Li L-L, Shimozawa M, Tambaro S, Mayer J, Zhang B, Li M, Wang J, Liu H, Yang A, Chen X, Liu J, Winblad B, Han H, Jiang T, Wang W, Nilsson P, Guo W, Lu B. An App knock-in rat model for Alzheimer's disease exhibiting Aβ and tau pathologies, neuronal death and cognitive impairments. Cell Res. 2022;32:157–75. https://doi.org/10.1038/s41422-021-00582-x.
50. Baglietto-Vargas D, Cai L, Forner S, Martini A, Trujillo-Estrada L, Cunha C, Mortazavi A, Wood MA, Green KN, Macgregor GR, Tenner A, LaFerla F. O1-01-04: HAβ-KI: a knock-in mouse model for sporadic Alzheimer's disease. Alzheimers Dement. 2018;14 https://doi.org/10.1016/j.jalz.2018.06.2333.
51. Jean YY, Baleriola J, Fà M, Hengst U, Troy CM. Stereotaxic infusion of oligomeric amyloid-beta into the mouse hippocampus. J Vis Exp. 2015; https://doi.org/10.3791/52805.
52. Castillo-Rangel C, Marín G, Diaz-Chiguer DL, Zarate-Calderon CJ, Viveros-Martinez I, Caycho-Salazar FDMDJ, Torres-Pasillas JG, Hernández-Contreras KA, Vega-Quesada LA, Cano-Ortega Y, García-Hernández L, Vásquez-Celaya L. Animal models in Alzheimer's disease: biological plausibility and mood disorders. Neurol Perspect. 2023;3:100110. https://doi.org/10.1016/j.neurop.2023.100110.
53. Defriani E. Parkinson disease: pathophysiology and histopathological overview. Sriwijaya J Neurol. 2023;1:34–8. https://doi.org/10.59345/sjn.v1i2.81.
54. El Khiat A, El-Mansoury B, Ali DA, El Koutbi M, Kaoutar K, Hamdan YA, El Houate B, Tamegart L, Smimih K, Hakam A, Kissani N, Gamrani H. Symptomatology and clinical features of Parkinson's disease. In: Experimental and clinical evidence of the neuropathology of Parkinson's disease. IGI Global; 2023. p. 14–24. https://doi.org/10.4018/978-1-6684-5156-4.ch002.

55. Tong J. The pathogenesis and treatment of Parkinson's disease, theoretical and natural. Science. 2024;47:15–9. https://doi.org/10.54254/2753-8818/47/2024PJ0101.
56. Slézia A, Hegedüs P, Rusina E, Lengyel K, Solari N, Kaszas A, Balázsfi D, Botzanowski B, Acerbo E, Missey F, Williamson A, Hangya B. Behavioral, neural and ultrastructural alterations in a graded-dose 6-OHDA mouse model of early-stage Parkinson's disease. Sci Rep. 2023;13:19478. https://doi.org/10.1038/s41598-023-46576-0.
57. Lal R, singh A, watts S, Chopra K. Experimental models of Parkinson's disease: challenges and opportunities. Eur J Pharmacol. 2024;980:176819. https://doi.org/10.1016/j.ejphar.2024.176819.
58. Yamakado H, Takahashi R. Experimental animal models of prodromal Parkinson's disease. J Parkinsons Dis. 2024;14:S369–79. https://doi.org/10.3233/JPD-230393.
59. Carvalho MM, Campos FL, Coimbra B, Pêgo JM, Rodrigues C, Lima R, Rodrigues AJ, Sousa N, Salgado AJ. Behavioral characterization of the 6-hydroxidopamine model of Parkinson's disease and pharmacological rescuing of non-motor deficits. Mol Neurodegener. 2013;8:14. https://doi.org/10.1186/1750-1326-8-14.
60. Thirugnanam T, Santhakumar K. Chemically induced models of Parkinson's disease. Comp Biochem Physiol C Toxicol Pharmacol. 2022;252:109213. https://doi.org/10.1016/j.cbpc.2021.109213.
61. Santoro M, Fadda P, Klephan KJ, Hull C, Teismann P, Platt B, Riedel G. Neurochemical, histological, and behavioral profiling of the acute, sub-acute, and chronic <scp>MPTP</scp> mouse model of Parkinson's disease. J Neurochem. 2023;164:121–42. https://doi.org/10.1111/jnc.15699.
62. Yang W, Hao W, Meng Z, Ding S, Li X, Zhang T, Huang W, Xu L, Zhang Y, Yang J, Gu X. Molecular regulatory mechanism and toxicology of neurodegenerative processes in MPTP/probenecid-induced progressive Parkinson's disease mice model revealed by transcriptome. Mol Neurobiol. 2021;58:603–16. https://doi.org/10.1007/s12035-020-02128-5.
63. Simola N, Morelli M, Carta AR. The 6-Hydroxydopamine model of parkinson's disease. Neurotox Res. 2007;11:151–67. https://doi.org/10.1007/BF03033565.
64. Masini D, Plewnia C, Bertho M, Scalbert N, Caggiano V, Fisone G. A guide to the generation of a 6-hydroxydopamine mouse model of Parkinson's disease for the study of non-motor symptoms. Biomedicine. 2021;9:598. https://doi.org/10.3390/biomedicines9060598.
65. Mat Taib CN, Mustapha M. MPTP-induced mouse model of Parkinson's disease: a promising direction of therapeutic strategies. Bosn J Basic Med Sci. 2020; https://doi.org/10.17305/bjbms.2020.5181.
66. Pasquali L, Caldarazzo-Ienco E, Fornai F. MPTP neurotoxicity: actions, mechanisms, and animal modeling of Parkinson's disease. In: Handbook of neurotoxicity. New York: Springer; 2014. p. 237–75. https://doi.org/10.1007/978-1-4614-5836-4_170.
67. Cannon JR, Tapias V, Na HM, Honick AS, Drolet RE, Greenamyre JT. A highly reproducible rotenone model of Parkinson's disease. Neurobiol Dis. 2009;34:279–90. https://doi.org/10.1016/j.nbd.2009.01.016.
68. Devine MJ, Gwinn K, Singleton A, Hardy J. Parkinson's disease and α-synuclein expression. Mov Disord. 2011;26:2160–8. https://doi.org/10.1002/mds.23948.
69. Rui Q, Ni H, Li D, Gao R, Chen G. The role of LRRK2 in neurodegeneration of Parkinson disease. Curr Neuropharmacol. 2018;16:1348–57. https://doi.org/10.2174/1570159X16666180222165418.
70. Ishihara-Paul L, Hulihan MM, Kachergus J, Upmanyu R, Warren L, Amouri R, Elango R, Prinjha RK, Soto A, Kefi M, Zouari M, Sassi SB, Yahmed SB, El Euch-Fayeche G, Matthews PM, Middleton LT, Gibson RA, Hentati F, Farrer MJ. *PINK1* mutations and parkinsonism. Neurology. 2008;71:896–902. https://doi.org/10.1212/01.wnl.0000323812.40708.1f.
71. Castelo Rueda MP, Raftopoulou A, Gögele M, Borsche M, Emmert D, Fuchsberger C, Hantikainen EM, Vukovic V, Klein C, Pramstaller PP, Pichler I, Hicks AA. Frequency of heterozygous Parkin (PRKN) variants and penetrance of Parkinson's disease risk markers in the population-based CHRIS cohort. Front Neurol. 2021;12 https://doi.org/10.3389/fneur.2021.706145.

72. Cano-Cano F, Martín-Loro F, Gallardo-Orihuela A, del C. González-Montelongo M, Ortuño-Miquel S, Hervás-Corpión I, de la Villa P, Ramón-Marco L, Navarro-Calvo J, Gómez-Jaramillo L, Arroba AI, Valor LM. Retinal dysfunction in Huntington's disease mouse models concurs with local gliosis and microglia activation. Sci Rep. 2024;14:4176. https://doi.org/10.1038/s41598-024-54347-8.
73. Alharbi KS. Europinidin mitigates 3-NPA-induced Huntington's disease symptoms in rats: a comprehensive analysis of oxidative stress, mitochondrial enzyme complex activity, pro-inflammatory markers and neurotransmitter alterations. Biomedicine. 2024;12:625. https://doi.org/10.3390/biomedicines12030625.
74. Han JY, Seo J, Choi Y, Im W, Ban J-J, Sung J-J. CRISPR-Cas9 mediated genome editing of Huntington's disease neurospheres. Mol Biol Rep. 2023;50:2127–36. https://doi.org/10.1007/s11033-022-08175-6.
75. Creus-Muncunill J, Haure-Mirande JV, Mattei D, Bons J, Ramirez AV, Hamilton BW, Corwin C, Chowdhury S, Schilling B, Ellerby LM, Ehrlich ME. TYROBP/DAP12 knockout in Huntington's disease Q175 mice cell-autonomously decreases microglial expression of disease-associated genes and non-cell-autonomously mitigates astrogliosis and motor deterioration. J Neuroinflammation. 2024;21:66. https://doi.org/10.1186/s12974-024-03052-4.
76. Ikenoshita S, Matsuo K, Yabuki Y, Kawakubo K, Asamitsu S, Hori K, Usuki S, Hirose Y, Bando T, Araki K, Ueda M, Sugiyama H, Shioda N. A cyclic pyrrole-imidazole polyamide reduces pathogenic RNA in CAG/CTG triplet repeat neurological disease models. J Clin Invest. 2023;133 https://doi.org/10.1172/JCI164792.
77. Vasilkovska T, Salajeghe S, Vanreusel V, Van Audekerke J, Verschuuren M, Hirschler L, Warnking J, Pintelon I, Pustina D, Cachope R, Mrzljak L, Muñoz-Sanjuan I, Barbier EL, De Vos WH, Van der Linden A, Verhoye M. Longitudinal alterations in brain perfusion and vascular reactivity in the zQ175DN mouse model of Huntington's disease. J Biomed Sci. 2024;31:37. https://doi.org/10.1186/s12929-024-01028-3.
78. Bertoglio D, Weiss AR, Liguore W, Martin LD, Hobbs T, Templon J, Srinivasan S, Dominguez C, Munoz-Sanjuan I, Khetarpal V, Verhaeghe J, Staelens S, Link J, Liu L, Bard JA, McBride JL. In vivo cerebral imaging of mutant huntingtin aggregates using ^{11}C-CHDI-180R PET in a nonhuman primate model of Huntington disease. J Nucl Med. 2023;64:1581–7. https://doi.org/10.2967/jnumed.123.265569.
79. Martin-Solana E, Diaz-Lopez I, Mohamedi Y, Ventoso I, Fernandez J-J, Fernandez-Fernandez MR. Progressive alterations in polysomal architecture and activation of ribosome stalling relief factors in a mouse model of Huntington's disease. Neurobiol Dis. 2024;195:106488. https://doi.org/10.1016/j.nbd.2024.106488.
80. Al-Khayri JM, Ravindran M, Banadka A, Vandana CD, Priya K, Nagella P, Kukkemane K. Amyotrophic lateral sclerosis: insights and new prospects in disease pathophysiology, biomarkers and therapies. Pharmaceuticals. 2024;17:1391. https://doi.org/10.3390/ph17101391.
81. Zhang Q, Lin J, Wang Y, Chen L, Ding Y, Zheng F, Song H, Lv A, Li Y, Guo Q, Lin M, Hu W, Xu L, Zhao W, Fang L, Cui M, Fu Z, Chen W, Zhang J, Wang Z, Wang N, Fu Y. Detection of pTDP-43 via routine muscle biopsy: a promising diagnostic biomarker for amyotrophic lateral sclerosis. Brain Pathol. 2024;34 https://doi.org/10.1111/bpa.13261.
82. Gao Y. Abnormal pathological mechanisms of TDP-43 protein and its associated pathway in association with ALS. MedScien. 2024;1 https://doi.org/10.61173/wq3pmm96.
83. Rizea RE, Corlatescu A-D, Costin HP, Dumitru A, Ciurea AV. Understanding amyotrophic lateral sclerosis: pathophysiology, diagnosis, and therapeutic advances. Int J Mol Sci. 2024;25:9966. https://doi.org/10.3390/ijms25189966.
84. Duranti E, Villa C. Muscle involvement in amyotrophic lateral sclerosis: understanding the pathogenesis and advancing therapeutics. Biomolecules. 2023;13:1582. https://doi.org/10.3390/biom13111582.
85. Lescouzères L, Patten SA. Promising animal models for amyotrophic lateral sclerosis drug discovery: a comprehensive update. Expert Opin Drug Discovery. 2024;19:1213–33. https://doi.org/10.1080/17460441.2024.2387791.

86. Zhou L, Xu R. Invertebrate genetic models of amyotrophic lateral sclerosis. Front Mol Neurosci. 2024;17 https://doi.org/10.3389/fnmol.2024.1328578.
87. Zhu L, Li S, Li X-J, Yin P. Pathological insights from amyotrophic lateral sclerosis animal models: comparisons, limitations, and challenges. Transl Neurodegener. 2023;12:46. https://doi.org/10.1186/s40035-023-00377-7.
88. Moreno-Jiménez L, Benito-Martín MS, Sanclemente-Alamán I, Matías-Guiu JA, Sancho-Bielsa F, Canales-Aguirre A, Mateos-Díaz JC, Matías-Guiu J, Aguilar J, Gómez-Pinedo U. Murine experimental models of amyotrophic lateral sclerosis: an update. Neurología (English Edition). 2024;39:282–91. https://doi.org/10.1016/j.nrleng.2021.07.004.
89. Zhou L, Xie M, Wang X, Xu R. The usage and advantages of several common amyotrophic lateral sclerosis animal models. Front Neurosci. 2024;18 https://doi.org/10.3389/fnins.2024.1341109.
90. Moriyama H, Yokota T. Recent progress of antisense oligonucleotide therapy for superoxide dismutase 1-mutated amyotrophic lateral sclerosis: focus on tofersen. Genes (Basel). 2024; https://doi.org/10.20944/preprints202409.2440.v1.
91. Caldi Gomes L, Hänzelmann S, Hausmann F, Khatri R, Oller S, Parvaz M, Tzeplaeff L, Pasetto L, Gebelin M, Ebbing M, Holzapfel C, Columbro SF, Scozzari S, Knöferle J, Cordts I, Demleitner AF, Deschauer M, Dufke C, Sturm M, Zhou Q, Zelina P, Sudria-Lopez E, Haack TB, Streb S, Kuzma-Kozakiewicz M, Edbauer D, Pasterkamp RJ, Laczko E, Rehrauer H, Schlapbach R, Carapito C, Bonetto V, Bonn S, Lingor P. Multiomic ALS signatures highlight subclusters and sex differences suggesting the MAPK pathway as therapeutic target. Nat Commun. 2024;15:4893. https://doi.org/10.1038/s41467-024-49196-y.
92. Ciaccio G, Kovalev S, McDermott A, Wong J, Sadiq S. Validating a novel CSF-mediated animal model of sporadic ALS (P6-11.018). Neurology. 2024;102 https://doi.org/10.1212/WNL.0000000000204674.
93. Guerrero EN, Wang H, Mitra J, Hegde PM, Stowell SE, Liachko NF, Kraemer BC, Garruto RM, Rao KS, Hegde ML. TDP-43/FUS in motor neuron disease: complexity and challenges. Prog Neurobiol. 2016;145–146:78–97. https://doi.org/10.1016/j.pneurobio.2016.09.004.
94. Cheng JL, Cook AL, Talbot J, Perry S. How is excitotoxicity being modelled in iPSC-derived neurons? Neurotox Res. 2024;42:43. https://doi.org/10.1007/s12640-024-00721-3.
95. Shaikh M, Doshi G. Epigenetic aging in major depressive disorder: clocks, mechanisms and therapeutic perspectives. Eur J Pharmacol. 2024;978:176757. https://doi.org/10.1016/j.ejphar.2024.176757.
96. Akil H, Gordon J, Hen R, Javitch J, Mayberg H, McEwen B, Meaney MJ, Nestler EJ. Treatment resistant depression: a multi-scale, systems biology approach. Neurosci Biobehav Rev. 2018;84:272–88. https://doi.org/10.1016/j.neubiorev.2017.08.019.
97. Elias E, Zhang AY, Manners MT. Novel pharmacological approaches to the treatment of depression. Life. 2022;12:196. https://doi.org/10.3390/life12020196.
98. Anmella G, Meehan A, Ashton M, Mohebbi M, Fico G, Ng CH, Maes M, Berk L, De Prisco M, Singh AB, Malhi GS, Berk M, Dodd S, Hidalgo-Mazzei D, Grande I, Pacchiarotti I, Murru A, Vieta E, Dean OM. Exploring clinical subgroups of participants with major depressive disorder that may benefit from adjunctive minocycline treatment. Clin Psychopharmacol Neurosci. 2024;22:33–44. https://doi.org/10.9758/cpn.23.1098.
99. White AG, Elias E, Orozco A, Robinson SA, Manners MT. Chronic stress-induced neuroinflammation: relevance of rodent models to human disease. Int J Mol Sci. 2024;25:5085. https://doi.org/10.3390/ijms25105085.
100. Lin S-S, Zhou B, Chen B-J, Jiang R-T, Li B, Illes P, Semyanov A, Tang Y, Verkhratsky A. Electroacupuncture prevents astrocyte atrophy to alleviate depression. Cell Death Dis. 2023;14:343. https://doi.org/10.1038/s41419-023-05839-4.
101. Song N, Gao Y, Lu S, Yang S, Yuan C, Sun W. NAc-DBS corrects depression-like behaviors in CUMS mouse model via disinhibition of DA neurons in the VTA. Mol Psychiatry. 2023; https://doi.org/10.21203/rs.3.rs-2869497/v1.
102. Bobkova NV, Chuvakova LN, Kovalev VI, Zhdanova DY, Chaplygina AV, Rezvykh AP, Evgen'ev MB. A mouse model of sporadic Alzheimer's disease with elements of major depression. Mol Neurobiol. 2024; https://doi.org/10.1007/s12035-024-04346-7.

103. Wang J, Wang Y, Chen Y, Zhang J, Zhang Y, Li S, Zhu H, Song X, Hou L, Wang L, Wang Y, Zhang Z, Rong P. Transcutaneous auricular vagus stimulation attenuates LPS-induced depression-like behavior by regulating central α7nAChR/JAK2 signaling. Mol Neurobiol. 2024; https://doi.org/10.1007/s12035-024-04438-4.
104. Jiang H, Long X, Wang Y, Zhang X, Chen L, Yang X, Zhao B, Zhang Y, Chai Y, Bao T. Acupuncture ameliorates depression-like behaviors through modulating the neuroinflammation mediated by TLR4 signaling pathway in rats exposed to chronic restraint stress. Mol Neurobiol. 2024;61:2606–19. https://doi.org/10.1007/s12035-023-03737-6.
105. Naughton SX, Yang E-J, Iqbal U, Trageser K, Charytonowicz D, Masieri S, Estill M, Wu H, Raval U, Lyu W, Wu Q, Shen L, Simon J, Sebra R, Pasinetti GM. Permethrin exposure primes neuroinflammatory stress response to drive depression-like behavior through microglial activation in a mouse model of gulf war illness. J Neuroinflammation. 2024;21:222. https://doi.org/10.1186/s12974-024-03215-3.
106. Głombik K, Kukla-Bartoszek M, Curzytek K, Basta-Kaim A, Budziszewska B. Contribution of changes in the orexin system and energy sensors in the brain in depressive disorder - a study in an animal model. Pharmacol Rep. 2024;76:51–71. https://doi.org/10.1007/s43440-023-00559-0.
107. Xie Y, Xie Z, Wei Q, Wang Y, Li T, Wei J, Wu Y. Preparation of static training animal models and research status analysis. J Acupunct Tuina Sci. 2023;21:330–6. https://doi.org/10.1007/s11726-023-1391-3.
108. Sarrazin DH, Gardner W, Marchese C, Balzinger M, Ramanathan C, Schott M, Rozov S, Veleanu M, Vestring S, Normann C, Rantamäki T, Antoine B, Barrot M, Challet E, Bourgin P, Serchov T. Prefrontal cortex molecular clock modulates development of depression-like phenotype and rapid antidepressant response in mice. Nat Commun. 2024;15:7257. https://doi.org/10.1038/s41467-024-51716-9.
109. Floris G, Godar SC, Braccagni G, Piras IS, Ravens A, Zanda MT, Huentelman MJ, Bortolato M. The sinking platform test: a novel paradigm to measure persistence in animal models. Neuropsychopharmacology. 2024;49:1373–82. https://doi.org/10.1038/s41386-024-01827-0.
110. Szuhany KL, Simon NM. Anxiety disorders. JAMA. 2022;328:2431. https://doi.org/10.1001/jama.2022.22744.
111. Kapoor A., Goel S. Prediction of anxiety disorders using machine learning techniques. 2022 IEEE Bombay Section Signature Conference (IBSSC), IEEE; 2022. p. 1–6. https://doi.org/10.1109/IBSSC56953.2022.10037459.
112. Rapee RM, Creswell C, Kendall PC, Pine DS, Waters AM. Anxiety disorders in children and adolescents: a summary and overview of the literature. Behav Res Ther. 2023;168:104376. https://doi.org/10.1016/j.brat.2023.104376.
113. Cruz APM, Castro-Gomes V, Landeira-Fernandez J. An animal model of trait anxiety: carioca high freezing rats as a model of generalized anxiety disorder. Pers Neurosci. 2024;7:e6. https://doi.org/10.1017/pen.2023.6.
114. Watanabe S, Al Omran A, Shao AS, Liang J. Social isolation model: a noninvasive rodent model of stress and anxiety. J Vis Exp. 2022; https://doi.org/10.3791/64567.
115. Nisar R, Batool Z, Haider S. Electric foot-shock induces neurobehavioral aberrations due to imbalance in oxidative status, stress hormone, neurochemical profile, and irregular cortical-beta wave pattern in rats: a validated animal model of anxiety. Life Sci. 2023;323:121707. https://doi.org/10.1016/j.lfs.2023.121707.
116. Zhao H, Zhou M, Liu Y, Jiang J, Wang Y. Recent advances in anxiety disorders: focus on animal models and pathological mechanisms. Animal Model Exp Med. 2023;6:559–72. https://doi.org/10.1002/ame2.12360.
117. Ennaceur A, Chazot PL. Preclinical animal anxiety research—flaws and prejudices. Pharmacol Res Perspect. 2016;4 https://doi.org/10.1002/prp2.223.
118. Ganella DE, Kim JH. Developmental rodent models of fear and anxiety: from neurobiology to pharmacology. Br J Pharmacol. 2014;171:4556–74. https://doi.org/10.1111/bph.12643.
119. Moreira FA, Gobira PH, Viana TG, Vicente MA, Zangrossi H, Graeff FG. Modeling panic disorder in rodents. Cell Tissue Res. 2013;354:119–25. https://doi.org/10.1007/s00441-013-1610-1.

120. Mitten EH, Souders A, Marron Fernandez de Velasco E, Wickman K. Stress-induced anxiety-related behavior in mice is driven by enhanced excitability of ventral tegmental area GABA neurons. Front Behav Neurosci. 2024;18 https://doi.org/10.3389/fnbeh.2024.1425607.
121. Peng X, Mao Y, Tai Y, Luo B, Dai Q, Wang X, Wang H, Liang Y, Guan R, Liu C, Guo Y, Chen L, Zhang Z, Wang H. Characterization of anxiety-like behaviors and neural circuitry following chronic moderate noise exposure in mice. Environ Health Perspect. 2023;131 https://doi.org/10.1289/EHP12532.
122. Lages YV, Balthazar L, Krahe TE, Landeira-Fernandez J. Pharmacological and physiological correlates of the bidirectional fear phenotype of the carioca rats and other bidirectionally selected lines. Curr Neuropharmacol. 2023;21:1864–83. https://doi.org/10.2174/1570159X20666221012121534.
123. Goulart VG, Rocha-Mendonça H, Maisonnette S, Pandolfo P, Landeira-Fernandez J, Campello-Costa P. Differential expression of glutamatergic receptor subunits in the hippocampus in carioca high- and low-conditioned freezing rats. Mol Cell Neurosci. 2021;116:103666. https://doi.org/10.1016/j.mcn.2021.103666.
124. Macêdo-Souza C, Maisonnette SS, Filgueiras CC, Landeira-Fernandez J, Krahe TE. Cued fear conditioning in carioca high- and low-conditioned freezing rats. Front Behav Neurosci. 2020;13 https://doi.org/10.3389/fnbeh.2019.00285.
125. Perusini JN, Fanselow MS. Neurobehavioral perspectives on the distinction between fear and anxiety. Learn Mem. 2015;22:417–25. https://doi.org/10.1101/lm.039180.115.
126. Ritov G, Ardi Z, Richter-Levin G. Differential activation of amygdala, dorsal and ventral hippocampus following an exposure to a reminder of underwater trauma. Front Behav Neurosci. 2014;8 https://doi.org/10.3389/fnbeh.2014.00018.
127. Danduga RCSR, Kola PK. Elevated plus maze for assessment of anxiety and memory in rodents. Methods Mol Biol. 2024;2761:93–6. https://doi.org/10.1007/978-1-0716-3662-6_8.
128. Walf AA, Frye CA. The use of the elevated plus maze as an assay of anxiety-related behavior in rodents. Nat Protoc. 2007;2:322–8. https://doi.org/10.1038/nprot.2007.44.
129. Walf AA, Frye CA Using the elevated plus maze as a bioassay to assess the effects of naturally occurring and exogenously administered compounds to influence anxiety-related behaviors of mice. 2009. p. 225–246. https://doi.org/10.1007/978-1-60761-303-9_12.
130. Kraeuter A-K, Guest PC, Sarnyai Z. The elevated plus maze test for measuring anxiety-like behavior in rodents. Methods Mol Biol. 2019;1916:69–74. https://doi.org/10.1007/978-1-4939-8994-2_4.
131. Roy V, Chapillon P, Jeljeli M, Caston J, Belzung C. Free versus forced exposure to an elevated plus-maze: evidence for new behavioral interpretations during test and retest. Psychopharmacology (Berl). 2009;203:131–41. https://doi.org/10.1007/s00213-008-1378-2.
132. Okonogi T, Nakayama R, Sasaki T, Ikegaya Y. Characterization of peripheral activity states and cortical local field potentials of mice in an elevated plus maze test. Front Behav Neurosci. 2018;12 https://doi.org/10.3389/fnbeh.2018.00062.
133. Yamamori Y, Roiser J, Robinson O. 242. Translational computational assessment of approach-avoidance conflict in human anxiety. Biol Psychiatry. 2023;93:S190–1. https://doi.org/10.1016/j.biopsych.2023.02.482.
134. Oberrauch S, Sigrist H, Sautter E, Gerster S, Bach DR, Pryce CR. Establishing operant conflict tests for the translational study of anxiety in mice. Psychopharmacology (Berl). 2019;236:2527–41. https://doi.org/10.1007/s00213-019-05315-y.
135. Gencturk S, Unal G. Rodent tests of depression and anxiety: construct validity and translational relevance. Cogn Affect Behav Neurosci. 2024;24:191–224. https://doi.org/10.3758/s13415-024-01171-2.
136. Resstel LBM, Souza RF, Guimarães FS. Anxiolytic-like effects induced by medial prefrontal cortex inhibition in rats submitted to the Vogel conflict test. Physiol Behav. 2008;93:200–5. https://doi.org/10.1016/j.physbeh.2007.08.009.
137. Lisboa SF, Guimarães FS, Resstel LBM. Anxiety-behavior modulated by ventral medial prefrontal cortex of rats submitted to the vogel conflict test involves a local NMDA receptor and nitric oxide. J Behav Brain Sci. 2011;01:181–7. https://doi.org/10.4236/jbbs.2011.13024.

138. Cuccheddu T, Floris S, Serra M, Luisa Porceddu M, Sanna E, Biggio G. Proconflict effect of carbon dioxide inhalation in rats. Life Sci. 1995;56:PL321–PL324. https://doi.org/10.1016/0024-3205(95)00093-3.
139. Xu Y-X, Liu G-Y, Ji Z-Z, Li Y-Y, Wang Y-L, Wu X-Y, Liu J-L, Ma D-X, Zhong M-K, Gao C-B, Xu Q. Restraint stress induced anxiety and sleep in mice. Front Psych. 2023;14 https://doi.org/10.3389/fpsyt.2023.1090420.
140. Yan L, Yang F, Wang Y, Shi L, Wang M, Yang D, Wang W, Jia Y, So K-F, Zhang L. Stress increases hepatic release of lipocalin 2 which contributes to anxiety-like behavior in mice. Nat Commun. 2024;15:3034. https://doi.org/10.1038/s41467-024-47266-9.
141. Toth I, Neumann ID. Animal models of social avoidance and social fear. Cell Tissue Res. 2013;354:107–18. https://doi.org/10.1007/s00441-013-1636-4.
142. White SW, Squires GD, Smith SJ, Wright GM, Sufka KJ, Rimoldi JM, Gadepalli RS. Anxiolytic-like effects of an mGluR 5 antagonist and a mGluR 2/3 agonist, and antidepressant-like effects of an mGluR 7 agonist in the chick social separation stress test, a dual-drug screening model of treatment-resistant depression. Pharmacol Biochem Behav. 2023;227–228:173588. https://doi.org/10.1016/j.pbb.2023.173588.
143. Jiang J, Tan S, Feng X, Peng Y, Long C, Yang L. Distinct ACC neural mechanisms underlie authentic and transmitted anxiety induced by maternal separation in mice. J Neurosci. 2023;43:8201–18. https://doi.org/10.1523/JNEUROSCI.0558-23.2023.
144. Luvsannyam E, Jain MS, Pormento MKL, Siddiqui H, Balagtas ARA, Emuze BO, Poprawski T. Neurobiology of schizophrenia: a comprehensive review. Cureus. 2022; https://doi.org/10.7759/cureus.23959.
145. MacDonald A, Demro C. Schizophrenia: presentation, affect and cognition, pathophysiology and etiology. In: Krueger RF, Blaney PH, editors. Oxford textbook of psychopathology. New York: Oxford University Press; 2023. p. 332–51. https://doi.org/10.1093/med-psych/9780197542521.003.0014.
146. Oliveras I, Cañete T, Sampedro-Viana D, Río-Álamos C, Tobeña A, Corda MG, Giorgi O, Fernández-Teruel A. Neurobehavioral profiles of six genetically-based rat models of schizophrenia- related symptoms. Curr Neuropharmacol. 2023;21:1934–52. https://doi.org/10.2174/1570159X21666230221093644.
147. Santos-Silva T, dos Santos Fabris D, de Oliveira CL, Guimarães FS, Gomes FV. Prefrontal and hippocampal parvalbumin interneurons in animal models for schizophrenia: a systematic review and meta-analysis. Schizophr Bull. 2024;50:210–23. https://doi.org/10.1093/schbul/sbad123.
148. Białoń M, Wąsik A. Advantages and limitations of animal schizophrenia models. Int J Mol Sci. 2022;23:5968. https://doi.org/10.3390/ijms23115968.
149. Benvenutti R, Gallas-Lopes M, Marcon M, Reschke CR, Herrmann AP, Piato A. Glutamate NMDA receptor antagonists with relevance to schizophrenia: a review of zebrafish behavioral studies. Curr Neuropharmacol. 2022;20:494–509. https://doi.org/10.2174/1570159X19666210215121428.
150. Delgado-Sallent C, Gener T, Nebot P, López-Cabezón C, Puig MV. Neural substrates of cognitive impairment in a NMDAR hypofunction mouse model of schizophrenia and partial rescue by risperidone. Front Cell Neurosci. 2023;17 https://doi.org/10.3389/fncel.2023.1152248.
151. Ju J, Liu L, Yang X, Men S, Hou S-T. Distinctive effects of NMDA receptor modulators on cerebral microcirculation in a schizophrenia mouse model. Biochem Biophys Res Commun. 2023;653:62–8. https://doi.org/10.1016/j.bbrc.2023.02.040.
152. Lee G, Zhou Y. NMDAR hypofunction animal models of schizophrenia. Front Mol Neurosci. 2019;12 https://doi.org/10.3389/fnmol.2019.00185.
153. Petty A, Cui X, Tesiram Y, Kirik D, Howes O, Eyles D. Enhanced dopamine in prodromal schizophrenia (EDiPS): a new animal model of relevance to schizophrenia. NPJ Schizophr. 2019;5:6. https://doi.org/10.1038/s41537-019-0074-z.
154. Mahoney HL, Bloom CA, Justin HS, Capraro BM, Morris C, Gonzalez D, Sandefur E, Faulkner J, Reiss S, Valladares A, Ocampo A, Carter B, Lussier AL, Dinh LP, Weeber E, Gamsby J, Gulick D. DISC1 and reelin interact to alter cognition, inhibition, and neuro-

genesis in a novel mouse model of schizophrenia. Front Cell Neurosci. 2024;17 https://doi.org/10.3389/fncel.2023.1321632.
155. Avloniti M, Evangelidou M, Gomini M, Loupis T, Emmanouil M, Mitropoulou A, Tselios T, Lassmann H, Gruart A, Delgado-García JM, Probert L, Kyrargyri V. IKKβ deletion from CNS macrophages increases neuronal excitability and accelerates the onset of EAE, while from peripheral macrophages reduces disease severity. J Neuroinflammation. 2024;21:34. https://doi.org/10.1186/s12974-024-03023-9.
156. Benítez-Fernández R, Josa-Prado F, Sánchez E, Lao Y, García-Rubia A, Cumella J, Martínez A, Palomo V, de Castro F. Efficacy of a benzothiazole-based <scp>LRRK2</scp> inhibitor in oligodendrocyte precursor cells and in a murine model of multiple sclerosis. CNS Neurosci Ther. 2024;30 https://doi.org/10.1111/cns.14552.
157. Papazian I, Kourouvani M, Dagkonaki A, Gouzouasis V, Dimitrakopoulou L, Markoglou N, Badounas F, Tselios T, Anagnostouli M, Probert L. Spontaneous human CD8 T cell and autoimmune encephalomyelitis-induced CD4/CD8 T cell lesions in the brain and spinal cord of HLA-DRB1*15-positive multiple sclerosis humanized immune system mice. Elife. 2024;12 https://doi.org/10.7554/eLife.88826.
158. Furube E, Ohgidani M, Yoshida S. Systemic inflammation leads to changes in the intracellular localization of KLK6 in oligodendrocytes in spinal cord white matter. Neurochem Res. 2023;48:2645–59. https://doi.org/10.1007/s11064-023-03929-5.
159. Vejar S, Pizarro IS, Pulgar-Sepúlveda R, Vicencio SC, Polit A, Amador CA, del Rio R, Varas R, Orellana JA, Ortiz FC. A preclinical mice model of multiple sclerosis based on the toxin-induced double-site demyelination of callosal and cerebellar fibers. Biol Res. 2024;57:48. https://doi.org/10.1186/s40659-024-00529-7.
160. Gingele S, Möllenkamp TM, Henkel F, Schröder L, Hümmert MW, Skripuletz T, Stangel M, Gudi V. Automated analysis of gray matter damage in aged mice reveals impaired remyelination in the cuprizone model. Brain Pathol. 2024;34 https://doi.org/10.1111/bpa.13218.
161. Fu X-X, Qu H, Wang J, Cai H-Y, Jiang H, Chen H-H, Han S. Novel nano-carriers with N-formylmethionyl-leucyl-phenylalanine-modified liposomes improve effects of C16-angiopoietin 1 in acute animal model of multiple sclerosis. Drug Deliv. 2023;30 https://doi.org/10.1080/10717544.2023.2241664.
162. Yamanaka G, Hayashi K, Morishita N, Takeshita M, Ishii C, Suzuki S, Ishimine R, Kasuga A, Nakazawa H, Takamatsu T, Watanabe Y, Morichi S, Ishida Y, Yamazaki T, Go S. Experimental and clinical investigation of cytokines in migraine: a narrative review. Int J Mol Sci. 2023;24:8343. https://doi.org/10.3390/ijms24098343.
163. Zhang S, Azubuine J, Schmeer C. A systematic literature review on the role of glial cells in the pathomechanisms of migraine. Front Mol Neurosci. 2023;16 https://doi.org/10.3389/fnmol.2023.1219574.
164. Kuburas A, Russo AF. Shared and independent roles of CGRP and PACAP in migraine pathophysiology. J Headache Pain. 2023;24:34. https://doi.org/10.1186/s10194-023-01569-2.
165. Alabbad S, Figueredo N, Yuan H, Silberstein S. Developments in targeting calcitonin gene-related peptide. Expert Rev Neurother. 2024;24:477–85. https://doi.org/10.1080/14737175.2024.2332754.
166. Xiao S, Lu G, Liu J, Su W, Li C, Liu Y, Meng F, Zhao J, Gao N, Chang Y, Guo X, Yu S, Liu R. Brain-wide mapping of c-Fos expression in nitroglycerin-induced models of migraine. J Headache Pain. 2024;25:136. https://doi.org/10.1186/s10194-024-01837-9.
167. Al-Omari A, Gaszner B, Zelena D, Gecse K, Berta G, Biró-Sütő T, Szocsics P, Maglóczky Z, Gombás P, Pintér E, Juhász G, Kormos V. Neuroanatomical evidence and a mouse calcitonin gene–related peptide model in line with human functional magnetic resonance imaging data support the involvement of peptidergic Edinger–Westphal nucleus in migraine. Pain. 2024; https://doi.org/10.1097/j.pain.0000000000003294.
168. Pietrobon D, Brennan KC. Genetic mouse models of migraine. J Headache Pain. 2019;20:79. https://doi.org/10.1186/s10194-019-1029-5.
169. Vuralli D, Bolay H. Animal models of headache. In: Handbook of animal models in neurological disorders. Elsevier; 2023. p. 305–15. https://doi.org/10.1016/B978-0-323-89833-1.00051-3.

170. Laborc KF. Activation of the trigeminal system in the animal models of migraine. Szegedi Tudományegyetem. 2023; https://doi.org/10.14232/phd.11677.
171. Spekker E, Fejes-Szabó A, Nagy-Grócz G. Models of trigeminal activation: is there an animal model of migraine? Brain Sci. 2024;14:317. https://doi.org/10.3390/brainsci14040317.
172. Cropper HC, Conway CM, Wyche W, Pradhan AA. Glial activation in pain and emotional processing regions in the nitroglycerin mouse model of chronic migraine. Headache. 2024;64:973–82. https://doi.org/10.1111/head.14740.
173. Pellesi L. The human NTG model of migraine in drug discovery and development. Expert Opin Drug Discovery. 2023;18:1077–85. https://doi.org/10.1080/17460441.2023.2236545.
174. Sureda-Gibert P, Romero-Reyes M, Akerman S. Nitroglycerin as a model of migraine: clinical and preclinical review. Neurobiol Pain. 2022;12:100105. https://doi.org/10.1016/j.ynpai.2022.100105.
175. Wattiez A-S, Wang M, Russo AF. CGRP in animal models of migraine. Handb Exp Pharmacol. 2018;255:85–107. https://doi.org/10.1007/164_2018_187.
176. Vitale M, Tottene A, Zarin Zadeh M, Brennan K, Pietrobon D. Mechanisms of initiation of cortical spreading depression. J Headache Pain. 2023;24:105. https://doi.org/10.1186/s10194-023-01643-9.
177. Zheng F. Canonical transient receptor potential channel 3 contributes to cerebral blood flow changes associated with cortical spreading depression in mice. Int J Mol Sci. 2023;24:12611. https://doi.org/10.3390/ijms241612611.
178. Zhang J, Zhao Y, Jing Y, Han L, Ma X, Yu L, Diao T. The plasma fibrinogen levels in the nitroglycerin-induced chronic migraine rat model and its association between migraine-associated vestibular dysfunction. Front Neurol. 2023;14 https://doi.org/10.3389/fneur.2023.980543.
179. Babuta M, Nagesh PT, Datta AA, Remotti V, Zhuang Y, Mehta J, Lami F, Wang Y, Szabo G. Combined insults of a MASH diet and alcohol binges activate intercellular communication and neutrophil recruitment via the NLRP3-IL-1β axis in the liver. Cells. 2024;13:960. https://doi.org/10.3390/cells13110960.
180. Lee MR, Moore CF, Weerts EM. Oxytocin decreases alcohol self-administration in male baboons. Transl Psychiatry. 2024;14:369. https://doi.org/10.1038/s41398-024-03076-7.
181. Reinwald JR, Schmitz CN, Skorodumov I, Kuchar M, Weber-Fahr W, Spanagel R, Meinhardt MW. Psilocybin-induced default mode network hypoconnectivity is blunted in alcohol-dependent rats. Transl Psychiatry. 2023;13:392. https://doi.org/10.1038/s41398-023-02690-1.
182. Tonetto S, Weikop P, Thomsen M. Nutritional ketosis as treatment for alcohol withdrawal symptoms in female C57BL/6J mice. Sci Rep. 2024;14:5092. https://doi.org/10.1038/s41598-024-55310-3.
183. Wagner J, Park LM, Mukhopadhyay P, Matyas C, Trojnar E, Damadzic R, Jung J, Bell AS, Mavromatis LA, Hamandi AM, Rosoff DB, Vendruscolo LF, Koob GF, Pacher P, Lohoff FW. PCSK9 inhibition attenuates alcohol-associated neuronal oxidative stress and cellular injury. Brain Behav Immun. 2024;119:494–506. https://doi.org/10.1016/j.bbi.2024.04.022.
184. Ghani S, Alkhlaif Y, Mann J, Moncayo L, Ulker E, Caillaud M, Barik M, Ditre JW, Miles MF, Damaj MI. Surgical incision pain induced an increase in alcohol consumption in mice. Alcohol. 2024;117:1–9. https://doi.org/10.1016/j.alcohol.2024.03.005.
185. Smith R, Taylor S, Bilek E. Computational mechanisms of addiction: recent evidence and its relevance to addiction medicine. Curr Addict Rep. 2021;8:509–19. https://doi.org/10.1007/s40429-021-00399-z.
186. Koob GF, Vendruscolo L. Theoretical frameworks and mechanistic aspects of alcohol addiction: alcohol addiction as a reward deficit/stress surfeit disorder. Curr Top Behav Neurosci. 2025;71:17–77. https://doi.org/10.1007/7854_2023_424.
187. Vierkant V, Xie X, Wang X, Wang J. Experimental models of alcohol use disorder and their application for pathophysiological investigations. Curr Protoc. 2023;3 https://doi.org/10.1002/cpz1.831.
188. Kuhn BN, Kalivas PW, Bobadilla A-C. Understanding addiction using animal models. Front Behav Neurosci. 2019;13 https://doi.org/10.3389/fnbeh.2019.00262.

189. Vanderschuren LJMJ, Ahmed SH. Animal models of the behavioral symptoms of substance use disorders. Cold Spring Harb Perspect Med. 2021;11:a040287. https://doi.org/10.1101/cshperspect.a040287.
190. Valles G, Huebschman JL, Chow E, Kelly C, Guo Y, Smith LN. Jugular vein catheter design and cocaine self-administration using mice: a comprehensive method. Front Behav Neurosci. 2022;16 https://doi.org/10.3389/fnbeh.2022.880845.
191. Venniro M, Shaham Y. An operant social self-administration and choice model in rats. Nat Protoc. 2020;15:1542–59. https://doi.org/10.1038/s41596-020-0296-6.
192. Bazovkina DV, Kondaurova EM, Adonina SN, Bazhenova EY, Kulikov AV. The effect of subchronic alcoholization on the behavior and monoaminergic systems of the brains of mice with a predisposition to depression-like behavior. Neurochem J. 2023;17:402–11. https://doi.org/10.1134/S1819712423030042.
193. Spanagel R. Animal models of addiction. Dialogues Clin Neurosci. 2017;19:247–58. https://doi.org/10.31887/DCNS.2017.19.3/rspanagel.
194. McKendrick G, Graziane NM. Drug-induced conditioned place preference and its practical use in substance use disorder research. Front Behav Neurosci. 2020;14 https://doi.org/10.3389/fnbeh.2020.582147.
195. Cajiao-Manrique M d M, Maldonado R, Martín-García E. A male mouse model of WIN 55,212–2 self-administration to study cannabinoid addiction. Front Pharmacol. 2023;14 https://doi.org/10.3389/fphar.2023.1143365.
196. Strickland JC, Smith MA. Animal models of social contact and drug self-administration. Pharmacol Biochem Behav. 2015;136:47–54. https://doi.org/10.1016/j.pbb.2015.06.013.
197. Li S, Samreen S, Ullah MB, Riaz FA, Awwad SWT. Global dynamics and computational modeling approach for analyzing and controlling of alcohol addiction using a novel fractional and fractal–fractional modeling approach. Sci Rep. 2024;14:5065. https://doi.org/10.1038/s41598-024-54578-9.
198. Neziri S, Köseoğlu AE, Deniz Köseoğlu G, Özgültekin B, Özgentürk NÖ. Animal models in neuroscience with alternative approaches: evolutionary, biomedical, and ethical perspectives. Animal Model Exp Med. 2024;7:868–80. https://doi.org/10.1002/ame2.12487.
199. Schneider E, Samsa LA, Dubljević V. A political and ethical landscape of brain organoid research. Mol Psychol Brain Behav Soc. 2024;2:3. https://doi.org/10.12688/molpsychol.17521.2.
200. Zhang T, Zhang N, Feng R, Wang H, Bi F, Wang S. Promoting the welfare of animals utilized in neuroscience research. Brain-X. 2024;2 https://doi.org/10.1002/brx2.70002.
201. Erler A. Human brain organoid transplantation: testing the foundations of animal research ethics. Neuroethics. 2024;17:20. https://doi.org/10.1007/s12152-024-09556-3.
202. Bozal CB, Bárcenas CA. Investigación preclínica en animales. Pautas actuales //Preclinical animal research. Current guidelines. Rev Asoc Odontol Argent. 2024:1–3. https://doi.org/10.52979/raoa.1120801.1284.
203. Khabade AM, Agarwal SS, Mahajan HB, Mandlik SK. Navigating regulatory requirements for stem cells therapy- a review on regulations of US, European Union, Japan and India. Int J Drug Regul Aff. 2024;12:66–81. https://doi.org/10.22270/ijdra.v12i3.704.
204. Bayne K, Turner PV. Animal welfare standards and international collaborations. ILAR J. 2019;60:86–94. https://doi.org/10.1093/ilar/ily024.
205. Francesca G, Ceruti S, Martini S, Picozzi M, Cosentino M, Marino F. Educating and training in research integrity (RI): a study on the perceptions and experiences of early career researchers attending an institutional RI course. J Acad Ethics. 2024;22:413–30. https://doi.org/10.1007/s10805-023-09497-1.
206. Carmichael ST. Medical school regulatory compliance burdens for the physician-scientist. Ann Neurol. 2024;96:417–22. https://doi.org/10.1002/ana.27020.

GPSR Compliance

The European Union's (EU) General Product Safety Regulation (GPSR) is a set of rules that requires consumer products to be safe and our obligations to ensure this.

If you have any concerns about our products, you can contact us on ProductSafety@springernature.com

In case Publisher is established outside the EU, the EU authorized representative is:

Springer Nature Customer Service Center GmbH
Europaplatz 3
69115 Heidelberg, Germany

Batch number: 10755222

Printed by Printforce, the Netherlands